Coastal and Marine Environments

The Handbook of Natural Resources, Second Edition

Series Editor:
Yeqiao Wang

Volume 1
Terrestrial Ecosystems and Biodiversity

Volume 2
Landscape and Land Capacity

Volume 3
Wetlands and Habitats

Volume 4
Fresh Water and Watersheds

Volume 5
Coastal and Marine Environments

Volume 6
Atmosphere and Climate

Coastal and Marine Environments

Edited by
Yeqiao Wang

CRC Press is an imprint of the
Taylor & Francis Group, an **informa** business

CRC Press
Taylor & Francis Group
6000 Broken Sound Parkway NW, Suite 300
Boca Raton, FL 33487-2742

First issued in paperback 2022

© 2020 by Taylor & Francis Group, LLC
CRC Press is an imprint of Taylor & Francis Group, an Informa business

No claim to original U.S. Government works

ISBN 13: 978-1-03-247440-3 (pbk)
ISBN 13: 978-1-138-33963-7 (hbk)

DOI: 10.1201/9780429441004

This book contains information obtained from authentic and highly regarded sources. Reasonable efforts have been made to publish reliable data and information, but the author and publisher cannot assume responsibility for the validity of all materials or the consequences of their use. The authors and publishers have attempted to trace the copyright holders of all material reproduced in this publication and apologize to copyright holders if permission to publish in this form has not been obtained. If any copyright material has not been acknowledged, please write and let us know so we may rectify in any future reprint.

Except as permitted under U.S. Copyright Law, no part of this book may be reprinted, reproduced, transmitted, or utilized in any form by any electronic, mechanical, or other means, now known or hereafter invented, including photocopying, microfilming, and recording, or in any information storage or retrieval system, without written permission from the publishers.

For permission to photocopy or use material electronically from this work, please access www.copyright.com (http://www.copyright.com/) or contact the Copyright Clearance Center, Inc. (CCC), 222 Rosewood Drive, Danvers, MA 01923, 978-750-8400. CCC is a not-for-profit organization that provides licenses and registration for a variety of users. For organizations that have been granted a photocopy license by the CCC, a separate system of payment has been arranged.

Trademark Notice: Product or corporate names may be trademarks or registered trademarks, and are used only for identification and explanation without intent to infringe.

Publisher's Note
The publisher has gone to great lengths to ensure the quality of this reprint but points out that some imperfections in the original copies may be apparent.

Library of Congress Cataloging-in-Publication Data

Names: Wang, Yeqiao, editor.
Title: Handbook of natural resources / edited by Yeqiao Wang.
Other titles: Encyclopedia of natural resources.
Description: Second edition. | Boca Raton: CRC Press, [2020] | Revised edition of: Encyclopedia of natural resources. [2014]. | Includes bibliographical references and index. | Contents: volume 1. Ecosystems and biodiversity — volume 2. Landscape and land capacity — volume 3. Wetland and habitats — volume 4. Fresh water and watersheds — volume 5. Coastal and marine environments — volume 6. Atmosphere and climate. |
Summary: "This volume covers topical areas of terrestrial ecosystems, their biodiversity, services, and ecosystem management. Organized for ease of reference, the handbook provides fundamental information on terrestrial systems and a complete overview on the impacts of climate change on natural vegetation and forests. New to this edition are discussions on decision support systems, biodiversity conservation, gross and net primary production, soil microbiology, and land surface phenology. The book demonstrates the key processes, methods, and models used through several practical case studies from around the world" — Provided by publisher.
Identifiers: LCCN 2019051202 | ISBN 9781138333918 (volume 1 ; hardback) | ISBN 9780429445651 (volume 1 ; ebook)
Subjects: LCSH: Natural resources. | Land use. | Climatic changes.
Classification: LCC HC85 .E493 2020 | DDC 333.95—dc23
LC record available at https://lccn.loc.gov/2019051202

Visit the Taylor & Francis Web site at
http://www.taylorandfrancis.com

and the CRC Press Web site at
http://www.crcpress.com

Contents

SECTION II Marine Environment

SECTION III Coastal Change and Monitoring

Preface

Coastal and Marine Environments is the fifth volume of *The Handbook of Natural Resources, Second Edition (HNR)*. This volume consists of 36 chapters authored by 46 contributors from 10 countries. The contents are organized in three sections: *Terrestrial Coastal Environment* (14 chapters); *Marine Environment* (13 chapters); and *Coastal Change and Monitoring* (9 chapters).

Coastal environments include areas of continental shelves, islands or partially enclosed seas, estuaries, bays, lagoons, beaches, and terrestrial and aquatic ecosystems within watersheds that drain into coastal waters. Coastal zone is the most dynamic interface between land and sea and represents the most challenging frontier between human civilization and environmental conservation. Worldwide, over 38% of the human population live in the coastal zones. An increasing proportion of the global population lives within the coastal zones of all major continents that require increasing attention to agricultural, industrial, and other human-related effects on coastal habitats and water quality, and their impacts on ecological dynamics, ecosystem health, and biological diversity. Climate change and sea level rise impose significant impacts and uncertainty on coastal ecosystems and infrastructures.

Coastal environments contain a wide range of natural habitats, such as sand dunes, barrier islands, tidal wetlands and marshes, mangrove forests, coral reefs, and submerged aquatic vegetation that provide foods, shelters, and breeding grounds for terrestrial and marine species. Coastal habitats also provide irreplaceable services, such as filtering pollutants and retaining nutrients, maintaining water quality, protecting shoreline, and absorbing flood waters. Coastal habitats are facing intensified natural and anthropogenic disturbances by direct impacts, such as hurricane, tsunami, harmful algae bloom, storm surge, and cumulative impacts, such as climate change, sea level rise, oil spill, and urban development. Inventory and monitoring of coastal environments become one of the most challenging tasks of the society in resource management and humanity administration.

With the challenges and concerns, the 36 chapters in this volume cover topics in *Terrestrial Coastal Environment*, including coral reef biology, ecology and conservation, quantifying reef ecosystem services, fishery conservation and management, mangrove forests, aquaculture, aquaforests and aquaforestry, coastal erosion and shoreline change, coast and marine tourism management, and science communication for natural resource managers; in *Marine Environment*, including archaeological oceanography, arctic hydrology, bathymetry assessment, bathymetry features and hypsography, seafloor mapping, marine benthic productivity, marine mammals, marine protected areas, marine resource management, maritime transportation and ports, market, trade and seafood, ocean's role in water cycle, and climate change impacts and adaptation for coastal transportation infrastructure; and in *Coastal Change Monitoring*, including submerged aquatic vegetation, tidal effects in salt marsh mapping, spatial and temporal change of mangrove-salt marsh ecotone, wetland change in Yellow River delta and Yangtze River estuary, coastal natural disasters, remote sensing of coastal estuarine waters and remote sensing of coastal environments.

The chapters provide updated knowledge and information in general environmental and natural science education and serve as a value-added collection of references for scientific research and management practices.

Yeqiao Wang
University of Rhode Island

About The Handbook of Natural Resources

With unprecedented attentions to the changing environment on the planet Earth, one of the central focuses is about the availability and sustainability of natural resources and the native biodiversity. It is critical to gain a full understanding about the consequences of the changing natural resources to the degradation of ecological integrity and the sustainability of life. Natural resources represent such a broad scope of complex and challenging topics.

The Handbook of Natural Resources, Second Edition (HNR), is a restructured and retitled book series based on the 2014 publication of the *Encyclopedia of Natural Resources (ENR)*. The *ENR* was reviewed favorably in February 2015 by CHOICE and commented as *highly recommended for lower-division undergraduates through professionals and practitioners*. This *HNR* is a continuation of the theme reference with restructured sectional design and extended topical coverage. The chapters included in the *HNR* provide authoritative references under the systematic relevance to the subject of the volumes. The case studies presented in the chapters cover diversified examples from local to global scales, and from addressing fundamental science questions to the needs in management practices.

The Handbook of Natural Resources consists of six volumes with 241 chapters organized by topical sections as summarized below.

Volume 1. Terrestrial Ecosystems and Biodiversity
Section I. Biodiversity and Conservation (15 Chapters)
Section II. Ecosystem Type, Function and Service (13 Chapters)
Section III. Ecological Processes (12 Chapters)
Section IV. Ecosystem Monitoring (6 Chapters)

Volume 2. Landscape and Land Capacity
Section I. Landscape Composition, Configuration and Change (10 Chapters)
Section II. Genetic Resource and Land Capability (13 Chapters)
Section III. Soil (15 Chapters)
Section IV. Landscape Change and Ecological Security (11 Chapters)

Volume 3. Wetlands and Habitats
Section I. Riparian Zone and Management (13 Chapters)
Section II. Wetland Ecosystem (8 Chapters)
Section III. Wetland Assessment and Monitoring (9 Chapters)

Volume 4. Fresh Water and Watersheds
Section I. Fresh Water and Hydrology (16 Chapters)
Section II. Water Management (16 Chapters)
Section III. Water and Watershed Monitoring (8 Chapters)

Volume 5. Coastal and Marine Environments
Section I. Terrestrial Coastal Environment (14 Chapters)
Section II. Marine Environment (13 Chapters)
Section III. Coastal Change and Monitoring (9 Chapters)

Volume 6. Atmosphere and Climate
Section I. Atmosphere (16 Chapters)
Section II. Weather and Climate (16 Chapters)
Section III. Climate Change (8 Chapters)

With the challenges and uncertainties ahead, I hope that the collective wisdom, the improved science, technology and awareness and willingness of the people could lead us toward the right direction and decision in governance of natural resources and make responsible collaborative efforts in balancing the equilibrium between societal demands and the capacity of natural resources base. I hope that this *HNR* series can help facilitate the understanding about the consequences of changing resource base to the ecological integrity and the sustainability of life on the planet Earth.

Yeqiao Wang
University of Rhode Island

Acknowledgments

I am honored to have this opportunity and privilege to work on *The Handbook of Natural Resources, Second Edition (HNR)*. It would be impossible to complete such a task without the tremendous amount of support from so many individuals and groups during the process. First and foremost, I thank the 342 contributors from 28 countries around the world, namely, Australia, Austria, Brazil, China, Cameroon, Canada, Czech Republic, Finland, France, Germany, Hungary, India, Israel, Japan, Nepal, New Zealand, Norway, Puerto Rico, Spain, Sweden, Switzerland, Syria, Turkey, Uganda, the United Kingdom, the United States, Uzbekistan, and Venezuela. Their expertise, insights, dedication, hard work, and professionalism ensure the quality of this important publication. I wish to express my gratitude in particular to those contributors who authored chapters for this HNR and those who provided revisions from their original articles published in the *Encyclopedia of Natural Resources*.

The preparation for the development of this *HNR* started in 2017. I appreciate the visionary initiation of the restructure idea and the guidance throughout the preparation of this *HNR* from Irma Shagla Britton, Senior Editor for Environmental Sciences and Engineering of the Taylor & Francis Group/CRC Press. I appreciate the professional assistance and support from Claudia Kisielewicz and Rebecca Pringle of the Taylor & Francis Group/CRC Press, which are vital toward the completion of this important reference series.

The inspiration for working on this reference series came from my over 30 years of research and teaching experiences in different stages of my professional career. I am grateful for the opportunities to work with many top-notch scholars, colleagues, staff members, administrators, and enthusiastic students, domestic and international, throughout the time. Many of my former graduate students are among and/or becoming world-class scholars, scientists, educators, resource managers, and administrators, and they are playing leadership roles in scientific exploration and in management practice. I appreciate their dedication toward the advancement of science and technology for governing the precious natural resources. I am thankful for their contributions in HNR chapters.

As always, the most special appreciation is due to my wife and daughters for their love, patience, understanding, and encouragement during the preparation of this publication. I wish my late parents, who were past professors of soil ecology and of climatology from the School of Geographical Sciences, Northeast Normal University, could see this set of publications.

Yeqiao Wang
University of Rhode Island

Aims and Scope

Land, water, and air are the most precious natural resources that sustain life and civilization. Maintenance of clean air and water and preservation of land resources and native biological diversity are among the challenges that we are facing for the sustainability and well-being of all on the planet Earth. Natural and anthropogenic forces have affected constantly land, water, and air resources through interactive processes such as shifting climate patterns, disturbing hydrological regimes, and alternating landscape configurations and compositions. Improvements in understanding of the complexity of land, water, and air systems and their interactions with human activities and disturbances represent priorities in scientific research, technology development, education programs, and administrative actions for conservation and management of natural resources.

The chapters of *The Handbook of Natural Resources, Second Edition (HNR)*, are authored by world-class scientists and scholars. The theme topics of the chapters reflect the state-of-the-art science and technology, and management practices and understanding. The chapters are written at the level that allows a broad scope of audience to understand. The graphical and photographic support and list of references provide the helpful information for extended understanding.

Public and private libraries, educational and research institutions, scientists, scholars, resource managers, and graduate and undergraduate students will be the primary audience of this set of reference series. The full set of the HNR and individual volumes and chapters can be used as references in general environmental science and natural science courses at different levels and disciplines, such as biology, geography, Earth system science, environmental and life sciences, ecology, and natural resources science. The chapters can be a value-added collection of references for scientific research and management practices.

Editor

Yeqiao Wang, PhD, is a professor at the Department of Natural Resources Science, College of the Environment and Life Sciences, University of Rhode Island. He earned his BS from the Northeast Normal University in 1982 and his MS degree from the Chinese Academy of Sciences in 1987. He earned the MS and PhD degrees in natural resources management and engineering from the University of Connecticut in 1992 and 1995, respectively. From 1995 to 1999, he held the position of assistant professor in the Department of Geography and the Department of Anthropology, University of Illinois at Chicago. He has been on the faculty of the University of Rhode Island since 1999. Among his awards and recognitions, Dr. Wang was a recipient of the prestigious Presidential Early Career Award for Scientists and Engineers (PECASE) in 2000 by former U.S. President William J. Clinton, for his outstanding research in the area of land cover and land use in the Greater Chicago area in connection with the Chicago Wilderness Program.

Dr. Wang's specialties and research interests are in terrestrial remote sensing and the applications in natural resources analysis and mapping. One of his primary interests is the land change science, which includes performing repeated inventories of landscape dynamics and land-use and land-cover change from space, developing scientific understanding and models necessary to simulate the processes taking place, evaluating consequences of observed and predicted changes, and understanding the consequences of change on environmental goods and services and management of natural resources. His research and scholarships are aimed to provide scientific foundations in understanding of the sustainability, vulnerability and resilience of land and water systems, and the management and governance of their uses. His study areas include various regions in the United States, East and West Africa, and China.

Dr. Wang published over 170 refereed articles, edited *Remote Sensing of Coastal Environments* and *Remote Sensing of Protected Lands*, published by CRC Press in 2009 and 2011, respectively. He served as the editor-in-chief for the *Encyclopedia of Natural Resources* published by CRC Press in 2014, which was the first edition of *The Handbook of Natural Resources*.

Contributors

Gordon N. Ajonina
CWCS Coastal Forests and Mangrove Programme
Cameroon Wildlife Conservation Society, and Institute of Fisheries and Aquatic Sciences
University of Douala
Yabassi, Douala, Cameroon

Stephanie I. Andersona
Graduate School of Oceanography
University of Rhode Island
Narragansett, Rhode Island

Regina Asariotis
Policy and Legislation Section, TLB/DTL, United Nations Conference on Trade and Development (UNCTAD)
Geneva, Switzerland

Frank Asche
Institute for Sustainable Food Systems and Fisheries and Aquatic Sciences
School of Forest Resources and Conservation
University of Florida
Gainesville, Florida

Robert D. Ballard
Graduate School of Oceanography
University of Rhode Island
Narragansett, Rhode Island

Austin Becker
Emmett Interdisciplinary Program in Environment and Resources (E-IPER)
Stanford University
Stanford, California

Michael L. Brennan
Center for Ocean Exploration
University of Rhode Island
Narragansett, Rhode Island

Richard Burroughs
Department of Marine Affairs
University of Rhode Island
Kingston, Rhode Island

Carrie J. Byron
Gulf of Maine Research Institute
Portland, Maine

Steven X. Cadrin
University of Massachusetts Dartmouth
New Bedford, Massachusetts

Anthony Daniel Campbell
Department of Natural Resources Science
University of Rhode Island
Kingston, Rhode Island

and

Yale School of Forestry and Environmental Studies
New Haven, Connecticut

Jennifer Caselle
Marine Science Institute
University of California
Santa Barbara, California

Don P. Chambers
College of Marine Science
University of South Florida
St. Petersburg, Florida

Lin Chen
Key Laboratory of Wetland Ecology and Environment
Northeast Institute of Geography and Agroecology
Chinese Academy of Sciences
Changchun, China

and

University of Chinese Academy of Sciences
Beijing, China

Sean D. Connell
Southern Seas Ecology Laboratories
School of Earth and Environmental Science
University of Adelaide
Adelaide, South Australia, Australia

Heidi M. Dierssen
Department of Biology
Norwegian University of Science and Technology
Trondheim, Norway

and

Department of Marine Sciences/Geography
University of Connecticut
Groton, Connecticut

Joshua Drew
Department of Environmental and Forest Biology
State University of New York College of Environmental Science and Forestry
Syracuse, New York

M. Tomedi Eyango
Institute of Fisheries and Aquatic Sciences
University of Douala
Yabassi, Douala, Cameroon

Graham E. Forrester
Department of Natural Resources Science
University of Rhode Island
Kingston, Rhode Island

C. Michael Hall
Department of Management
University of Canterbury
Christchurch, New Zealand

and

Centre for Tourism
University of Eastern Finland
Savonlinna, Finland

and

University of Oulu
Oulu, Finland

Kate Henderson
Department of Environmental and Forest Biology
College of Environmental Science and Forestry
State University of New York
Syracuse, New York

Jinrong Hu
Yuen Yuen Research Centre for Satellite Remote Sensing
Institute of Space and Earth Information Science
Chinese University of Hong Kong
Hong Kong, China

and

Laboratory of Coastal Zone Studies
Shenzhen Research Institute
Shenzhen, China

Darryl J. Keith
Atlantic Ecology Division
U.S. Environmental Protection Agency (EPA)
Narragansett, Rhode Island

Jinwoo Kim
The Ohio State University
Columbus, Ohio

Zhong Lu
Cascades Volcano Observatory
U.S. Geological Survey (USGS)
Vancouver, Washington

Ariel E. Lugo
International Institute of Tropical Forestry, Forest Service
U.S. Department of Agriculture (USDA-FS)
Río Piedras, Puerto Rico

Katharine McDuffie
Metcalf Institute for Marine and Environmental Reporting
Department of Natural Resources Science
University of Rhode Island
Kingston, Rhode Island

Emma McKinley
School of Earth and Ocean Sciences
Cardiff University
Cardiff Wales, United Kingdom

Ernesto Medina
International Institute of Tropical Forestry, Forest Service
U.S. Department of Agriculture (USDA-FS)
Río Piedras, Puerto Rico

and

Center for Ecology
Venezuelan Institute for Scientific Research
Caracas, Venezuela

Sunshine Menezes
Metcalf Institute for Marine and Environmental Reporting
Department of Natural Resources Science
University of Rhode Island
Kingston, Rhode Island

Alyssa B. Novak
Department of Earth and Environment
Boston University
Boston, Massachusetts

Chunying Ren
Key Laboratory of Wetland Ecology and Environment
Northeast Institute of Geography and Agroecology
Chinese Academy of Sciences
Changchun, China

Wilfrid Rodriguez
Smithsonian Environmental Research Center
Edgewater, Maryland

Cathy A. Roheim
Department of Agricultural Economics and Rural Sociology
University of Idaho
Moscow, Idaho

Bayden D. Russell
Southern Seas Ecology Laboratories
School of Earth and Environmental Science
University of Adelaide
Adelaide, South Australia, Australia

Frederick T. Short
University of New Hampshire
Durham, New Hampshire

C. K. Shum
Division of Geodetic Science
School of Earth Sciences
The Ohio State University
Columbus, Ohio

Martin D. Smith
Nicholas School of the Environment and Department of Economics
Duke University
Durham, North Carolina

Bretton Somers
Department of Geography and Anthropology
Louisiana State University
Baton Rouge, Louisiana

Matthew L. Stutz
Department of Chemistry, Physics, and Geoscience
Meredith College
Raleigh, North Carolina

Takehiko Takano
Tohoku Gakuin University
Sendai, Japan

Albert E. Theberge, Jr.
Central Library
National Oceanic and Atmospheric Administration (NOAA)
Silver Spring, Maryland

Kathleen J. Vigness-Raposa
Marine Acoustics, Inc.
Middletown, Rhode Island

H. Jesse Walker
Department of Geography and Anthropology
Louisiana State University
Baton Rouge, Louisiana

Yeqiao Wang
Department of Natural Resources Science
University of Rhode Island
Kingston, Rhode Island

Yuanzhi Zhang
Yuen Yuen Research Centre for Satellite Remote Sensing
Institute of Space and Earth Information Science
Chinese University of Hong Kong
Hong Kong, China

and

Laboratory of Coastal Zone Studies
Shenzhen Research Institute
Shenzhen, China

I

Terrestrial Coastal Environment

1

Aquaculture

Carrie J. Byron
Gulf of Maine Research Institute

Introduction

Aquaculture is the process of cultivating aquatic organisms in a controlled setting and may include some or all of the processes of breeding, rearing, and harvesting. Aquaculture can be done on land in tanks, freshwater habitats such as ponds, brackish water habitats such as estuaries, or marine habitats such as in the coastal zone and open ocean. Mariculture is often used to describe brackish or marine aquaculture. Sea ranching describes the process of rearing early life stages in controlled systems and releasing them back to the ocean often for the purpose of restocking a population.[1] Tuna ranching is a relatively recent development whereby young wild fish are captured and reared in net pens to marketable sized[2,3]

Global Perspective

Aquaculture is increasing on a global scale. However, the increase in aquaculture is not uniform across the globe.[4] In 2010, global aquaculture production equaled 63.6 million tonnes.[4] Asia comprised 89% (53,301,157 million tonnes) of the world aquaculture production by volume, with China as the leading nation comprising 61.4% (36,734,215 million tonnes) of global production. The remaining global aquaculture production came from the Americas (4.3%; 2,576,428 million tonnes), Europe (4.2%; 2,523,179 million tonnes), Africa (2.2%; 1,288,320 million tonnes), and Oceana (0.3%; 183,516 tonnes).[4]

The National Aquaculture Act of 1980 states that it is "in the national interest, and it is the national policy, to encourage aquaculture development in the U.S." Despite this national policy, the United States is not one of the top 10 leading nations in aquaculture production. Aquaculture production has grown most strongly in developing countries, particularly in Asia.[4]

Freshwater is the dominant habitat type for aquaculture production comprising 62% of global production and 58.1% of global value of products produced.[4] Of all the animal species produced in freshwater, 91% (33.9 million tonnes) of the production are freshwater finfishes such as carps, tilapia, and catfish.[4] Marine aquaculture comprised 30.1% of global production and 29.2% of global value.[4] Of all the animal species produced in marine water, three-quarters are molluscs (75.5%; 13.9 million tonnes) such as clams, cockles, oysters, and mussels.[4] Brackish water habitats comprised only 7.9% of global aquaculture production and 12.8% of its value.[4] Of all the animal species produced in brackish water, more than half are crustaceans (57.7%; 2.7 million tonnes) such as shrimp.[4]

Most aquaculture species are produced for human consumption. Some ornamental fish, such as clownfish, are produced for aquarium trade.[5,6] Some algae and small fin-fish species are produced for consumption by other farmed animals including shellfish, fish, chickens, pigs, and other mammals. Of the animal species being produced through aquaculture for human consumption, there has been a proportionally large increase in freshwater finfish and molluscs over the past four decades.[4] More than half (56.6%; 33.7 million tonnes) of global aquaculture production is freshwater finfish and nearly one-quarter (23.6%; 14.2 million tonnes) is molluscs.[4] Crustaceans (9.6%; 5.7 million tonnes), diadromous fish (6.0%; 3.6 million tonnes), marine fish (3.1%; 1.8 million tonnes), and other assorted animals (1.4%; 814,3000 tonnes) such as sea cucumbers and softshell turtles comprise the rest of the aquaculture production by species type.[4]

Species and Habitats

Finfish are vertebrate fish species and include freshwater, diadromous, and marine fish. Diadromous fish such as salmon live part of their life in freshwater and another part in salt water. Most freshwater finfish are grown on land in tanks or raceways (long narrow troughs). U.S. freshwater aquaculture is dominated by channel catfish (*Ictalurus punctatus*) but is shadowed by the dramatic increase in the production of pangas catfish (*Pangasius* spp.) in Vietnam in recent years.[4,7] Trout, tilapia, and some bass species are also grown in land-based facilities often using a recirculating aquaculture system designed to reduce and reuse water. Land-based aquaculture facilities require high amounts of electricity to maintain operations, and efficient systems that cause little pollution are generally required to maintain these facilities, which have a little impact on the environment.[8] Recirculating aquaculture systems clean and reuse effluent water within the facility instead of discharging dirty water into the environment.

Although some marine species can also be raised on land, most are raised in the ocean in net pens or cages (Figure 1.1). Atlantic salmon represents many of the environmental and social debates surrounding the efficacy of finfish farming. Salmon is farmed around the world with the two leading producers of farmed Atlantic salmon being Norway and Chile.[4] Criticisms for finfish aquaculture concern environmental and human health. Escaped farm salmon may breed with wild salmon, thereby polluting "wild" genetic stocks with genes from domesticated stocks, which may not be derived from local populations.[9] The mixed-gene offspring have lower survival in natural habitats than their wild counterparts, thereby further hindering the persistence of an already threatened and endangered species.[10] Another environmental concern is the fecal waste from farmed fish as well as the uneaten fish food on farms, which can pollute surrounding habitats.[11,12] Additionally, like any dense monoculture of species,

FIGURE 1.1 Floating salmon net pen. Notice the netting to keep out predators, the pneumatic feeder in the center of the cage, and the floating feed-supply house to the left and behind the cage.

disease also becomes more prevalent. Sea lice are obligate ectoparasites that attach to and graze on the skin of the fish, thereby causing shallow lesions that can permit bacterial infections or lead to death, which devalue the seafood product.[13,14]

The commercially formulated pellet feed that many commercially aquacultured fish are fed is composed of proteins, lipids, and carbohydrates that vary in proportion with each farm-raised species.[15,16] The protein meal is derived from a variety of sources including fish caught specifically to produce fish meal and terrestrial-based crops such as soya and maize. Little is known regarding the impact of terrestrial-derived nutrients in marine food webs.[17]

Advances in technology and scientific knowledge are minimizing the impacts of finfish culture in the coastal environment. The risk of escapees from pens has also declined over the years with increased net-pen technology. Technologies for feeding the fish are becoming more advanced where very little food goes uneaten, thereby reducing waste to the environment. Some countries, such as the United States, require pens to remain fallow for a period of time between stocking cycles to further reduce the loading of fecal waste to an area and to mitigate the spread of diseases and pests, such as sea lice.[18,19] Furthermore, innovative approaches to aquaculture such as integrated multitrophic aquaculture and utilizing carrying capacity for an ecosystem approach to aquaculture hold promise for further reductions on environmental impact in the future.[20–23]

Crustacean species that are most commonly grown using aquaculture are primarily Penaeid marine shrimp and freshwater prawns.[4] In 2010, the fastest growing crustacean production was the white-leg shrimp (*Penaeus vannamei*). Thailand, China, and Indonesia are the major exporting countries of shrimp, whereas the United States and Japan are the primary importers of shrimp.[4] Like salmon, there are several environmental and social issues surrounding shrimp farming including the destruction of mangrove habitats for the construction of rearing ponds, salinization of groundwater, depletion of wild fish stocks for formulated feeds, depletion of wild broodstocks, and disease outbreaks[24] (http://www.asc-aqua.org; http://www.wildlife.org). Shrimp farming is highly unregulated in the global market and poses several environmental threats. Recent technologies hold promise for a more sustainable method of culturing shrimp in the future. Pilot studies of native shrimp species grown in Aquapod™ net pens in the open ocean on natural productivity, which serves as a food source without herbicides, pesticides, commercial fish oil, or processed fish meal, have shown that there is minimal impact on the environment and that it alleviates pressure on fish stocks and wild broodstocks[5,7] (http://www.olazul.org). Other finfish species can also be grown in various designs of large pens in the open ocean, which present additional engineering and economic challenges but would alleviate environmental impacts in crowded and heavily used coastal areas[25–29] (http://www.amac.unh.edu).

Bivalve molluscs such as clams, cockles, oysters, and mussels are commonly grown worldwide (Figures 1.2 and 1.3).[4] Unlike finfish farming, bivalves grown in coastal habitats filter feed directly on ambient plankton and do not require additional additive feeds. Bivalve aquaculture in coastal environments has the potential to alter the community structure of the ecosystem, nutrient dynamics in the water column, and deposition of sediment.[30–39] Nitrogen is the primary nutrient of concern in coastal habitats where bivalves live naturally and are grown. In extreme cases, excessive nitrogen can lead to large algae blooms and subsequently anoxia in the water column, which can kill finfish and shellfish.[8] As bivalves consume algae and other particulate organic matter in the water, nitrogen gets assimilated into their hard shell, thereby removing it from water and making it unavailable to other organisms, which is a method of cleansing the water in high nutrient systems.[40–42] Bioextraction describes the process of using bivalves to clean the water by extracting excessive nutrients. Some economists are exploring ways to trade nitrogen credits in a way that mirrors the carbon credit trading system[43] (http://www.motu.org.nz/research/detail/nutrient_trading).

Globally, cultivation of algae is dominated by marine macroalgae or seaweeds.[4] Seaweeds are algae and can include both small microscopic species such as *Isochrysis* sp. primarily used to feed aquaculture-grown bivalve shellfish and macroalgae, commonly called kelp. In 2010, algae production equaled 19 million tonnes and came primarily from eight countries: China, Indonesia, the Philippines,

FIGURE 1.2 Oyster aquaculture. This photo demonstrates one of many techniques for raising oysters in coastal waters. Notice the rack-and-bag system. The rack, or cage, sits on the bottom under the water and is attached to a rope and surface buoy. The oyster farm is not visible from the surface of the water, other than a few buoys marking the boundaries of the lease area.

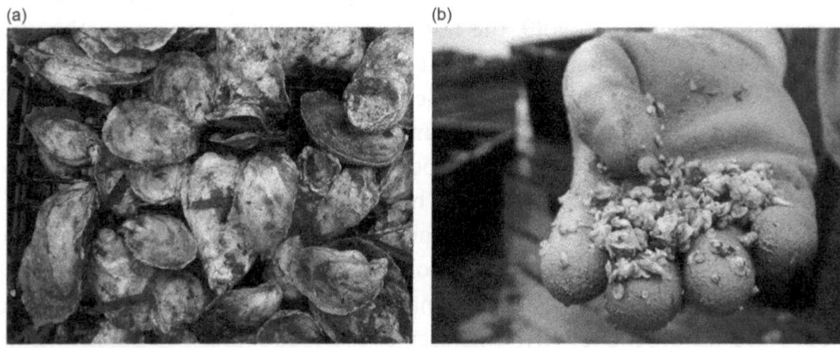

FIGURE 1.3 **(See color insert.)** Cultured oysters (**a**) almost grown to market size in bags; (**b**) juvenile oyster spat, sometimes called oyster seed.
Source: Figure 3B by Trisha Towanda, 2008.

the Republic of Korea, Democratic People's Republic of Korea, Japan, Malaysia, and the United Republic of Tanzania.[4] Aquatic algae production by volume increased at an average annual rate of 9.5% in the 1990s, which is comparable to rates for farmed aquatic animals.[4] The most-cultivated algae is the seaweed called Japanese kelp (*Saccharina japonica*) consisting of 98.9% of global production of algae cultured.[4] One of the many methods for cultivation involves growing kelp spores into seedlings in land-based facilities before being transplanted to long ropes or other rack-and-line structures in coastal waters. Seaweeds are used in several food products, fertilizers, and animal feeds.

Integrated Multitrophic Aquaculture

Integrated multitrophic aquaculture (IMTA) is an extension of polyculture, whereby the culture of fed organisms are combined with the culture of organisms that extract dissolved inorganic nutrients (seaweeds) or particulate organic matter (shellfish) or both, thereby linking trophic levels.[44] The term trophic level describes the position an organism occupies in the food chain. One of the goals of IMTA is to reduce organic pollution to the environment from fish culture. Another goal is to improve the growth of all the IMTA species due to a rich environment in terms of food. Species from lower trophic levels such as mussels and algae ingest particulate and dissolved nutrients from the fish farm, thereby "cleansing" the water[25] while getting additive nutrition. Mussels may also be capable of the infectious pressure of the bacteria *Vibrio anguillarum*.[45] Sea cucumbers (Echinodermata: Holothuroidea) can also be integrated into IMTA and have a growing market in aquaculture, independent of IMTA.[46] Coupled aquaculture-agriculture systems have been practiced throughout time primarily on a small scale in developing countries and have undoubtedly led to the initial conceptualization of IMTA.[40,47]

Management

Marine Spatial Planning

The coastal environment, where much aquaculture activity takes place, is a busy place that is used by several industries and activities (e.g., shipping, fishing, recreational boating, renewable energy, protected habitat areas, etc.). Conflict between aquaculture and other coastal activities presents a hurdle for the aquaculture industry, often making it difficult to obtain permits for operation. Effective spatial planning of coastal habitats is a way to mitigate these types of user conflicts. Marine spatial planning allows for optimizing the use of the coastal zone. For example, by coupling IMTA aquaculture farms with wind turbine farms, the unfishable and unnavigable spaces between turbines can be efficiently used for food production)[15,48–51]

Ecosystem Approach to Aquaculture

The ecosystem approach to aquaculture is defined as "a strategic approach to the development and management of the sector aiming to integrate aquaculture within the wider ecosystem such that it promotes sustainability of interlinked social-ecological systems" and is guided by three main principles:[52] i) "Aquaculture development and management should take account of the full range of ecosystem functions and services, and should not threaten the sustained delivery of these to society;" ii) "Aquaculture should improve human well-being and equity for all relevant stakeholders;" and iii) "Aquaculture should be developed in the context of other sectors, policies and goals." The principles of the ecosystem approach to aquaculture follow closely from the principles of ecosystem-based management, which emphasize the interdependencies of ecological and social goals within the same system (http://www.ebmtools.org/about_ebm.html).

Carrying Capacity

There are four types of carrying capacity that have been adapted to describe coastal aquaculture where a farm is located in an ecosystem. The first type is physical carrying capacity, which is the total area of marine farms that can be accommodated in the available physical space.[53] The definition of physical carrying capacity is limited in that it only considers available space and not the function or impacts of the farm in the ecosystem. The second type is production carrying capacity, which is the stocking density at which harvests are maximized.[53] Every successful farmer knows the production carrying capacity for the farm because it is directly linked to sales and profits. The definition of production carrying capacity is limited to the farm site and does not consider the larger ecosystem in which the farm is located. To extend this definition, consider ecological carrying capacity, which is the stocking or farm density above which would cause unacceptable ecological impacts.[53] And similarly, social carrying capacity is the level of farm development that causes unacceptable social impacts.[53] Both ecological and social carrying capacity recognize that the farm is simply part of the larger ecosystem, which is important when taking an ecosystem approach to aquaculture. As useful as these two definitions may be, they are also somewhat vague. How does one define what is unacceptable in an ecosystem? McKindsey et al.[54] suggest that ecological and social carrying capacity must be considered together. Society, represented by a group of stakeholders, needs to decide what is acceptable.[20]

Once acceptability limits are set, carrying capacity can be calculated.[21,22] Ecological carrying capacity can be calculated using mass-balance ecosystem modeling as described by Byron et al.[20–22] Social carrying capacity can be calculated using economic modeling, spatial modeling, and/or resource valuation. These techniques are still in development and are not yet well established. Ultimately, ecosystem carrying capacity and social carrying capacity can be used as techniques for implementing an ecosystem approach to aquaculture.

The aquaculture industry is growing, with most of that growth occurring in Asia. The United States National Policy on Aquaculture encourages further sustainable growth in U.S. coastal waters but currently contributes only a very small fraction to global production. Now is an appropriate time to set measures in place to ensure sustainable growth within the capacity of the ecosystem and society.

References

1. Schraga, T.S.; Cloern, J.E.; Dunlavey, E.G. Primary production and carrying capacity of former salt ponds after reconnection to San Francisco Bay. Wetlands (Wilmington, NC) **2008**, *28* (3), 841–851.
2. De Stefano, V.; Van Der Heiden, P.G.M. Bluefin tuna fishing and ranching: a difficult management problem. New Medit. **2007**, *2*, 59–64.
3. Longo, S.B.; Clark, B. The commodification of bluefin tuna: the historical transformation of the mediterranean fishery. J. Agrarian Change **2012**, *12* (2–3), 204–226.
4. *FAO.* The State of the World Fisheries and Aquaculture 2012*; FAO: Rome, Italy 2012; 209.*
5. Bren. Spatial planning and bio-economic analysis for sustainable offshore shrimp aquaculture. Donald Bren School of Environmental Science and Management & University of California, Santa Barbara. 2010; 33. https://www.box.com/s/3aa1bb89b09cbd7329cb.
6. Molina, L.; Segade, A. Aquaculture as a potential support of marine aquarium fish trade sustainability. In *Management of Narual Resources, Sustainable Development and Ecological Hazards III.* Brebbia, C., Zubir, S. WIT Press: Southhampton, UK, 2012.
7. Olivares, A.E.V. *Design of a Cage Culture System for Farming in Mexico*; The United Nations University, Fisheries Training Programme; Reykjavik, Iceland, 2003; 47 pp.
8. Howarth, R.; Anderson, D.; Cloern, J.; Elfring, C.; Hopkinson, C.; Lapointe, B.; Malone, T.; Marcus, N.; McGathery, K.; Sharpley, A.; Walker, D. Nutrient pollution of coastal rivers, bays, and seas. Issues in Ecology. 2000. http://www.esa.org/science_resources/issues/TextIssues/issue7.php.
9. Houde, A.L.S.; Fraser, D.J.; Hutchings, J.A. Fitness-related consequences of competitive interactions between farmed and wild Atlantic salmon at different proportional representations of wild-farmed hybrids. ICES J. Mar. Sci. **2010a**, *67* (4), 657–667.
10. Houde, A.L.S.; Fraser, D.J.; Hutchings, J.A. Reduced antipredator responses in multi-generational hybrids of farmed and wild Atlantic salmon (*Salmo salar* L.). Conserv. Genet. **2010b**, *11* (3), 785–794.
11. Hargrave, B.T. Empirical relationships describing benthic impacts of salmon aquaculture. Aquaculture Environ. Interact. **2010**, *1* (1), 33–46.
12. Piedecausa, M.A.; Aguado-Gimenez, F.; Valverde, J.C.; Llorente, M.D.H.; Garcia-Garcia, B. Influence of fish food and faecal pellets on short-term oxygen uptake, ammonium flux and acid volatile sulphide accumulation in sediments impacted by fish farming and non-impacted sediments. Aquaculture Res. **2012**, *43* (1), 66–74.
13. Krkosek, M. Host density thresholds and disease control for fisheries and aquaculture. Aquaculture Environ. Interact. **2010**, *1* (1), 21–32.
14. Revie, C.; Dill, L.; Finstad, B.; Todd, C.D. Salmon aquaculture dialog working group report on sea lice. commissiond by the salmon aquaculture dialogue. 2009. http://www.nina.no/archive/nina/PppBasePdf/temahefte/039.pdf

15. Buck, B.H.; Krause, G.; Rosenthal, H. Extensive open ocean aquaculture development within wind farms in Germany: the prospect of offshore co-management and legal constraints. Ocean Coastal Manag. **2004**, *47* (3–4), 95–122.
16. Southgate, P.C. Foods and Feeding. In *Aquaculture. Farming Aquatic Animals and Plants.* Lucas, J.S., Southgate, P.C. Wiley-Blackwell: Oxford, UK, 2012; 188–213.
17. Spivak, A.C.; Canuel, E.A.; Duffy, J.E.; Richardson, J.P. Nutrient enrichment and food web composition affect ecosystem metabolism in an experimental seagrass habitat. PLoS ONE **2009**, *4* (10), e7473.
18. Brauner, C.J.; Sackville, M.; Gallagher, Z.; Tang, S.; Nendick, T.L.; Farrell, A.P. Physiological consequences of the salmon louse (*Lepeopthteirus salmonis*) on juvenlie pink salmon (*Onchorhynchus gorbuscha*): implications for wild salmon ecology and management, and for salmon aquaculture. Philos. Trans. R. Soc. **2012**, *367* 1770–1779.
19. USASAC. *Annual Report of the U.S. Atlantic Salmon Assessment Committee 22.* Prepared for U.S. Section to NASCO: Portland, ME, 2010.
20. Byron, C.; Bengtson, D.; Costa-Pierce, B.; Calanni, J. Integrating science into management: ecological carrying capacity of bivalve shellfish aquaculture. Mar. Policy **2011a**, *35* (3), 363–370.
21. Byron, C.; Link, J.; Bengtson, D.; Costa-Pierce, B. Calculating carrying capacity of shellfish aquaculture using mass-balance modeling: Narragansett Bay, Rhode Island. Ecol. Model. **2011b**, 222, 1743–1755.
22. Byron, C.; Link, J.; Costa-Pierce, B.; Bengtson, D. Modeling ecological carrying capacity of shellfish aquaculture in highly flushed temperate lagoons. Aquaculture **2011c**, *314* (1–4), 87–99.
23. Neori, A.; Chopin, T.; Troell, M.; Buschmann, A.H.; Kraemer, G.P.; Halling, C.; Shpigel, M.; Yarish, C. Integrated aquaculture: rationale, evolution and state of the art emphasizing seaweed biofiltration in modern mariculture. Aquaculture **2004**, *231* (1–4), 361–391.
24. Hopkins, J.S.; Devoe, M.R.; Holland, A.F. Environmental impacts of shrimp farming with special reference to the situation in the continental United States. Estuaries **1995**, *18* (1A), 25–42.
25. Chambers, M.D.; Howell, W.H. Preliminary information on cod and haddock production in submerged cages off the coast of New Hampshire, USA. ICES J. Mar. Sci. **2006**, *63* (2), 385–392.
26. Fredriksson, D.W.; DeCew, J.; Swift, M.R.; Tsukrov, I.; Chambers, M.D.; Celikkol, B. The design and analysis of a four-cage grid mooring for open ocean aquaculture. Aquacultural Eng. **2004**, *32* (1), 77–94.
27. Fredriksson, D.W.; Muller, E.; Baldwin, K.; Robinson Swift, M.; Celikkol, B. Open ocean aquaculture engineering: system design and physical modeling. Mar. Technol. Soc. J. **2000**, *34* (1), 41–52.
28. Fredriksson, D.W.; Swift, M.R.; Eroshkin, O.; Tsukrov, I.; Irish, J.D.; Celikkol, B. Moored fish cage dynamics in waves and currents. J. Ocean Eng. **2005**, *30* (1), 28–36.
29. Howell, W.H.; Chambers, M.D. Growth performance and survival of Atlantic halibut (*Hippoglossus hippoglossus*) grown in submerged net pens. Bulletin of the Aquaculture Assoc. Canada **2005**, *9*, 35–37.
30. Callier, M.D.; Richard, M.; McKindsey, C.W.; Archambault, P.; Desrosiers, G. Responses of benthic macrofauna and biogeochemical fluxes to various levels of mussel biodeposition: an in situ "benthocosm" experiment. Mar. Pollut. Bull. **2009**, *58* (10), 1544–1553.
31. Forrest, B.M.; Keeley, N.B.; Hopkins, G.A.; Webb, S.C.; Clement, D.M. Bivalve aquaculture in estuaries: Review and synthesis of oyster cultivation effects. Aquaculture **2009**, *298* (1–2), 1–15.
32. McKindsey, C.; Landry, T.; O'Beirn, FX.; Davies, IM. Bivalve aquaculture and exotic species: a review of ecological considerations and management issues. J. Shellfish Res. **2007**, *26* (2), 281–294.
33. McKindsey, C.W.; Lecuona, M.; Huot, M.; Weise, A.M. Biodeposit production and benthic loading by farmed mussels and associated tunicate epifauna in Prince Edward Island. Aquaculture **2009**, *295* (1–2), 44–51.

34. Nakamura, Y.; Kerciku, F. Effects of filter-feeding bivalves on the distribution of water quality and nutrient cycling in a eutrophic coastal lagoon. J. Mar. Syst. **2000**, *26* (2), 209–221.
35. Nugues, M.M. Benthic community changes associated with intertidal oyster cultivation. Aquaculture Res. **1996**, 27, 913–924.
36. Sara, G. Aquaculture effects on some physical and chemical properties of the water column: a meta-analysis. Chem. Ecol. **2007a**, *23* (3), 251–262.
37. Sara, G. Ecological effects of aquaculture on living and non-living suspended fractions of the water column: a meta-analysis. Water Res. **2007b**, *41* (15), 3187–3200.
38. Sara, G. A meta-analysis on the ecological effects of aquaculture on the water column: dissolved nutrients. Mar. Environ. Res. **2007c**, *63* (4), 390–408.
39. Smaal, A.; van Stralen, M.; Schuiling, E. The interaction between shellfish culture and ecosystem processes. Canad. J. Fisheries Aquatic Sci. **2001**, *58*, 991–1002.
40. Costa-Pierce, B.A. *Ecological Aquaculture: The Evolution of the Blue Revolution.* John Wiley & Sons Blackwell Science, Oxford, 2008.
41. Stadmark, J.; Conley, D.J. Mussel farming as a nutrient redution measure in the Baltic sea: consideration of nutrient biogeochemical cycles. Mar. Pollut. Bull. **2011**, *62* (7), 1385–1388.
42. White, C.; Halpern, B.S.; Kappel, C.V. Ecosystem service tradeoff analysis reveals the value of marine spatial planning for multiple ocean uses. Proc. Natl. Acad. Sci. **2012**, *109* (12), 4696–4701.
43. Kerr, S.; Lauder, G.; Fairman, D. Towards design for a nutrient trading programme to improve water quality in lake Rotorua 07-03. Motu Work. Paper **2007**.
44. Chopin, T. Integrated multi-trophic aquaculture. What it is, and why you should care… and don't confuse it with polyculture. Northern Aquaculture **2006**, *4*.
45. Pietrak, M.R.; Molloy, S.D.; Bouchard, D.A.; Singer, J.T.; Bricknell, I. Potential role of *Mytilus edulis* in modulating the infectious pressure of *Vibrio anguillarum* 02β on an integrated multi-trophic aquaculture farm. Aquaculture **2012**, *326–329*, 36–39.
46. FAO. *Sea Cucumbers. A Global Review of Fisheries and Trade.* FAO Fisheries and Aquaculture Technical Paper: Rome, Italy, 2008; 516.
47. Murshed-E-Jahan, K.; Pemsl, D.E. The impact of integrated aquaculture-agriculture on small-scale farm sustainability and farmers livelihoods: Experience from Bangladesh. Agricultural Syst. **2011**, *104*, 392–402.
48. Buck, B.H.; Ebeling, M.;Michler-Cieluch, T. Mussel cultivation as a co-use in offshore wind farms: potentials and economic feasibility. Aquaculture Econ. Manag. **2010**, *14* (4), 1365–7305.
49. Buck, B.H.; Krause, G.; Michler-Cieluch, T.; Buchholz, B.M.; Busch, C.M.; Fisch, J.A.; Geisen, R.; Zielinski, M. Meeting the quest for spatial efficiency: progress and prospects of extensive aquaculture within offshore wind farms. Helgoland Mar. Res. **2008**, *62*, 269–281.
50. Michler-Ceiluch, T.; Krause, G.; Buck, B. Marine aquaculture within offshore wind farms: Social aspects of multiple use planning. GAIA Ecol. Perspect. Sci. Soc. **2009a**, *18* (2), 158–162.
51. Michler-Ceiluch, T.; Krause, G.; Buck, B. Reflections on integrating operation and maintanence activities of offshore wind farms and mariculture. Open Coastal Manag. **2009b**, *52* (1), 57–68.
52. Soto, D.; Manjarez, J.A., Eds.; *Building an Ecosystem Approach to Aquaculture. FAO/Universitat de les Illes Balears Expert Workshop. 7–11 May 2007;* FAO Fisheries and Aquaculture Proceedings: No. 14. Rome, FAO., Palma de Mallorca, Spain, **2008**.
53. *Inglis, G.J.; Hayden, B.J.; Ross, A.H.* An Overview of Factors Affecting the Carrying Capacity of Coastal Embayments for Mussel Culture. *Client Report CHC00/69: NIWA, Christchurch,* **2002**.
54. McKindsey, C.W.; Thetmeyer, H.; Landry, T.; Silvert, W. Review of recent carrying capacity models for bivalve culture and recommendations for research and management. Aquaculture **2006**, *261* (2), 451–462.

2

Aquaforests and Aquaforestry: Africa

Gordon N. Ajonina and M. Tomedi Eyango
University of Douala

Introduction

As water covers over 70% of the Earth's surface, all over the world from the mountain to the sea and particularly visible in tropical savanna, urban and periurban areas, the natural vegetation associated with waterways which is mostly represented by forests at various levels of inundation from permanent (swamp forests) to dry riparian forests is credited to be among the most species-rich ecosystems.[1] These forests which we now group as "aquaforests," "aquatic forests," "blue forests," or forested wetlands are important areas for global sources and quality and stream environment.[2] These forests are widely found in many large landscape in the world especially peat forests in the temperate zones, the Amazonian swamp forests, swamp forests in Malaysia, and the peat swamps of Indonesia—to name a few examples. In Africa, these forests abound with the most species-rich forests at least for woody plants in the water areas of West Central Africa (from Mt. Cameroon South East into Gabon) and in Northeast Madagascar.[1] In proportion to their area within a watershed, they perform more ecological and productive functions than do adjacent uplands.[3,4] They harbor a diversified flora and physical structure.[5] Ecologists have also examined the value of these forests, especially riparian forests, as habitats for many animals and recognized them as a priority area for conservation of terrestrial mammals,[6,7] fisheries as well as birdlife.[5,8,9] They are marginal among the wooded vegetation and because of their advanced state of degradation. Despite the fact that there is growing recognition of the ecological, hydrologic, biogeochemical, sociocultural, economic, and aesthetic importance of these forests in Africa, they have remained insufficiently studied and managed as key vegetation formation for biodiversity protection and sustained livelihoods for surrounding communities. Aquaforests have often been ignored or excluded from vegetation studies in favor of upland forests.

This entry sets to raise the status of such forests especially in Africa and to demonstrate their capacity and potentials to satisfy mankind's needs if sustainably managed under an emerging management regime of "aquaforestry." Hence, the entry is divided into two major parts: part 1—aquaforests deals with the definitions, characteristics, typology, distribution, structure, and functions of aquaforests including major benefits, threats, and drivers of degradation and loss of aquaforests; part 2—aquaforestry covers the essentials of strategies for sustainable management of aquaforests within integrated river basin approach for sustained production of environmental goods and services to satisfy mankind under the new management regime of "aquaforestry" or "aquatic forestry."

Aqua-Forests

Definitions

"Aquaforests," "aquatic forests," "blue forests," or "forested wetlands" is an umbrella term that we use to refer to diverse range of ecosystems with vegetation (trees, shrubs) and marshes that is permanently or temporally covered by either fresh or salt water or bordering water courses. Riverine landscapes are heterogeneous, dynamic, and biologically and spatially complex.[10] Within riverine landscapes, riparian forests are transitional between terrestrial and aquatic ecosystems and are distinguished by gradients in biophysical conditions, ecological processes, and biota. They are portions of terrestrial ecosystems that influence exchanges of energy and matter between aquatic and uplands ecosystems[4] and contribute to the diversity and function of both terrestrial and aquatic ecosystems.[11] At all latitudes, swamps, riparian, or streamside forests are recognized as distinct terrestrial ecosystems.[12] They occur in areas of intense land-water interaction and are known to support concentrated and diverse assemblages of wildlife and plant species landscape mosaic, spatial heterogeneity, complex environmental gradients, and unique natural disturbance regimes[13] which instigate a wide variety and abundance of resources and substrate.[11] These forests are adapted to life on banks of river estuaries and lagoons where salt-water flooding follows a daily cycle or to freshwater flooding of a permanent or seasonal nature in forest areas.[14]

General Characteristics of Aquaforests

Certain general characteristics of these forest habitats can be identified:

- *Critical role of hydrology in system sustenance and integrity.* As pointed out by Brinson (1990),[15] these forests own their dynamics, structure, and composition (species and habitat diversity and food webs) to river processes of inundation, sediments dynamics (transport of sediments), and biogeochemistry and nutrient cycling hence the erosive forces of water. The characteristic plant species, plant communities, and associated aquatic or semiaquatic animal species are intrinsically linked to the role of water as both an agent of natural disturbance and as a critical requirement of biota survival. Alternating environmental stress[16] such as periodic flooding is a form of disturbance to which many of the taxa occurring in riparian communities appear well adapted by sprouting from remaining root or trunk (e.g. *Pterocarpus santalinoides, Cola laurifolia, Szygium guineense, Cynometra megalophylla*).[17] According to Acker et al. (2003),[11] flooding apparently promotes complexity at the life form, species composition, and stand structural levels, thus the significance of channel constraint for severity of floods may be different in mountainous versus relatively flat terrain. Frequent floods are known to maintain the native riparian vegetation in a mosaic of different successional stages. [18,19]

Sediments deposited by floods can create initial seedbeds for germination but also injure seedlings.[20] Research on the effects of flooding have shown that flooding increases wood and leaf decomposition rates and sites that are flooded generally have a lower accumulation of organic matter.[21] As a result, these forests tend to be very diverse in species composition and physical structure.[22]

- *Fragile and vulnerable ecosystems.* They resemble island species in their vulnerability to environmental stresses. They have a limited range (often a single watershed, lake, or river system) and low population numbers.
- *Transition ecosystems.* These ecosystems are seemingly transitional ecosystems (ecotones) between terrestrial systems and open water systems but most heavily influenced by terrestrial ones. Where they occur, they attract species from both systems and are often very productive regions with high species richness.[23]
- *High litter production.* Mangroves and other aquaforests produce considerable litter such as leaves, fruits, and deadwood. While some of this detritus stay in the forests, the rest end up in estuaries. There, biological degradation by bacteria, fungi, and larger organisms results in detrital complexes that appear to be the most important source of energy for maintaining estuarine fisheries and supporting tropical marine coastal ecosystems.
- *High levels of primary productivity.* The high levels of primary production coupled with their characteristic habitat complexity provide the support base for some of the world's highest levels of biodiversity. Many support important populations of wildlife and plants including endangered species, terrestrial mammals such as primates, and often a spectacular concentration of birds. Numerous aquaforest areas form part of international flyways for migratory birds.
- *Aquaforests as biodiversity vegetation hotspots.* Hotspots of biodiversity are areas particularly rich in species, rare, endemic, threatened species, or some combination of these attributes.[24] These forests can be considered as hotspots of biodiversity for several reasons:
  - They are among the most vulnerable forest formations at all latitudes, yet of high ecological importance.
  - Their high species richness is relatively in small areas (patches and narrow corridors). Natta (2003)[17] collected one-third of the estimated species in the Benin flora in just 20 ha of riparian forests.
  - Their numerous species are rare, threatened, or with superior adaptability to a specific habitat.
  - They are habitat for an endemic plant species in a fire-prone environment.
  - They are a vital ecosystem for numerous wild and birdlife species.
  - They protect many water courses all over the landscape.

Typology and Distribution of Aquaforests

Aquaforests are associated with natural or artificial water bodies or courses (streams, rivers, lakes, tidewaters, etc.) from mountain (limit of tree-line) to the sea. They can be classified based on landscape units; dominant vegetation, hydroperiod, and salinity broadly into (see Figures 2.1 and 2.2):

- *Inland aquaforests*
  - Upland watersheds and catchment forests
  - Swamp forests and marshes
  - Riverine/riparian/fringing/gallery forests
  - Periodically inundated/floodplain forests/flooded forests
- *Coastal and marine aquaforests*
  - Transitional swamps
  - Intertidal forests (mangrove forests)

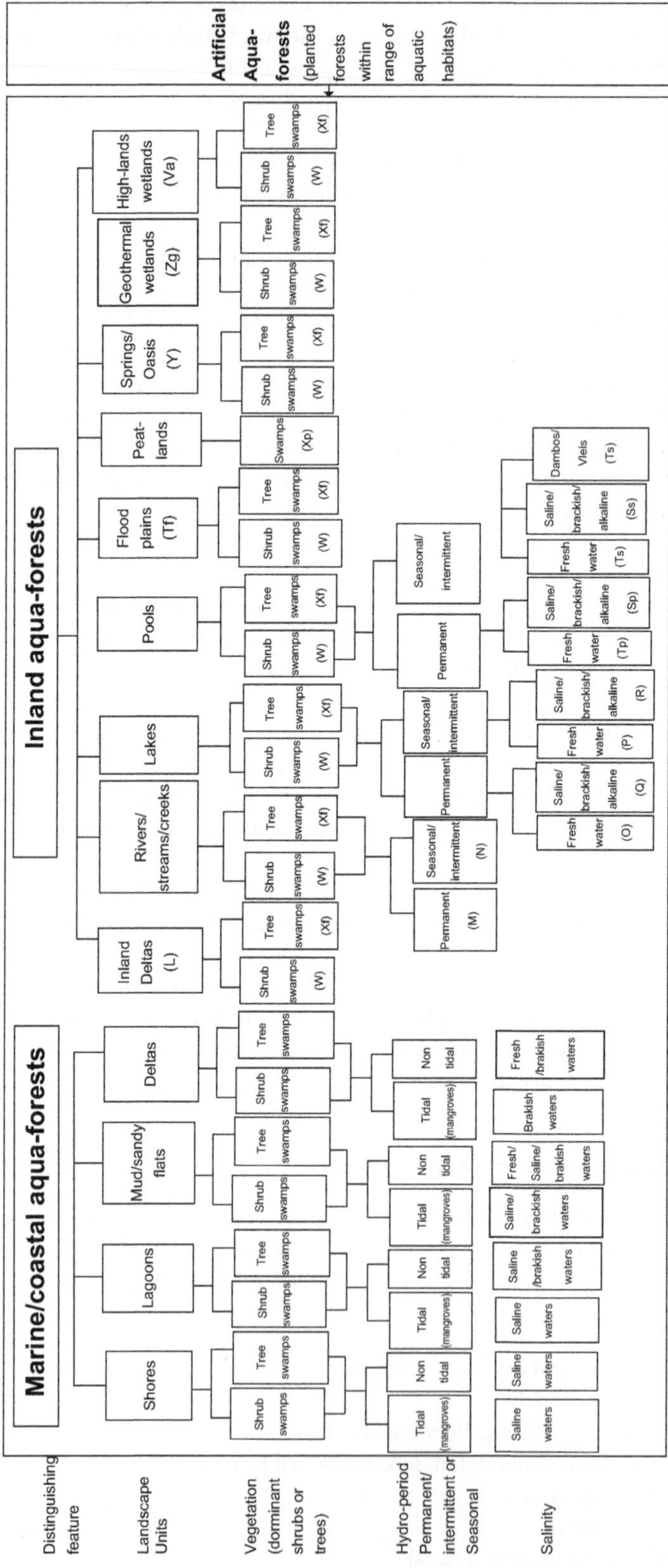

FIGURE 2.1 Attempted classification of aquaforests following Ramsar classification of wetlands habitats and codes.

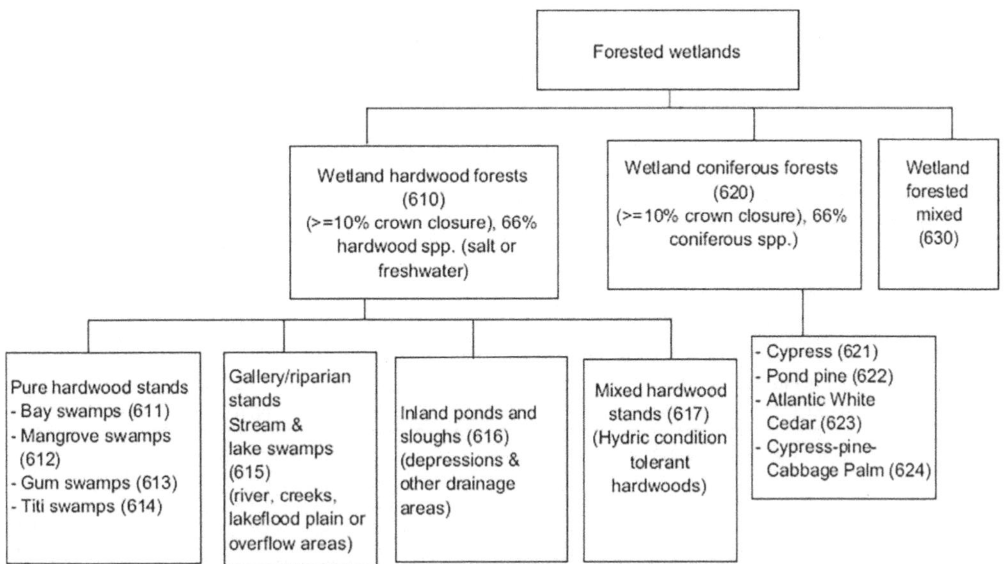

FIGURE 2.2 Classification following The Florida Land Use Cover and Forms Classification System (FLUCCS) and codes.

- *Artificial aquaforests*
    - Planted forests around rivers or lakes, e.g., *Raphia* forests
    - Planted mangroves

Swamp forests are found in still waters around lake margins and certain parts of floodplains. These typically consist of *Ficus* species, borassus palms, and *Syzgium.*[25] Extensive areas are found in the Congo basin and the Niger Delta. The Guineo-Congolian swamp forests have a diversified endemic flora. Cameroon's swamp forests along the Nyong River are unusual and important representatives of this type of habitat.

Mangrove is a broad name used to describe over 50 species from five families worldwide, which have been adapted to live in sheltered low-wave action intertidal areas within the tropical and subtropical areas. They are more intensive in the deltas of large rivers, but also occur along small bays and lagoons. The major element of mangrove community is generally comprised of a pure stand of a particular mangrove species; each stand may have a unique associated fauna. Table 2.1 shows the global distribution of mangrove forests now occupying less than 140,000 km^2, of which most (33.4%) is in Southeast Asia and 15% in South America. In Africa, mangroves are found discontinuously from Senegal to Angola on west coast and from Somalia to South Africa on the east coast. Extensive stands are also found in Madagascar.

Structure and Function of Aquaforest Ecosystems

A cross-section through any aquatic forest habitat will display both its horizontal and vertical structure (see Figure 2.3).

Horizontal Structure

A horizontal structure will display a water body within the land mass with different substrates due to various levels of inundation from submerged (swamp forest) within open water through temporarily inundated to dry forests on firm ground.

TABLE 2.1 World Distribution of Mangrove Areas per Regions

Region	Area (km²)	Proportion of Global (%)
East and South Africa	7.917	5.2
Middle East	624	0.4
South Asia	10.344	6.8
Southeast Asia	51.049	33.4
East Asia	215	0.1
Australia	10.171	6.7
Pacific Ocean	57.17	3.8
North and Central America	22.402	14.7
South America	23.882	15.7
West and Central Africa	20.04	13.2
Total	134.325	100.0

Source: Adapted from Spalding et al. (2010).

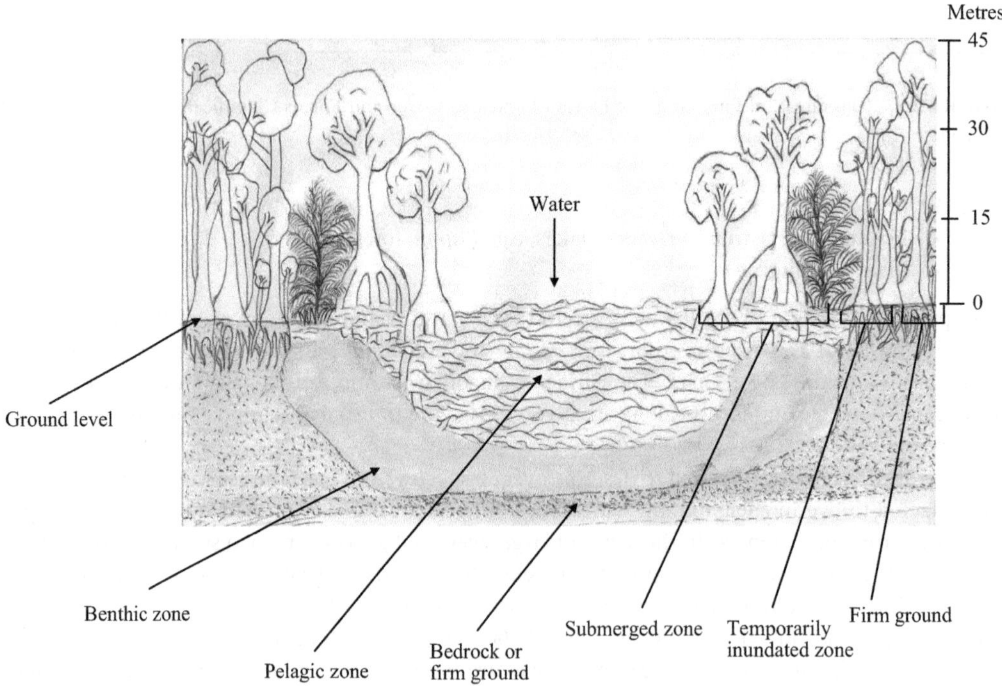

FIGURE 2.3 Cross-section through aquatic forest habitat showing horizontal and vertical layers.

Vertical Structure

The vertical structure of any aquaforest will show the characteristic three tree-layered strata of any matured forest depending on the ecological region on land boundary (the upper tree stratum: 30–45 m, the middle tree stratum: 23–30 m and lower tree stratum: 9–23 m), and a shrub layer: <9 m also consisting of young trees with a herb layer also with tree seedlings with an admixture of herbs which may attain 1 meter.[26]

For the deep water part such as any water body, a vertical section through it will reveal two distinct zones: the benthic zone which consisting of the layers closer to the bottom of the water body above the bottom sediment varying from a few cm to several 1000s of meters as in the open ocean.

The biotas that dwell in this zone (termed benthos) consist of organisms that have adapted to tolerate low temperatures and high pressure, as well as low oxygen levels found in this zone. Many of them have bottom dwelling adaptations. Since light cannot penetrate this depth, this zone lacks the ability to photosynthesis as its energy source. The main energy source of this zone consists of organic materials that drift down the upper layers, and this region is dominated by detritivores and scavengers. The second layer is the pelagic zone (Greek meaning: "open sea") which includes the free water column that interacts with the surface layers of a water body directly with the atmosphere. Physical and chemical properties of this zone vary greatly because of the vastness of this area, which extends from the uppermost waters down to the deeper layers near the benthic zone of a water column. As the depth increases, favorable life-sustaining features of the pelagic zone reduce, resulting in a decrease in the biota, as well.

A riverine or aquaforest is an ecosystem with dynamics, structure, and composition (species and habitat diversity and food webs) that depend on river processes of erosion, sediment transport, flooding inundation, and alluvial deposition.[15] These processes are briefly described as follows:

- *Nutrient flow.* Rivers and streams move nutrients and minerals from watershed and lowland areas to the sea and all of the points between. For example, floodplains are extraordinarily productive agricultural areas as rivers transport organic material and sediments over the floodplain soil. The flooding also releases terrestrial nutrients into the aquatic system, generating increased aquatic plant production and food for spawning fish and their young.[27] These also play a role in sustaining dryland gazing systems.
- *Species distribution.* The distribution of both terrestrial and aquatic species can depend on one another. Quite a few African amphibians, reptilian, bird, and mammal species are found near riverine or floodplain habitats, such as buffaloes, certain antelopes, snakes, and cane rats.[28] Similarly, terrestrial wildlife depends on ephemeral rivers in arid areas, for both water and the associated plant life. The distributions of some floodplain tree species depend on animal activity such as hippopotamus that influence the distribution of *Acacia albida* along the Luanga and Zambezi river valleys in Zambia. The hippo and other wildlife eat the seeds, enhancing germination and fostering dispersion.
- *Food webs.* Riparian habitats provide leaves, fruits, fruits, seeds, flowers, and branches to river and streams. A number of freshwater fishes and invertebrates depend on this terrestrial plant material for shelter and food. This is important as Lowel-McConnell[29] remarked that certain lakes (Ntomba and Mai-Ndombe in the Congo basin and the Great Lakes—Lakes Tanganyika, Malawi, and Victoria) are particularly important for the evolution of Africa's endemic fishes, interestingly the richness of the fauna especially in Lake Ntomba appears to be due to food and nutrients from the surrounding forest, the water in the lakes being impoverished in nutrients.

Aquaforests Relationships with Upland Communities

Forested wetlands occur in various climatic, geomorphic, edaphic, hydroperiod, and geographic settings and have similar as well as contrasting features with surrounding upland plant communities with regard to species composition, structure, and ecosystem diversity.

In most savanna regions, narrow bands of vegetation in the vicinity of the river contrast strongly with open forests and savanna woodlands which otherwise dominate the landscape.[5] Numerous studies confirm not only the great richness of these forests over upland forests.[30] It is now accepted that these forests harbor more species of birds[8] and mammals[6] than upland habitats. Roché[31] found bird species richness in riverine flood plains to be twice that in adjacent uplands while[32] documented that the majority of bird species in arid regions of Great Basin (USA) are associated with riparian habitats. Also in the intensive natural landscape of Australia's tropical savanna, species richness and abundance of birds were significantly greater in riparian forests than in the matched riparian forests areas and riparian vegetations allow many species to extend their distributions into lower rainfall areas.[5] The physical and vegetation characteristics of the streamside area differ from those upslope because of frequent inundation, soil saturation, and physical disturbance of streamside vegetation due to flood flows, mass soil movement, etc.[23,33]

On the contrary, along the Helena River and its tributaries and in the Grampians (Western Australia), riparian forests were generally less species rich than adjacent upland communities.[16,34] According to these authors, their results were corroborated by other studies in other parts of Australia and South Africa,[35] but not in Europe. Riparian forests in the Oregon Coast Range are said to have lower tree density than upland stands.[36,37] Reasons for such richness pattern have not been fully investigated.

Light penetration that depends upon topographic position, slope, and vegetation characteristics plays a major role in the ecology of the riparian forest and protection of the light environment is generally seen as an important function of buffers that aims to protect riparian habitat values.[38] The ecotone between the riparian and upland vegetation may be very narrow or gradual depending on the fact that gradients are sharp or not. In the latter case, it could be the site of highest species richness because of the presence of species from both communities.[16,39] In Benin, field observations by Natta (2003)[17] show that riparian forests tree flora is not only more diverse but also has marked differences in structure, abundance, and composition with surrounding upland plant communities (e.g., in Yarpao, Pénéssoulou, Idadjo, etc.).

Benefits of Aquaforests

A large body of work has demonstrated that such forests, especially riparian forests, play a critical role in regulating interactions between terrestrial and aquatic components of temperate zones landscapes[23,40] however, there have been relatively few studies dedicated to them in the tropics[41] in general. Aquatic ecosystems provide innumerable ecological and economic benefits (see Table 2.2), for example:

- *Provide habitat corridors.* As tropical forests become more fragmented due to deforestation, riparian forests play a crucial role in providing habitat corridors between forest patches to increase landscape connectivity.[42]
- *Protection.* Aquaforests serve as "green-shields" helping to stabilize shores, prevent erosion, and protect many coastal areas from storms through their damping of wind and wave action and their retention of the bottom sediments by plant roots. Trees lining the sides of rivers help to prevent erosion of the banks, while those lining the shores of lakes play a critical role in the ecological balance of the lakes. Wetlands also control floods by absorbing floodwaters and releasing runoff evenly.
- *Wastewater treatment and reduction of sediment loading.* Forested wetlands provide wastewater treatment playing a central role in water purification (are therefore ecofilters) preventing eutrophication by absorbing nutrients and retaining sediments and toxicants and also aiding in flood control and protection of coastal areas. If they are destroyed, these functions must be achieved artificially at a significant cost!
- *Perenniating water supply.* Aqua forests help to ensure a year-round water supply by their role in groundwater recharging (i.e., the movement of water from the wetlands into the underground aquifers) and discharging (i.e., when underground water moves upward into the wetland). It has been shown that where wetlands have been lost the hydrological cycle is affected and water supply during the dry season has been lost.[25]
- *As fisheries and other aquatic fauna habitats enhance fishery production. Forested wetlands including mangrove forests serve as feeding, breeding, hatching, and nursery grounds for*

TABLE 2.2 Ecosystem Services in Aqua-Forest Habitats

	Provisioning
Food	Production of fish, wild game, fruits, and grains
Fresh water	Storage and retention of water for domestic, industrial, and agricultural use
Fiber & fuel	Production of timber and fuel wood, peat, fodder
Biochemical	Extraction of pharmaceuticals, natural products including salts, medicines, and other materials from biota
Genetic materials	Genes for resistance to plant pathogens, ornamental species, etc.
Others	Extraction of minerals and other natural products, space for ports/ transportation
	Regulating
Climate regulation	Source of and sink for greenhouse gases; influence local and regional temperature, precipitation, and other climatic processes
Water regulation (hydrological flows)	Groundwater recharge/discharge
Water purification & waste treatment	Retention, recovery and removal/filtration of excess nutrients, other pollutants, and waste water
Erosion regulation	Retention of soils and sediments
Natural hazard regulation	Flood control, storm protection
Pollination	Habitats for pollinators
Others	Waste disposal
	Cultural
Spiritual & inspirational	Source of inspiration; many religions attach spiritual and religious values to aspects of wetland ecosystems
Recreational	Opportunities for recreational and touristic activities
Aesthetic	Many people find beauty or aesthetic value in aspects of wetland ecosystems
Educational	Opportunities for formal and informal education and training
	Supporting
Soil formation	Sediment retention and accumulation of organic matter
Nutrient cycling	Storage, recycling, processing and acquisition of nutrients
	Option Use Values
	Future unknown and speculative benefits. Currently unknown potential future uses of aquaforest biodiversity

Source: Adapted from Millennium Ecosystem Assessment Ecosystems and Human Well Being (2005) and Forest Trends (2010).

numerous commercially important fin and shell fish species. African river basins with associated wetlands are highly productive ecosystems with an estimated 40% of fish in Africa coming from riverine and floodplain fisheries.[43] Fish also represent an irreplaceable source of protein (about 25%) for a continent in which there is often a widespread and chronic shortage of animal protein. Most of African lakes and rivers and a number of African wetlands support tribes whose livelihood has historically depended upon fishing.

- *Aquaforest products and resources.* Wetlands are the source of a variety of products which can be harvested sustainably such as fish, fodder, timber, nontimber forest products (e.g., resins and medicines), agricultural crops, and wildlife resources (e.g., meat, skins, honey, and both bird and turtle eggs). Worldwide, people use the products of mangroves and the mangrove environment for subsistence.

- *Energy generation.* Most aquaforests provide vast sources of energy, especially mangroves provide fuel wood for cooking and fish preservation.
- *Tourisms and other cultural values.* Aquaforests and wetlands offer great potential for tourism because they are important to both terrestrial and aquatic life. In one spot, tourists can view both large mammals and an astonishing abundance of waterfowl.
- *Climate change indicators and mitigation.* Wetlands often provide visible indicators of climate change when water courses (streams, rivers, and lakes) begin to dry off. They are also considerable areas for carbon capture and storage. The mangrove ecosystems have the capacity to store great amounts of organic carbon; significant enrichments of organic carbon have been reported even at several meters of depth.[44,45]

Threats to Aquaforests, Drivers, and Consequences of Degradation and Loss

Despite the growing recognition of the ecological, hydrologic, biogeochemical, sociocultural, economic, and aesthetic importance of these forests, they are degraded or lost at an alarming rate due to factors mostly anthropogenic:

- Pollution

  Pollution from farms, cities, and factories can affect adjacent aquatic areas as well as areas further downstream. Agricultural threats include both pesticide and fertilizer runoff. The chemicals in pesticides runoff become more concentrated and toxic as they work their way up the food chain. They accumulate in the bodies of fish and other higher-level organisms. Fertilizer runoff increases the nutrient loading in waters, thereby causing eutrophication and algal blooms. This impacts fisheries and drinking water and reduces biological diversity.[46] Municipal pollution also increases eutrophication as the greater the number of people living along a river, the greater the amount of nitrates in the river.[47] Industrial pollution mainly threatens coastal waters as the oceans are utilized by most nations worldwide as a vast dumping ground for waste. It is estimated that more than 90% of all chemical refuse and other material entering coastal waters remain there in sediments, wetlands, fringing reefs, and other coastal ecosystems. The most widespread and serious chemical pollutants are hydrogenated hydrocarbons (such as pesticides, herbicides, and PDBs), heavy metals, petroleum products, and fertilizers.[48] These substances can cause tumors and disease in estuarine and coastal fish. In addition, plastic and other debris, such as pieces of nets, entangle and kill a variety of aquatic animals.
- Aquaforest site conversion

  For agricultural production

  The rapidly increasing human population in Africa and other developing nations puts heavy demand on increased food production, leading to agricultural conversion of critical wetlands.

  Conversion not only damages the environment but seldom works mainly due to the development of acid sulfate soil conditions. Many mangrove areas have soil containing large amounts of pyrate sulfur that when exposed to air, oxidize to release sulfuric acid. The soil then becomes extremely acidic and very high in soluble salts, a condition which often leads to nutrient and fertilizer problems resulting in the impairment and even failure of most crops especially rice, coconut, etc. Even if drained land does produce reasonable yields for several years, long-term sustainability on the site is virtually impossible because of the resulting environmental degradation. Wetland drainage can have additional effects of disrupting the water supply of nearby towns affecting the quality of water and altering the microclimate.

  For aquaculture production

  Improperly sited aquaculture projects can greatly damage freshwater ecosystems. If cultured stocks escape to the wild, they can affect native gene pools. In addition, the potential spread of pests and diseases poses a serious risk to wild stocks since the spread of farmed species happens all too often. From experience, Bartley and Minchin (1995)[49] noted that complete containment of exotic species in aquaculture facilities is nearly impossible. Unfortunately, economic analyses of aquaculture often neglect the economic cost of conversion. For example, for mangrove conversion to aquaculture ponds, Dixon (1989)[50] noted that usually only the marketable mangrove forestry products are considered, ignoring the economic benefit mangroves provide to fisheries both within mangroves and in nearly coastal and estuarine ecosystems. Further, this neglects the fact that some products or services such as nutrient flows to estuaries do not have market prices. Finally, the decision as to whether or not mangroves should be converted is made by comparing the minimum partial estimate with the total expected benefit from conversion. No wonder mangroves are being lost as such a rapid pace. In fact, large areas of mangrove have been cleared worldwide for the establishment of shrimp mariculture. In addition to the destruction of mangroves, shrimp farming reduces the area of estuaries due to the diking of sand and mudflats to create ponds and water intake channels and changes estuarine flow by channelization and by controlling the flow of water into the ponds. Waterfowl numbers decline due to habitat degradation and in some cases through routine shooting of those birds believed to eat shrimp.

- Deforestation from within and adjoining areas

  Due to grazing, dams, embankments or other urban infrastructural development projects

  Deforestation damages both freshwater and marine ecosystems. The deforestation of riparian areas due to grazing, dams, embankments, or other urban infrastructural development projects can cause stream temperatures to fluctuate widely, reduce dissolved oxygen concentrations and reduce the contribution of terrestrial nutrients, including leaf letter to aquatic food webs. The erosion that results from deforestation can change the shape of a stream or river harming the quality of spawning grounds and habitat. Sedimentation can also smother spawning grounds and reduce habitat complexity of rocky substrates. Sediments harm freshwater food webs in a variety of ways. They reduce the rate of light penetration and hence rate of photosynthesis. They cover benthic algae, reducing foraging efficiency in herbivorous fish. They reduce the nutritional value of detritus (organic waste).[51] They also interfere with the feeding apparatus of filter-feeding organisms. Sedimentary particles themselves can abrade the bodies of aquatic organisms. Fortunately, maintaining strip of riparian forest can do much to minimize these effects.

  Wood bioenergy issues

  Vast areas of aquaforests are also devastated to meet rising energy demands in both rural and urban areas especially in developing countries where wood is a major source of energy. Throughout the coasts of West and Central Africa, mangrove forests are largely cut to preserve fish catch through smoking.[52–54]

  Fire

  Riparian plants are not dependent upon fire for renewal, but fire can influence the composition and structure of riparian ecosystems,[55] in particular the understory in combination or not with other factors such as flooding.[56] In Central New Mexico (USA), the suppression of flooding along RFs has also increased forest floor litter and woody debris which may have contributed to the increased frequency and severity of fires.[21]

- Water diversion

  Dams and water diversions are known to modify surface flow rates, flood periodicity, and sediment and nutrient transport often to the detriment of riparian plants.[20] When channeling occurs within riparian systems, removal of sediments and nutrients from surface runoff is less effective.[57] Large dams are constructed for various reasons such as to increase agricultural production from large-scale irrigation, generate hydroelectricity, control floods, and assure water supplies. The benefits of large dams are usually not as large as expected and their adverse effects are often severely underestimated. These include the following:

- *Habitat degradation and reduced biodiversity.* Dams directly impact riverine channel characteristics and habitat quality and greatly modify the river's influence downstream. They alter the timing and volume of river flow which reduces the sediment and nutrient transport downstream, alters temperature and salinity and at times even increases the river's acidity due to plant decomposition. Mangrove forest systems are influenced by reduction of freshwater inflows. The loss of such inflows affects the three most important factors in maintaining mangrove ecosystems: a sufficient amount of water, a sufficient supply of nutrients, and stability of the substrate. Mangrove forests degrade without periodic pulses of freshwater. They can also be degraded by saline fronts caused by dams. Channel mouth closure due to sedimentation is a growing threat to lagoonal mangroves in West and Central Africa especially in Ghana.[58]
- *Declining fisheries and long-established indigenous patterns of floodplain agriculture including livestock production.* Studies have demonstrated that fish yield is directly related to river and catchment size with rivers with extensive floodplains having a higher yield than those with smaller floodplains.[59] In addition to degrading riverine habitat, dams directly block fish migration, both of marine fish that spawn in rivers and riverine fish that spawn in marine realms. Dams can also disrupt the hydrological and chemical cues needed to introduce migratory and spawning behavior in fish. Finally, the profound changes caused by dams exacerbate the spread of exotic fish species. On the impact of dams on floodplain fisheries, construction of the Kainji Dam in Nigeria resulted in the loss of 50% of the fish catches in lands extending to 200 km downstream of the dam.[60] Dams also impact lacustrine and coastal fisheries due to irregular freshwater input.

- Encroachments from urbanization

  Many aquaforests areas have been encroached by increased urbanization from burgeoning population pressures and so are lost at a rapid rate. Urbanization on lands adjacent to intact riparian woodland has substantial impacts on riparian bird species richness, density, and community composition.[61,62] Likewise, invasion of the naturalized shrubs potentially alters competitive hierarchies and disturbance regimes in riparian systems.[56]
- Climate change

  Wetlands are one of the most vulnerable ecosystems to global warming. Effects are more visible when water courses (streams, rivers, and lakes) begin to dry off. Although considerable areas for carbon capture and storage can also rapidly emit carbon dioxide when degraded or destroyed.
- *Policy deficiencies, overlapping jurisdiction, and inadequate planning.* Above all else, aquatic systems require integrated planning. Too often, however, governmental agencies consider aquatic systems only for a single purpose rather than for multiple uses benefiting a variety of sectors. A number of government agencies (fisheries, tourism, urban sanitation and water, agriculture, and forestry) can have responsibilities over aquatic areas and resources leading to confusion, inertia, and ineffective planning. Overlapping jurisdiction can occur horizontally (within the same level of government) and vertically (across different levels of government from local authorities to regional). In the case of wetlands, for example, despite increasing efforts to conserve these areas, many wetlands are lost because of competing government priorities. Governments can have a stated commitment to wetlands conservation while at same time their national agricultural policy favors wetlands drainage!

  Subsidies according to Shumway (1999)[46] often lead to waste and misuse of aquatic resources. These include agricultural subsidies for pesticides, fertilizers, and water, forestry subsidies which can exacerbate deforestation and erosion in marginal and vulnerable areas and tax relief for overseas investors with little incentives to sustainably manage a given resource.

- *Introduction of exotic species.* According to Shumway (1999)[46] citing other sources, invasion of exotics is believed to be the second greatest threat to global biodiversity after habitat loss. Islands, coastal estuaries and lakes are at particular risk. Some are deliberately introduced with good intentions of filling apparently "empty" niches and producing more big fish, to control unwanted organisms to improve sport fishing and to control fish production in ponds through the introduction of predators and also with the introduction of the now extensive West-Central African invasive mangrove palm *(Nypa fruticans)* from Asia to Nigeria in 1906. Others have been introduced accidentally through translocation especially as most African river systems are connected by tunnels, pipes, and canals for the sake of development infrastructure. In Africa, two South American species have caused considerable havoc to wetlands systems: water hyacinth *(Eichhornia crassipes)* and *Salvinia molesta.* Both species rapidly colonize water bodies, forming a dense floating mat of interlocking plants.[23] These species crowd out native vegetation, reduce light penetration, limit water column mixing, and increase detrital inputs. Kaufman and Ochumba (1993)[63] report that a single water hyacinth plant can produce 140 million daughter plants per year enough to cover 140 ha with a weight of 28,000 tons, the seeds can surviving for 30 years in mud, making the plants almost impossible to eradicate completely.
- *Over exploitation of aquaforest resources.* Overfishing and postharvest losses due to lack of infrastructure for handling/processing facilities which causes waste. In most African countries between 15–25% the catch is lost due to spoilage, wastage at the time of capture, insect infestation, or improper handling and storage. Some countries have postharvest losses of up to 50% particularly those where smoking and drying are the normal preservation methods.[46]

Managing Aquaforests (Aquaforestry or Aquatic Forestry)

Definitions and Scope of Aquaforestry

Given the aforementioned ecological, hydrologic, biogeochemical, sociocultural, economic, and aesthetic importance of aquaforests and their threats to degradation and loss, there is a need to manage these forests sustainably to perpetuate their goods and services to mankind through desired land-use management practices of aquaforestry. We therefore use the term "aquaforestry" to include all sustainable land-use management practices aimed at perpetuating the ecological services of aquaforests within the framework of the integrated basin approach. This is not much different from the traditional forestry practices but needs to be highly ecological, integrative, and participative in all its dimensions to make it produce the desired land use management change for aquaforests. The challenging task of sustainable aquaforest management is being confronted by the need to assess these resources (aquaforest inventory and mensuration) the outcome of which can produce management practices that may range from conservation, sustainable utilization to regeneration/restoration of degraded habitats.

Aquaforest Inventory and Mensuration

Effective management strategies must be devised to preserve and restore riparian corridors[64] and shores[65] which rely on accurate understanding of the structure and dynamics of aquaforests communities.[66] Forest inventories and mensuration is a means to obtain reliable information on the stock and dynamics of forest resources through various remote sensing and ground survey methods. Because of diversity of landscapes and tree forms in aquaforests, the traditional forest inventory and mensuration procedures[67,68] need to be adapted to aquaforests and especially for mangrove vegetation[69–71] with characteristic stilt roots.

Aquaforest Conservation

Cost-effective conservation of biodiversity requires that maximum biodiversity should be protected in a minimum area.[72] In proportion to their area within a watershed, aquaforests perform more ecological and productive functions than do adjacent uplands.[4] The vital ecological, hydrological, and biogeochemical vital functions of these forests bordering waterways can be protected through conservation that can preserve a large range of plants, animals, water sources, soils, and watersheds. In Amazon and the Congo basins, numerous streams and rivers provide huge potentials for increasing the conservation value of deforested and fragmented landscape through the protection of linear remnants along waterways.[73] Not only do they constitute a natural habitat or the last refuge for many species, but they also lodge many endemic species and extinction-menaced species[74] and usually act as routes for movement of terrestrial plants and animals across the landscape.[42] It should be noted that simply protecting aquaforests in a buffer zone may not be adequate to ensure their existence in the long term.[75] Instead, the management of these forests must be a component of good watershed or landscape management. Therefore, awareness has to be raised to various stakeholders of such forests as unique physical and natural systems in their own right and warranting special management and protection. Integrated management of these forests that optimize their values as habitat for native plants and animals requires planning and acting with all stakeholders at both site-specific and watershed levels. This can also be achieved within the framework of the emerging engagement of private sector and other land-use developers in various payments for ecosystem services arrangements.

Sustainable Utilization of Aquaforests

It is important that all human interactions within and outside aquaforests should embrace the sustainable exploitation concept. This range from consumptive uses (harvests of timber and nontimber forest resources) to nonconsumptive uses (cultural, tourisms, etc.). Appropriate technologies should be employed, especially the reduced impact logging approach.[76] In the Congo basin logging practiced,[77] aquaforests are preserved in logging concessions and classified under sensitive habitats barred from logging activities while efficient wood energy utilization technology is becoming widespread in Africa to reduce the traditional open-cooking ovens. In the mangrove zones of central Africa such improved fish smoking ovens or cook stoves could reduce as much as 60% wood used.[78]

Regenerating and Restoring Degraded Aquaforests

Aquasilviculture

Where aquaforests are degraded or lost altogether, they may be regenerated by natural regeneration methods and/or artificial regeneration methods through replanting (reforestation) or planting (afforestation) where they did not exist before. The silvicultural techniques will differ from one type of forest to the other and will require appropriate adaptation to the edaphic and hydrological peculiarities of aquaforests. For example, a *Rhizophora* dominated mangrove can be regenerated only through viviparous propagules not through cuttings as this widely spread mangrove genus in West-Central Africa does not coppice. Lagoon, estuaries, or coastal shores are opened to facilitate tidal movements to facilitate the establishment of mangrove stands.

Wetlands Farming

The natural rise and fall of rivers and streams can be used to advantage in farming. IUCN notes that if properly managed, natural wetland agriculture (i.e., agriculture that takes advantage of the normal flooding cycle rather than draining the site) can yield substantial benefits to rural communities. Natural wetlands farming occurs in the inland Delta in Mali,[79] the flood plains of the larger rivers in the Sudan and Sahel Savanna zones and estuarine swamps of the Upper Guinea coast. In the large floodplains, complex systems of flood-advance and flood-retreat agriculture have been developed.[25]

Conclusion

Aquatic forest ecosystems provide critical services for man and wildlife. They are important both to flora and to fauna living in water as well as those living on land. Aquatic biodiversity though threatened is still abundant especially in Africa with one of the world's richest treasures and equally importantly to African nations and peoples. We have the chance to conserve some of this wealth if action is taken now and alternatives are provided. It is not too late. Integrated approaches to conservation and development are necessary. The goal of aquaforestry or aquatic forestry is to seek to perpetuate aquaforests for sustained production of ecological services derived from this special and fragile forested habitats.

Acknowledgments

The author thanks the students and lecturers of the Institute of Fisheries and Aquatic Sciences of the University of Douala at Yabassi, Cameroon where aquaforests and aquaforestry have been taught as a series of courses and who greatly contributed to shaping the ideas presented in this entry. The authors also acknowledge support of the Cameroon Wildlife Conservation Society (CWCS) in collaboration with the Cameroon Ministry of Forestry & Wildlife (MINFOF) and Ministry of Environment, Nature Protection and Sustainable Development (MINEPDED) and funding partners through its Coastal forests and Mangrove Conservation Programme, Cameroon. We also thank Mr. Lissouck Benard and Ms. Jeanette Wiwa of Government High School Mouanko for the drawings in Figure 2.3.

Life forms (LF) following Raunkaier (1934), Schnell (1971) and Keay & Hepper (1954–1972) are

- Phanerophytes (Ph): megaphanerophytes (MPh: >30m), mesophanerophytes (mPh: 8–29m), microphanerophytes (mph: 2–7 m), and nanophanerophytes (nph <2m)
- Therophytes (Th: plant survive as seeds)
- Hemicryptophytes (Hc: perennating buds at soil surface)
- Chamaephytes (Ch: perennating buds near soil surface)
- Lianas (L)
- Cryptophytes (Ch): Geophytes (Ge: perennating buds underground), Helophytes (Hel: perennating buds underground in marshes), hydrophytes (hyd: perennating buds in water)
- Epiphytes (Ep)
- Parasites (Par)

Phytogeographic types (PT) following White (1986) and Keay & Hepper (1954–1972) are

- Species widely distributed in the tropics (Cosmopolitan – cosmo, pantropical – pan, Afro-American – AA and Paleotropical – Paleo)
- Species widely distributed in Africa (Tropical Africa –TA, Pluri Regional in Africa – PRA)
- Regional species in Africa (Sudanian – S, Guinean – G, Sudano-Guinean – SG, Sudano-Zambesian – SZ, Guineo Congolian – GC)

APPENDIX 1 Typical Aquaforest Higher Plant Species of West and Central Africa

						Habitat					
						Forest zone			Savanna zone		
No.	Species	Family	Habit	Life forms*	Geo affinity*	River banks/ fringing forests	Fresh water swamps	Tidal swamps/mangroves	River banks/ fringing forests	Fresh water swamps	Tidal swamps/ mangroves
1	*Acacia ataxacantha*	Mimosaceae	tree	mph	S				X		
2	*Aeschynomene elaphroxylon*	Papilionaceae	tree	mph	SG	X	X		X	X	
3	*Afzelia africana*	Caesalpiniaceae	tree	mPh	SG	X			X		
4	*Albizia malacophylla*	Mimosaceae	tree	mPh	SG				X		
5	*Albizia zygia*	Mimosaceae	tree	mPh	SG				X		
6	*Alchornea cordifolia*	Euphorbiaceae	tree	mph	TA	X	X		X	X	
7	*Allanblackia floribunda*	Guittiferae	tree	mPh	GC	X					
8	*Allophylus africanus*	Sapindaceae	shrub	nph	Pan	X	X	X	X	X	X
9	*Alstonia boonei*	Apocynaceae	tree	mPh	GC	X	X				
10	*Alstonia congensis*	Apocynaceae	tree	MPh	GC	X	X				
11	*Ancistrophyllum sp.*	Palmae	liana	mph	GC		X				
12	*Ancylobotrys amoena*	Apocynaceae	liana	nph	SG				X		
13	*Andıra inermis*	Papilionaceae	tree	mph	SG				X	X	
14	*Annona senegalensis*	Annonaceae	shrub	nph	TA	X			X		
15	*Anogeissus leiocarpus*	Combretaceae	tree	mph	SG	X			X		
16	*Anthocleista nobilis*	Loganiaceae	tree	mPh	GC		X				
17	*Anthocleista schweinfurthii*	Loganiaceae	tree	mPh	GC		X				
18	*Anthocleista vogelii*	Loganiaceae	tree	mPh	SG				X	X	
19	*Antiaris africana*	Moraceae	tree	mPh	SG				X		
20	*Antidesma venosum*	Euphorbiaceae	tree	mph	TA	X	X		X	X	
21	*Aphania sengalensis*	Sapindaceae	tree	mph	TA	X	X	X	X	X	X
22	*Avicennia germinans*	Avicenniaceae	tree	mPh	GC			X			X
23	*Berlinia grandiflora*	Caesalpiniaceae	tree	mPh	GC	X			X		

(Continued)

APPENDIX 1 *(Continued)* Typical Aquaforest Higher Plant Species of West and Central Africa

						Habitat					
						Forest zone			Savanna zone		
No.	Species	Family	Habit	Life forms*	Geo affinity*	River banks/ fringing forests	Fresh water swamps	Tidal swamps/mangroves	River banks/ fringing forests	Fresh water swamps	Tidal swamps/ mangroves
24	*Berlinia grandiflora*	Caesalpiniaceae	tree	mPh	SG				X		
25	*Borassus aethiopum*	Palmae	palm	mPh	TA				X		
26	*Bosqueia angolense*	Moraceae	tree	MPh	GC	X			X		
27	*Brachystegia eurycoma*	Caesalpiniaceae	tree	MPh	GC	X					
28	*Breonadia salicina*	Rubiaceae	tree	mph	SG				X		
29	*Bridelia micrantha*	Euphorbiaceae	tree	mPh	SG				X		
30	*Calamus sp.*	Palmae	liana	mph	GC		X				
31	*Campa procera*	Meliaceae	tree	mPh	GC	X					
32	*Cassia sieberiana*	Caesalpiniaceae	tree	mph	SG				X		
33	*Ceiba pentandra*	Bombacaceae	tree	mPh	SG				X		
34	*Celtis integrifolia*	Ulmaceae	tree	mPh	SG				X		
35	*Clausena ainisata*	Rutaceae	shrub	nph	SG				X		
36	*Cleistopholis glauca*	Annonaceae	tree	mPh	GC	X					
37	*Cleistopholis patens*	Annonaceae	tree	mPh	GC	X	X				
38	*Clerodendrum thyrsoideum*	Verbenaceae	liana	mph	SG						
39	*Cola laurifolia*	Stercubaceae	tree	mPh	GC	X			X		
40	*Combretum paniculatum*	Combretaceae	shrub	nph	TA	X			X		
41	*Combretum tomentosum*	Combretaceae	tree	mPh	SG	X			X		
42	*Cordia myxa*	Boraginaceae	tree	mph	SG				X		
43	*Cordia sinensis*	Boraginaceae	tree	mph	Pan	X	X		X	X	
44	*Crataeva adansonii*	Capparaceae	tree	nph	TA	X	X		X	X	
45	*Cynometra megalophylla*	Caesalpiniaceae	tree	mPh	GC	X	X				

(Continued)

APPENDIX 1 *(Continued)* Typical Aquaforest Higher Plant Species of West and Central Africa

						Habitat					
						Forest zone			Savanna zone		
No.	Species	Family	Habit	Life forms*	Geo affinity*	River banks/ fringing forests	Fresh water swamps	Tidal swamps/mangroves	River banks/ fringing forests	Fresh water swamps	Tidal swamps/ mangroves
46	*Cynometra sanagaensis*	Caesalpiniaceae	tree	mPh	GC	X	X				
47	*Cynometra vogelii*	Caesalpiniaceae	tree	mph	SG				X		
48	*Dalbergia melanoxylon*	Papilionaceae	tree	mph	Pan	X			X		
49	*Dalbergia sissoo*	Papilionaceae	tree	mph	Pan	X			X		
50	*Detarium senegalense*	Caesalpiniaceae	tree	mph	SG				X		
51	*Dialium dingklagei*	Caesalpiniaceae	tree	mPh	GC	X	X				
52	*Dialium guineense*	Caesalpiniaceae	tree	mph	SG	X			X		
53	*Diospyros mespiliformis*	Ebenaceae	tree	mPh	SG				X		
54	*Dracaena arborea*	Agavaceae	tree	mPh	GC	X	X				
55	*Endodesmia calophylloides*	Hypericaceae	tree	mPh	GC		X				
56	*Entada manii*	Mimosaceae	liana	mph	GC	X					
57	*Erythrophleum suaveolens*	Caesalpiniaceae	tree	mPh	SG	X			X		
58	*Eugenia nigerina*	Myrtaceae	shrub	nph	SG				X		
59	*Ficus abutilifolia*	Moraceae	shrub	nph	GC				X		
60	*Ficus asperifolia*	Moraceae	shrub	nph	GC	X					
61	*Ficus capreif olia*	Moraceae	shrub	nph	GC				X		
62	*Ficus dicranostyla*	Moraceae	shrub	nph	GC				X		
63	*Ficus exasperata*	Moraceae	shrub	nph	GC				X		
64	*Ficus glumosa*	Moraceae	shrub	nph	GC				X		
65	*Ficus natalensis*	Moraceae	shrub	nph	GC				X		
66	*Ficus ovata*	Moraceae	shrub	nph	GC				X		
67	*Ficus polita*	Moraceae	shrub	nph	GC				X		
68	*Ficus sur*	Moraceae	shrub	nph	GC				X		

(Continued)

APPENDIX 1 (*Continued*) Typical Aquaforest Higher Plant Species of West and Central Africa

						Habitat					
						Forest zone			Savanna zone		
No.	Species	Family	Habit	Life forms*	Geo affinity*	River banks/ fringing forests	Fresh water swamps	Tidal swamps/mangroves	River banks/ fringing forests	Fresh water swamps	Tidal swamps/ mangroves
69	*Ficus sycomorus*	Moraceae	shrub	nph	GC				X		
70	*Ficus thonningii*	Moraceae	shrub	nph	GC				X		
71	*Ficus trichopoda*	Moraceae	tree	mPh	GC				X	X	
72	*Ficus vallis-choudae*	Moraceae	tree	mPh	GC				X		
73	*Ficus vogelii*	Moraceae	tree	mPh	GC				X		
74	*Flacourtia indica*	Flacourtiaceae	tree	mph	SG				X		
75	*Gambeya africana*	Sapotaceae	tree	MPh	GC	X					
76	*Garcinia livingstonei*	Guittiferae	tree	mph	SG				X		
77	*Garcinia ovalifolia*	Guittiferae	tree	mph	SG				X		
78	*Gardenia imperialis*	Rubiaceae	tree	mph	GC	X	X				
79	*Guibourtia capallifera*	Caesalpiniaceae	tree	mph	SG				X		
80	*Guibourtia demeusei*	Caesalpiniaceae	tree	MPh	GC	X	X				
81	*Hippocratea africana*	Hippocrate aceae	liana	mph	SG				X		
82	*Holarrhena floribunda*	Apocynaceae	tree	mph	GC	X			X		
83	*83 Ipomoea carnea*	Convolvulaceae	shrub	nph	Pan	X	X		X	X	
84	*84 Irvingia smithii*	Simaroubaceae	tree	mPh	PRA	X					
85	*Jasminum dichotomum*	Oleaceae	liana	mph	SG				X		
86	*86 Kaya senegalensis*	Meliaceae	tree	MPh	SG				X		
87	*Keayodendron bridelioides*	Euphorbiaceae	tree	mPh	GC	X	X				
88	*Keetia Cornelia*	Rubiaceae	shrub	nph	SG				X		
89	*Faguncularia racemosa*	Combretaceae	tree	mPh	G			X			X
90	*Lecaniodiscus cupanioides*	Sapindaceae	shrub	nph	SG				X		
91	*Leptoderris brachyptera*	Papilionaceae	liana	LmPh	GC	X					

(*Continued*)

APPENDIX 1 *(Continued)* Typical Aquaforest Higher Plant Species of West and Central Africa

						Habitat					
						Forest zone			Savanna zone		
No.	Species	Family	Habit	Life forms*	Geo affinity*	River banks/ fringing forests	Fresh water swamps	Tidal swamps/mangroves	River banks/ fringing forests	Fresh water swamps	Tidal swamps/ mangroves
92	*Loeseneriella africana*	Hippocrateaceae	liana	mPh	SG				X		
93	*Lonchocarpus cyanescens*	Papilionaceae	shrub	mph	SG				X		
94	*Lonchocarpus grijfonianus*	Papilionaceae	tree	mPh	GC		X				
95	*Lonchocarpus sericeus*	Papilionaceae	tree	mPh	SG				X		
96	*Lophira alata*	Ochnaceae	tree	MPh	SG		X				
97	*Macrosphyra longistyla*	Rubiaceae	shrub	nph	SG				X		
98	*Malacantha alnifolia*	Sapotaceae	tree	nph	TA				X		
99	*Manilkara multinervis*	Sapotaceae	tree	mPh	TA	X					
100	*Militia excelsa*	Moraceae	tree	mPh	SG				X		
101	*Mimosa pigra*	Mimosaceae	shrub	nph	Pan	X					
102	*Mitragyna ciliata*	Rubiaceae	tree	MPh	GC	X	X				
103	*Mitragyna inermis*	Rubiaceae	shrub	nph	SG				X		
104	*Mitragyna stipulosa*	Rubiaceae	tree	mPh	SG		X				
105	*Morelia senegalensis*	Rubiaceae	shrub	mph	SG	X	X		X		
106	*Napoleonaea vogelii*	Lecythidaceae	tree	mph	G	X					
107	*Nauclea pobeguinii*	Rubiaceae	tree	mph	G		X				
108	*Neocarya macrophylla*	Chrysobal anaceae	tree	nph	SG				X		
109	*Olax subsorpioides*	Olacacea	shrub	nph	SG				X		
110	*Oncoba spinosa*	Flaco urtiaceae	tree	mph	SG				X		
111	*Ouratea glaberrima*	Ochnaceae	shrub	nph	GC	X					
112	*Oxytenanthera abyssinica*	Poaceae	bamboo	mPh	SG				X		
113	*Pachystela pobeguiniana*	Sapotaceae	shrub	nph	TA				X		

(Continued)

APPENDIX 1 *(Continued)* Typical Aquaforest Higher Plant Species of West and Central Africa

						Habitat					
						Forest zone			Savanna zone		
No.	Species	Family	Habit	Life forms*	Geo affinity*	River banks/ fringing forests	Fresh water swamps	Tidal swamps/mangroves	River banks/ fringing forests	Fresh water swamps	Tidal swamps/ mangroves
114	*Parinari congensis*	Chrysobalanaceae	tree	MPh	GC	X					
115	*Paullinia pinnata*	Sapindaceae	liana	nph	TA				X		
116	*Pericopsis laxiflora*	Papilionaceae	tree	mph	SG				X		
117	*Phoenix reclinata*	Palmae	palm	mPh	SG	X	X		X	X	
118	*Phyllanthus muellerianus*	Euphorbiaceae	shrub	nph	SG				X		
119	*Phyllanthus reticulatus*	Euphorbiaceae	liana	nph	SG	X			X		
120	*Phyllanthus welwitschianus*	Euphorbiaceae	shrub	nph	SG	X			X		
121	*Plagiosiphon longitubus*	Caesalpiniaceae	tree	mPh	GC	X	X				
122	*Psychotria calva*	Rubiaceae	shrub	nph	GC	X					
123	*Pterocarpus santalinoides*	Papilionaceae	tree	mPh	PRA	X					
124	*Raphia hockeri/vinifera*	Palmae	palm	mPh	SG	X	X		X	X	
125	*Raphia sudanica*	Palmae	palm	mPh	SG				X	X	
126	*Rhizophora harrisonii*	Rhizophoraceae	tree	mPh	TA			X			X
127	*Rhizophora mangle*	Rhizophoraceae	shrub	nph	AA			X			X
128	*Rhizophora racemosa*	Rhizophoraceae	tree	mPh	TA			X			X
129	*Ricinodendron heudelotii*	Euphorbiaceae	tree	mPh	GC	X					
130	*Rinorea spp.*	Violaceae	liana	mPh	SG				X		
131	*Rotula aquatica*	Boraginaceae	shrub	nph	Pan	X	X	X	X	X	X
132	*Romea thomsonii*	Connaraceae	shrub	mph	SG				X	X	
133	*Rytigynia senegalensis*	Rubiaceae	shrub	nph	SG				X		
134	*Saba comorensis*	Apocynaceae	liana	mPh	SG	X			X		

(Continued)

APPENDIX 1 *(Continued)* Typical Aquaforest Higher Plant Species of West and Central Africa

						Habitat					
						Forest zone			Savanna zone		
No.	Species	Family	Habit	Life forms*	Geo affinity*	River banks/ fringing forests	Fresh water swamps	Tidal swamps/mangroves	River banks/ fringing forests	Fresh water swamps	Tidal swamps/ mangroves
135	*Saba senegalensis*	Apocynaceae	liana	mPh	SG	X					
136	*Sacoglottis gabonensis*	Humiriaceae	tree	mPh	GC		X				
137	*Salix subserrata*	Salicaceae	shrub	nph	Pan	X	X	X	X	X	X
138	*Santaloides afzelii*	Connaraceae	shrub	nph	SG				X		
139	*Scyphocephalium mannii*	Myristericacee	tree	mPh	GC		X				
140	*Sesbania sesban*	Papibonaceae	shrub	nph	Pan	X	X		X	X	
141	*Sorindeia grandifolia*	Anacardiaceae	tree	mPh	GC	X					
142	*Spathodea campanulata*	Bignoniaceae	tree	mPh	GC	X					
143	*Spathodea tomentosa*	Bignoniaceae	tree	mPh	SG				X		
144	*Spondianthus preussii*	Euphorbiaceae	tree	mPh	GC	X	X				
145	*Spondias mombin*	Anacardiaceae	tree	mPh	GC	X			X		
146	*Sterculia tragacantha*	Sterculiaceae	tree	mPh	SG	X	X		X	X	
147	*Strombosia grandifolia*	Olacacea	tree	mPh	GC	X					
148	*Strophanthus sarmentosus*	Apocynaceae	liana	mPh	SG	X			X		
149	*Symphonia globulifera*	Guittiferae	tree	mPh	GC		X				
150	*Synsepalum brevipes*	Sapotaceae	tree	mph	GC	X					
151	*Syzygium guineense*	Myrtaceae	tree	mPh	TA	X	X				
152	*Taccazea apiculata*	Aslepiadaceae	liana	mPh	GC	X					
153	*Talbotiella batesii*	Caesalpiniaceae	tree	mPh	GC	X	X				
154	*Terminalia glaucescens*	Combretaceae	tree	mPh	SG				X		
155	*Terminalia laxiflora*	Combretaceae	tree	mPh	SG				X		

(Continued)

APPENDIX 1 ***(Continued)*** Typical Aquaforest Higher Plant Species of West and Central Africa

						Habitat					
						Forest zone			Savanna zone		
No.	Species	Family	Habit	Life forms*	Geo affinity*	River banks/ fringing forests	Fresh water swamps	Tidal swamps/mangroves	River banks/ fringing forests	Fresh water swamps	Tidal swamps/ mangroves
156	*Tetracera alnifolia*	Dilleniaceae	liana	mph	SG				X		
157	*Uapaca guineensis*	Euphorbiaceae	tree	mPh	GC	X					
158	*Uapaca heudelotii*	Euphorbiaceae	tree	mPh	GC	X					
159	*Uapaca togoensis*	Euphorbiaceae	tree	mPh	SG				X		
160	*Uvaria chamae*	Annonaceae	shrub	nph	SG				X		
161	*Vitex chrysocarpa*	Verbenaceae	tree	mPh	SG				X		
162	*Vitex doniana*	Verbenaceae	tree	mPh	SG				X		
163	*Vitex madiensis*	Verbenaceae	tree	mPh	SG				X		
164	*Voacanga africana*	Apocynaceae	tree	mPh	SG				X		
165	*Xylopia aethiopica*	Annonaceae	tree	mPh	GC	X			X		
166	*Xylopia aurantiiodora*	Annonaceae	tree	mph	GC	X	X				
167	*Zanthoxylum zanthoxyloides*	Rutaceae	shrub	nph	SG				X		
168	*Ziziphus spina-christi*	Rhamnaceae	shrub	nph	SG				X		
	Total					**75**	**43**	**9**	**109**	**19**	**9**
	As % of total					**44.6**	**25.6**	**5.4**	**64.9**	**11.3**	**5.4**

Source: Compiled from Arbornnier (2004); Thikakul (1985); Gledhill (1972); Etukudo, et al. (1994).

APPENDIX 2 Typical Aquaforest Higher Fauna Species of West and Central Africa

				Type of Aquaforests		
No.	Scientific Name	Family	Common Name	Fringing Forest	Fresh Water	Mangroves/ Coastal
1	*Albula vulpes*	Albulidae	Sea banana	0	1	1
2	*Alectis alexandrinus*	Carangidae	Jack fish	0	0	1
3	*Alestes sp.*	Alestidae	Alestes	0	1	1
4	*Alopochen aegyptiacus*	Anantidae	Ouette of Egypt	0	1	0
5	*Antennarius striatus*	Antenarridae	Frog fish	0	1	1
6	*Aplocheilichthys spilauchen*	Poeciliidae	Poeciliids	1	0	0
7	*Aristichthys nobilis*	Cyprinidae	Big carp head	0	1	0
8	*Arius spp.*	Bagridae	Cat fish	0	1	1
9	*Barbus sp.*	Gobeiidae	River barber	1	1	1
10	*Batrachoides liberiensis*	Batrechoididae	Toadfish	0	0	1
11	*Brachydeuterus auritus*	Haemulidae	Grunds	0	0	1
12	*Brycinus longipinnus*	Alestidae	African tetras	1	1	1
13	*Bufo regularis*	Bufonidae	Toad	1	0	0
14	*Carassius auratus*	Cyprinidae	Gilded (bronzed) carasin	0	1	0
15	*Carcharhinus leucas*	Carcharhinidae	Shark	0	0	1
16	*Chrysichtys nigrodigitatus*	Bagridae	Cat fish	0	1	0
17	*Cirrhina molitorella*	Cyprinidae	Mud carp	0	1	0
18	*Clarias gariepinus*	Claridae	Cat fish	1	1	0
19	*Clupea harengus*	Clupeidae	Herring	0	0	1
20	*Coryphaena hippurus*	Coryphaenidae	Sea bream	0	0	1
21	*Crocodilus fuscus*	Crocodylidae	Caiman	1	1	0
22	*Crocodylus sp.*	Crocodylidae	Crocodile	0	1	1
23	*Ctenopharyngodon idella*	Cyprinidae	Herbivorous carp	0	1	0
24	*Cynoglosis spp.*	Soleidae	Sole	0	1	1
25	*Cynoglossus senegalensis*	Soleidae	Sole	0	0	1
26	*Cyprinus carpio*	Cyprinidae	Common carp	0	1	0
27	*Dasyatis spp.*	Rhinobatidae	Skate	0	1	1
28	*Dermocherys coriacea*	Dermochelyidae	Tortoise	0	1	1
29	*Dicentractchus labrax*	Serranidae	Common bar	0	0	1
30	*Dorminator lebretonis*	Eleotridae	Sleeper	0	1	1
31	*Dreissena polymorpha*	Dreissenidae	Streaked mussel	0	1	0
32	*Drepana africana*	Drepaneidae	Disk	0	0	1
33	*Elops lacerta*	Elopidae	Ladyfish	0	1	1
34	*Engraulis encrasicolus*	Engraulidae	Anchovies	0	0	1
35	*Ephippion guttifer*	Tetraodontidae	Puffer	1	0	0
36	*Epinephelus sp.*	Serranidae	Grouper	0	0	1
37	*Ethmalosa fimbriata*	Clupeidae	Wadding	0	1	1
38	*Fluviatilis sp.*	Fluviatilis	Ecrevisse	0	1	1
39	*Fodiator acutes*	Exocotidae	Wheel fish	0	0	1
40	*Galeoides decadactylus*	Polynemidae	Threadfins	0	0	1

(*Continued*)

APPENDIX 2 (*Continued*) Typical Aquaforest Higher Fauna Species of West and Central Africa

				Type of Aquaforests		
No.	Scientific Name	Family	Common Name	Fringing Forest	Fresh Water	Mangroves/ Coastal
41	*Gymnarcus niloticus*	Gymnarchidae	Aba – aba, Frankfish	0	0	1
42	*Gymnura micrura*	Gymnuridae	Butterfly-rays	0	0	1
43	*Hemichromis faciatus*	Cichlidae	Panther fish	0	1	0
44	*Hemiramphus balao*	Hemiramphidae	Halfbeaks	0	0	1
45	*Hepsetus odoe*	Hepsetidae	Pikes	0	1	1
46	*Hetérobrancus longifilis*	Claridae	Cat fish	1	1	0
47	*Heterotis niloticus*	Osteoglossidae	Kanga	0	1	0
48	*Hippoglossus Hippoglossus*	Pleuronectidae	Halibut	0	0	1
49	*Hippoglossus stenolepis*	Pleuronectidae	Halibut	0	0	1
50	*Hydrochoeris hydrochoeris*	Hydrochoeridae	Water pig	0	1	0
51	*Hydrocynus forskalii*	Alestidae	Tiger fish	0	0	1
52	*Hypleurochilus langi*	Blennidae	Blennie	0	1	1
53	*Hypophtalmichthys molitrix*	Cyprinidae	Silver carp	0	1	0
54	*Inia geoffrensis*	Platanistidae	Fresh	0	0	1
55	*Labeo sp.*	Cyprinidae	Labeo	0	1	0
56	*Lates niloticus*	Centropomidae	Captain	0	1	0
57	*Lepomis gibbosus*	Centrachidae	Perche sun	0	1	0
58	*Lisha africana*	Clupeidae	Clupeids	0	1	1
59	*Litjanus agennes*	Lutjaidae	Snapper	1	1	1
60	*Liza sp.*	Mugilidae	Mullet	0	1	1
61	*Malapterus electricus*	Malapteruriade	Electricial fish	0	1	1
62	*Mugil cephalus*	Mugilidae	Mullet	0	1	1
63	*Mylopharyngodon piceus*	Cichlidae	Black carp	0	0	1
64	*Myocastor coypus*	Myocastoridae	Coypu	0	0	1
65	*Myrophis plumbeus*	Ophichthyidae	Snake eel	0	1	1
66	*Oncorhynchus mykiss*	Salmonidae	Rainbow trout	0	1	0
67	*Ondatra zibethicus*	Muridae	Muskrat	0	1	0
68	*Oreochromis mosambicus*	Cichlidae	Tilapia	0	1	0
69	*Oreochromis niloticus*	Cichlidae	Nile Tilapia	0	1	0
70	*Oxyura jamaicensis*	Anantidae	Fuzz Erismature	0	1	0
71	*Pangasius gigas*	Pangasiides	Mekong's fish	1	1	1
72	*Parachanna abscura*	Channidae	Viper fish	1	1	0
73	*Paraconger arisona*	Ophichthyidae	Conger	0	1	1
74	*Penaeus sp.*	Peneidae	Shrimp	0	1	1
75	*Periphtalmus sp.*	Gobeiidae	Periophtalm	0	0	0
76	*Poecilia latipina*	Poeciliidae	Molly	0	0	1
77	*Poecilia reticulata*	Poeciliidae	Guppy	0	0	1
78	*Polydacrylus sp.*	Polynemidae	Small African threadfin	0	0	1

(*Continued*)

APPENDIX 2 (*Continued*) Typical Aquaforest Higher Fauna Species of West and Central Africa

				Type of Aquaforests		
No.	Scientific Name	Family	Common Name	Fringing Forest	Fresh Water	Mangroves/ Coastal
79	*Polypterus senegalus*	Polyteridae	Senegal bichir	0	0	1
80	*Pomedasys jubilini*	Sparidae	Sompat grunt	0	0	1
81	*Procyon lotor*	Procyonidae	Racoon	0	1	0
82	*Psettodes belcheri*	Psettodiae	Psettodis	0	0	1
83	*Psettus sebas*	Drepaneidae	African moony	0	0	1
84	*Pseudotolithus brachygnathus*	Sciaenidae	Drums	0	1	1
85	*Pseudotolithus elongatus*	Sciaenidae	Bobo croaker	0	1	1
86	*Pseudotolithus senegalensis*	Sciaenidae	Bar	0	0	1
87	*Pteroscion peli*	Sciaenidae	Fried fish	0	1	1
88	*Rana sp.*	Ranidae	Frog	0	1	0
89	*Rhinobatos rhinobatos*	Rhinobatidae	Guitar skate	0	0	1
90	*Salmo salar*	Salmonidae	Salmon	0	0	1
91	*Sardinella maderensis*	Clupeidae	Clupeids	0	1	1
92	*Sarotherodon galilaeus*	Cichlidae	Tilapia	0	1	0
93	*Schilber mystus*	Schilbedae	Cat fish	0	1	0
94	*Scomber scombrus*	Scombridae	Mackerel	0	0	1
95	*Sphyraena barracuda*	Sphyraenidae	Barracuda	0	0	1
96	*Sphyraena dubia*	Hepsetidae	Brochet	0	0	1
97	*Sphyraena piscatorium*	Spyraenidae	Guinean barracuda	0	0	1
98	*Strongylura senegalensis*	Belonidae	Orphies	1	1	1
99	*Synodontis sp.*	Mochokidae	Synodontis	0	1	0
100	*Tilapia guineensis*	Cichlidae	Tilapia	0	1	0
101	*Trichechus senegalensis*	Trichechidae	Manatee	1	1	1
102	*Trichiurus lepturus*	Trichiuridae	Cutlassfish	0	0	1

Source: Compiled from Stiassny et al. (2007).

References

1. Nilsson, C.; Jansson, R.; Zinko, U. Long term responses of river-margin vegetation to water-level regulation. Science **1997**, *276*, 798–800.
2. Trimbe, S.W. Decreased rates of alluvial sediment storage in the Coon Creek basin, Wiscosin, 1975–1993. Science **1999**, *285*, 1244–1246.
3. Gentry, A.H. Tropical forest biodiversity: distributional patterns and their conservational significance. Oikos **1992**, *63*, 19–28.
4. National Research Council (NRC). *Riparian Areas: Functions and Strategies for Management. Summary. WSTB CRZFSM;* National Academy Press: Washington, 2002.
5. Woinarski, J.C.Z.; Brock, C.; Armstrong, M.; Hempel, C.; Cheal, D.; Brennan, K. Bird distribution in riparian vegetation in the extensive natural landscape of Australia's tropical savanna: a broad-scale survey and analysis of a distributional data base. J. Biogeography **2000**, *27*, 843–868.
6. Doyle, A.T. Use of riparian and upland habitats by small mammals. J. Mammal. **1990**, *71*, 14–23.

7. Darveau, M.; Labbé, P.; Beauchesne, P.; Bélanger, L.; Huot, J. The use of riparian forest strips by small mammals in a boreal balsam fir forest. Forest Ecol. Manag. **2001**, *143*, 95–104.
8. Larue, P.; Bélanger, L.; Huot, J. Riparian edge effects on boreal balsam fir bird communities. Canadian J. Forest Resources **1995**, *25*, 555–566.
9. Saab, V. Importance of spatial scale to habitat use by breeding birds in riparian forests: a hierarchical analysis. Ecol. Appl. **1999**, *9*, 135–151.
10. Ward, J.V.; Tockner, K.; Arscott, D.B.; Claret, C. Riverine landscape diversity. Freshwater Biol. **2002**, *47*, 517–539.
11. Acker, A.S.; Lienkaemper, G.; McKee, W.A.; Swanson, F.J.; Mlller, S.D. Composition, complexity, and tree mortality in riparian forests in the central Western Cascades of Oregon. Forest Ecol. Manag. **2003**, *173*, 293–308.
12. Cordes, L.D.; Hughes, F.M.R.; Getty, M. Factors affecting the regeneration and distribution of riparian woodlands along a northern praire river: the Red Deer River Alberta, Canada. J. Biogeography **1997**, *24*, 675–695.
13. Robinson, C.T.; Tockner, K.; Ward, J.V. The fauna of dynamic riverine landscape. Freshwater Biol. **2002**, *47*, 661–677.
14. Gledhill, D. *West African trees*. Longman. 1972.
15. Brinson, M.M. Riparian forests. In *Forested Wetlands Ecosystems of the World 15;* Lugo, A.E., Brinson, M., Brown, S. Eds.; Elsevier, Amsterdam **1990**, 87–141
16. Hancock, C.N.; Ladd P.G.; Froend, R.H. Biodiversity and management of riparian vegetation in Western Australia. Forest Ecol. Manag. **1996**, *85*, 239–250.
17. Natta, A.K. *Ecological Assessment of Riparian Forests in Benin: Phytodiversity, Phytosociology and Spatial Distribution of Tree Species:* PhD thesis Wageningen University: Netherlands, 2003.
18. Salo, J. Kalliola, R.; Häkkinen, I.; Mäkinen, Y.; Puhakka, M.; Coley, P.D. River dynamics and the diversity of Amazon lowlands forests. Nature **1986**, *322*, 254–258.
19. Naiman, R.J.; Décamps, H.; Polleck, M. The role of riparian corridors in maintaining regional biodiversity. Ecol. Appl. **1993**, *3*, 209–212.
20. Levine, C.M.; Stromberg, J.C. Effects of flooding on native and exotic plant seedlings: implications for restoring southwestern riparian forests by manipulating water and sediment flows. J. Arid Environ. **2001**, *49*, 111–131.
21. Ellis, L.M.; Molles, M.C., Jr.; Crawford, C.S. Influence of experimental flooding on litter dynamics in a Rio Grande riparian of central New Mexico. Restoration Ecol. **1999**, *7*, 1–13.
22. Gregory, S.V.; Swanson, F.J.; MsKee, W.A.; Cummins, K.W. An ecosystem perspective of riparaian zones: focus on links between land and water. BioScience **1991**, *41*, 540–551.
23. Harper, D.M.; Mavuti, K.M. Freshwater wetlands and marshes. *In*: *East African Ecosystems and their Conservation;* McClanahan, T.R., Young, Y.P., Eds.; Oxford University Press: New York, NY, 1996; 217–239 pp.
24. Reid, W.V. Biodiversity hotspots. Tree **1998**, *13* (7), 275–280.
25. IUCN. *Wetland Conservation: A Review of Current Issues and Required Action;* Dugan, PJ., Ed.; Gland: Switzerland, 1990.
26. Hopkins, B. *Forest and Savanna. An Introduction to Terrestrial Ecology with Special Reference to West Africa;* Heinemann: London, 1974.
27. Scudder, T. The need and justification for maintaining transboundary flood regimes: The Africa Case. Natural Resources Journal: University of New Mexico School of Law **1991**, *31*, 75–107.
28. Cooper, S.D. Rivers and streams. In *East African Ecosystems and their Conservation*; In: McClanahan, T.R., Young, T.P., Eds.; Oxford University Press: New York, NY, 1996; 133–170 pp.
29. Lowel-McConnell, R.H. Ecological studies in tropical fish communities, Cambridge University Press: Cambridge, 1987.
30. Melick, D.R.; Ashton, D.H. The effects of natural disturbances on warm temperate rainforests in South-eastern Australia. Aust. J. Ecol. **1991**, *39*, 1–30.

31. Roché, J. The use of historical data in the ecological zonation of rivers: the case of the 'tern zone'. Vie et Milieu **1993,** *43,* 27–41.
32. Wakentin, I.G.; Reed, J.M. Effects of habitat type and degradation on avian species richness in great Basin riparian habitats. Great Basin Nat. **1999**, *59,* 205–212.
33. Fetherston, K.L.; Naiman, R.J.; Billy, R.E. Large woody debris, physical process and riparian forest development in montane river networks of the Pacific Northwest. Geomorphology **1995**, *13,* 133–144.
34. Enright, N.J.; Miller, B.P.; Crawford, D. Environmental correlates of vegetation paterns and species richness in the northern Grampians, Victoria. Aust. J. Ecol. **1994**, *19,* 159–168.
35. Cowling, R.M.; Holmes, P.M. Flora and vegetation. In: *The Ecology of Fynbos: Nutrient, Fire and Diversity;* Cowling, R.M., Ed.; Oxford University Press: Cape Town, 1992.
36. Gregory, S.V.; Beschta, R.; Swanson, F.J.; Sedell, J.; Reeves, G.; Everest F. *Abundance of Conifers in Oregon's Riparian Forests*, COPE Report, Vol 3. No. 2. College of Forestry, Oregon State University: Corvallis, OR, 1990, 5–6 pp.
37. Nierenberg, T.; Hibbs, D.E. Characterization of unmanaged riparian areas in the central coast range of Western Oregon. Forest Ecol. Manag. **2000**, *129,* 195–206.
38. Dignan, P.; Bren, L. A study of logging on the understorey light environment in riparian buffer strips in a south-east Australian forests. Forest Ecol. Manag. **2003**, *172,* 161–172.
39. Tilman, D. *Plant Strategies and the Dynamics and Structure of Plant Communities.* Princeton University Press: Princeton, 1988.
40. Gilliam, J.W. Riparian Wetlands and Water Quality. J. Environ. Qual. **1994**, 25, 896–900.
41. Groffman, P.M.; McDowell, W.H.; Myers, J.C.; Merriam, J.L. Soil microbial biomass and activity in tropical riparian forests. Soil Biol. Biochem. **2001**, *33,* 1339–1348.
42. Forman, R.T.T.; Godron, M. *Landscape Ecology.* John Wiley and Sons: New York, 1986.
43. FAO. *Review of the State of World Fishery Resources;* FAO Fisheries Circular no. 710. (Rev. 2). FAO: Rome, 1981.
44. Bouillon, S.; Dahdouh-Guebas, F.; Rao, A.V.V.S.; Koedam, N.; Dehairs, F. Sources of organic carbon in mangrove sediments: variability and possible ecological implications. Hydrobiology **2003**, *495,* 33–39.
45. Donato, D.C.; Kauffman, J.B.; Murdiyarso, D.; Sofyan Kurnianto, S.; Melanie Stidham, M.; MarkkuKanninen, M. Mangroves among the most carbon-rich forests in the tropics. Nat. Geosci. **2011**, 4, doi: 10.1038/NGEO1123.
46. Shumway, C.A. *Forgotten waters: Freshwater and marine ecosystems in Africa. Strategies for Biodiversity Conservation and Sustainable Development;* Boston University: USA, 1999.
47. Cole, J.J.; Peierls, B.I.; Caraco, N.E.; Pace, M.I. Nitrogen loading of rivers as a human-driven process. In *Humans as Components of Ecosystems;* McDonnell, M., Pickett, S., Eds., Springer-Verlag, New York, 1993; 141–157 pp.
48. National Research Council (NRC). Understanding Riverine biodiversity: A Research Agenda for the Nation; Academy of Sciences: Washington, D.C, 1995.
49. Bartley, D.; Minchin, D. Precautionary approach to introductions and transfers of aquatic organisms. In Precautionary approach to fisheries, Part 2, FAO Fish. Tech. Pap. **1995**, *350* (2), 159–189.
50. Dixon, J.A. *Valuation of mangroves.* Tropical Coastal Area Management. ICLARM Newsl. **1989**, *4* (3), 1–6.
51. Graham, A. Siltation of stone-surface periphyton in rivers by clay-sized particles from low concentrations in suspension. Hydrobiologia **1990**, *199,* 107–115.
52. Ajonina, G.N.; Usongo, L. Preliminary quantitative impact assessment of wood extraction on the mangroves of Douala-Edea Forest Reserve, Cameroon. Trop. Biodivers. **2001**, *7* (2–3), 137–149.

53. Ajonina, P.U.; Ajonina, G.N.; Jin, E.; Mekongo, F.; Ayissi, I.; Usongo, L. Gender roles and economics of exploitation, processing and marketing of bivalves and impacts on forest resources in the Douala-Edaa Wildlife Reserve, Cameroon. Int. J. Sustainable Develop. World Ecol. **2005**, *12*, 161–172.
54. Feka, N.Z.; Ajonina, G.N. Drivers causing decline of mangrove in West-Central Africa, a review. Int. J. Biodivers. Sci. Ecosystem Serv. Manag. 2011, doi:10.1080/21513732.201 1.634436.
55. Kellman, M.; Meave, J. Fire in the tropical gallery forests of Belize. J. Biogeography **1997**, *24* (1), 23–34.
56. Ellis, L.M. Short-term response of woody plants to fire in a Rio Grande riparian forest of Central New Mexico, USA. Biol. Conserv. **2001**, *97*, 159–170.
57. Norris, V.O.L. The use of buffer zones to protect water quality: a review. Water Resources Manag. **1993**, *7*, 257–272.
58. Ajonina, G. *Rapid Assessment of Mangrove status and Conditions for use to Assess Potential for Marine Payment for Ecosystem Services in Amanzuri and Surrounding Areas in the Western Coastal Region of Ghana, West Africa*, Consultancy report to Coastal Resource Center, 2011.
59. Ruwa, R.K. Intertidal wetlands. In *East African Ecosystems and their Conservation*; McClanahan, T.R., Young, Y.P., Eds.; Oxford Univ. Press: New York, N.Y, 1996; 101–130 pp.
60. Scudder, T. Recent experiences with river basin development in the tropics and subtropics. Nat. Resources Forum **1994**, *18*, 101–113.
61. Katibath, E.F. A brief history of riparian forests in the Central Valley of California. In *California Riparian Systems: Ecology, Conservation and Productive Management*; Warner R.E., Hendrix, K.M., Eds.; University of California Press: Berkeley, 1984; 23–29 pp.
62. Rottenborn, S.C. Predicting the impacts of urbanization on riparian bird communities. Biol. Conserv. **1999**, *88*, 289–299.
63. Kaufman, L.; Ochumba, P. Evolutionary and conservation biology of ciclid fishes as revealed by faunal remnants in northern Lake Victoria. Conserv. Biol. **1993**, *7*, 719–730.
64. Taylor, J.R.; Cardonne, M.A.; Mitsch, W.J. Bottomland hardwood forests: their functions and values. In *Cumulative Impacts: Illustrated by Bottomland Hardwood Wetlands Ecosytems;* Gosselink, J.P., Lee, L.C., Muir, T.A., Eds.; Lewis Publishers: Chelsea MI, 1990, 13–86 pp.
65. Spalding, M.; Kainyuma, M.; Collins, L. *World Atlas of Mangroves;* Earthscan, London, 2010.
66. Sagers, C.L.; Lyon, J. Gradient analysis in a riparian landscape: contrasts in amongst forests layers. Forest Ecol. Manag. **1997**, *96*, 13–26.
67. Loetsch, I. Zohier, F.; Haller, K. *Forest Inventory Vol.* 2; Second Edition BIV: Germany, 1973.
68. Husch, B.; Beers, T.W.; Keuhaw, J.A., Jr. *Forest Mensuration*, 4th Ed.; John Wiley & Sons, 2003. pp.
69. Pool, D.G.; Snedaker, S.C.; Lugo, A.E. Structure of mangrove forests in Florida, Puerto Rico, Mexico and Costa Rica. Biotropica **1977**, *9*, 195–212.
70. Cintron, G.; Schaeffer-Novelli, Y. Methods of studying mangrove structure. In *The Mangrove Ecosystem: Research Methods*; Snedaker, S., Snedaker, J.G., Eds.; UNESCO Publication, 1984.
71. Kjerfve, B.; Macintosh, D.J. The impact of climatic change on mangrove ecosystems. In *Mangrove Ecosystems Studies in Latin America and Africa;* Kjerfve, B.B., Lacerda, L.D., Diop S. H., Eds.; UNESCO: Paris, **1997**; 1–7 pp.
72. Williams, P.H.; Humpries, C.J.; Vane-Wright, R.I. Measuring biodiversity: Taxonomic relatedness for conservation priorities. Aust. Syst. Bot. **1991**, *4*, 665–679.
73. De Lima, M.G.; Gascon, C. The conservation value of linear forest remnants in central Amazonia. Biol. Conserv. **1999**, *91*, 241–247.
74. Roggeri, H. *Zones Humides Tropicales D'eau Douce. Guide de Connaissances Actuelles Et De La Gestion Durable.* WIW, Union Europeene, Université de Leiden. Kluwer Academic Publishers: Pays-Bas, 1995.
75. Hibbs, D.E.; Bower, A.L. Riparian forests in the Oregon Coast Range. Forest Ecol. Manag. **2001**, *154*, 201–213.

76. Blas, D.E.; Perez, M.R. Prospects for reduced impact logging in Central African logging concessions. Forest Ecology and Management **2008**, *256*, 1509–1516.
77. Cerutti, P.O.; Nasi, R.; Tacconi, L. Sustainable forest management in Cameroon needs more than approved forest management plans. Ecol. Soc. **2008**, *13* (2), 36.
78. Feka, N.Z.; Chuyong, G.B.; Ajonina, G.N. Sustainable utilization of mangroves using improved fish smoking systems: A management perspective from the Douala-Edea Wildlife Reserve, Cameroon. Trop. Conserv. Sci. **2009**, *4*, 450–468.
79. Wetlands International. *Planting Trees to Eat Fish: Field Experiences in Wetlands and Poverty Reduction;* Wetlands International, Wageningen: the Netherlands, 2009.

Bibliography

Arbornnier, M. *Trees, Shrubs and Lianas of West African Dry Zones;* CIRAD, MNHN, Paris, 2004.

Etukudo, I.G.; Akpan-Ebe, I.N.; Udofia, A.; Attah, V.I. *Elements of Forestry;* Usanga & Sons. UYO: Nigeria, 1994.

Forest Trends. Payments for Ecosystem Services: Getting Started in Marine and Coastal Ecosystems A Primer. 2010.

Keay, R.W.J.; Hepper, F.N. *Flora of West tropical Africa,* 2nd Ed.; Millbank: London, 1954–1972; Vol. I-III pp.

Millennium Ecosystem Assessment. *Ecosystems and Human Well-Being: Synthesis;* World Resources Institute: Washington, D.C. 2005.

Raunkaier, C. *The Life Forms of Plants and Statistical Plant Geography.* Clarendon Press: London, 1934.

Schnell, R. *Introduction à la Phytogéographie des Pays Tropicaux;* Volume II. Milieux, les groupements végétaux. Edition Gauthier–Villars : Paris, France, 1971; 503–951 pp.

Stiassny, M.L.J.; Teugels, G.G.; Hopkins, C.D. *Poissons d'eaux douces et saumâtres de basse Guinée, ouest de l'Afrique centrale/The Fresh and Fishes of Lower Guinea, West-Central Africa.* Volume 1. IRD, Publications scientifiques du Museum, MRAC: Paris, 2007.

Swanson, F.J.; Johnson, S.L; Gregory, S.V.; Acker, A.S. Flood disturbance in a forested Mountain Landscape. Bioscience **1998**, *48*, 681–689.

Thikakul, S. *Manual of Dendrology-Cameroon.* Groupe Poulin Quebec: Canada, 1985.

White, F. *La végétation de l'Afrique. Mémoire accompagnant la carte de végétation de l'Afrique.* UNESCO/AET-FAT/UNSO, 1986.

3

Coastal and Estuarine Waters: Light Behavior

Darryl J. Keith
U.S. Environmental Protection Agency (EPA)

Introduction

Remote sensing of near-coastal ocean and estuaries is very important as 30–70% of the world's population lives in along the coast.[1,2] Water quality parameters (e.g., water clarity, chlorophyll concentration) are derived using remote-sensing systems from direct solar radiation entering the water column and either being absorbed or scattered within the water column. Some of the radiation is then reflected back to the atmosphere as water-leaving radiance. The ratio of the radiance reflected out to the direct solar radiation incident on the sea surface is termed reflectance. When the reflectance is passively recorded by a sensor, it is referred to as remote sensing reflectance. The remote-sensing reflectance (R_{rs}) represents the ocean color signature of the water. R_{rs} can be used to deduce the concentrations of chlorophyll, colored dissolve organic matter (CDOM), or mineral particles within the near-surface water; the bottom depth and type in shallow waters; and other ecosystem information such as net primary production, phytoplankton functional groups, or phytoplankton physiological state. This entry discusses the inherent and apparent optical properties that affect the retrieval of color from coastal and estuarine waters. The most common example of remote sensing, and the one primarily discussed in this entry, is the use of sunlight that has been backscattered within the water and passively returned to the sensor.

Remote Sensing Basics

Coastal and inland waters contain a wide variety of optically active constituents that combine to create an optically complex system. Particulate and dissolved materials such as phytoplankton, detritus, suspended sediments, and CDOM vary over orders of magnitude in concentration to affect the underwater light regime and water quality. Water quality parameters are derived using remote-sensing systems through direct solar radiation entering the water column and either being absorbed or scattered within the water column. Remote-sensing systems can be *active* or *passive. Active remote sensing* means that a signal of known characteristics (e.g., a laser beam) is sent from the sensor platform—an aircraft or satellite—to the sea surface, and the return signal is then detected after a time delay determined by the distance from the platform to the ocean and by the speed of light. The Light Detection and Ranging (LIDAR) system that is usually flown on low-altitude aircraft and satellites is an example of an active system. In *passive remote sensing*, we simply observe the light that is naturally emitted or reflected by the water body. The night-time detection of bioluminescence from aircraft is an example of the use of emitted light at visible wavelengths. To derive water-quality parameters, coastal scientists primarily rely on passive remote-sensing systems. The most common example of passive remote sensing, and the one primarily discussed in this entry, is the use of sunlight that has been backscattered within the water and returned to the sensor. This light can be used to deduce the concentrations of chlorophyll, CDOM, or mineral particles within the near-surface water; the bottom depth and type in shallow waters; and other ecosystem information such as net primary production, phytoplankton functional groups, or phytoplankton physiological state.

Optical Properties of Coastal and Estuarine Waters

The solar irradiance (E_d) that enters the Earth's atmosphere undergoes absorption and scattering by air molecules and aerosols. Radiance generated in the atmosphere by scattering (L_a) along the path between the sea surface and the sensor contains information about atmospheric aerosols and other atmospheric parameters. Aerosols are suspended liquid and particulate matter that include smoke, water, and hydrogen sulfate droplets, dust, ashes, pollen, spores, and other forms of atmospheric suspensions.[3] As a result, two radiative components reach the surface of the Earth: direct solar radiation that maintains the directionality of the sun's radiation as it exists outside the atmosphere and diffuse solar radiation (generally referred to as skylight or path radiance) that reaches the water surface from multiple angles due to molecular scattering (Rayleigh scattering) and aerosol scattering (Mie scattering) (Figure 3.1). When direct sunlight (L_t) enters a water body, it is either absorbed or scattered (Figure 3.1). The portion of L_t that is not absorbed by materials within the water column has its radiance altered, depending on the absorption and scattering properties of the water body and on the types and concentrations of the various constituents within the particular water body. The altered radiance (L_u) is then reflected (scattered) back out as water—leaving radiance (L_w). The total up welling radiance (L_u) received by a sensor

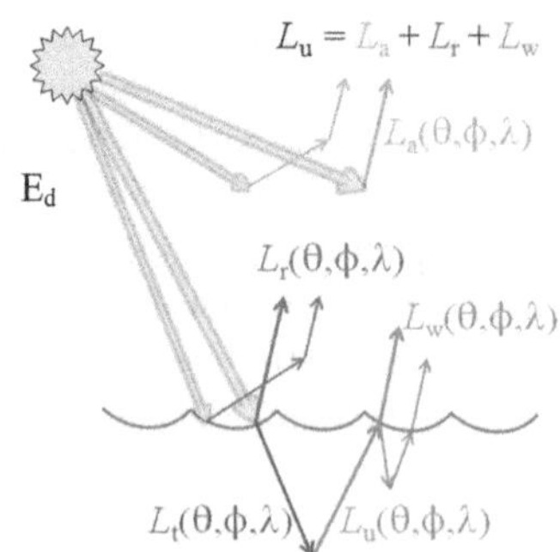

FIGURE 3.1 Contributions to the total upwelling radiance above the sea surface, L_u, where θ = zenith angle, Φ = azimuth angle, and λ = wavelength. Thick arrows are the sun's unscattered beam; thin arrows are atmospheric path radiance L_a; dashed is surface-reflected radiance L_r; dotted is water-leaving radiance L_w. The curve arrow is total radiance (L_t) that enters the sea surface from direct sunlight. L_u, below the sea surface, is total upwelling radiance produced by scattering properties of the water column that contributes to the magnitude of L_w. E_d represents solar irradiance. Thick arrows represent single-scattering contributions; thin arrows illustrate multiple scattering contributions.

Source: Reproduced with permission from http://wwwoceanop-ticsbook.infobasedonMobley.[33]

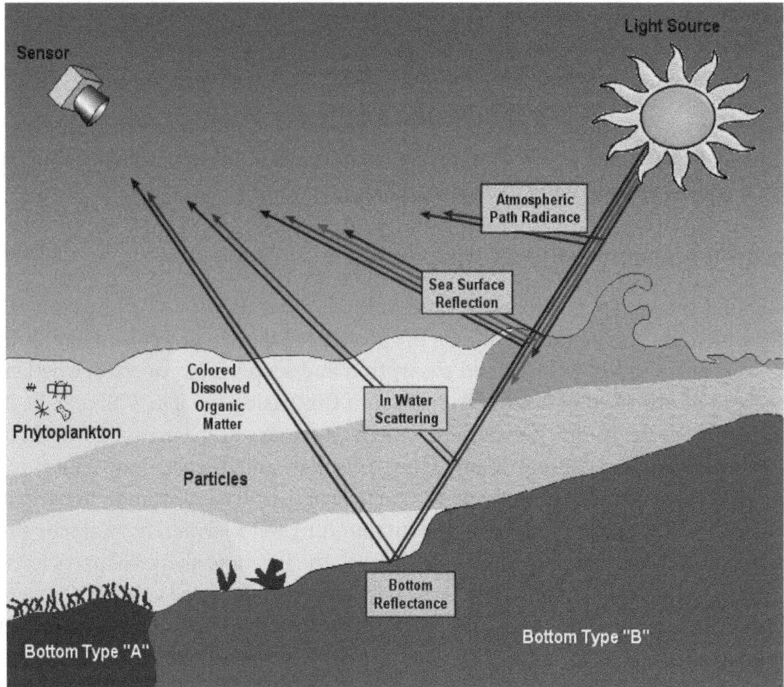

FIGURE 3.2 Optical components and pathways of radiance and reflectance based on the absorption and scattering properties of coastal waters.
Source: Reproduced with permission from Arnone, R.A. from "Hyperspectral Imager for Coastal Ocean (HICO)" presented at the NASA Gulf Workshop in 2009.

above the sea surface is the sum of the L_w that carries information about the water column, radiance reflected by the sea surface (L_r) contains information about the wave state of the sea surface, and radiance generated in the atmosphere by scattering (L_a). The ratio of the radiance reflected out (in units of Watts/m^2) to the solar irradiance (Watts/m^2) is the *reflectance.* When the reflectance is passively recorded by a sensor, it is called remote sensing reflectance (R_{rs}, Figure 3.2). The R_{rs} represents the "ocean color signature" of the water integrating the spectral absorption and backscattering properties of all the materials present in the water column.[4]

Where Do Estuarine and Coastal Waters Get Their Color Signature

The waters get their colors from the mixture of pigments in phytoplankton, CDOM, and minerals. If≈we are able to know how different substances (e.g., phytoplankton, suspended sediments) alter sunlight (either by wavelength-dependent absorption, scattering, or fluorescence), then we can attempt to deduce from the ocean color signature what substances must have been present in the water, and their concentrations. By working the analytical process "backward" in an inverse manner from the sensor to the estuary or coastal ocean, the remote-sensing community has been able to:

- Map chlorophyll concentrations
- Measure inherent optical properties (IOPs) such as absorption and backscatter
- Determine phytoplankton physiology, phenology, and functional groups
- Study ocean carbon fixation and cycling
- Monitor ecosystem changes resulting from climate change
- Map coral reefs, sea grass beds, and kelp forests

- Map shallow-water bathymetry and bottom type for military operations
- Monitor water quality for recreation and environmental protection
- Detect harmful algal blooms and pollution events

Ocean color remote sensing has completely revolutionized the ability of oceanographers, the environmental monitoring community, and coastal managers to understand estuaries and the coastal ocean at local to global spatial scales and daily to decadal temporal scales.

Inherent and Apparent Optical Properties

The absorption and scattering properties of fresh and sea water are described by their IOPs. IOPs are properties of the medium and do not depend on the position of the sun or ambient light field in the water column. A volume of water has well-defined absorption and scattering properties whether or not there is any light there to be absorbed or scattered. Because of this basic principle, IOPs can be measured in a laboratory setting on a water sample, as well as in situ in an estuary or the ocean. The IOPs depend on the composition, morphology, and concentration of the particulate and dissolved substances in the medium. Composition refers to what materials make up the particle or dissolved substance, in particular the index of refraction of that material relative to that of the surrounding water. Different materials absorb light differently as a function of wavelength. Morphology refers to the sizes and shapes of particles. Particles with different shapes scatter light differently even if the particles have the same volume. Conversely, particles with different volumes scatter light differently even if they have the same shape. Concentration refers to the number of particles in a given volume of water, which is described by the particle size distribution. Given these complexities, it is not unusual that the magnitude of IOPs can vary by orders of magnitude.[5]

Apparent optical properties are those properties that depend both on the IOPs of the media as well as change in solar elevation (the angle between the horizon and the sun's disk, which alters the radiance distribution).[5,6] Apparent optical properties are R_{rs}, diffuse absorption, and diffuse backscattering.

Absorption Properties of Coastal and Estuarine Waters

There are only two things that can happen when light enters the water, it is either absorbed or scattered. Therefore, in order to understand the behavior of light as it passes into a water body, we need some measure of the extent to which water absorbs and scatters the incoming radiation. In estuarine and coastal waters, the total absorption of light is partitioned into absorption by the water itself, phytoplankton pigments, detritus particles, and CDOM (Figures 3.3 and 3.4). Coastal and estuarine waters are considered to be optically complex because CDOM and mineral particles can make a significant

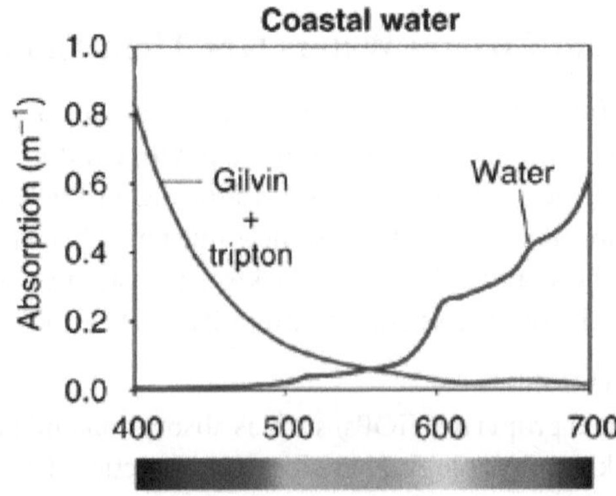

FIGURE 3.3 Light absorption spectra of pure water (dashed line) and CDOM plus suspended matter (tripton). **Source:** Reproduced with permission from http://www.oceanop-ticsbook.infobasedonStomp et al.[34]

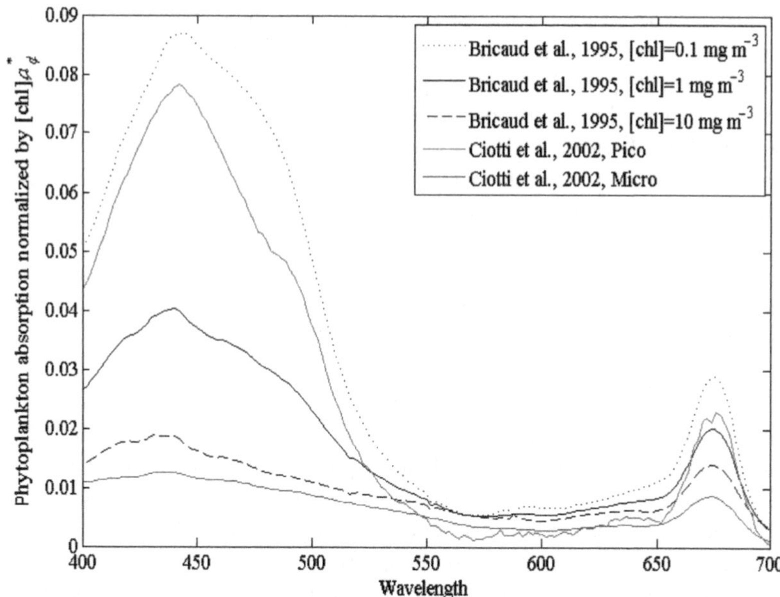

FIGURE 3.4 Phytoplankton absorption spectra normalized by chlorophyll, which illustrates two dominant peaks: a primary peak in the blue region (at 440 nm) and a secondary absorption peak in the red region of the spectra (at 675 nm).

Source: Reproduced with permission from http://www.oceanopticsbook.info based on Bricaud et al.,[35] and Ciotti et al.[14]

(or even the dominant) contribution to water color and brightness. Both CDOM and mineral particles absorb light most strongly in the blue part of the spectrum and exponentially decay with very low or no absorption in the red region (Figure 3.3). When added to water (which absorbs mostly red light), they produce a green color that is easily confused with the effect of phytoplankton pigments.

Pure Water

Pure water is a blue liquid whose color is derived from the fact that it absorbs very weakly in the blue and green regions of the spectrum.[7] However, absorption increases in the yellow and orange portions of the spectrum and is very significant in the red region (Figure 3.3).[8]

Colored Dissolved Organic Matter

Coastal waters get their color primarily due to the presence of CDOM (which is also known in scientific literature as yellow substance and gilvin), which imparts a brown to yellow-brown color to these waters. CDOM is the dissolved product of the decomposition of plant and animal matter from both terrestrial and marine sources that is composed of humic and fulvic acids. These acids form a water-soluble and chemically complex group of compounds termed "humic substances." As rainfall runoff and melting snow percolate through soils in coastal watersheds, humic substances are extracted and flow into rivers and streams and ultimately to estuaries and nearshore coastal waters.[4,7] Another likely origin for CDOM is from the decomposition of phytoplankton.

Tripton

Tripton (also known as suspended matter) includes both inorganic [e.g., mineral particles (colloids)] and organic (e.g., fecal pellets and cell fragments) materials that are suspended in estuarine and coastal waters. The composition of tripton generally reflects the geologic structure and composition of the adjacent watershed and coastal areas. Kirk[7] suggested that the yellow-brown color characteristic of suspended matter (tripton) from coastal waters is primarily due to particulate material either coated with humic compounds bound to particles or as free particles of humus. Characteristically, these substances do not absorb light strongly but scatter quite intensely. Spectrally, the absorption spectra for tripton is similar to that of CDOM with low absorption at the red region of the spectrum and rises increasingly as wavelength decreases toward the blue and ultraviolet end of the spectrum (Figure 3.3). When light is transmitted through the water column, constituents (e.g., tripton, phytoplankton) in the water selectively absorb light at selective wavelengths. The resulting spectrum represents the unique absorption signature for that constituent. In the case of tripton, the spectral signature is dependent on particle shape, particle size distributions, and refractive index. Of all the optical constituents, detrital particles remain relatively understudied.[9,10]

Phytoplankton

Phytoplankton have a major effect on ocean color as they are a major absorber of light in estuarine, coastal, and open ocean environments. Phytoplankton possess chlorophyll, a pigment that allows them to harvest the sunlight and through the process of photosynthesis produce energy. Chlorophylls cause two dominant peaks in the absorption spectra: a primary peak in the blue region (at 440 nm) and a secondary absorption peak in the red region of the spectra (at 675 nm) (Figure 3.4). Other pigments that are present (depending on species and taxa) will cause the broadening of the blue peak and the appearance of additional absorption maxima (Figure 3.5). Spectra of phytoplankton absorption varies in magnitude and shape due to different cellular pigment compositions and pigment packaging.[11–16] Specific

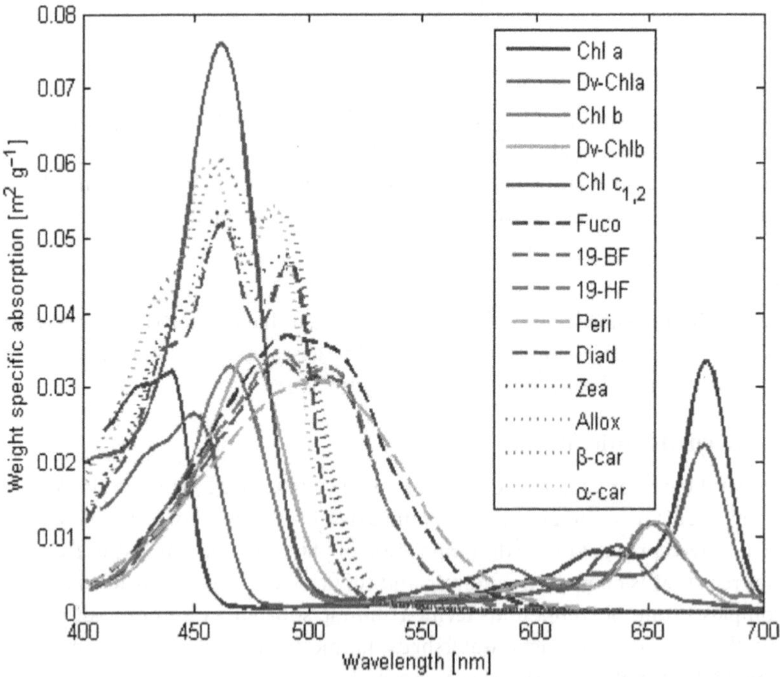

FIGURE 3.5 Absorption spectra showing the primary and secondary absorption peaks for key phytoplankton pigment groups including Chlorophyll a (Chl a), Divinyl chlorophyll (Dv-Chl a), Chlorophyll b, Divinyl chlorophyll b (Dv-Chl b), Chlorophyll c1 and c2 (chl c1,2), 19'-Butanoyloxyfucoxanthin (19-BF), 19'-Hexanoxyfucoxanthin (19-HF), Fucoxanthin (Fuco), Peridinin (Peri), Diadino (Diad), Zeaxanthin (Zea), Alloxanthin (Allox), Beta-carotene (β-car), and Alpha-carotene (α-car).

Source: Reproduced with permission from http://www.oceanopticsbook.info based on Bidigare et al.[11]

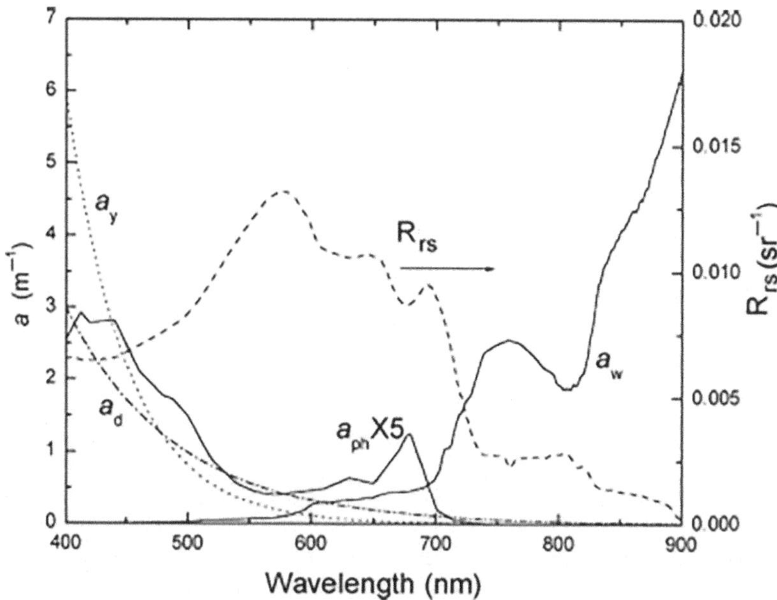

FIGURE 3.6 Typical curve shapes for coastal remote-sensing reflectance (R_{rs}), phytoplankton pigment absorption (a_{ph}), C_{DOM} (a_y or a_{CDOM}), detrital absorption (a_d), and pure water absorption (a_w).
Source: Reproduced with permission from the Journal of Coastal Research based on Sydor.[16]

pigment-protein complexes present in the cell will cause changes in absorption spectra and magnitudes. Furthermore, the increase in cellular pigment concentration and cell size (i.e., packaging effect) will flatten the specific absorption spectra.[17–20] As the concentration of phytoplankton changes in a water body, changes in the different regions of the absorption spectra allow coastal scientists to create algorithms (step-by-step procedures) that can estimate chlorophyll abundance from R_{rs}.

In the open ocean, as phytoplankton concentrations increase, the water-leaving reflectance has been observed to "green" in a predictable fashion that allowed for the creation of empirical relationships from the ratio of blue-to-green wavelengths of light.[10] In contrast, in the productive turbid waters of estuarine and coastal environments, R_{rs} in the blue-green end of the spectrum are significantly affected by absorption due to CDOM, tripton, and phytoplankton pigments (Figure 3.6). For this reason, satellite remote-sensing algorithms for chlorophyll developed from spectral data from the open ocean regularly fail in the coastal ocean and estuaries because they overestimate the chlorophyll content. In remote-sensing terms, these waters are classified as Case 2[21] as distinct from Case 1 open ocean waters.

To quantify chlorophyll concentrations in estuarine and coastal environments, algorithms have been developed which use information in spectral regions away from the influences of CDOM and tripton absorption and which are based on the properties of the phytoplankton reflectance peak in the red and near infrared portion of the spectrum near 700 nm.[22–24] The algorithms commonly ratio remotely sensed reflectances at 705 and 670 nm.[22,23,25]

Scattering Properties of Coastal and Estuarine Waters

Coastal and estuarine waters appear bright (highly reflective) when viewed from space due to the presence of sediments resuspended from the sea bottom by wave action, tidal currents, and storms; sediments brought in by river discharge; and phytoplankton. Scattering properties of phytoplankton are important since they are directly related to R_{rs} calculations. The manner in which incident light penetrates these waters depends not only on the scattering properties of the media but also on the angular

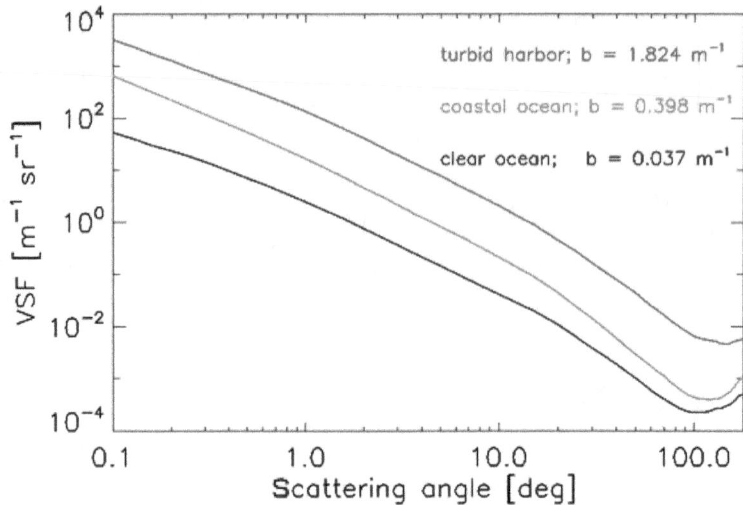

FIGURE 3.7 Log-log plots of measured volume scattering function at 514 nm from the harbor at San Diego, San Pedro Channel off the California coast, and the Tongue of the Ocean, Bahama Islands.
Source: Reproduced with permission from http://www.oceanopticsbook.info based on Petzold.[26]

distribution of the scattered flux resulting from the primary scattering process.[7] The angular distribution has a characteristic shape that is described in terms of the *volume scattering function* (VSF). The VSF describes the distribution of light scattered by a suspension of particles in a direction (forward or backward) at a specified wavelength (Figure 3.7).[26] A measure of the overall magnitude of the scattered light is given by the scattering coefficient, which is the integral of the VSF over all angles. A quantitative definition for the scattering coefficient is found in Kirk.[7] Scattering and backscattering coefficients of phytoplankton, as well as the VSF, are derived from either theoretical models (Mie theory) or direct measurements of the above-mentioned properties.[27–31] Coefficients are highly dependent on the size, shape, and refractive index of all components of the phytoplankton cell.[31,32]

Conclusion

Estuaries and coastal waters represent the dynamic interface between land and the open ocean. These waters are physically and biologically complex environments from which water quality information can be derived based on the absorption and scattering characteristics of light within this medium. All the apparent, inherent, and specific optical properties that were described in the preceding sections are wavelength dependent. It is the particular wavelength dependencies of these properties that define ocean color signatures in coastal and estuarine waters. The spectral qualities of the optical properties of pure water, CDOM, tripton, and phytoplankton determine the exact nature of the interdependence between individual constituents of the water and the R_{rs} signal that emerges from coastal and estuarine waters.

Acknowledgments

I would like to thank Kenneth Rocha, Kristin Hychka, and Anne Kuhn for their review and comments of the manuscript. This is contribution number ORD-002860 of the Atlantic Ecology Division, National Health and Environmental Effects Research Laboratory, Office of Research and Development, U.S. Environmental Protection Agency. Any mention of trade names or commercial products does not constitute endorsement or recommendation for use.

References

1. *UNEP.* United Nations Environmental Programme Annual Report; *UNEP: Nairobi, Kenya, 2007.*
2. Wilson, S.G.; Fischetti, T.R. *Coastline Population Trends in the United States: 1960 to 2008;* U.S. Census Bureau: Washington, D.C., 2010.
3. *Sathyendranath, S., Ed.; IOCCG. Remote Sensing of Ocean Colour in Coastal, and Other Optically-Comoplex Waters, Reports of the International Ocean-Colour Coordinating Group, No. 3; IOCCG: Dartmouth, Canada, 2000.*
4. *Coble, P.; Hu, C.; Gould, R.W., Jr.; Change, G.; Wood, A.M. Colored dissolved organic matter in the coastal ocean* – An optical tool for coastal zone environmental assessment and management. Oceanography **2004,** *17* (2), 50–59.
5. Mobley, C.D.; Stramski, D.; Bisset, W.P.; Boss, E. Optical modeling of ocean waters: Is the case-1 case-2 still useful? Oceanography **2004,** *17* (2), 60–67.
6. Prisendorfer, R.W. Application of radiative transfer theory to light measurements in the sea. Union Geod. Geophys. Inst. Monogr. **1961,** *10,* 11–30.
7. Kirk, J.T.O. *Light and Photosynthesis in Aquatic Ecosystems;* Cambridge University Press: Great Britain, 1994; p. 509.
8. Pope, R.M.; Fry, E.S. Absorption spectrum (380–700 nm) of pure water. II Integrating cavity measurements. Appl. Opt. **1997,** *36* (33), 8710–8723.
9. Bukata, R.P.; Jerome, J.H.; Kondratyev, K.Y.; Pozdnyakov, D.V. *Optical Properties and Remote Sensing of Inland and Coastal Waters;* CRC Press: Boca Raton, FL, 1995; pp. 362.
10. Schofield, O.; Arnone, R.; Bissett, W.P.; Dickey, T.D.; Davis, C.O.; Finkel, Z.; Oliver, M.; Moline, M.A. Watercolors in the coastal zone—What can we see? Oceanography **2004,** *17* (2), 24–31.
11. Bidigare, R.R.; Ondrusek, M.E.; Morrow, J.H.; Kiefer, D. In vivo absorption properties of algal pigments. In *Ocean Optics X,* Proceedings of SPIE, 1990; Vol. 1302, 290–302.
12. Bricaud, A.; Stramski, D. Spectral absorption coefficients of living phytoplankton and nonalgal biogenous matter: A comparison between the Peru upwelling area and the Sargasso Sea. Limnol. Oceanogr. **1990,** *35* (3), 562–582.
13. Hoepffner, N.; Sathyendranath, S. Effect of pigment composition on absorption properties of phytoplankton. Mar. Ecol. Prog. Ser. **1991,** *73* (1), 11–23.
14. Ciotti, A.M.; Lewis, M.R.; Cullen, J.J. Assessment of the relationships between dominant cell size in natural phytoplankton communities and the spectral shape of the absorption coefficient. Limnol. Oceanogr. **2002,** *47* (2), 404–417.
15. Bricaud, A.; Claustre, H.; Ras, J.; Oubelkheir, K. Natural variability of phytoplanktonic absorption in oceanic waters: Influence of the size structure of algal populations. J. Geophys. Res. Oceans **2004,** *109* (C11).
16. Sydor, M. Use of hyperspectral remote sensing reflectance in extracting the spectral volume absorption coefficient for phytoplankton in coastal water: Remote sensing relationships for the inherent optical properties of coastal water. J. Coast. Res. **2006,** *22* (3), 587–594.
17. Duysens, L.M.N. The flattening effect of the absorption spectra of suspensions as compared to that of solutions. Biochim. Biophys. Acta **1956,** *19* (1), 1–12.
18. Kirk, J.T.O. A theoretical analysis of the contribution of algal cells to the attenuation of light within natural waters, III. Cylindrical and spheroidal cells. New Phytol. **1976,** *77* (2), 341–358.
19. Morel, A.; Bricaud, A. Theoretical results concerning light-absorption in a discrete medium, and application to specific absorption of phytoplankton. Deep-Sea Res. A **1981,** *28* (11), 1375–1393.
20. Johnsen, G.; Nelson, N.B.; Jovine, R.V.M.; Prezelin, B.B. Chromoprotein-dependent and pigment-dependent modeling of spectral light-absorption in 2 dinoflagellates, prorocentrumminimum and heterocapsa-pygmaea. Mar. Ecol. Prog. Ser. **1994,** *114* (3), 245–258.
21. Morel, A.; Prieur, L. Analysis of variations in ocean color. Limnol. Oceanogr. **1977,** *22* (4), 709–722.

22. Gitelson, A. The peak near 700 nm on radiance spectra of Algae and water - Relationships of its magnitude and position with chlorophyll concentration. Int. J. Remote Sens. **1992,** *13* (17), 3367–3373.
23. Gitelson, A.A.; Kondratyev, K.Y. On the mechanism of formation of maximum in the reflectance spectra near 700 nm and its application for remote monitoring of water quality. In *Transactions Doklady of the USSR Academy of Sciences: Earth Science Sections;* 1991; Vol. 306, 1–4.
24. Schalles, J.F. Optical remote sensing techniques to estimate phytoplankton chlorophyll a concentrations in coastal waters with varying suspended matter and CDOM concentrations. In *Remote Sensing of Aquatic Coastal Ecosystem Processes: Science and Management Applications;* Richardson, L., Ledrew, E., Eds.; Springer: the Netherlands, 2006; 27–79.
25. *Dekker, A.G.* Detection of Optical Water Quality Parameters for Eutrophic Waters by High Resolution Remote Sensing, Ph.D. Thesis; *Vrije Universiteit: Amsterdam, 1993; p. 222.*
26. 26 Petzold, T.J. Volume scattering functions for selected ocean waters. In *Light in the Sea;* Tyler, J.E., Ed.; Hutchinson & Ross: Dowden, 1977; 150–174.
27. 27 Bricaud, A.; Morel, A.; Prieur, L. Optical-efficiency factors of some phytoplankters. Limnol. Oceanogr. **1983,** *28* (5), 816–832.
28. Volten, H.; de Haan, J.F.; Hoovenier, J.W.; Schreurs, R.; Vassen, W. Laboratory measurements of angular distributions of light scattered by phytoplankton and silt. Limnol. Oceanogr. **1998,** *43* (6), 1180–1197.
29. Witkowski, K.; Krol, T.; Zielinski, A.; Kuten, E. A lightscattering matrix for unicellular marine phytoplankton. Limnol. Oceanogr. **1998,** *43,* 859–869.
30. Vaillancourt, R.D.; Brown, C.W.; Guillard, R.R.L.; Balch, W.M. Light backscattering properties of marine phytoplankton: Relationships to cell size, chemical composition and taxonomy. J. Plankton Res. **2004,** *26* (2), 191–212.
31. Sullivan, J.M.; Twardowski, M.S. Angular shape of the oceanic particulate volume scattering function in the backward direction. Appl. Opt. **2009,** *48* (35), 6811–6819.
32. *Jonasz, M.; Fournier, G.R.* Light Scattering by Particles in Water: Theoretical and Experimental Foundations*; Academic Press, 2007.*
33. Mobley, C. Overview of Optical Oceanography-Reflectances, 2010. http://www.oceanopticsbook.info.
34. Stomp, M.; Huisman, J.; Vörös, L.; Pick, F.R.; Laamanen, M.; Haverkamp, T.; Stal, L.J. Colorful coexistence of red and green picobacteria in lakes and seas. Ecol. Lett. **2007,** *10* (4), 290–298.
35. Bricaud, A.; Babin, M.; Morel, A.; Claustre, H. Variability in the chlorophyll-specific absorption coefficients of natural phytoplankton: analysis and paramentarization. J. Geophys. Res. **1995,** *100* (C7), 13211–3332.

4

Coastal and Estuarine Waters: Optical Sensors and Remote Sensing

Darryl J. Keith
U.S. Environmental Protection Agency

Introduction

There are two basic types of optical multispectral or hyperspectral sensors used in imaging estuarine, coastal, and open ocean systems: whiskbroom (across-track scanning; Figure 4.1) and pushbroom scanners (along track scanning; Figure 4.2).

Scanners

Whiskbroom Scanning (Across-Track Scanning)

In a whiskbroom scanner system, a scan mirror rotates in front of a telescope to continuously sweep the Earth beneath the aircraft or satellite perpendicular to the direction of flight. Airborne scanners typically sweep large angles (between 90° and 120°), while satellites, because of their higher altitudes, only need to sweep fairly small angles (10–20°) to cover a broad region. At each ground cell or pixel (a two-dimensional array of individual picture elements arranged in rows or columns that has an intensity value and a location address), the incoming reflected or emitted radiation is detected independently and separated into ultraviolet, visible, near-infrared, and thermal spectral components based on their constituent wavelengths. Each pixel represents an area on the earth's surface. A bank of internal *detectors,* each sensitive to a specific range of wavelengths, detects and measures the energy for each spectral band and converts it into an electrical signal. The electrical signal is converted to digital data and recorded for subsequent computer processing. The ground resolution of each pixel is defined by the instantaneous field of view (IFOV) based on sensor characteristics and flight altitude. The sweep of the mirror (angular field of view) records a scan line of pixels. A scan line of pixels is equivalent to the imaged swath. As the platform moves forward over the Earth, successive scans build up a two-dimensional image of the earth's surface. A collection of scan lines in a single direction is generally defined as a mission flight line. For aircraft remote sensing, adjacent parallel flight lines are flown with adequate overlap to provide complete coverage of the desired area. Many well-known aircraft instruments are whiskbroom

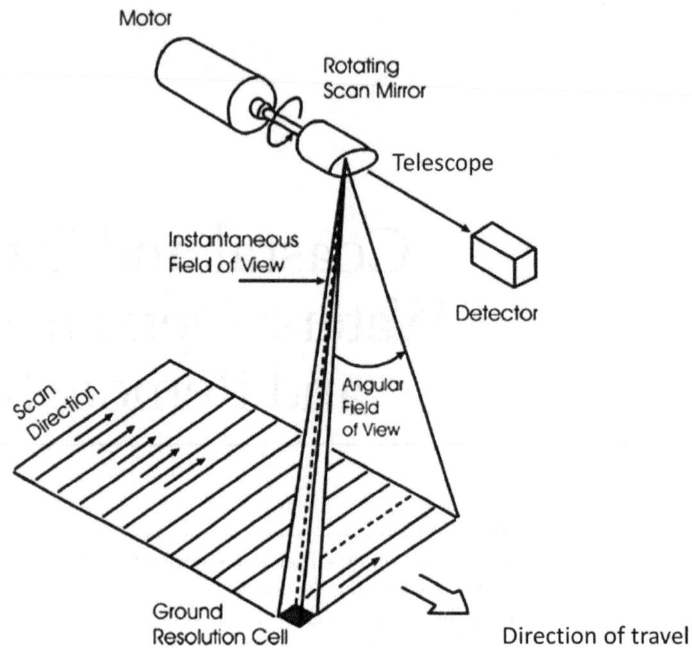

FIGURE 4.1 Whiskbroom (across track) scanner system.
Source: Modified from Canadian Centre for Remote sensing.

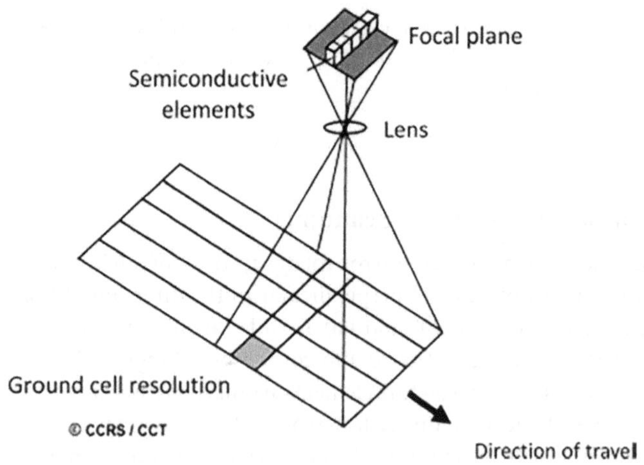

FIGURE 4.2 Along-track scanner system.
Source: Modified from Canadian Centre for Remote Sensing.

scanners such as the Airborne Ocean Color Imager (AOCI), the Calibrated Airborne Multispectral Scanner (CAMS), and the Airborne Visible/Infrared Imaging Spectrometer (AVIRIS).[1–3]

Pushbroom Scanners (Along-Track Scanning)

The pushbroom or along-track scanner is an imaging system that is "pushed" along in the flight track direction (i.e., along track) to map a scene. The system uses a linear array of solid semiconductive elements located at the focal plane of the image formed by the lens systems to collect spectral data.

The semiconductive elements are individual detectors that measure the energy for a single pixel. A separate linear array is required to measure each spectral band or channel. For each scan line, the energy detected by each detector of each linear array is sampled electronically and digitally recorded. The size and IFOV of the detectors determine the spatial resolution of the system. Since all pixels for a given scan line are projected onto a detector array at the same time, the pushbroom design allows for a longer dwell-time over each pixel that increases the signal-to-noise performance. Common pushbroom systems for aircraft applications include the CASI (Compact Airborne Spectrographic Imager),[4–7] HYDICE (Hyperspectral Digital Imagery Collection Experiment),[8] PHILLS (Hyperspectral Imager for Low Light Spectroscopy),[9] and HICO (Hyperspectral Imager for the Coastal Ocean)[10] sensors.

Sensors

Multispectral Imaging Sensors

Multispectral imaging is remote sensing that obtains optical representations in two or more ranges of frequencies or wavelengths. Multispectral imaging sensors capture image data from at least two or more wavelengths across the electromagnetic spectrum. On the sensor, each channel is sensitive to radiation within a narrow wavelength band resulting in a multilayer image (Figure 4.3) that contains both the brightness and spectral (color) information of the pixels sampled. In a multilayer image, data from each wavelength forms an image that carries some specific spectral information about the pixels in that image. By "stacking" images together from the same area, a multilayer image is formed composed of individual component images. Please note that multispectral images do not produce the spectrum of a pixel because wavelengths may be separated by filters or by the use of instruments that are sensitive to particular bands in the spectrum. Multispectral images are the main type of images acquired by most spaced-based or airborne radiometer systems. Radiometers are devices that measure the flux of electromagnetic radiation. Satellites may carry many radiometers in order to acquire data from selected portions of the electromagnetic spectrum. For example, one radiometer may acquire data from wavelengths in the red–green–blue (RGB) [700–400 nanometers (nm)] portion of the visible spectrum, a second radiometer may acquire data from wavelengths in the near infrared (700–3000 nm), and another might acquire data from mid-infrared to thermal region (greater than 3000 nm).

Hyperspectral Imaging Sensors

Hyperspectral systems offer the high spatial and spectral resolution needed to provide reflectance data from the numerous bays and estuaries that because of their spatial dimensions are smaller than the resolution capability of the multispectral ocean color satellite sensors. Hyperspectral sensors collect

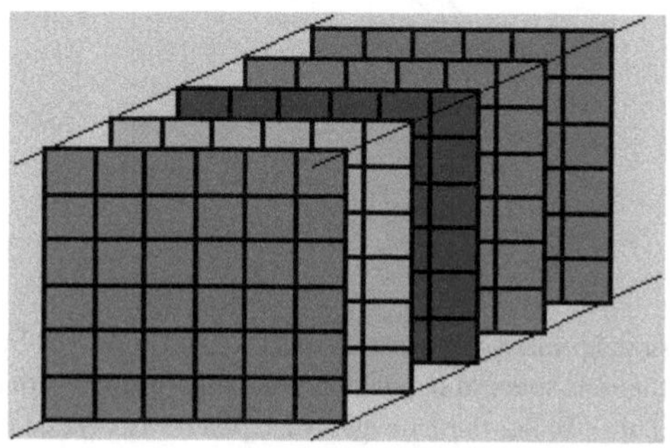

FIGURE 4.3 An illustration of a theoretical multilayer image consisting of spectral data from five layers.
Source: Modified from the Centre for Remote Imaging, Sensing, and Processing, http:/www.crisp.nus.edu.sg.

information from about a hundred or more contiguous spectral bands across the visible wavelengths from the near ultraviolet to near infrared regions of the electromagnetic spectrum. Information for these bands is combined to form a three-dimensional hyperspectral data cube (Figure 4.4a) for processing and analysis. The precision of these sensors is typically measured in spectral resolution, which is the width of each band of the spectrum that is captured. Hyperspectral sensors usually have greater than 10 nm resolution. Hyperspectral sensors image narrow spectral bands over a continuous spectral range and produce the spectra of all pixels in the scene (Figure 4.4b). Hyperspectral remote sensing is used in a wide array of applications. Although originally developed for mining, geology, and astronomy applications, hyperspectral imaging has been successfully used in oceanographic research for about 30 years.[11] Hyperspectral technology has expanded from hand-held radiometers to submerged sensors for measurements of inherent and apparent optical properties in estuarine/coastal and ocean waters.[11] Hyperspectral airborne and space-based detectors are now used to collect high spectral and spatial resolution measurements of radiance and reflectance from estuarine, coastal, and open ocean environments. The primary advantage to hyperspectral imaging is that each pixel has the entire spectrum acquired during overflights. The primary disadvantages are cost, complexity, and the need for fast computers, sensitive detectors, and large data-storage capacities for analyzing these data. Significant data storage capacity (exceeding hundreds of megabytes) is necessary since hyperspectral cubes are large multidimensional datasets.

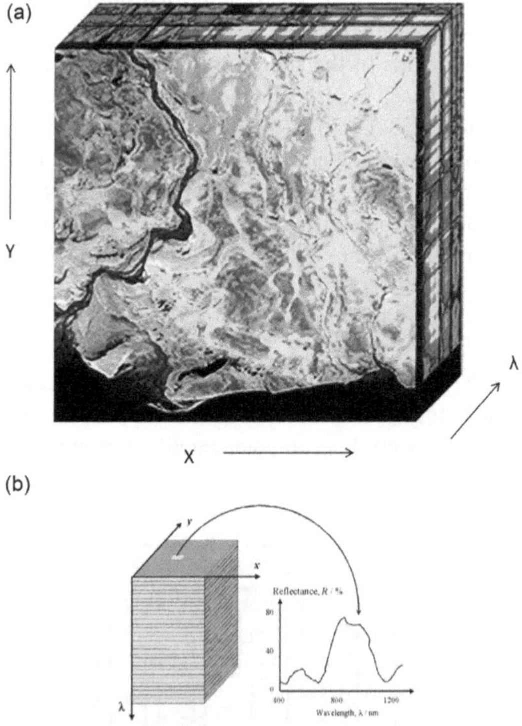

FIGURE 4.4 (**a**) Two-dimensional projection of a hyperspectral data cube. The hyperspectral image data usually consists of over a hundred contiguous spectral bands, forming a three-dimensional (two spatial dimensions and one spectral dimension) image cube. (**b**) For these images, each pixel is associated with a complete spectrum of the imaged area. In this figure, X and Y represent location coordinates and λ represents all wavelengths associated with each pixel in the image.

Source: Modified from aviris.jpl.nasa.gov.

Platforms

Satellite and Space-Based Remote Sensing

Satellites have been an integral key to retrieving reflectance data from the coastal and open ocean. Remote sensing of ocean color began in 1978 with the launch of the Coastal Zone Color Scanner (CZCS). Through the years, governmental organizations around the world have developed increasingly advanced remote-sensing scanners that have provided higher spatial and temporal resolutions of the earth's surface. There are two types of orbits for Earth observation satellites, polar orbiting and geostationary. Polar-orbiting satellites typically operate at an altitude between 700 and 800 km, with a revisit time of 2–3 days, whereas geostationary satellites operate in time scales of hours, which could theoretically provide data on the diurnal variation in phytoplankton abundance and productivity. A summary of the operational characteristics of current satellite-based and other space-based systems that have provided data for coastal and estuarine monitoring is found in Table 4.1.

Aircraft Remote Sensing

Aircraft are now widely used to remotely sense estuaries and coastal waters. In general, an airborne system can provide considerably higher spatial resolution data (e.g., less than a meter to tens of meters) than space-based systems at relatively low costs and can be flown when atmospheric (i.e., cloud-free), environmental, and solar conditions are acceptable to study a specific phenomenon.[12] Virtually every class of aircraft from small single-engine propeller planes to large multi-engine commercial and specialized military platforms can be used to conduct remote sensing. Single-engine propeller planes are typically un-pressurized and operate at altitudes below 25,000 ft, with a range less than 1000 miles. Turbine or turbo-charged twin engine propeller aircraft operate at altitudes up to 35,000 ft. Commercial jets flying above 40,000 ft can map large areas (at lower resolutions) but are relatively expensive to operate. Special-purpose high-altitude platforms, such as the NASA ER-2 aircraft flying at 65,000 ft, collect spectral and image data at regional scales. The major benefits of airborne remote sensing, compared to satellite-based systems, are that the user can define the deployment schedule and operational characteristics of the remote-sensing system. The deployment can also be coordinated with a field program to acquire in situ measurements to monitor biological and physical processes that occur over temporal scales, and at spatial resolutions that cannot be sampled by most satellite instruments.[13,14]

Post-flight Data Processing

Post-flight data processing is an extremely important process for removing atmospheric and seabed effects to convert the digital radiances into reflectance data. There are several commercially available software packages to process aircraft and satellite remote-sensing data such as ERDAS IMAGINE (Leica Geosystems GIS & Mapping) and ENVI (Research Systems International). Data are provided to a data user in a standard image processing format (e.g., ERDAS, ENVI, and GEOTIFF) or in a generic scientific data format such as HDF (Hierarchal Data Format). The HDF is an efficient structure for storing multiple sets of scientific, image, and ancillary data in a single data file. The data may be sent to a user as raw radiance files with no processing, image files with radiometric calibrations applied, or as radiometrically calibrated and atmospherically corrected digital image files geoferenced to a map projection. Complex atmospheric correction procedures and models [such as MODTRAN 4.0 (MODerate resolution TRANSsmittance)] are employed to compute the ocean color signal by determining the magnitude of and removing atmospheric scattering and absorption effects between the water surface and the sensor[13,15,16] However, a commonly used and simple approach is the "clear water pixel" or "dark pixel" subtraction technique, which assumes that the sensor has a spectral band for which clear water is essentially a black body (i.e., no reflectance). Therefore, any radiance measured by the instrument in this band is due to atmospheric backscatter and can be subtracted from all pixels in the image.[17,18] In the shallow waters of estuarine and coastal systems, the seabed reflects part of the incident light in a way that is highly dependent on the bottom material and roughness. The reflected light is spectrally different than

TABLE 4.1 Operational Characteristics of Current Ocean-Color Satellite Sensors (IOCCG, 2012; http://www.ioccg.org/sensors/current.html)

Sensor	Agency	Satellite	Launch Date	Swath (km)	Spatial Resolution (m)	Bands	Spectral Coverage (nm)	Orbit
COCTS	CNSA	HY-1B	11 April	2400	1100	10	402–12,500	Polar
CZI	(China)	(China)	2007	500	250	4	433–695	
GOCI	KARI/KORDI (South Korea)	COMS	26 June 2010	2500	500	8	400–865	Geostationary
HICO	ONR and DOD Space Test Program	International Space Station	18 Sept. 2009	50 km Selected coastal scenes	100	124	380–1000	51.6°, 15.8 orbits p/d
MERIS	ESA	ENVISAT			300			
MERSI	CNSA (China)	FY-3A (China)	27 May 2008	2400	250/1000	20	402–2155	Polar
MERSI	CNSA (China)	FY-3B (China)	5 Nov. 2010	2400	250/1000	20	402–2155	Polar
MODIS-Aqua	NASA (USA)	Aqua (EOS-PM1)	4 May 2002	2330	250/500/1000	36	405–14,385	Polar
MODIS-Terra	NASA (USA)	Terra (EOS-AM1)	18 Dec. 1999	2330	250/500/1000	36	405–14,385	Polar
OCM-2	ISRO (India)	Oceansat-2 (India)	23 Sept. 2009	1420	360/4000	8	400–900	Polar
POLDER-3	CNES (France)	Parasol	18 Dec. 2004	2100	6000	9	443–1020	Polar
VIIRS	NOAA/NASA (USA)	NPP	28 Oct. 2011	3000	370/740	22	402–11,800	Polar

that of deep water, which allows scientists to obtain useful information about the nature of the seabed. The maximum depth at which a sensor receives any significant signal varies as a function of spectral wavelength and the clarity of the water. In some coastal waters, the bottom is detected to less than 10 m. In highly turbid waters, the bottom would not be visible as the depth of light penetration is less than a meter.[19] Once the data have been corrected for atmospheric and seabed effects, the standard method is to then georeference the imagery that links specific pixel locations in the image to their corresponding location on a mapped surface for which the mapped coordinates are well known.

Remote Sensing for Management of Coastal and Estuarine Environments

Increased population and development have contributed significantly to the environmental pressures in coastal regions. These pressures have resulted in substantial physical changes to the coastline, declines in water quality, and biological/chemical changes to waters with the addition of high volumes of nutrients (primarily nitrogen and phosphorous) from urban, nonpoint source runoff. Remote sensing provides near-synoptic, local, regional, and global views of the indicators of environmental condition and change that can be routinely monitored. Unfortunately, the application of ocean color remote sensing techniques and data to address coastal environmental issues has been limited to a small segment of the scientific community and there has not been a general transfer of the capability to the environmental management community for operational decision-making. Schaeffer et al.[20] indicated that four attributes may be responsible for the reluctance of the management community to incorporate remote sensing data into their decision-making: cost, data accuracy, satellite mission continuity, and obtaining upper level management approval to include remote sensing data in their work. Results from a survey of selected U.S. Environmental Protection Agency managers indicated there was an impression that all satellite imagery could only be obtained with a financial commitment.[20] Survey respondents did not know that most satellite data were available at no cost from NASA and the European Space Agency (ESA). Respondents wanted assurance that satellite products could be validated, with accuracy or error estimates identified, for their particular water body of interest.[20] Operational remote sensing of coastal and estuarine environments is in an evolutionary phase where the availability of real-time (or near real-time) data, coupled with advances in algorithm development and image processing, will soon revolutionize our ability to monitor coastal and estuarine systems. The ability to monitor and predict changes in biological and physical processes characteristic of the coastal environment, over large areas and long time periods, is only possible through satellite and aircraft observations. Even with uncertainties in accuracy estimates, the detection of change in a particular water body is possible if the data product was derived with a consistent methodology.[20–22] When incorporated into specific monitoring plans, remotely sensed information has been shown to provide an added value component to data analysis for better decision support.[23] For example, high-resolution optical systems on space-based platforms (e.g., Landsat and Satellite Pour l'Observation de la Terre (SPOT) satellites) provide a critical source of spatial information that gives coastal managers the ability to assess environmental conditions at a given point in time. When collected over multiple years, this information can be used to better understand the cumulative effects of human development, including impacts of changes in land cover on coastal water quality or ecosystem health.[24–27] The survey further indicated that mission continuity was identified as a critical element for acceptance as funding issues may jeopardize the life expectancy of satellite programs.[20] Finally, the need for experienced personnel and to educate environmental managers was identified as an important requirement to create an organizational commitment to use remote sensing data.[20]

Conclusion

Optical sensors flown on space-based platforms and aircraft can remotely estimate the underwater light field of coastal and estuarine waters and link these optical properties to their in situ biological, chemical, and geological constituents. With this information, a wealth of new and exciting products that support coastal and estuarine monitoring are being created and could be effectively used to continually address new questions and challenges faced by decision-makers and managers.

Acknowledgments

I would like to thank Kenneth Rocha, Kristin Hychka, and Anne Kuhn for their review and comments of the manuscript. This is contribution number ORD-002860 of the Atlantic Ecology Division, National Health and Environmental Effects Research Laboratory, Office of Research and Development, U.S. Environmental Protection Agency. Any mention of trade names or commercial products does not constitute endorsement or recommendation for use.

References

1. Bagheri, S.; Peters, S.W.M. Retrieval of Marine Water Constituents Using Atmospherically Corrected AVIRIS Hyperspectral Data, 12th Aviris Workshop, JPL, Pasadena, CA, 2003.
2. Porter, W.M.; Enmark, H.T. A system overview of the Airborne Visible/Infrared Imaging Spectrometer (AVIRIS). *Proc SPIE, Imaging Spectrometer II;* San Diego, CA, 1987; Vol. 834, 22–31.
3. Richardson, L.L.; Buison, D.; Lui, C.J.; Ambrosia, V. The detection of algal photosynthetic accessory pigments using Airborne Visible-Infrared Imaging Spectrometer (AVIRIS) spectral data. Mar. Tech. Soc. J. **1994,** *28* (28), 10–21.
4. Anger, C.D.; Mah, S.; Babey, S.K. Technological enhancements to the Compact Airborne Spectrographic Imager (CASI). In *Proceedings of the First International Airborne Remote Sensing Conference and Exhibition;* Strasbourg, France, 1994; Vol. 2, 205–213.
5. Anger, C.D.; Achal, S.; Ivanco, T.; Mah, S.; Price, R.; Busler, J. Extended operational capabilities of CASI. In *Proceedings of the Second International Airborne Remote Sensing Conference and Exhibition-Technology, Measurement, and Analysis,* Environmental Research Institute of Michigan: Ann Arbor, MI, 1996; Vol. 2, 124–133.
6. Clark, C.D.; Ripley, H.T.; Green, E.P.; Edwards, A.J.; Mumby, PJ. Mapping and measurement of tropical coastal environments with hyperspectral and high spatial resolution data. Int. J. Remote Sens. **1997,** *18* (2), 237–242.
7. Hoogenboom, H.J.; Dekker, A.G.; De Haan, J.F. Retrieval of chlorophyll and suspended matter in inland waters from CASI data by matrix inversion. Can. J. Remote Sens. **1998,** *24* (2):144–152.
8. Rickard, L.J.; Basedow, R.W.; Zalewski, E.F.; Silverglate, P.R.; Landers, M. HYDICE: An airborne system for hyperspectral imaging. In *Proc SPIE, Imaging Spec-55 trometry of the Terrestrial Environment*; Vane, G., Ed.; San Diego, CA, 1993; Vol. 1937, 173–179, doi: 10.1117/12.157055.
9. Davis, C.O.; Bowles, J.; Leathers, R.A.; Korwan, D.; Downes, V.; Snyder, W.A.; Rhea, W.J.; Chen, W.; Fisher, J.; Bissett, W.P.; Reisse, R.A. Ocean PHILLS hyperspectral imager: Design, characterization, and calibration. Opt. Expr. **2002,** *10* (4), 210–221.
10. Corson, M.; Davis, C.O. The Hyperspectral Imager for the Coastal Ocean (HICO) provides a new view of the Coastal Ocean from the International Space Station. AGU EOS, **2011,** *92* (19), 161–162.
11. Chang, G.; Mahoney, K.; Briggs-Whitmire, A.; Kohler, D.D.R.; Mobley, C.D.; Lewis, M.; Moline, M.A.; Boss, E.; Kim, M.; Philpot, W.; Dickey, T.D. The new age of hyperspectral oceanography. Oceanography **2004,** *17* (2), 16–23.
12. Myers, J.S.; Miller, R.L. Optical airborne remote sensing. In *Remote Sensing of Coastal Aquatic Environments Technologies, Techniques, and Applications* (*Remote Sensing and Digital Image Processing*); Miller, R.L., Del Castillo, C.E., McKee, B., Eds.; Springer: Dordrecht, the Netherlands, 2005.
13. Miller, R.L.; Cruise, J.C.; Otero, E.; Lopez, J.M. Monitoring suspended particulate matter in puerto rico: Field measurements and remote sensing. Water Resour. Bull. **1994,** *30* (2), 271–282.

14. Miller, R.L.; Twardowski, M.S.; Moore, C.; Cassagrande, C. The Dolphin: Technology to support remote sensing algorithm development and applications. Backscatter **2003,** *14* (2), 8–12.
15. Richter, R.; Schläpfer, D. Geo-atmospheric processing of airborne imaging spectrometry data. Part 2: Atmospheric/ Topographic correction. Int. J. Remote Sens. **2002,** *23* (13), 2631–2649.
16. Lavender, S.J.; Nagur, C.R.C. Mapping coastal waters with high resolution imagery: Atmospheric correction of multiheight airborne imagery. J. Opt. A Pure Appl. Opt. **2002,** *4* (4), S50–S55.
17. Gordon, H.R.; Morel, A. Remote assessment of ocean color for interpretation of satellite visible imagery: A review. In *Lecture Notes on Coastal and Estuarine Studies*; Springer-Verlag: Berlin, 1983; Vol. 4.
18. Siegel, D.A.; Wang, M.; Maritorena, S.; Robinson, W. Atmospheric correction of satellite ocean color imagery: the black pixel assumption. Appl. Opt. **2000,** *39* (21), 3582–3591.
19. Sathyendranath, S., Ed.; IOCCG. *Remote Sensing of Ocean Colour in Coastal, and Other Optically-Comoplex Waters*, Reports of the International Ocean-Colour Coordinating Group, No. 3; IOCCG: Dartmouth, Canada, 2000, 140 pp.
20. Schaeffer, B.A.; Schaeffer, K.G.; Keith, D.; Lunetta, R.S.; Conmy, R.; Gould, R. Barriers to adopting satellite remote sensing for water quality management. Int. J. Remote Sens. **2013,** *34* (21), 7534–7544.
21. Stumpf, R.P.; Culver, M.E.; Tester, P.A.; Tomlinson, M.; Kirkpatrick, G.J.; Pederson, B.A.; Truby, E.; Ransibrahmanakul, V.; Soracco, M. Monitoring *Karenia brevis* blooms in the Gulf of Mexico using satellite ocean color imagery and other data. Harmful Algae **2003,** *2* (2), 147–160.
22. Hu, C.; Muller-Karger, F.E.; Taylor, C.; Carder, K.L.; Kelbe, C.; Johns, E.; Heil, C.A. Red Tide detection and tracing using MODIS fluorescence data: A regional example in SW Florida coastal waters. Remote Sens. Environ. **2005,** *97* (3), 311–321.
23. Arnone, R.A.; Parsons, A.R. Real-time use of ocean color remote sensing for coastal monitoring. In *Remote Sensing of Coastal Aquatic Environments Technologies, Techniques, and Applications (Remote Sensing and Digital Image Processing);* Miller, R.L., Del Castillo, C.E., McKee, B., Eds.; Springer: Dordrecht, the Netherlands, 2005; 317–337.
24. Wilson, E.H.; Sadler, S.A. Detection of forest type using multiple dates of Landsat TM imagery. Remote Sens. Environ. **2002,** *80* (3), 385–396.
25. Wang, Y.; Bonynge, G.; Nugranad, J.; Traber, M.; Ngusaru, A.; Tobey, J.; Hale, L.; Bowen, R.; Makota, V. Remote sensing of mangrove change along the Tanzania Coast. Mar. Geod. **2003,** *26* (1–2), 35–48.
26. Turner, W.; Spector, S.; Gardiner, N.; Fladeland, M.; Sterling, E.; Steininger, M. Remote sensing for biodiversity science and conservation. Trends Ecol. Evol. **2003,** *18* (3), 306–314.
27. Hurd, J.D.; Civco, D.L.; Wilson, E.M.; Arnold, C.L. Coastal area land-cover change analysis for connecticut. In *Remote Sensing of Coastal Environments;* Wang, Y., Ed.; CRC Press: Boca Raton, FL, 2010; 333–353.

5

Coastal Environments

Yuanzhi Zhang
and Jinrong Hu
Chinese University of Hong Kong Shenzhen Research Institute

Introduction

The coastal zone, broadly defined as near-coast waters and the adjacent land area, forms a dynamic interface of land and water of high ecological diversity and critical economic importance.[1] Coasts are often highly scenic and contain abundant natural resources. Coastal environments contain a wide range of natural habitats, such as sand dunes, barrier islands, tidal wetlands and marshes, mangrove forests, coral reefs, and submerged aquatic vegetation that provides food, shelter, and breeding grounds for terrestrial and marine species. Figure 5.1 shows a schematic view of coastal environments of coastal embayment. The majority of the world's population lives close to the sea. As many as 3 billion people (50% of the global total) live within 60 km of the shoreline.[2] The coastal zone is therefore perhaps the most critical zone for the world economy, culture, and future survival. However, the coasts also face many environmental challenges including natural hazards and human-induced impacts.

Coastal environments are characteristic of the coastal zones. This entry is designed as an introduction to the coastal environments to increase awareness and knowledge about coast. Beaches, barrier islands, salt marshes, mangrove swamps, estuaries, coral reefs are some examples of coastal environments and some typical coastal issues are described in this entry.

Characteristic Landforms and Environments of Coastal Zone

The coastal areas are the most beautiful areas on Earth. For much of Earth's existence, coastal zones have been its most dynamic and changing areas. Coastal zones are continually changing because of the dynamic interaction between the oceans and the land. Coastal environments are also highly dynamic and complex. Currents, waves, and tides as the main coastal energy source are fundamental to defining coastal environments. Coastal regions have distinctive land-forms that represent a balance among forces from the ocean, land, and atmosphere.[3] The shape of coastal landforms is a response of the materials

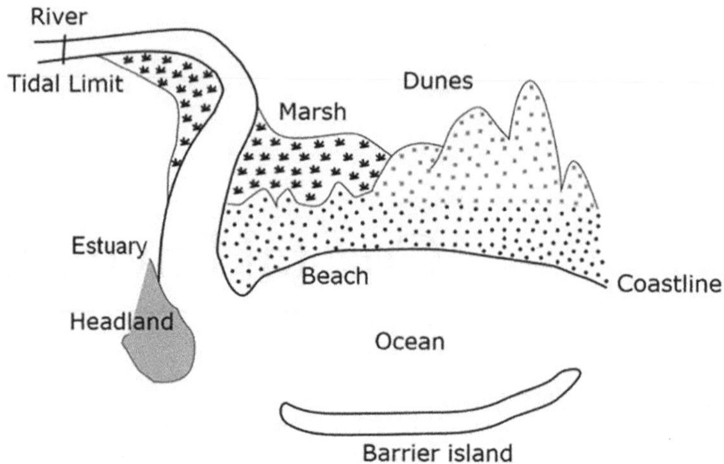

FIGURE 5.1 A schematic coastal environment of a coastal embayment, which includes different types of coastal environments: Barrier island, beach, dunes, estuary, marsh, and headland.

that are available to the processes acting on them. In this section, some characteristic land-forms and environments of coastal zone are introduced.

Beaches

A beach is a geological landform along the shoreline of an ocean, sea, lake, or a river. It usually consists of loose particles that are often composed of rock such as sand, gravel, shingle, pebbles, or cobblestones. The particles comprising the beach are occasionally biological in origin, such as mollusc shells or coralline algae. Beaches typically occur in areas along the coast where wave or current action deposits and reworks sediments.

Beaches are the result of wave action by which waves or currents move sand or other loose sediments of which the beach is made as these particles are held in suspension. Alternatively, sand may be moved by saltation (a bouncing movement of large particles). Beach materials come from erosion of rocks offshore, as well as from headland erosion and slumping producing deposits of scree.

Beaches also play an important role as a habitat for coastal plants and animals. Some small animals burrow into the sand and feed on material deposited by the waves. Crabs, insects, and shorebirds feed on these beach dwellers. The endangered Piping Plover and some tern species rely on beaches for nesting. Sea turtles also lay their eggs on ocean beaches. Sea grasses and other beach plants grow on undisturbed areas of the beach and dunes.

Barrier Islands

Barrier islands, a coastal landform and a type of barrier system, are relatively narrow strips of sand that parallel the mainland coast. Barrier islands often form as chains of long, low, narrow offshore deposits of sand and sediment, running parallel to a coast but separated from it by bays, estuaries, or lagoons. Unlike stationary landforms, barrier islands build up, erode, migrate, and rebuild over time in response to waves, tides, currents, and other physical processes in the open ocean environment.

The world's barrier islands measure about 13,000 miles (21,000 km) in length.[4] They are found along all continents except Antarctica and in all oceans, and they make up roughly 10% of the Earth's continental shorelines. The northern hemisphere is home to 74% of these islands.

Barrier islands play an enormous role in mitigating ocean swells and other storm events for the water systems behind on the mainland side of the barrier island.[5] This effectively creates a unique environment of relatively low-energy brackish water. Multiple wetland systems, such as lagoons, estuaries, and/or marshes, can result from such conditions depending on the surroundings. Without barrier islands, these wetlands cannot exist and will be destroyed by daily ocean waves and tides as well as ocean storm events. One of the most prominent examples is that of the Louisiana barrier islands.

Salt Marshes

Salt marshes are coastal wetlands rich in marine life.[6] They are sometimes called tidal marshes, because they occur in the zone between low and high tides. Salt marshes are composed of a variety of plants: rushes, sedges, and grasses. Salt marsh plants cannot grow where waves are strong, but they thrive along low-energy coasts. They also occur in areas called estuaries, where freshwater from the land mixes with sea water. A distinctive feature of salt marshes is the color; the plants grow in various shades of gray, brown, and green.

Salt marshes serve many important functions.[7] They buffer stormy seas, slow shoreline erosion, and are able to absorb excess nutrients before they reach the oceans and estuaries. High concentrations of nutrients can cause oxygen levels low enough to harm wildlife, such as the "Dead Zone" in the Gulf of Mexico. Salt marshes also provide vital food and habitat for clams, crabs, and juvenile fish, as well as offering shelter and nesting sites for several species of migratory waterfowl.

Mangrove Forests

Mangroves are various kinds of trees growing up to a medium height and shrubs that grow in saline coastal sediment habitats in the tropics and subtropics, mainly between latitudes 25°N and 25°S.[8] Mangroves provide enormously important and economically valuable ecosystem services to coastal communities throughout the tropics. Their important ecosystems provide wood, food, fodder, medicine, and honey for humans and habitats for many animals such as crocodiles, snakes, tigers, deer, otters, dolphins, and birds. A wide range of fish and shellfish also depend on mangroves as the swamps help filter sediment and pollution from water upstream and stop it from disturbing the delicate balance of ecosystems like coral reefs.

Mangrove ecosystems should be better protected,[8] the UN's food agency has warned, as it published new figures showing that 20% of the world's mangrove area has been destroyed since 1980. The main causes of the destruction of mangrove swampland include population pressure, conversion for shrimp and fish farming, agriculture, infrastructure and tourism, as well as pollution and natural disasters.

Estuaries

An estuary is a partly enclosed coastal body of water with one or more rivers or streams flowing into it, and with a free connection to the open sea.[9] Estuaries are protected from the full force of the ocean by mudflats, sandpits, and barrier islands. Estuaries are sometimes called bays, lagoons, harbors, or sounds. All these examples are estuaries if fresh water mixes with salt water.

Estuaries play an important role in the economy. They attract tourists who like fishing, boating, and other water sports. They are an important part of the shipping industry because many industrial ports are located in estuaries. Estuaries are also a critical part of the commercial fishing industry. It is estimated that over 75% of all fish caught by commercial fishing operations lived in an estuary for at least part of their life cycle.

Coral Reefs

Reefs[10] built by coral and associated organisms occur extensively in tropical waters and are widespread between latitudes 30°N and 30°S in the western parts of the Pacific, Indian, and Atlantic oceans. They are well developed in the Caribbean Sea, the Red Sea, and Indonesia. The distribution[11] of coral reefs is influenced by environmental factors such as light (symbiotic zooxanthellae in corals require light to photosynthesize), sea-surface temperature, and carbonate saturation state (closely related to temperature). Corals are limited to waters, where sea-surface temperature rarely drops below 17–18°C or exceed 33–34°C for prolonged periods. Coral species diversity decreases with latitude in response to sea-surface temperature and currents of dispersal.[12] Today, richly diverse coral reefs are found in the tropics along coastlines, on the margins of volcanic islands, and on isolated coral atolls.

Coral reefs[13] are the most diverse and beautiful of all marine habitats; they are home to 25% of all marine species. However, many of the world's coral reefs have been severely damaged by natural processes and human activities in recent decades. Corals face serious risks from various diseases, including black-band, white-band, and yellow-band diseases that have been reported from many localities worldwide. Black-band disease is primarily caused by cyanobacteria, but the causes of white-band disease and yellow-band disease are unknown. When corals are stressed, they often expel the algal symbionts that are critical to their health in a process commonly known as coral bleaching. One known cause of coral bleaching is an increase in ocean temperatures. Regional increases in sea-surface temperatures occur during El Niño events, and ocean temperatures worldwide may be changing as a result of global warming.

The susceptibility of corals to disease may also be on the rise as a result of human activities. Many human activities are known to directly and indirectly harm coral reefs. Oil spills and pollutants can threaten entire reefs. Excessive nutrients from land sources, such as sewage outfall and agricultural fertilizers, promote the growth of algae that can smother corals. Such algae also thrive when fish that graze on them are overharvested. Other organisms harmful to corals, such as the crown-of-thorns starfish, multiply when the species that prey on them are removed. The collection of live corals and other reef organisms can directly degrade large areas of reef.

Coastal Hazards

The highly variable environments[14] found along coastlines are shaped by a combination of different processes including waves, tides, storms, and in geologically recent times, humankind's changes to the shoreline environment. Most of the time, these processes operate at levels that are predictable, expected, and form part of the daily rhythm of life on the coasts, but other times these processes can impart dramatic changes very quickly, and sometimes pose great hazards to coastal residents. To appreciate and mitigate the hazards posed by living along coastlines, the processes that affect the interaction of the land, water, and atmosphere in this critical, dynamic, and ever-changing environment must be understood. There are several major factors that influence the development and potential of hazards along the coastline, including waves, tides, storms, and human-induced changes to the shoreline. These are each introduced in turn in this section.

Origin of Coastal Hazards

Waves

All coasts are affected to a certain degree by wave activity,[15] and waves provide the energy, which drives many of the coastal processes that create many of the world's most spectacular coasts. Waves are superficial undulations of the water surface produced by winds blowing over the sea.

Waves provide energy "powers" to many coastal processes. Coastlines are zones along which water is continually making changes. Waves can both erode rock and deposit sediment. Because of the continuous nature of ocean currents and waves, energy is constantly being expended along coastlines and they are thus dynamically changing systems, even over short (human) time scales.

Tides

Tides,[16] the rhythmic, twice-daily rise and fall of ocean waters, are caused by gravitational attraction between the Moon (and to a lesser degree, the Sun) and the Earth.

In some places, currents induced by tides are the most significant factor controlling development of the beach and shoreline environment. These places include tidal inlets, passages between islands and the mainland, and areas with exceptionally large tidal ranges. Tides are responsible for depositing deltas on the lagoonal and oceanward sides of tidal inlets, and for moving large amounts of sediment in regions with high tidal ranges. Tides also affect erosion and coastal sediment processes, since they control to what height waves can influence the land. Storm surges (high energy waves) that occur during spring tides can be particularly damaging to the shoreline and/or human property.

Storms

Storms[17] can cause some of the most dramatic and rapid changes to the coastal zones and represent one of the major, most unpredictable hazards to people living along coastline. Storms include hurricanes, which form in the late summer and fall, and extra tropical lows, which form in the late fall through spring. High winds blowing over the surface of the water during storms bring more energy to the coastline and can cause more rapid rates of erosion. One of the most famous hurricanes that had a significant impact on coastal communities in the United States was Hurricane Katrina in August 2005. According to the statistics, around 80% of New Orleans and some areas of neighboring communities were flooded, after which it took several weeks for the land surface to show up again. The hurricane left about 1800 people dead, which made it one of the deadliest U.S. tropical storms. The property damage was estimated over $100 billion, which was considered as the costliest hurricane in the USA.

Tsunami

Tsunami[18] is a series of traveling ocean waves of extremely long length generated by disturbances associated primarily with earthquakes occurring below or near the ocean floor. In the deep ocean, their length from wave crest to wave crest may be a hundred miles or more but with a wave height of only a few feet or less. They cannot be felt aboard ships nor can they be seen from the air in the open ocean. In deep water, the waves may reach speeds exceeding 500 miles per hour. Large tsunamis have been known to rise over 100 feet, while tsunamis 10–20 feet high can be very destructive and cause many deaths and injuries.

Tsunamis are a threat to life and property to anyone living near the ocean. For example, in 1992 and 1993 over 2000 people were killed by tsunamis occurring in Nicaragua, Indonesia, and Japan. Property damage was nearly $1 billion. The 1960 Chile earthquake generated a Pacificwide tsunami that caused widespread death and destruction in Chile, Hawaii, Japan, and other areas in the Pacific. Two of the most famous tsunamis in recent years are the 2004 Indian Ocean tsunami and March 11, 2011 tsunami in northeastern Japan. The December 26, 2004 Indian Ocean tsunami was caused by an earthquake that was thought to have had the energy of 23,000 atomic bombs. By the end of the day, more than 150,000 people were dead or missing and millions more were homeless in 11 countries, making it perhaps the most destructive tsunami in history. On March 11, 2011, a devastating tsunami with 33 feet high waves, triggered by the biggest earthquake on record in Japan, destroyed huge areas of the country's northeastern coast, causing more than 1000 deaths.

Human-Inducted Hazards

The human influence on coastlines looms large. Man is a major factor in coastal changes at various scales. Coastal zones are relatively fragile ecosystems, and disordered urbanization and development of infrastructure, alone or in combination with uncoordinated industrial, tourism-related, fishing and agricultural activities, can lead to rapid degradation of coastal habitats and resources. Coastal land use, pollution, and sea-level rise will be introduced in this section.

Coastal Land-Use

Coastal zones are an important area for human habitation, industry, location of centers of energy production, military activities, fisheries, bird life, and recreation. The fast development of coastal regions[19] will inevitably lead to the creation of vast built-up areas (such as construction of ports and tourist facilities) at the expense of natural habitats (e.g., dunes, saltmarshes) and as a result, will damage or destroy a substantial part of the natural coastline's habitats. In France, for example, 15% of natural areas on the coast have disappeared since 1976 and are continuing to do so at the rate of 1% a year. Italy, which had around 700,000 ha of coastal marshes at the end of the last century, had no more than 192,000 ha in 1972 and has less than 100,000 ha today. Some estimates suggest that about one-third of the coastal dunes in northwestern Europe and three quarters in the western Mediterranean have disappeared. Such large-scale habitat destruction will inevitably lead to a decline in species distribution and abundance.

Coastal Pollution: Red Tide

Pollution of the coastal zone[20] is primarily a result of contaminant load being discharged into receiving waters, resulting in deleterious effects such as harm to plants and animals, hazards to human health, hindrance to marine activities (including fishing), impairment of quality for use of sea water, and reduction of amenities. Despite the ability of coastal zones to reduce the harmful effects of some contaminants, coasts are also vulnerable to pollution since wastewater is often discharged directly or indirectly into sheltered and shallow coastal waters with poor mixing.

The most important contaminants[20] in the coastal zone are organic matter, synthetic organic compounds (e.g., PCBs and pesticides such as DDT and residues), and microbial organisms, nutrients (mainly nitrogen and phosphorus), and so on. This is the main reason for the steadily increase of harmful algal bloom (HAB) in coastal waters in recent years. Algal blooms[21] may cause harm through the production of toxins or by their accumulated biomass, which can affect co-occurring organisms and alter food-web dynamics. Impacts include human illness and mortality following consumption of or indirect exposure to HAB toxins, substantial economic losses to coastal communities and commercial fisheries, and HAB-associated fish, bird, and mammal mortalities. For example, in recent years, the frequent toxic red tide in the Gulf of Mexico has increasingly becoming a threat to the sea turtle population.

Coastal Pollution: Oil Spill

In recent years, oil spills caused by human error or carelessness have become one of the most serious disasters for the coastal and ocean ecosystems. Oil spills often result in both immediate and long-term environmental damage. Some of the environmental damage caused by an oil spill can last for decades after the spill occurs. Here, we only point out their effects on coastal environments, for example, the 2010 Gulf of Mexico oil spill and the 2011 Bohai Bay oil spill. The 2010 Gulf of Mexico oil spill led to at least 2500 km^2 of water covered with oil, and a large number of fish, birds, marine life, and plants were seriously affected. In November 2010, the U.S. government reported that 6104 birds, 609 turtles, and 100 dolphins were killed by this oil spill. The China State Oceanic Administration reported that the 2011 Bohai Bay oil spill resulted in 5500 km^2 of water pollution, roughly equivalent to 7% of the Bohai Sea area.

Sea-Level Rise

World sea-level is known to be rising, threatening an increase in storm and flood damage along many low-lying, populated coasts. The global rise may be as much as 1 m by the year 2050. The reasons for this rise in sea-level are complex, but one factor seems to be an increase in atmospheric CO_2 and trace gases, leading to increased heat absorption, the so-called "Greenhouse Effect."[22] Anthropogenic CO_2 emissions are currently responsible for more than half of the enhanced greenhouse effect.[23]

Conclusions

Coastal environments contain a wide range of natural habitats such as beaches, barrier islands, salt marshes, mangrove forests, coral reefs, and submerged aquatic vegetation that provides food, shelter, and breeding grounds for terrestrial and marine species. The coast and its adjacent areas on and off shore are an important part of a local ecosystem as the mixture of fresh water and salt water in estuaries provides many nutrients for marine life. However, coasts also face many environmental challenges including natural hazards and human-induced impacts. Currently, more and more issues face coastal managers: coastal storms and coastal habitat loss caused by land use, pollution, sea-level rise, and some other issues are becoming increasingly prominent. Management and protection of coastal environments to achieve the sustainable development of coastal zone are arduous and long-term tasks for humankind.

The coast, which is shaped by a variety of different forces, is a complex dynamic system and the coastal environments are also diverse. Due to space limitations, this entry cannot include detailed descriptions for each coastal environment but hopefully it can help the reader gain a general understanding about coastal environments.

References

1. McLean, R.F.; Tsyban, A.; Burkett, V.; Codignotto, J.O.; Forbes, D.L.; Mimura, N.; Beamish, R.J.; Ittekkot, V. Coastal zones and marine ecosystems. In *Climate Change 2001: Impacts, Adaptation and Vulnerability*; Cambridge University Press: United Kingdom, 2001.
2. Woodroffe, C.D. *Introduction. Coasts: Form, Process and Evolution,* 1st Ed.; Cambridge University Press: Cambridge, United Kingdom, 2003.
3. *Kusky, T. M. Introduction. The Coast: Hazardous Interactions within the Coastal Environment; Facts on File, Inc.: New York, 2008.*
4. Stutz, M.L.; Pilkey, O.H. Open-Ocean Barrier Islands: Global Influence of Climatic, Oceanographic, and Depositional Settings. J. Coast. Res. **2011,** *27* (2), 207–222.
5. Stone, G.W.; McBride, R.A. Louisiana barrier islands and their importance in wetland protection: forecasting shoreline change and subsequent response of wave climate. J. Coast. Res.**1998,** *14,* 900–915.
6. http://www.dep.state.fl.us/coastal/habitats/saltmarshes.htm (accessed March 2012).
7. http://water.epa.gov/type/wetlands/marsh.cfm (accessed March 2012).
8. http://www.guardian.co.uk/environment/2008/feb/01/endangeredhabitats.conservation (accessed March 2012).
9. http://www.nhptv.org/natureworks/ (accessed March 2012)
10. *Bird, Eric C.F. Coral Reefs and Atolls. Coasts: An Introduction to Coastal Geomorphology, 3rd Ed.; Blackwell: New York, 1984; 252.*
11. Woodroffe, C.D. *The Coastal Zone. Coasts: Form, Process and Evolution*; 1st Ed.; Cambridge University Press: Cambridge, United Kingdom, 2003; 190–191.
12. Yonge, C.M. The biology of reef-building corals. Scientific Reports of the Great Barrier Reef Expedition 1928–1929, British Museum (Natural History), **1940,** *1,* 353–891.
13. http://pubs.usgs.gov/fs/2002/fs025-02/ (accessed March 2012).
14. *Kusky, T.M. Introduction. The Coast: Hazardous Interactions within the Coastal Environment; Facts on File, Inc. Press: New York, 2008; 34.*
15. Bascom, W.N. Ocean waves. Scient. Am. **1959,** *201,* 74–84.
16. Skinner, B.J.; Porter S.C. *The Oceans and Their Margins. The Dynamic Earth: An Introduction to Physical Geology,* 4th Edition; John Wiley & Sons, Inc.: New York, 2000; 378–379.
17. Kusky, T.M. Introduction. *The Coast: Hazardous Interactions within the Coastal Environment*; Facts on File, Inc. Press: New York, 2008; 45.
18. http://www.nws.noaa.gov/om/brochures/tsunami.htm (accessed March 2012)

19. Gehu, J.M. European dune and shoreline vegetation. *Nature and Environment* Series No. 32. Council of Europe, Strasbourg, 1985.
20. Europe's Environment – The Dobris Assessment; European Environment Agency: 1995; Contamination and coastal pollution.
21. http://www.whoi.edu/website/redtide/home (accessed March 2012).
22. Carter, R.W.G. *Coastal Issues. Coastal Environments: An Introduction to the Physical, Ecological and Cultural Systems of Coastlines*; Academic Press: 1989.
23. *The Supplementary Report to the IPCC Scientific Assessment of Climate Change*; Houghton, J.T., Callander, B.A., and Varney, S.K., Eds.; Cambridge University Press, Cambridge, 1992.

6

Coastal Erosion and Shoreline Change

Matthew L. Stutz
Meredith College

Introduction

The shoreline defines the boundary between land and sea. The position of the shoreline is in constant flux, existing in a dynamic equilibrium between shifting coastal sediment, fluctuating waves and currents, and sea level. Shorelines in most places worldwide are experiencing erosion (i.e., landward movement) and may erode even more rapidly in the next century if global sea level continues to rise at its present pace. Coastal societies have a long history of attempting to hold back the sea; however, future attempts are likely to be more widespread, more extreme, more expensive, and in some cases more futile.

This synopsis of shoreline change and coastal erosion first addresses the underlying causes of shoreline change and its coastal impacts—sea level change, storms, and sediment supply. It also describes some of the common responses to coastal erosion—hard stabilization, beach nourishment, and relocation. The conclusion considers the future implications of global climate change and rising sea level on shoreline change and coastal societies.

Herein, the general term "shoreline change" will be used to apply to any movement of the shoreline landward or seaward. "Coastal erosion" is a term that denotes a landward movement of the shoreline due to loss of sediment, increased wave activity, and/or a rise in sea level. "Accretion" is used to describe a seaward movement of the shoreline.

Causes of Shoreline Erosion

The shoreline exists in a balance between three elements of the coastal environment—sea level, waves and currents, and sediment. A change or shift in any of these factors results in shoreline change. The direction of the change—landward or seaward—depends upon the specific nature of the change in the coastal environment. In general, an influx of new sediment, reduction in wave intensity, or fall in sea level produces shoreline accretion. Removal of sediment, increase in wave intensity, or rise in sea level usually results in shoreline erosion.

Sea Level

The *effective sea level* for any given coastal location is influenced by both *eustatic* and *relative* sea level. Eustatic sea level is governed by the volume of water in the oceans, and is "global." Relative sea level is governed by the height of the land itself, and is generally "local." A rise in either eustatic or relative sea level results in the inundation of land and shoreline erosion.

Eustatic Sea Level

Eustatic sea level rise results from an increase in ocean volume. Ocean volume increases due to the addition of water to the ocean, principally through the melting of land-based glacial ice. At the end of the Pleistocene Ice Age when vast ice sheets covered North America and Europe, eustatic sea level was 130 m lower (Figure 6.1),[1] and continental shorelines were located at the edge of today's continental shelf—in some cases more than 100 km seaward. Eustatic sea level rose a total of 0.2 m during the last century and is presently rising by 3 mm each year according to satellite altimeter measurements and coastal tide gauges (Figure 6.2). The Intergovernmental Panel on Climate Change (IPCC) attributes about half of the present eustatic sea level rise to glacial melting and the other half to the thermal expansion of ocean water.[2] The present rate of sea level rise is widely projected to accelerate, rising another 0.5 to 2 m in the coming century. This estimate is highly uncertain due to the complexity of ice sheet behavior in Antarctica and Greenland, the two largest stores of glacial ice. The Antarctic and Greenland ice sheets presently hold enough ice to add about 70 m and 7 m, respectively, to sea level if completely melted.

Relative Sea Level

Relative sea level rise results from localized vertical land movement. Tectonic land movement is a significant source of relative sea level rise. The March 2011 Japanese earthquake caused some coastal areas to instantaneously sink several feet, flooding some coastal villages now during normal daily high tides.[3] Relative sea level rise also affects deltas because as their thick sediment deposits compact, the delta surface gradually subsides. Compaction is naturally offset by continuous new sediment deposition.

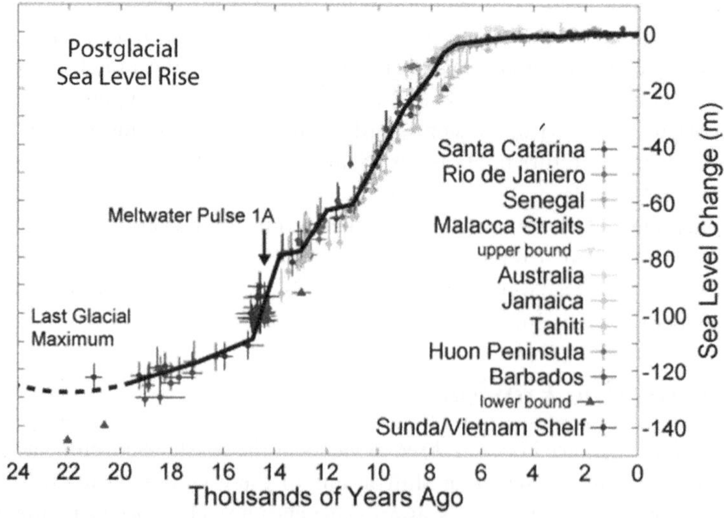

FIGURE 6.1 Global eustatic sea level rose rapidly in the late Pleistocene and early Holocene Epochs but has been nearly stable for the past 6000 years.
Source: Adapted from Rohde, Wikimedia Commons, http://commons.wikimedia.org/wiki/File:Post-Glacial_Sea_Level.png, accessed March 2012.

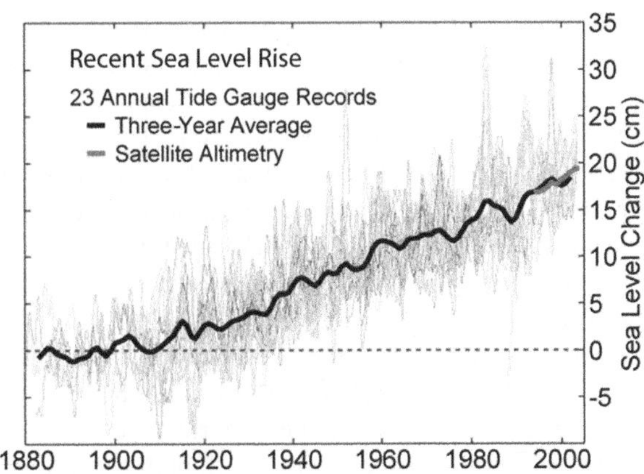

FIGURE 6.2 Modern tide gauge records show a rate of eustatic sea level rise of 2 mm/yr between 1900 and 2000. Satellite altimetry measured since 1990 indicates a rate of 3 mm/yr.
Source: Adapted from Rohde, Wikimedia Commons, http://commons.wikimedia.org/wiki/File:Recent_Sea_Level_Rise.png, accessed March 2012.

However, upstream dams and levees starve many deltas today, such as the Nile Delta, of essential new sediment buildup. Wetland drainage and petroleum extraction—often accompanying urban or agricultural development of deltas—accelerates compaction and is a significant cause of subsidence on the Mississippi Delta. The Mississippi Delta is sinking as much as 9 mm each year and loses about one football field of area *every hour* to erosion.[4] The city of New Orleans is as much as 3 m (10 ft) below sea level due to this long-term subsidence. The impact of Hurricane Katrina was more extreme as a result of the long-term subsidence. One last cause of relative sea level rise is *isostasy*, the uplift or subsidence of the earth's crust in response to the addition or removal of overlying weight. The massive Pleistocene ice sheets covering North America and Eurasia depressed the crust underneath and pushed up a bulge in the crust around the edges. When the ice sheets melted, the isostatic effect reversed and is still ongoing. The land around Hudson Bay, Canada, and Scandinavia is actually rising, causing relative sea level to fall. Coasts throughout much of the United States and western Europe, located on the former ice age bulge, are now gradually sinking.[1,5] The addition of more than 100 m of sea water over the continental shelf also contributes to subsidence through *hyrdroisostasy*.[1]

Waves and Currents

Waves and currents are chiefly influenced by storm activity and daily tides. The impacts of storms on shorelines can range from minor and temporary to catastrophic and permanent. The magnitude and duration of storms largely determine what type of impact the storm has. The height of tides at a given coastal location is relatively consistent and predictable; however, tidal currents in and around coastal inlets vary greatly due to the natural shifting of inlet shoals and to human modification to inlets.

Storms

Storms are meteorological events that cause temporarily elevated tide levels, or *storm surge*, and wave heights to impact the shoreline. The following storm impact classification illustrates the range of coastal impacts caused by storms.[6] Storm impacts begin with a narrowing of the dry beach as larger waves begin surging higher up the beach (the *run-up regime*) and may soon begin to scour the base of dunes (the *collision regime*). Most storms fall into these two categories, and their impacts are usually temporary. The sand eroded by beach and dune erosion is swept offshore and built up into offshore bars. The return of calm waves brings this sand gradually back to the beach over a period of days to months. A major storm may break through low, narrow sections of the dunes, washing water and sand inland (the *overwash regime*). A catastrophic storm can completely drown and wash over the dunes, causing wholesale landward shoreline retreat (the *inundation regime*). Shoreline erosion caused by overwash is generally permanent, because the sand is moved inland and cannot return to the beach by waves. At the inundation level, new inlets may cut completely through barrier islands or, in rare cases, cause barrier islands to disappear permanently (Figure 6.3).

Storm impacts on the shoreline are also closely related to the coastal topography. The Chandaleur Islands in Louisiana were almost completely destroyed by Hurricane Katrina. Whereas the Chandaleur Islands were uniformly low lying, narrow, and easily inundated, wider islands in the Gulf of Mexico with higher dunes survived the immense storm surge and waves of Katrina, experiencing some overwash but not complete destruction.[7]

Along rocky coasts, storm waves may pound the base of steep coastal bluffs, slowly weakening the bluffs and eventually causing a collapse of the bluff. Bluff erosion is more extreme in California during years with a strong *El Nino,* which brings more frequent storms and higher water levels to the Pacific U.S. coast. Sea level rise can also accelerate bluff erosion. All bluff erosion is permanent although the collapsed debris may temporarily protect the base of the bluff from further erosion. Future storms or sea level rise will eventually remove the debris and renew bluff erosion.

Tides and Currents

The daily rise and fall of ocean tides can generate strong tidal currents, particularly in coastal bays, estuaries, and lagoons. These environments are similar in that they are narrow passages that intensify the flow of tidal currents. In barrier island systems, *tidal inlets* are the narrow passes that separate two islands and connect the ocean with a quiescent barrier lagoon. Inlets are typically narrow unstable channels that allow ocean water to circulate into and out of lagoons as the tide rises and falls, respectively. Tidal currents capture and transport enormous amounts of sediment, forming large shoals just inside and outside tidal inlets. The rising (flood) tide flowing into the lagoon deposits a flood tidal delta just inside the lagoon. The falling (ebb) tide flowing out of the lagoon deposits an ebb tidal delta just seaward of the inlet.

Tidal inlet channels and their tidal deltas are in constant flux, but they are often cyclical and therefore predictable on a decadal timescale. Whenever the channel bends toward one side of the inlet, the ebb tidal delta also shifts its sand to the barrier island on that side of the inlet and the island begins accreting seaward. The island on the opposite side of the inlet will conversely erode rapidly. When the channel shifts back, the pattern reverses. Some tidal inlets migrate alongshore as opposed to shifting back and forth. The barrier island that the inlet is migrating toward is steadily consumed, while the updrift island is lengthened at the same time.

FIGURE 6.3 (See color insert.) Hurricane Irene generated significant erosion and overwash on the North Carolina Outer Banks, north of Cape Hatteras. Storm surge completely breached the island here and destroyed a section of Highway 12. Note that the highway makes a broad landward curve at this location, due to relocation of the highway following damage from previous storms.
Source: Adapted from United States Geological Survey Coastal and Marine Geology, August 31, 2011, http://coastal.er.usgs.gov/hurricanes/irene/post-storm-photos/20110830/20110831_150835d.jpg, accessed March 2012.

Sediment Supply

The action of waves, wind, and tidal currents causes a continuous movement of coastal sediment. Net loss of sediment along any shoreline reach always results in coastal erosion. A net gain of sediment results in shoreline accretion or a seaward advance of the shoreline. Storms or other events that cause periods of larger waves and higher water levels cause net sediment loss by sweeping sand from the beach and dunes offshore to form offshore bars. Fair-weather waves nudge that sand back to the beach slowly, resulting in net sediment gain and widening of the beach.

Longshore Transport

Most beaches experience a *longshore current*, which for swimmers means drifting slowly down the beach away from their beach towel as they swim in the surf. The longshore current is caused by waves approaching from any direction other than directly ashore—the wave crests are not parallel to the beach. Also, if waves are higher along one stretch of beach compared to an adjacent beach, the longshore current will flow along the beach from where the waves are high to where they are lower. Sand eroded from one beach can be transported along the beach by the resulting longshore drift.

Evidence of longshore transport is prevalent along *spits*, which are curved accumulations of sediment that extend across shallow bays. Along rocky coasts, where bays are too deep for spits to cross, sediment is nonetheless swept alongshore where it accumulates at the head of the bay. For many years, the accepted geologic wisdom was that all shorelines ultimately straighten out as waves erode sediment from headlands and fills in bays. The exception to this rule is *cuspate forelands* (e.g., Cape Hatteras, NC), which actually build seaward due to a large net gain of sediment from longshore transport. Because cuspate forelands jut out seaward, each side is sheltered from waves approaching from the opposite side. Thus, all waves and longshore transport converge toward the foreland, continuing its growth.[8]

Because beach sediment is often eroded from headlands and transported alongshore, the construction of seawalls along eroding headlands can also result in erosion down-drift because new sediment is blocked from entering the longshore drift. Delta shorelines are also starved of sediment if dams capture sediment upstream or flood-control levees prevent sediment from spreading over the delta. Heavy-mineral sand has long been mined directly from beaches and dunes where these dark minerals are concentrated in conspicuous layers (much like gold is concentrated in alluvial channels), resulting in a direct loss of sediment and subsequent erosion. All of these examples reveal that virtually all human intervention in the coastal zone is followed by reduced sediment supply and increased erosion somewhere.

Responses to Shoreline Erosion

Shorelines are transient features due to the always-shifting nature of waves, currents, and wind along the coast. Beaches are always eroding or accreting. No harm is done if erosion is occurring in an unpopulated area. The "problem" of erosion only exists where it threatens human interests such as buildings or roads. Three major categories of erosion control measures have been tried around the world—(1) hard stabilization, including seawalls, jetties, and groins; (2) soft stabilization, primarily beach nourishment; and (3) relocation.[9]

Hard Stabilization

Sea Walls

Sea walls are shore-parallel walls designed to stop further landward erosion. Waves still impact beaches in front of the seawall, which almost inevitably results in the narrowing or complete disappearance of the usable beach (Figure 6.4). Sea walls can be constructed from a wide variety of materials, including boulder revetments, concrete, steel, and wooden bulkheads, gabions, and sandbags as a

FIGURE 6.4 View of the Galveston sea wall that was built after the 1900 hurricane that killed more than 6000 residents. One of the drawbacks of sea walls is seen here as wave run-up reaches the base of the wall, leaving no visible beach for recreation.
Source: Adapted from Morton et al.[4]

few examples. For centuries, sea walls were the most preferred strategy to slow coastal erosion. Their negative effects on the beach environment—especially the loss of the dry "visible" beach—has led to a shift in strategy toward beach nourishment.[10] Sea walls are no longer permitted in many locations due to these concerns.

Jetties and Groins

Jetties and groins are engineered structures that extend offshore allowing them to block the longshore transport of sand. Jetties are employed at tidal inlets to improve navigation by blocking sediment from entering the channel (Figure 6.5). They also alter the flow of tidal currents and the size and position of the tidal deltas. They are frequently accompanied by dredging, which affects sediment transport in inlets with or without jetties. Instead of flowing to the next island, sediment will be trapped in the "hole" created by the dredging. Groins are built along beaches rather than inlets.

The primary problem with groins and jetties is that they interrupt the natural longshore transport of sediment. By trapping sediment on the updrift side, they cause sediment loss and erosion on the downdrift side. This may lead to the proliferation of more and more groins further downdrift. The jetties built at Ocean City, Maryland, in the 1930s and the groins built at Westhampton Beach, New York, in the 1960s demonstrate the impact of jetties and groins on the erosion of downdrift beaches.

FIGURE 6.5 View of jetties at Humboldt Bay, California. The jetties interrupt longshore sediment transport, causing a net buildup of sediment on the north jetty (left) and a net loss to the south.
Source: Adapted from Campbell, R. U.S. Army Corps of Engineers Digital Visual Library. Retrieved from Wikimedia Commons, http://commons.wikimedia.org/wiki/File:Humboldt_Bay_and_Eureka_aerial_view.jpg, accessed March 2012.

Beach Nourishment

Beach nourishment is the most commonly employed response to erosion today, largely due to the negative aspects of hard stabilization. Beach nourishment simply involves the emplacement of sand to widen the beach. Although nourished beaches keep the natural ability to adjust to the dynamic ocean, their drawback is that they must be renourished frequently to maintain their width—as often as every 3 to 4 years. Carolina Beach, North Carolina, has been renourished more than 20 times since its first beach nourishment project.[11]

For beach nourishment, sand is typically dredged and pumped from a nearby tidal inlet, shipping channel, or offshore location, or else trucked in a land source. Grading equipment is subsequently used to design it into a predetermined shape (Figure 6.6). In some instances, it is shaped into a wide berm a few feet above normal high tide. It may also be shaped into taller artificial dunes. Such an artificial dune was built along the North Carolina Outer Banks in the 1930s and 1940s, and these have been built extensively throughout the Netherlands today.

The limiting factor in beach nourishment is a sufficient supply of quality sand, which is not as readily available as one would think. Nourishment sand is usually of different quality than the natural beach sand, commonly being coarser or finer than the natural sand. It may also contain more mud, shells, or even rock depending on where the sand was obtained. Detailed geologic mapping of the continental shelf allows the identification of good-quality sand; however, the best sand is not always close to the beach, and lesser-quality sand is frequently used due to convenience and cost. Seafloor dredging may also alter wave patterns and currents, causing accelerated erosion of the nourished beach or adjacent beaches. In the Netherlands, sand cannot be dredged less than 2 km offshore to minimize the loss of beach sand offshore. Dredging sand from the seafloor also causes considerable damage to benthic habitats.

Relocation

Relocation is the movement of threatened structures to a more inland location. The most famous move in recent times occurred when the Cape Hatteras Lighthouse in North Carolina was moved in 1999 a half mile inland (Figure 6.7). It was not the first move for the lighthouse—it had moved to its former location in the late 1800s from its original site. During the lighthouse's lifetime, the shoreline has eroded almost 1500 ft. To the north of Cape Hatteras, North Carolina Highway 12 also feels the impact of persistent, annual erosion. Numerous sections of the highway are buried by overwash each year as a result of frequent winter storms and occasional hurricanes. Severe storms, such as Hurricanes Irene (2011) and Isabel (2003) often destroy the road completely and cut new inlets through the island. In these instances,

FIGURE 6.6 View of a nourished beach construction in Tybee Island near Savannah, Georgia. Sediment dredged from offshore is discharged from the pipeline and repositioned by the bulldozers.
Source: Adapted from Jordan, J. U.S. Army Corps of Engineers Digital Visual Library, http://eportal.usace.army.mil/sites/DVL/DVL%20Images/Forms/DispForm.aspx?ID=1146, accessed March 2012.

FIGURE 6.7 The Cape Hatteras Lighthouse was moved in 1999 from its former location (top of the photo) along the path through the center of the photo to the site cleared in the foreground. It is now 1600 ft from the shoreline. **Source:** Adapted from United States National Park Service, http://www.nps.gov/caha/historyculture/images/LH-move-path-ariel.jpg, accessed March 2012.

the road is commonly moved landward when it is rebuilt. The length of the road and the frequency of storms in North Carolina would make beach nourishment too expensive and impractical here.

More dramatic relocation efforts are already occurring worldwide and will likely become more prevalent in the future as sea level continues to rise. The Alaskan village of Shishmaref, home to about 500 mostly native Alaskans, will soon be moved to the mainland of Seward Peninsula in the Bering Strait.[12] The village has tried other measures of erosion control for several decades but none provided a permanent solution. In the south Pacific, the 2000 residents of the Carteret Islands (officially part of Papua New Guinea) have already abandoned their coral atoll. Other low-lying Pacific atolls, including Kiribati, may also soon follow suit.

The Future of Shoreline Erosion

World eustatic sea level will be 30 cm or 1 ft higher a century from now if the present rise of 3 mm/yr continues unchanged. There is a growing scientific consensus based on improved observational methods that the Greenland ice sheet—which would raise sea level by 7 m (more than 20 ft) if completely melted—is melting at an accelerating rate. It is unlikely that the entire ice sheet will melt within the next century; however, continued acceleration of the ice sheet's shrinkage could easily raise sea level by 1 to 2 m (3–6 ft) by 2100.[12] This scenario would expose all coastlines to the rate of submergence now seen only in Louisiana and a handful of other places where local relative sea level rise has been dramatically accelerated by humans.

What does this mean for coastal erosion rates? It is a certainty that due to the local variability of topography, rock types, and wave and storm patterns that coastal erosion rates will vary enormously as the rate of sea level rise increases. But it is also a certainty that all of the world's coastlines and a significant portion of the world's population will deal with increased environmental, economic, and political stresses due to sea level rise. All of recorded civilization has developed and expanded during a time of stable sea level and coastlines. The last century has seen global population grow by 5 billion people. In 2011, the world population officially eclipsed 7 billion, roughly half of whom live in low-lying coastal areas. In addition to the thousands of Shishmarefs of the world, sea level rise will impact the Shanghais of the world as well—making it more difficult for hundreds of millions to get fresh water, grow food on fertile low-lying deltas, and protect homes from floods.

The last period of Earth's history that witnessed a rate of sea level rise of more than 1 m/century occurred when the last remnants of the Pleistocene ice sheets melted between 11,000 and 6000 years ago (Figure 6.1). Sea levels rose more than 60 m (200 ft) over that period as coastlines worldwide retreated rapidly across the continental shelf, stopping roughly where they stand today. Based on the location of the 60 m depth contour on the continental shelf today, Atlantic coast shorelines would have eroded 15 times greater than average erosion rates today,[13] comparable to the extreme erosion rates in Louisiana.[4] In such a scenario, we would see an increasing number of vulnerable barrier island systems such as the North Carolina Outer Banks disintegrate much like the Chandeleur Islands.[14] Coastal marsh and wetland loss would greatly accelerate in places such as Chesapeake Bay and in already threatened and crowded deltas such as the Ganges, Mekong, Nile, Rhine, and Yangtze.

Conclusion

Although there is much uncertainty regarding future sea level rise and coastal erosion, the possibility of a catastrophic sea level rise is growing larger as we continue to see warmer temperatures in the Arctic and more rapid glacial melting. We may not yet grasp the full implications for coastal regions because we have no historical parallel with which to compare such a scenario. The traditional responses to coastal erosion such as sea walls and beach nourishment may not hold up to accelerating shoreline change. A long and methodical withdrawal from vulnerable shorelines is the most prudent approach for the future; however, this would require a level of long-term planning and cooperation—both political and economic and local, national, and global—that has only been rarely achieved in modern human society.

References

1. Woodroffe, C.J. Ed.; *Coasts: Form, Process, and Evolution*; Cambridge University Press: Cambridge, UK, 2003.
2. Solomon, S.; Qin, D.; Manning, M.; Chen, Z.; Marquis, M.; Averyt, K.B.; Tignor, M.; Miller, H.L. Eds.; *Climate Change 2007: The Physical Science Basis.* Contribution of Working Group I to the Fourth Assessment Report of the Intergovernmental Panel on Climate Change; Cambridge University Press: Cambridge, UK, 2007.
3. Gibbons, H. Japan Lashed by Powerful Earthquake, Devastating Tsunami, Sound Waves, United States Geological Survey, March 2011, http://soundwaves.usgs.gov/2011/03/ (accessed March 2012).
4. Morton, R.A.; Miller, T.M.; Moore, L.J. *National Assessment of Shoreline Change: Part 1, Historical Shoreline Changes and Associated Coastal Land Loss along the U.S. Gulf of Mexico.* Open file report 2004-1043; United States Geological Survey: St. Petersburg, FL, 2004, http://pubs.usgs.gov/of/2004/1043 (accessed March 2012).
5. Pirazzoli, P.A. *World Atlas of Holocene Sea-Level Changes;* Elsevier Oceanography Series; Elsevier: Amsterdam, 1991; Vol. 58.
6. Sallenger, A.H., Jr. Storm impact scale for barrier islands. J. Coastal Res. **2000**, *16* (3), 890–895.
7. Sallenger, A.H., Jr.; Wright, C.W.; Lillycrop, W.J.; Howd, P.A.; Stockdon, H.F.; Guy, K.K.; Morgan, K.L.M. Extreme changes to barrier islands along the central Gulf of Mexico coast during Hurricane Katrina. *In Science and the Storms—The USGS Response to the Hurricanes of 2005.* Farris, G.S., Smith, G.J., Crane, M.P., Demas, C.R., Robbins, L.L., Lavoie, D.L., Eds.; United States Geological Survey Circular 1306, 2007; 113–118, http://pubs.usgs.gov/circ/1306/ (accessed March 2012).

8. Ashton, A.; Murray, A.B.; Arnoult, O. Formation of coastline features by large-scale instabilities induced by highangle waves. Nature **2001**, *414*, 296–300.
9. Kaufman, W.; Pilkey, O.H. *The Beaches Are Moving: The Drowning of America's Shoreline;* Duke University Press: Durham, North Carolina, 1979.
10. Pilkey, O.H.; Wright, H.L., III. Seawalls versus beaches. J. Coastal Res. **1988**, *4*, 41–64.
11. Valverde, H.L.; Trembanis, A.C.; Pilkey, O.H. Summary of beach nourishment episodes on the U.S. East Coast Barrier Islands. J. Coastal Res. **1999,** *15* (4), 1100–1118.
12. Pilkey, O.H.; Young, R. *The Rising Sea;* Island Press: Washington, D.C., 2009.
13. CCSP. *Coastal Sensitivity to Sea-Level Rise: A Focus on the Mid-Atlantic Region.* A report by the U.S. Climate Change Science Program and the Subcommittee on Global Change Research. Titus, J.G. (Coordinating Lead Author), Anderson, K.E., Cahoon, D.R., Gesch, D.B., Gill, S.K., Gutierrez, B.T., Thieler, E.R., Williams, S.J. (Lead Authors); United States Environmental Protection Agency: Washington, D.C., U.S.A., 2009.
14. Riggs, S.R.; Ames, D.V.; Culver, S.J.; Mallinson, D.J. *The Battle for North Carolina's Coast: Evolutionary History, Present Crisis, and Vision for the Future;* University of North Carolina Press: Chapel Hill, NC, 2011.

7

Coastal Natural Disasters: Tsunamis on the Sanriku Coast

Takehiko Takano
Tohoku Gakuin University

Introduction

The Sanriku Coast is located on the Pacific side of the northeastern part of Japan's main island. Its coastal area was severely damaged when a huge tsunami with the height of more than 10 m arrived after the powerful earthquake on March 11, 2011, which shocked the world. Although this devastating tsunami was among the most severe natural disasters in world history, it was not the first one on the Sanriku Coast. The severely affected coastal area from Aomori to Miyagi is known as the "Sanriku Coast" or simply "Sanriku." The name "Sanriku" was originally used in the Meiji Era for the whole region of ancient Rikuzen (presently Miyagi), Rikuchu (Iwate), and Rikuou (Aomori), all of which contain the Japanese character "Riku." However, after the severe tsunami disaster that occurred in 1896, the name "Sanriku" began to mean only its coastal area. It was an impressive case that a severe natural disaster changed the meaning of a region's name.

History of Tsunami Disasters on the Sanriku Coast

Throughout history, the Sanriku Coast frequently suffered from tsunamis. According to "History of Earthquake and Tsunami in Miyagi Prefecture," there was a record of 16 tsunamis during the Edo Era from 1600 to 1868. Each of the historically known tsunamis, such as Jogan in 869, Keicho in 1611, Meiji in 1896, and Showa in 1933, were accompanied by severe damage with more than a thousand casualties (Table 7.1).

TABLE 7.1 Major Tsunami Disasters Affected the Sanriku Coast

Year	Name	Magnitude	Killed and Lost Lives	Tsunami Reach (m)
869	Jogan	8.3*	1,000*	
1611	Keicho	8.9*	7,800*	
1896	Meiji	8.2	21,959	38.2
1933	Showa	8.1	3,064	28.7
2011	East Japan	9.0	18,649	40.5

*Estimated value in the cases for Jogan and Keicho.
Source: Data from Japan Meteorological Agency, Central Disaster Prevention Council.

Although there was a lack of scientific observation data for the pre-modern Jogan and Keicho tsunamis, their outlines were described in ancient documents telling that the Jogan tsunami reached inland to a castle town about 4 km from the shoreline.[1] The description indicated that 1,000 people drowned, which reflects how severe the tsunami impact was in an ancient age when the population was far fewer than that of today. The Keicho tsunami is also considered to be the biggest one in Japan's history, of which the inundation wave was estimated to reach more distant places from the shoreline than the tsunamis in 1896 and 1933 (Figure 7.1).

After the Meiji Restoration in 1868, seismological and geophysical survey methods were introduced to Japan and reliable data was collected. The Japan Society of Civil Engineering published a comparative chart of the past major tsunamis in the modern age including the East Japan tsunami in 2011 (Figure 7.2). Figure 7.2 clearly shows that the 2011 tsunami was more intense than any of the other tsunamis in the modern age.

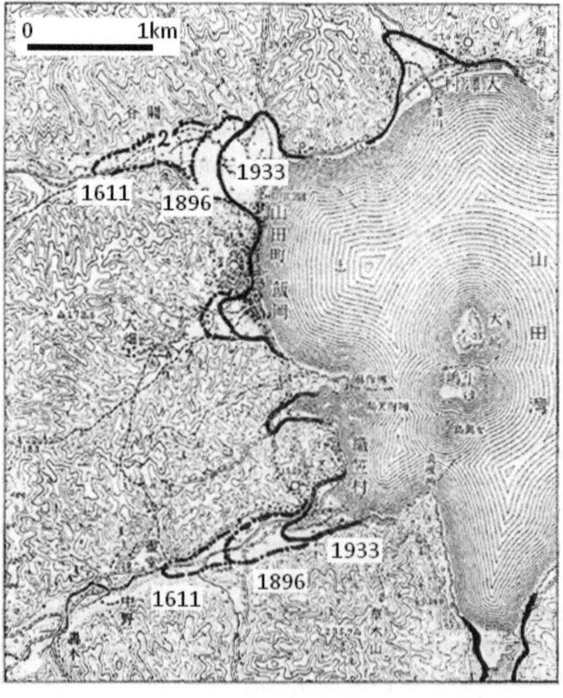

FIGURE 7.1 Inundated line of tsunami wave at Yamada Bay.
Source: Adapted from Imamura.[10]

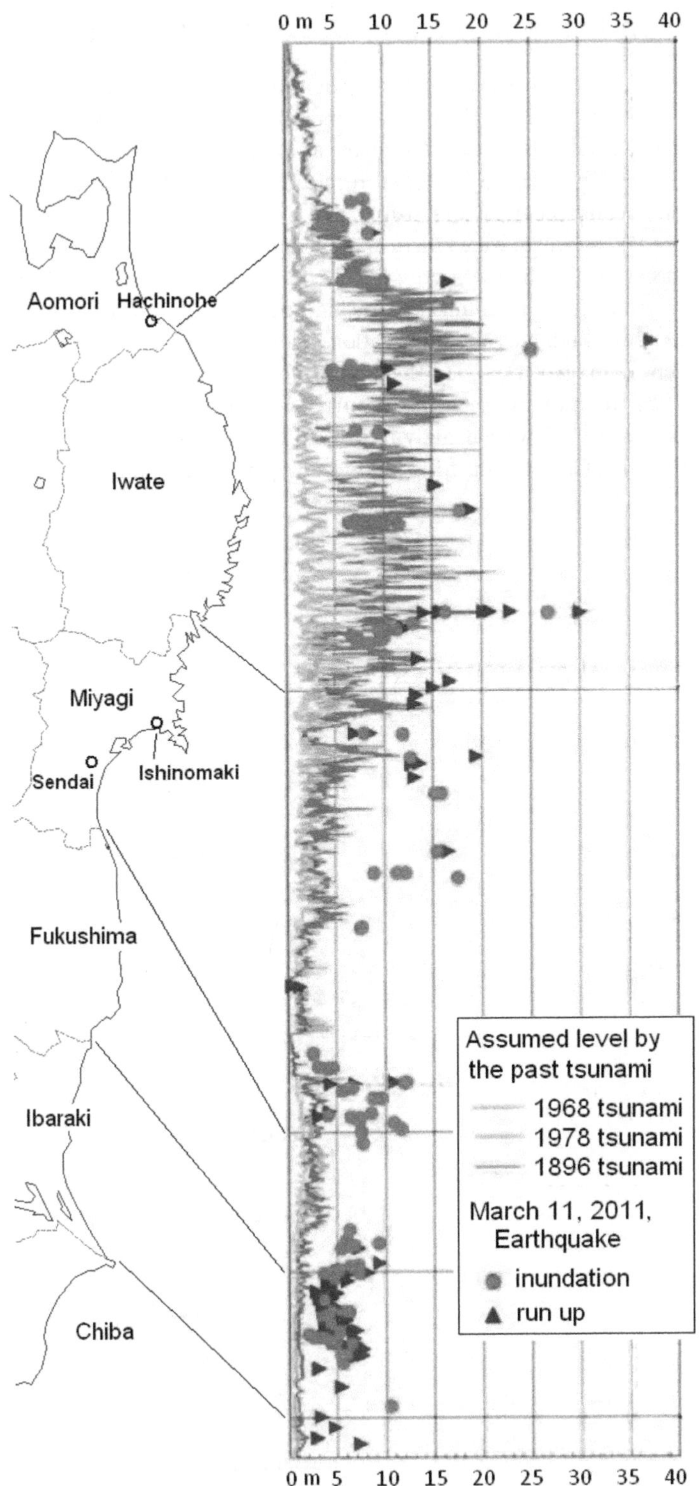

FIGURE 7.2 Comparison of tsunami levels.

Source: Data from http://www.hkd.mlit.go.jp/zigyoka/z_kowan/jishintsunami/pdf/110801_2_1.pdf, partly modified by author.

2011 East Japan Earthquake and the Sanriku Coast

On JST 14:46, March 11, 2011, a mega earthquake with a Richter scale magnitude of 9.0 severely shook the Pacific coast of East Japan with a JMA seismic intensity of 6 or more. The U.S. Geological Survey reported that "this magnitude places the earthquake as the fourth largest in the world since 1900 and the largest in Japan since modern instrumental recordings began 130 years ago."[2] It was one of the most intense natural phenomena in Japan's history, and could be rated as a "millennium" class disaster. Among the damages, 15,872 people were killed, 2,777 people were missing, 129,577 houses were entirely destroyed, and another 266,101 houses were severely damaged.[3] Almost every type of consequence associated with an earthquake disaster followed, including lifeline cut, infrastructure damages, collapse of residential and agricultural lands, and liquefaction in lowland areas.

The most severely damaged area was the Pacific coast of the northeastern part of the main island, where most of the human casualties occurred (Figure 7.3). The coastal areas of Iwate and Miyagi prefectures (Figure 7.4) were heavily impacted, because the huge tsunami arrived about 50 min after the earthquake. Countless fishing boats and aquaculture facilities were washed away and destroyed.

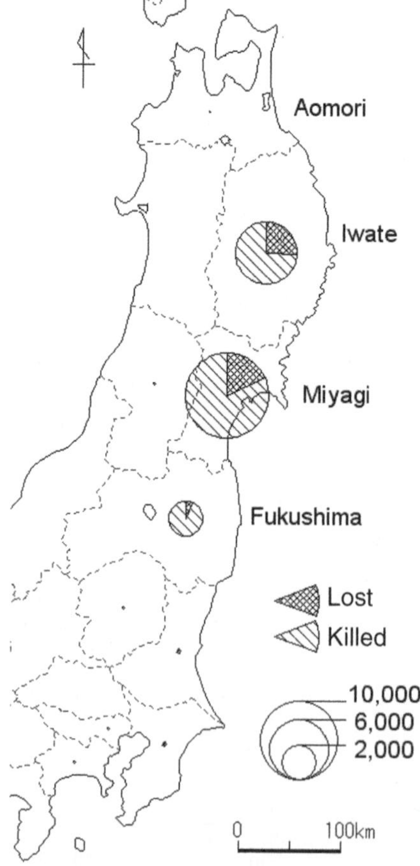

FIGURE 7.3 Casualties by prefecture.
Source: Data from National Police Agency, October 11, 2011.

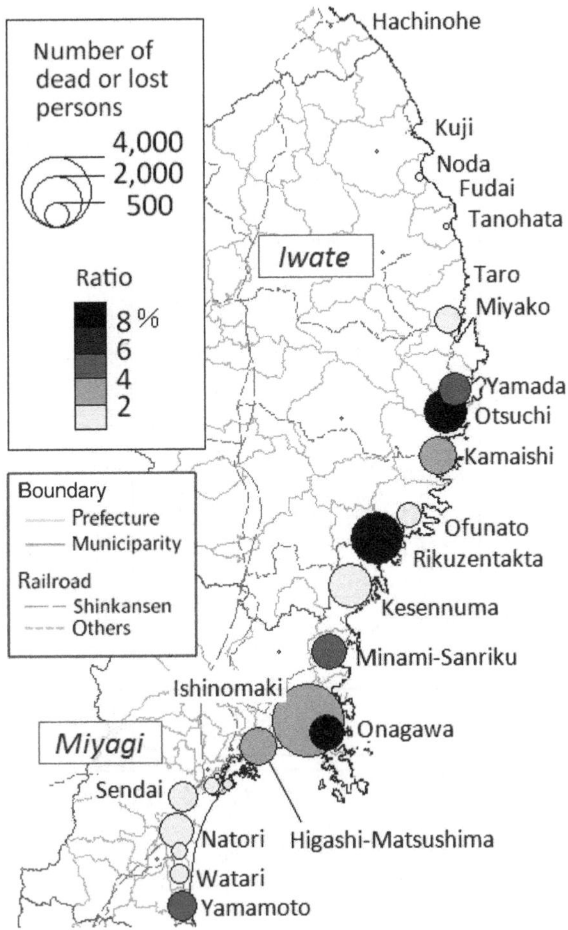

FIGURE 7.4 Casualties by municipality.
Source: Data from National Police Agency, October 11, 2011.

Mechanism of Tsunami on the Sanriku Coast

Tectonic Movement around the Japan Trench

Tsunamis along the Sanriku Coast are usually caused by tectonic movement of the plates around the Japan Trench, where the Pacific plate is creeping beneath the North American plate at a rate of 83 mm/yr.[4] The plate is actually considered to be divided into several micro plates and their activities can cause earthquakes and tsunamis. Smoothly descending micro plates are unlikely to cause major earthquakes and tsunamis. However, at the place where the micro plate strongly clings to the upper North American plates on which the Japanese archipelago lies, a sudden thrust of the plate can happen periodically. In the case where several micro plates move correlatively together, a greater magnitude of earthquake and an intensive scale of tsunami tend to occur.

The actual scale of a tsunami depends on which micro plates move. In the 2011 East Japan Earthquake, it was estimated that many micro plates offshore of East Japan (Figure 7.5) moved correlatively in only about 3 minutes, which caused a huge tsunami to arrive not only to the Sanriku Coast but also to the entire Pacific shore of East Japan.

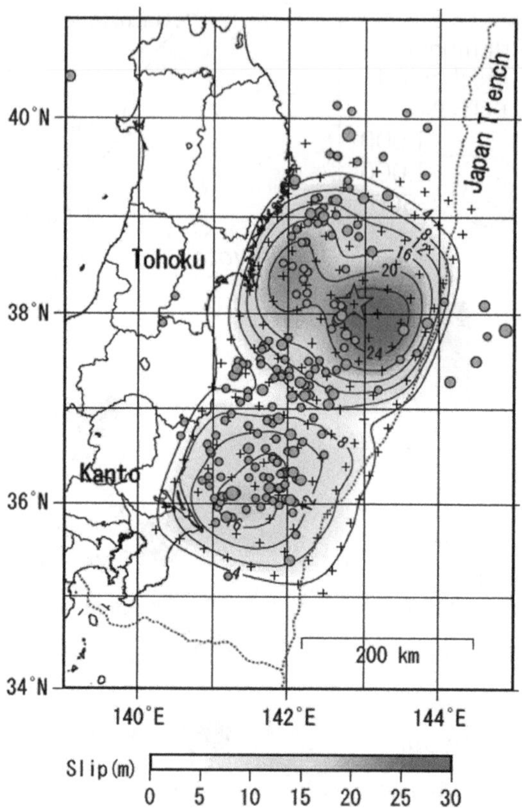

FIGURE 7.5 Hypo-central area of 2011 East Japan Earthquake.
Source: Technical Report of The Japan Meteorological Agency, No.133 (2012), http://www.jma.go.jp/jma/kishou/books/gizyutu/133/gizyutu_133.html.

Tsunamis from Distant Places

Many of the tsunamis that affect the Sanriku Coast are caused not only in and near the Japan Trench, but also distant places where the plates meet around the Pacific Ocean. Trenches such as Kamchatka, Aleutian, Peru, and Chile are possible places. In the early morning of May 24, 1960, a tsunami with a height of 2–6 m struck the Sanriku Coast without any registered earthquake. It was caused by an M 9.5 mega earthquake that occurred at the southern part of the Chile Trench the previous day. The report indicated that 142 human lives were lost, 46,000 houses were washed away or destroyed, and fishing boats and aquaculture facilities were swept out.

Shape of Shoreline

The irregular shape of the Sanriku Coast is well known as a typical example of a "Ria type" coast, where narrow peninsulas and inlets appear in turn (Figure 7.6). Ria is the regional name of Northwestern Spain of which the shape is sawtooth-like. Such a coastal shape can easily increase the height and power of a tsunami especially at the bottom of inlets, where settlements are often located in the narrow coastal flat lands. That came true in the 2011 tsunami disaster that followed the East Japan Earthquake.

FIGURE 7.6 "Ria type" of coastline in the Sanriku Coast (Otsuchi and Kamaishi).
Source: Data from Geospatial Information Authority of Japan, http://www.gsi.go.jp/common/000059842.pdf.

Countermeasures against Tsunami Disaster

Some of the suggested countermeasures against tsunami disaster are summarized as follows.

Seawalls, Breakwaters, and Water Gates

Seawalls, breakwaters, and water gates are the direct countermeasures to prevent or decrease the tsunami impact. Such structures have been constructed in almost every coastal settlement in Sanriku after suffering severe damages from two major tsunami disasters in 1896 and 1933. The fishery town of Taro is well known for its huge seawalls with a height of 10 m and total distance of about 2 km (Figure 7.7). The 1,350 m long wall (shown by the bold line in Figure 7.7) was constructed by 1958 and surrounds the old settlement near the western hill side. This was planned to divert tsunami flows eastward from the settlement. However, as the town expanded outside the original seawall, two other seawalls, shown by broken and dotted lines in Figure 7.7, were constructed by 1978. In addition, breakwaters were constructed at the mouth of the port. The great tsunami of 2011 arrived higher than those seawalls, which swept away both the new settlement outside the original wall as well as the old settlements. Even so, the seawalls were effective in delaying the arrival of the tsunami into the settlement area.[5]

FIGURE 7.7 Taro town. Bold line: constructed by 1958; broken line and dotted line: constructed by 1978.
Source: Data from Google Earth, July 20, 2009.

In Fudai village, about 1,500 people lived along the narrow flat land near the coast (Figure 7.8). The 15.5 m high seawall was constructed by 1967 at a fishing settlement of Otanabu, and a water gate with the same height was constructed by 1984 at the river mouth of Fudai (Figure 7.9). Both structures perfectly protected the inland settlement from the invasion of the 2011 tsunami. The height of the 15.5 m seawall was determined by a former chief of the village who believed villagers stating that the 1896 tsunami reached a height of about 15 m.

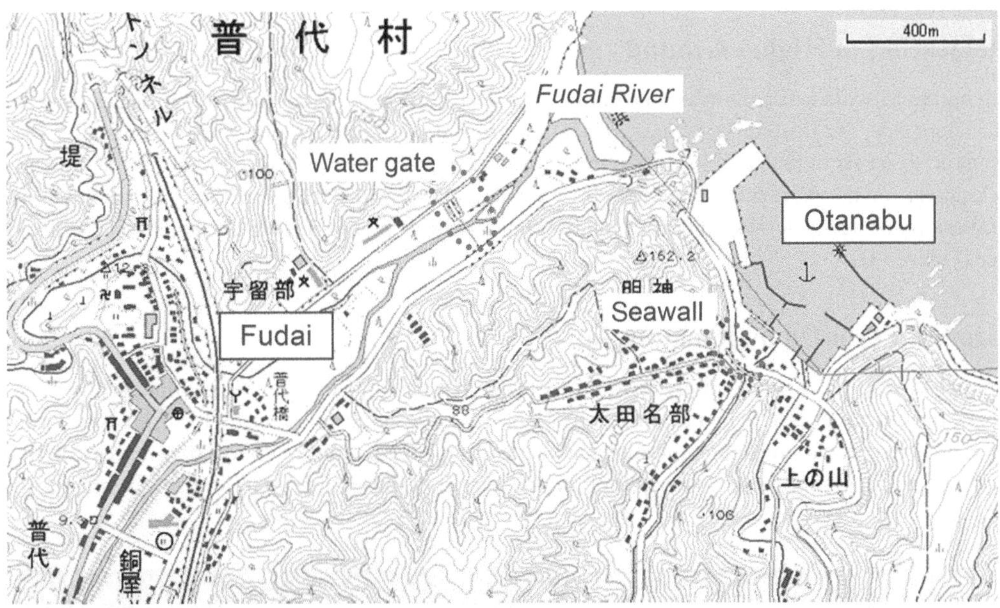

FIGURE 7.8 Topographical map of Fudai village.
Source: Data from Geospatial Information Authority of Japan.

FIGURE 7.9 The 15.5 m seawall at Otanabu port. The 2011 tsunami reached from the shore (left side), destroyed the factory (flat roof), but was interrupted by the seawall (upper right) before it could invade the settlement.
Source: Data from http://www.yomiuri.co.jp/national/news/20110403-OYT1T00599.htm.

The city of Kamaishi is known as the birthplace of Japan's modern iron manufacturing and is a major base port of the fishery on the Sanriku Coast. At the mouth of Kamaishi Bay with the maximum depth of 61 m, the world's biggest breakwater was constructed from 1978 to 2009 (Figure 7.10). The height is 4 m above sea level, and its total length of the north and south parts is 1,660 m. By the 2011 tsunami, about half of the 66 caissons were subsided below sea level. However, it proved that every caisson kept its original shape and that the tsunami level was decreased less than 7 m in the central part of the city. Where no breakwater existed, the tsunami level was estimated to reach up to about 13 m.[6]

Relocation to Higher Ground

An option to avoid tsunami invasion would be the relocation of settlements to higher places away from the shore side. On the Sanriku Coast, relocation projects were done in many settlements after severe tsunami disasters in 1896 and 1933.[7] Figure 7.11 shows a small fishery village named "Toni," which was relocated from the shore side to a higher ground inland after the 1933 tsunami. The 2011 tsunami flooded and destroyed newly developed houses in lower land, but the relocated older settlement had little damage (Figure 7.12).

Ground Level Up by Reclamation

Though relocation is surely an effective measure against tsunami invasion, it is difficult to realize in larger port towns and settlements with no available higher grounds to relocate. In such cases, "ground level up" by reclamation is a possible alternative measure. Needless to say, its cost may increase with the area and height of the ground to be uplifted, which may limit the feasible uplift level lower than the ideal

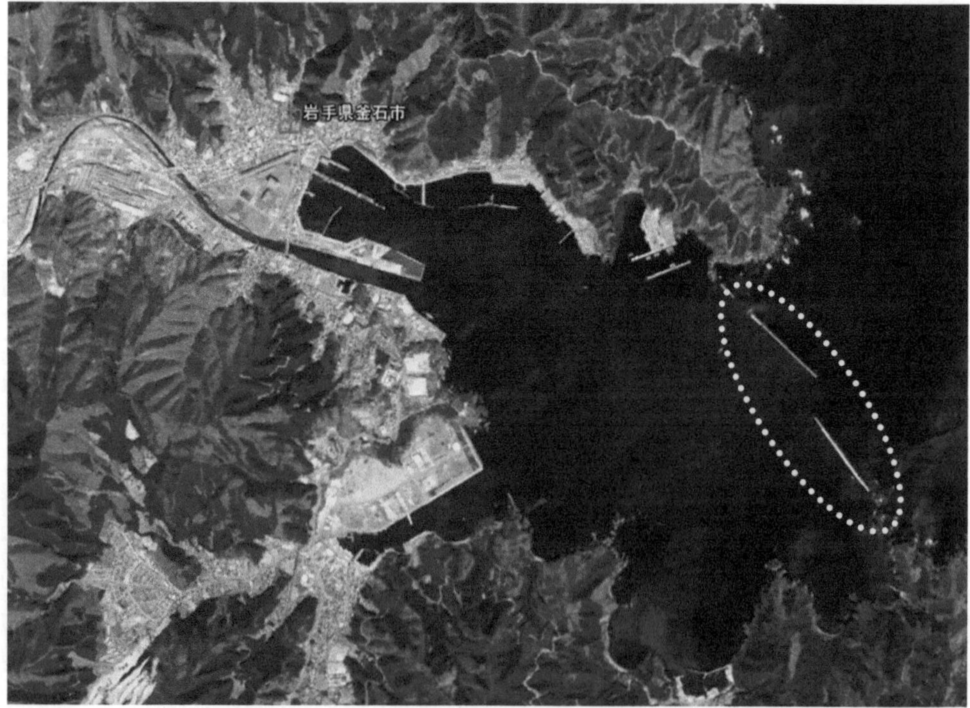

FIGURE 7.10 Kamaishi bay.
Source: Data from Google Earth, July 20, 2009.

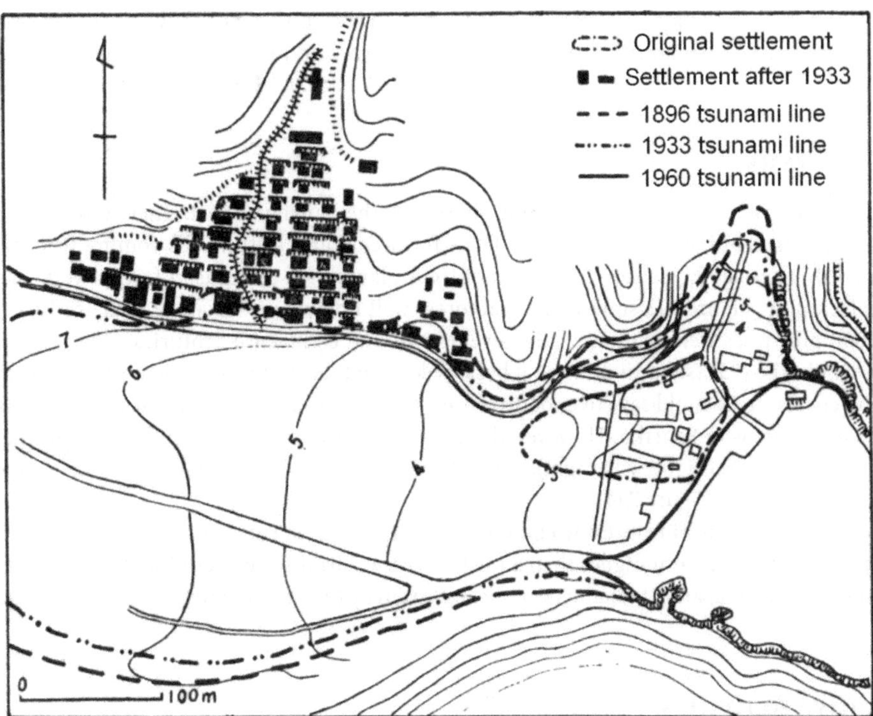

FIGURE 7.11 Movement of the settlement to higher land after the 1933 tsunami in Toni village.
Source: Data from http://dil.bosai.go.jp/disaster/2011eq311/pdf/gsi_chileeq1961_all.pdf, partly modified by the author.

FIGURE 7.12 Toni village just after the 2011 tsunami.
Source: Data from Google Earth, April 4, 2011.

safer level expected from past tsunamis. However, ground level up is an effective measure to reduce the height and pressure of tsunami flow to some extent, which could save the properties and lives of the inhabitants.

Coastal Forests

Coastal forests were created in many sandy shores on the Sanriku Coast. They were not originally created for the tsunami, but to protect rice fields from cold and salty winds coming from the north Pacific mainly after the Edo Era. The species of trees planted for the coastal forests were usually red pine or black pine. Figure 7.13 shows a coastal forest well known as "Takata Matsubara," where more than 70,000 pine trees were planted in the seventeenth and eighteenth centuries. "Matsubara" means "pine forest." In the 1933 tsunami disaster, it became known that the forest could decrease the tsunami's power and coastal trees could save lives of drowning people. As a result, coastal forests were established in many coastal areas in Sanriku. However, the 2011 tsunami was so powerful that it almost destroyed the entire coastal forest (Figure 7.13, right), and only one pine tree was left, which gave hope to the people of the devastated town (Figure 7.14).

It is concluded that the pine forest would not be effective against tsunamis higher than 3 m. The trees would fall down by a tsunami higher than 5 m.[8] After the 2011 disaster, activities to create new coastal forests more defensive against tsunamis have been started by the Forestry Agency[9] and other interested ecologist groups.

Preparative Activities

The physical and biological measures would not be sufficient against the unexpected scale of tsunamis. Preparative activities of the people such as hazard map making, evacuation drills, and leaving lessons for next generations by the monuments and folktales are important. The Sanriku Coast experienced smaller scale tsunamis other than the big tsunamis in 1896, 1933, and 1960. Historically, a major earthquake has occurred offshore the Miyagi Prefecture about every 30 years. Such a situation increased people's awareness in Sanriku about the earthquake and tsunami possibly coming in the future. The installation of new caution signs with past tsunami levels (Figure 7.15) and detailed and effective evacuation maps and practice drills have been performed by almost every community in the Sanriku Coast. The unexpected disaster in Sumatra and the Indian Ocean in December 2004 has further increased the awareness of tsunamis.

FIGURE 7.13 Rikuzentakata and its coastal forest before and after the 2011 tsunami.
Source: Data from Google Earth, July 20, 2010 – April 1, 2011.

FIGURE 7.14 Pine tree of hope (Rikuzentakata). This tree died due to salinity and was cut down, and returned to the original place as a monument after antiseptic treatment.
Source: Photographed by the author on July 2, 2013.

FIGURE 7.15 Tsunami warning sign in Taro town.
Source: Data from http://www.bo-sai.co.jp/tunamihyoujiban.htm

Having experienced the 2011 tsunami, planning activities to create villages or towns safer from coastal disasters have started in every Sanriku community, using almost all possible methods fitting local contexts. Those included the preservation of ruined structures by the tsunami as a memorial for the disaster (Figure 7.16). However, negotiation and agreement among the local people are necessary before the decision to preserve is made, because such monuments can serve as a traumatic reminder of the disaster.

FIGURE 7.16 **(See color insert.)** Ruined building in Minami-Sanriku town, Miyagi Prefecture. This building was the Disaster Prevention Office in town. Forty-two persons were lost in the 2011 tsunami, which reached up to 2 m higher than the roof top of the building. Many people requested to preserve the ruin as a memorial for the disaster. **Source:** Photo by the author.

Conclusion

The Sanriku Coast is a rare region in the world that has experienced severe tsunami disasters cyclically. We can learn much about coastal natural disasters and the effectiveness of countermeasures from the experiences of the Sanriku Coast.

References

1. Disaster Prevention Council: Tsunami Disasters in Sanriku Region (in Japanese), http://www.bousai.go.jp/kyoiku/kyokun/kyoukunnokeishou/rep/1896-meijisanrikuJISHINTSU-NAMI/pdf/1896-meiji-sanrikuJISHINTSUNAMI_05_chap1.pdf
2. USGS Updates Magnitude of Japan's 2011 Tohoku Earthquake to 9.0, http://www.usgs.gov/newsroom/article.asp?7ID=2727&from=rss_home#.UC86EKCa2VA (accessed March 2011).
3. National Police Agency, http://www.npa.go.jp/archive/keibi/biki/higaijokyo.pdf (accessed October 2012).
4. USGS. Poster of the Great Tohoku Earthquake (northeast Honshu, Japan) of March 11, 2011 – Magnitude 9.0, http://earthquake.usgs.gov/earthquakes/eqarchives/poster/2011/20110311.php (accessed November 2012).
5. http://www.gyokou.or.jp/100sen/100img/02tohoku/021.pdf; http://www.ajg.or.jp/disaster/files/201105_taro.pdf.
6. http://www.pa.thr.mlit.go.jp/kamaishi/bousai/bouhatei/bouhatei-kamaishi.html; http://www.nikkei.com/article/DGXNASFK3100F_R30C11A3000000/.

7. Shudo, N. *History of Tsunami in Sanriku Region;* Part 5, Move to higher place (in Japanese)
8. Geographical Survey Institute, 1960: "Report on the Survey of the Abnormal Tidal Waves Tsunami Caused by the Chilean Earthquake on May 24, 1960" (http://tsunami.dbms.cs.gunma-u.ac.jp/xml_tsunami/xmltext.php). Shudo, N. *History of Tsunami in Sanriku Region;* Part 4, Takata pine forest (in Japanese).
9. Forestry Agency, 2011: On the reclamation of disaster preventive coastal forest (in Japanese), http://www.rinya.maff.go.jp/j/tisan/tisan/pdf/chuukanhoukoku.pdf
10. Imamura, A. *Past Tsunamis of the Sanriku District;* Vol. 1, Bulletin of the Earthquake Research Institute, University of Tokyo: 1933; pp. 1–16 (in Japanese).

8

Coral Reef: Biology and History

Graham E. Forrester
University of Rhode Island

Introduction

Although they had been explored from above the surface for hundreds of years, scientific studies of coral reefs proliferated tremendously once SCUBA equipment became widely available in the 1960s. Like the millions of visitors that visit coral reefs each year, scientists have been captivated by the extraordinary beauty, abundance, and sheer variety of life to be found on coral reefs. This chapter summarizes some of the discoveries made by those who have ventured underwater to learn more about these spectacular places.

Corals and *Symbiodinium*

Corals belong to the class Anthozoa, within the phylum Cnidaria. The members of this group that build reefs (called hermatypic corals) are primarily members of the order Scleractinia. Their body form, shown in Figure 8.1 below, is called a polyp. Polyps consist of a ring of tentacles surrounding a mouth. The mouth forms the only opening to the body cavity (coelenteron). Most corals on the reef are colonial; they grow as interconnected groups of genetically identical polyps that spread in a single layer over the reef surface. It is the external skeleton of calcium carbonate (limestone) secreted by polyps around (theca in Figure 8.1) and under their tissues (basal plates in Figure 8.1) that slowly builds up under the veneer of living tissue to form reefs.

A key feature of most hermatypic corals is their mutualistic relationship with single-celled dinoflagellate algae in the genus *Symbiodinium* (sometimes referred to colloquially by the more general term zooxanthellae). The algal cells live in the tissues of the polyp, and their photosynthesis produces energy that appears to increase the polyps rate of growth, reproduction, and skeletal deposition. Some *Symbiodinium* can also live on the bottom separate from coral polyps, but those living inside polyps appear to benefit from an increased supply of nutrients and carbon dioxide, protection from UV damage, and maintenance within a relatively constant set of ambient conditions.[1] It is this symbiosis

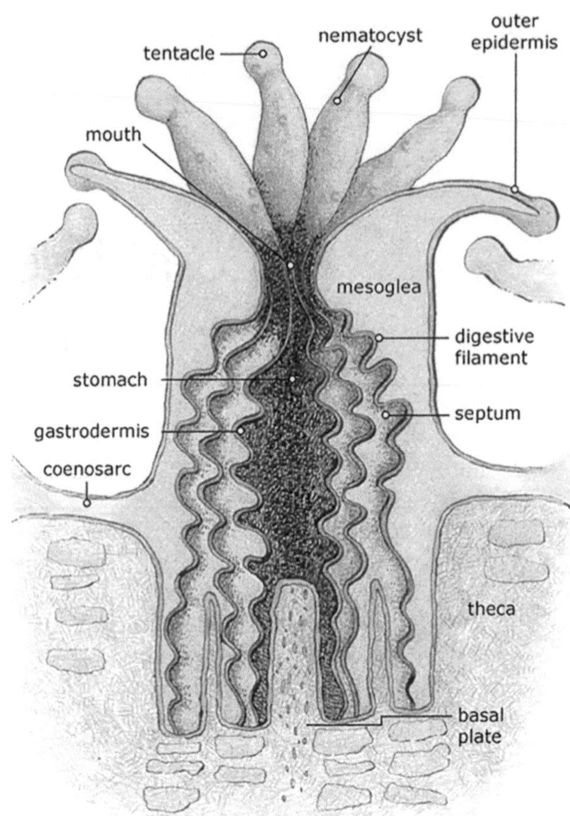

FIGURE 8.1 Cross section through a coral polyp. (Wikipedia: http://en.wikipedia.org/wiki/Coral.)

between *Symbiodinium* and coral polyps that allows the high primary productivity and enhanced calcification making reef growth possible. While hermatypic corals acquire much of the energy they need from their symbiotic algae, most also capture prey by extending their tentacles, as shown in Figure 8.1. Feeding mostly occurs at night, and typical prey include zooplankton, bacteria, and small organic particles.

Coral Reproduction

Corals have diverse reproductive strategies. Some corals can reproduce asexually, which occurs when a colony fragments to form genetically identical daughter colonies.[2] Corals display varied patterns of sexual reproduction and, while most can produce male and female gametes at the same time (simultaneous hermaphrodites), others are sequential hermaphrodites or have separate sexes (gonochores). Gametes may be fertilized internally so that the ciliated larvae develop internally, but the majority of species broadcast eggs and sperm into the water column to be fertilized.[3] Many broadcast spawning species only reproduce on a few nights each year, and the timing of these events is often predictable based on lunar cycles. On the Great Barrier Reef, spawning of up to 100 species is synchronized, and the clouds of gametes form massive slicks that are visible from the air.[4] How this precisely coordinated timing is achieved is not completely understood but appears to involve a combination of responses to environmental signals, genetic precision, and possibly communication among neighboring corals.[5]

Reef Growth

Coral reefs are biogenic habitats, so called because these rocky reefs that rise from the seabed are built by living organisms. The calcium carbonate skeletons deposited by growing corals and calcareous algae gradually accrete reefs that can form enormous limestone landscapes underwater.[6,7] Reef-building corals and algae are, therefore, called "ecosystem engineers"[8] because they create a place to live for so many other species. Coral growth rates are highly variable, e.g., the linear extension rate of two major Caribbean corals ranged from 0.6–0.9 cm/year for *Montastraea annularis* and 5–10 cm/year for *Acropora palmata*.[9] Some corals form growth rings in their limestone skeletons as they grow, much like tree rings.[10] These growth rings can be used to age coral colonies and reveal that colonies of slow-growing corals can live for hundreds of years. Reef growth is also fostered by the deposition of calcium and silica from sponges, which, along with calcareous algae also have an important role in cementing the various calcium carbonate skeletal deposits into a solid coherent matrix.[6]

The growth of reefs is continually opposed by processes that erode the reef to create coral rubble and sand. Most erosion is caused by organisms living in and around the reef and so is called bioerosion.[11] Some bioeroders are mobile species that roam above the reef, such as large parrotfishes that will bite chunks out of both live corals and dead corals covered by algae. Other important external bioeroders include sea urchins and molluscs, which erode the reef as they scrape its surface to consume algae. Another group of bioeroding species lives with the reef itself. These internal bioeroders include microscopic algae, fungi, and bacteria as well as larger organisms like sponges and molluscs that bore into the limestone. The rate of erosion is highly variable but can be very damaging. Sponges in the family Clionaidae can, e.g., remove up to 2.3 kg calcium carbonate from a square meter of reef per year.[12]

Reef History

Anthozoan corals have been forming extensive reefs in shallow coastal waters since the lower Cambrian period over 500 million years ago. The principal Paleozoic reef-forming corals were in the orders Tabulata and Rugosa, which deposited skeletons of calcite. These two groups are now extinct, and most reefs formed in the Mesozoic Era are dominated by Scleractinian corals. Scleractinian corals first appeared in the fossil record roughly 240 million years ago and deposit aragonite skeletons. Over geologic time, coral reefs have undergone mass extinctions and regenerations with the movement of continents, the rise and fall of sea level, as well as with changes in ocean acidity that may have triggered shifts in dominance between Scleractinian corals and other groups such as rudist bivalves. Although some of the coral reefs we can see today date back at least 50 million years, others are surprisingly recent. The earliest limestone that forms the Great Barrier Reef in Australia was laid down in the last 600,000 years and most of the extant reef was formed since the last glacial maximum 20,000 years ago, when sea level was 120 m lower than it is today.[13]

Reef Distribution and Limits to Growth

Large Scale

Currently, coral reefs are estimated to cover about 284,000 km^2 of the sea bed, of which most (261,000 km^2) are in the Indo-Pacific region. Reefs cover much smaller areas of the Caribbean (21,600 km^2) and Eastern Pacific (1,600 km^2). As illustrated in Figure 8.2, coral reefs are found in tropical waters, mostly within 30° north or south of the equator. They occur where temperatures are warm, growing best between 26°C and 28°C. Hermatypic corals are generally unable to form reefs when temperatures regularly drop below 18°C or rise above 36°C, which largely explains their absence from the tropical coasts of West Africa and the Americas where upwelling brings cool water to the surface. Coral reefs form in shallow areas where sufficient light penetrates to support photosynthesis by zooxanthellae. Turbidity of the water, thus, also

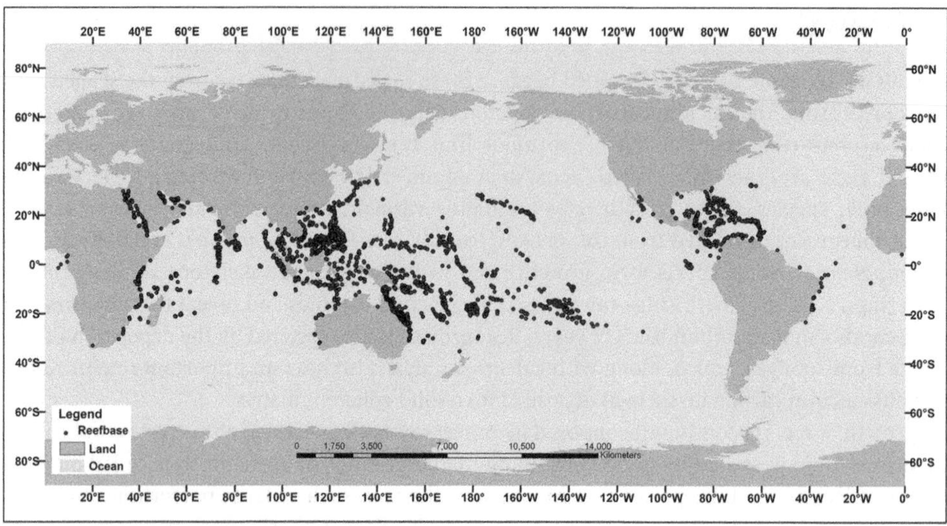

FIGURE 8.2 A map of the current global distribution of coral reefs. (NOAA Ocean Service Education: https://oceanservice.noaa.gov/education/kits/corals/media/supp_coral05a.html.)

plays a role in limiting the distribution of hermatypic corals primarily by reducing the light available for photosynthesis. Corals are also restricted by their tolerance for salinity levels typical of seawater (25–42 PSU) and their requirement for enough dissolved calcium carbonate in the surrounding seawater to precipitate their skeletons (aragaonite saturation greater than 3.2).[14]

Most of our knowledge of coral reefs comes from studying reefs less than 30 m deep, the comfortable limit for conventional diving on SCUBA. In recent years, however, advances in diving technology and underwater robotics have fueled a growing appreciation of the importance of mesophotic coral reefs, which extend from 30–40 m to about 150 m, the lower limit of light penetration sufficient for photosynthesis by zooxanthellae.[15]

Local Distribution

Charles Darwin[16] devised a simple and enduring classification of reefs, dividing them into fringing reefs, barrier reefs, and atolls. Fringing reefs grow on the sloping perimeter of most tropical islands and along many coastlines. Here, corals colonize rocky substrata and form a layer of reef that grows upward toward the water's surface and may spread offshore if conditions permit. Barrier reefs, in contrast, are separated from the coastline or island landmass by a lagoon at least 10 m deep and typically 1–10 km wide. The best known examples are the world's two largest reef systems, the Great Barrier Reef (over 2,900 reefs spanning 2,600 km off the coast of Australia) and the Mesoamerican Barrier Reef (stretching roughly 1,000 km from Mexico to Honduras). Another reef type, not classified by Darwin, is the patch reef, which as their name suggests grow as small isolated outcrops within lagoons. The third of Darwin's reef types is the atoll, which is a ring of reef that surrounds a lagoon. Underlying the lagoon is a volcanic basement. Atolls originally develop as fringing reefs around volcanic islands. Over time, as the island gradually collapses or sea level rises, the ring of reef is eventually all that remains.

Conclusion

Based on the partnership between coral polyps and single-celled algae, Scleractinian corals have been forming limestone reefs for over 200 million years. Their current distribution reflects this long history of evolution in response to changing water conditions and the configuration of ocean basins.

Their narrow tolerance of temperature and salinity and their need for light and dissolved calcium carbonate restrict corals to warm shallow seas. Here, continued reef growth depends on the balance between the deposition of limestone by corals and erosive processes that break up the reef.

References

1. Muller-Parker G, D'Elia CF. Interactions between corals and their symbiotic algae. In: Birkeland C editor. *Life and Death of Coral Reefs*; New York: Chapman and Hall; 1997.
2. Highsmith RC. Reproduction by fragmentation in corals. *Marine Ecology Progress Series* 1982:7(2): 207–226.
3. Szmant AM. Reproductive ecology of caribbean reef corals. *Coral Reefs* 1986:5(1): 43–53.
4. Harrison PL, Babcock RC, Bull GD, Oliver JK, Wallace CC, Willis BL. Mass spawning in tropical reef corals. *Science* 1984:223(4641): 1186–1189.
5. Levitan DR, Fogarty ND, Jara J, Lotterhos KE, Knowlton N. Genetic, spatial, and temporal components of precise spawning synchrony in reef building corals of the montastraea annularis species complex. *Evolution* 2011:65(5): 1254–1270.
6. Adey WH. Review-coral reefs: Algal structured and mediated ecosystems in shallow, turbulent, alkaline waters. *Journal of Phycology* 1998:34(3): 393–406.
7. Veron JEN. *Corals of the World.* Townsville, Australia: Australian Institute of Marine Science; 2000.
8. Jones CG, Lawton JH, Shachak M. Organisms as ecosystem engineers. *Oikos* 1994:69(3): 373–386.
9. Gladfelter EH, Monahan RK, Gladfelter WB. Growth rates of five reef-building corals in the Northeastern Caribbean. *Bulletin of Marine Science* 1978:28(4): 728–734.
10. Knutson DW, Buddemeier RW, Smith SV. Coral chronometers: Seasonal growth bands in reef corals. *Science* 1972:177(4045): 270.
11. Glynn PW. Bioerosion and coral reef growth: A dynamic balance. In: Birkeland C editor. *Life and Death of Coral Reefs*; New York: Chapman & Hall; 1997.
12. Neumann AC. Observations on coastal erosion in bermuda and measurements of the boring rate of the sponge, *cliona lampa*. *Limnology and Oceanography* 1966:11(1): 92–108.
13. Hopley D, Smithers SG, Parnell KE. *The Geomorphology of the Great Barrier Reef: Development, Diversity, and Change.* Cambridge: Cambridge University Press; 2007.
14. Kleypas JA, McManus JW, Menez LAB. Environmental limits to coral reef development: Where do we draw the line? *American Zoologist* 1999:39(1): 146–159.
15. Lesser MP, Slattery M, Mobley CD. Biodiversity and functional ecology of mesophotic coral reefs. *Annual Review of Ecology, Evolution, and Systematics* 2018:49: 49–71.
16. Darwin CR. *The Structure and Distribution of Coral Reefs.* Being the first part of the geology of the voyage of the beagle, under the command of capt. Fitzroy, r.N. During the years 1832 to 1836. London: Smith Elder and Company.

9

Coral Reef: Ecology and Conservation

Graham E. Forrester
University of Rhode Island

Introduction

Humans have a long history of interaction with coral reefs. Many tropical coastal communities have depended on coral reefs as a source of food, materials, and shoreline protection for millennia. More recently, millions of visitors have been drawn to snorkel and dive amidst their beautiful underwater scenery. Scientific studies of coral reefs proliferated tremendously once SCUBA equipment became widely available in the 1960s and have focused on the spectacular biodiversity supported by coral reefs. This chapter summarizes our knowledge of reef ecology and the effects humans are having on this ecosystem.

Biodiversity of Coral Reefs

Corals reefs are often called "rainforests of the sea" because both ecosystems have a complex biogenic habitat, are highly productive, and support spectacularly high biological diversity.[1–3] Globally, coral reefs harbor more species than any other marine ecosystem. Estimates of the total number of reef-associated species range from 600,000 to more than 9 million species.[4,5] Uncertainty over levels of diversity stems, in part, from the enormous number of small cryptic species that go largely unnoticed and are hard to survey using traditional methods but are now being catalogued by sequencing traces of their DNA from the environment (eDNA).[6,7] For well-studied groups, this biodiversity reaches its highest level in the Indo-Australian archipelago (the "coral triangle"), where there are over 500 species of stony (Scleractinian) coral and 2,600 fishes. Diversity gradually declines as one moves away from this area across the Western Pacific and Indian Oceans to the eastern Pacific, which supports less than 50 species of coral and 350 fishes.[8–10] The diversity of less well-known groups such as invertebrates and seaweeds may, however, also follow this pattern. For example, there are around 50 species of reef-associated mantis shrimp (members of the order Stomatopoda) in the coral triangle but less than five in the eastern Pacific.[11] The gradual closure of the Central American Seaway that separated the eastern Pacific and Caribbean Sea roughly 3.4 million years ago led to the evolution of a distinct Caribbean reef fauna and flora. Consequently, the two areas now have no hermatypic (reef-building) corals in common,[12] and the

Caribbean today has lower diversity than any of the major Indo-Pacific regions, with about 62 species of Scleractinian coral[13]. These patterns of diversity appear to result from the interplay of global, regional, and local factors, which combine to generate the extraordinary biological richness we see on coral reefs today.[3,14,15]

The Functioning of Reef Ecosystems

Coral reefs are highly productive ecosystems, particularly in comparison to the clear oligotrophic waters that surround them.[16] Globally, primary producers on coral reefs are estimated to fix roughly 700×10^{12} g carbon per year.[17] The unicellular algae in the genus *Symbiodinium* that reside within coral polyps are key primary producers on coral reefs and provide energy to corals in the form of reduced carbon. Photosynthesis by *Symbiodinium* accounts for a substantial portion of the carbon fixation that occurs on reefs. Up to half of the carbon fixed by *Symbiodinium* is exuded by the coral as mucus, some of which dissolves to provide food for bacteria in the water column and most of the remainder eventually reaches the sediments.[18] Relatively little of the carbon fixed with coral polyps carbon is, however, transferred directly up the food chain[17] because, with a few notable exceptions such as the crown-of-thorns starfish (*Acanthaster planci*) plus some butterflyfishes and parrotfishes, relatively few species consume coral polyps directly. Instead, turfs of filamentous and fleshy seaweeds are more important as the base of the reef food chain because they use sunlight to fix carbon but are heavily grazed by a variety of herbivorous fish and invertebrates. Reef food chains are also subsidized by offshore production, from a constant supply of zooplankton that are delivered to reefs by currents and consumed by resident planktivores.[19] While the total amount of carbon dioxide fixed by all photosynthetic organisms in the reef is high (between 2-3–6.0 g carbon per m^2 per day), the net production of organic matter by the community is much lower (0.01–0.29 g carbon per m^2 per day), which indicates that most of this material is retained within reefs. Because of this strong competition for food, and because the surrounding waters are nutrient poor, it is also assumed that nutrients are efficiently recycled with coral reef ecosystems, but the mechanisms for this are still uncertain.[20]

The Value of Reefs to Humans

Coral reef ecosystems provide a number of services to humans including the provision of raw materials, protection of coastlines, food from fish and shellfish, processing of nutrients, and venues for recreation and tourism.[21–23] Although coral reefs are estimated to underlay only about 0.1% of the global surface area of the ocean, they are of tremendous importance to the people who live along coasts fringed by reefs.[9] More than 100 countries have coral reefs along part of their coastline, and tens of millions of people in those countries depend partly on food from coral reefs.[24] Globally, species harvested from coral reefs account for 9%–12% of fisheries but, in some areas, they assume a much larger role in the provision of protein. For example, on many Pacific Islands and in coastal areas of Southeast Asia, the catch from corals reefs makes up at least 25% of the total fish catch[25] and noncommercial fishing provides a large fraction of meals for local residents.[26] Other important goods derived from coral reefs include live fish for the aquarium trade, coral limestone that is mined to produce sand, lime, mortar and cement for building, as well as seaweed harvested to make agar, and shells that are collected for jewelry.[23,25] Coral reefs are also valuable to local economies by attracting tourists and supporting their recreational activity. The average economic value of reef-related tourism is estimated to be $184 US per visitor, but estimates vary widely and appear to depend on factors such as the total number of visitors and the number of reefs accessible to divers.[27] The high biological diversity of coral reefs has also made them a priority target for the extraction of chemical compounds from tissues to be screened for medical applications.[28]

Other important ecosystem services provided by coral reefs have less direct, but no less important, value for coastal communities. For example, coral reefs are important to the cycling of organic and inorganic nutrients in tropical seas, including the transfer of surplus nitrogen produced by benthic microbes to the open water ecosystem.[25] Coral reefs also dissipate wave energy and so buffer shorelines from the potentially devastating effects of storms and tsunamis.[29] By dissipating waves and currents, coral reefs also create sheltered conditions inshore that promote the development of mangrove forests and seagrass beds. These habitats in turn provide their own benefits to people living nearby and in combination with corals reefs constitute a set of interlinked and mutually supporting habitats whose combined value is greater than that of each in isolation.[21,25]

Effects of Humans on Reefs

Despite their aesthetic, cultural, and economic value, coral reefs have been substantially impacted by human activity over the past few hundred years, as illustrated visually in Figure 9.1. Based on expert opinion and global monitoring efforts, 19% of the area of modern coral reefs has been effectively lost, and a further 15% is in serious threat of loss within 10–20 years.[30] One leading coral reef biologist has argued that within as little as 30–40 years, reefs may bear so little resemblance to those historically present that they may become the first major ecosystem to be effectively eliminated by human activity.[31]

These declines have a number of causes, some of which act in combination. Although people have fished on coral reefs for millennia, historical analyses suggest that fishing began to cause major changes to coral reef ecosystems in the 19th and 20th centuries.[32] Many large carnivores and herbivores, such as monk seals,[33] sea turtles,[34] dugongs, and manatees, had already become rare by the beginning of the 20th century. Likewise, large fishes, such as sharks,[35] groupers, snappers, and bumphead wrasses,[36,37] appear to have been depleted by overfishing before underwater research on coral reefs began in earnest in the 1970s. The use of dynamite, drive nets, and poisons like cyanide to catch fish is particularly

FIGURE 9.1 **(See color insert.)** Illustration of the loss of coral on Caribbean reefs in the past 30 years. (**a**) A Caribbean reef with roughly 40% of its surface covered with live coral (typical in the 1970s). (**b**) A similar Caribbean reef with less than 4% of its surface covered with live coral (typical in since 2000). (Photographs by the author.)

destructive and has a major impact on some pacific reefs because these methods destroy the underlying reef and many animals in addition to their intended targets.[38]

More intensive scientific scrutiny over the past 30 years has revealed progressive declines in the cover of corals[39,40] and dramatic "phase shifts" from coral- to seaweed-dominated states.[41] Some of these declines are attributable to localized effects of pollution and sedimentation, which coincide with the growth of human populations along tropical coastlines. In other cases, the loss of herbivores though overfishing and disease has allowed seaweeds to flourish and made it more likely that they will out-compete corals for space on the reef. In Jamaica, this phase shift from coral to seaweed dominance was triggered by hurricane damage, which caused widespread coral mortality. Whereas historically corals could recover from storm-related mortality, the absence of herbivores allowed seaweeds to take over the reef and prevent corals from repopulating.[41]

More recently, the most important recent threats to coral reefs have come from the emergence of new diseases and from the effects of global climate change, though there are signs that invasive species will have increasing impacts in the future. By 1965, scientists had identified just one coral disease. Since then, the number of reported diseases has grown exponentially and 18 different diseases had been reported by 2004.[42] Some new diseases have emerged through the expansion of pathogens to new areas, followed by the infection of new host species.[43] Similarly, the expansion of other species to areas outside their historical range has strong effects when those species become "invasive" and disrupt the resident community. Some coral reefs in Hawaii have been overgrown by invasive seaweeds, but perhaps the most notably damaging invasive species yet is the Pacific lionfish, which has spread rapidly around the Caribbean over the past 20 years. Lionfish consume smaller reef fishes and crustaceans, and are very efficient predators in their new environment.[44]

Climate change has exerted strong effects on coral reefs, and may now represent the greatest threat to their future integrity, because even modest increases in seawater temperature can induce coral bleaching, a generalized stress response in which the zooxanthellae are lost from the coral polyp.[45,46] As the ocean warms, episodes of higher than normal seawater temperature are becoming increasingly common and have triggered massive bleaching events, notably the global bleaching events of 1998, 2010, and 2015–2016.[47–49] Although bleaching-related stress can kill corals directly, some major bleaching events have coincided with epidemics of coral disease suggesting a synergistic effect where temperature stress and bleaching make corals more susceptible to disease.[50,51] Another threat to corals from climate change comes from the acidification of seawater. Elevated carbon dioxide emissions into the atmosphere have caused an increased flux of carbon dioxide into the ocean. Through a series of chemical reactions, the pH of surface waters has declined along with the availability of carbonate ions. As a result, the ability of corals and calcifying macroalgae to deposit calcium carbonate skeletons and form reefs will be progressively compromised.[52–54]

Conclusion

Arresting the declines in diversity and abundance of reef-dwelling species, and restoring some the ecosystem functions they provide, will require a concerted effort.[55] Local impacts, such as those from overfishing, can be addressed using marine protected areas and improved management of human activities in coastal regions. The growing impacts of climate change, though, require a worldwide commitment to reduce carbon emissions.

References

1. Stehli FG, Wells JW. Diversity and age patterns in hermatypic corals. *Systematic Biology* 1971:20(2): 115–126.
2. Huston M. Patterns of species diversity on coral reefs. *Annual Review of Ecology and Systematics* 1985:16(1): 149–177.

3. Bellwood DR, Hughes TP. Regional-scale assembly rules and biodiversity of coral reefs. *Science* 2001:292(5521): 1532–1535.
4. Reaka-Kudla ML, Wilson DE, Wilson EO. *Biodiversity 2: Understanding and Protecting Our Biological Resources*, Washington, DC: Joseph Henry Press; 1997.
5. Knowlton N, Brainard RE, Fisher R, Moews M, Plaisance L, Caley MJ. Coral reef biodiversity. In: Macintyre A editor. *Life in the World's Oceans: Diversity Distribution and Abundance*, Oxford, UK: Wiley-Blackwell; 2010.
6. Stat M, Huggett MJ, Bernasconi R, DiBattista JD, Berry TE, Newman SJ, Harvey ES, Bunce M. Ecosystem biomonitoring with edna: Metabarcoding across the tree of life in a tropical marine environment. *Scientific Reports* 2017:7(1): 12240.
7. Pearman JK, Leray M, Villalobos R, Machida R, Berumen ML, Knowlton N, Carvalho S. Cross-shelf investigation of coral reef cryptic benthic organisms reveals diversity patterns of the hidden majority. *Scientific Reports* 2018:8(1): 8090.
8. Veron JEN. *Corals in Space and Time: The Biogeography and Evolution of the Scleractinia*, Ithaca: Comstock/Cornell; 1995.
9. Spalding M, Ravilious C, Green EP. *World Atlas of Coral Reefs Berkeley*, Oakland, CA: University of California Press; 2001.
10. Zapata FA, Robertson DR. How many species of shore fishes are there in the tropical eastern pacific? *Journal of Biogeography* 2007:34(1): 38–51.
11. Reaka ML, Rodgers PJ, Kudla AU. Patterns of biodiversity and endemism on indo-west pacific coral reefs. *Proceedings of the National Academy of Sciences* 2008:105(Supplement 1): 11474–11481.
12. Veron JEN. *Corals of the World*, Townsville, Australia: Australian Institute of Marine Science; 2000.
13. Miloslavich P, Diaz JM, Klein E, Alvarado JJ, Diaz C, Gobin J, Escobar-Briones E, Cruz-Motta JJ, Weil E, Cortes J et al. Marine biodiversity in the caribbean: Regional estimates and distribution patterns. *Plos One* 2010:5(8): e11916.
14. Pandolfi J. Coral community dynamics at multiple scales. *Coral Reefs* 2002:21(1): 13–23.
15. Connolly SR, Hughes TP, Bellwood DR, Karlson RH. Community structure of corals and reef fishes at multiple scales. *Science* 2005:309(5739): 1363–1365.
16. Crossland CJ, Hatcher BG, Smith SV. Role of coral reefs in global ocean production. *Coral Reefs* 1991:10(2): 55–64.
17. Hatcher BG. Coral reef primary productivity: A beggar's banquet. *Trends In Ecology & Evolution* 1988:3(5): 106–111.
18. Wild C, Huettel M, Klueter A, Kremb SG, Rasheed MYM, Jorgensen BB. Coral mucus functions as an energy carrier and particle trap in the reef ecosystem. *Nature* 2004:428(6978): 66–70.
19. Hamner WM, Jones MS, Carleton JH, Hauri IR, Williams DM. Zooplankton, planktivorous fish, and water currents on a windward reef face: Great barrier reef, australia. *Bulletin of Marine Science* 1988:42(3): 459–479.
20. Silveira CB, Cavalcanti GS, Walter JM, Silva-Lima AW, Dinsdale EA, Bourne DG, Thompson CC, Thompson FL. Microbial processes driving coral reef organic carbon flow. *FEMS Microbiology Reviews* 2017:41(4): 575–595.
21. Barbier EB, Hacker SD, Kennedy C, Koch EW, Stier AC, Silliman BR. The value of estuarine and coastal ecosystem services. *Ecological Monographs* 2011:81(2): 169–193.
22. Barbier EB. Marine ecosystem services. *Current Biology* 2017:27(11): R507–R510.
23. Woodhead AJ, Hicks CC, Norström AV, Williams GJ, Graham NA. Coral reef ecosystem services in the anthropocene. *Functional Ecology* 2019: 33:1023–1034.
24. Salvat B. Coral reefs: A challenging ecosystem for human societies. *Global Environmental Change-Human and Policy Dimensions* 1992:2(1): 12–18.
25. Moberg F, Folke C. Ecological goods and services of coral reef ecosystems. *Ecological Economics* 1999:29(2): 215–233.

26. Grafeld S, Oleson KL, Teneva L, Kittinger JN. Follow that fish: Uncovering the hidden blue economy in coral reef fisheries. *PloS one* 2017:12(8): e0182104.
27. Brander LM, Van Beukering P, Cesar HSJ. The recreational value of coral reefs: A meta-analysis. *Ecological Economics* 2007:63(1): 209–218.
28. Quinn RJ, De Almeida Leone P, Guymer G, Hooper JNA. Australian biodiversity via its plants and marine organisms. A high-throughput screening approach to drug discovery. *Pure and Applied Chemistry* 2002:74: 519–526.
29. Beck MW, Losada IJ, Menéndez P, Reguero BG, Díaz-Simal P, Fernández F. The global flood protection savings provided by coral reefs. *Nature Communications* 2018:9(1): 2186.
30. Wilkinson CR. *Status of Coral Reefs of the World*, Townsville, Australia: Global Coral Reef Monitoring Network and Reef and Rainforest Research Center; 2008.
31. Sale PF. *Our Dying Planet: An Ecologist's View of the Crisis We Face*, Berkeley: University of California Press; 2011.
32. Pandolfi JM, Bradbury RH, Sala E, Hughes TP, Bjorndal KA, Cooke RG, McArdle D, McClenachan L, Newman MJH, Paredes G et al. Global trajectories of the long-term decline of coral reef ecosystems. *Science* 2003:301(5635): 955–958.
33. McClenachan L, Cooper AB. Extinction rate, historical population structure and ecological role of the caribbean monk seal. *Proceedings of the Royal Society B-Biological Sciences* 2008:275(1641): 1351–1358.
34. McClenachan L, Jackson JBC, Newman MJH. Conservation implications of historic sea turtle nesting beach loss. *Frontiers in Ecology and the Environment* 2006:4(6): 290–296.
35. Ward-Paige CA, Mora C, Lotze HK, Pattengill-Semmens C, McClenachan L, Arias-Castro E, Myers RA. Large-scale absence of sharks on reefs in the greater-caribbean: A footprint of human pressures. *Plos One* 2010:5(8): e11968.
36. Sadovy Y, Kulbicki M, Labrosse P, Letourneur Y, Lokani P, Donaldson TJ. The humphead wrasse, *cheilinus undulatus*: Synopsis of a threatened and poorly known giant coral reef fish. *Reviews in Fish Biology and Fisheries* 2003:13(3): 327–364.
37. McClenachan L. Documenting loss of large trophy fish from the florida keys with historical photographs. *Conservation Biology* 2009:23(3): 636–643.
38. McManus JW. Tropical marine fisheries and the future of coral reefs: A brief review with emphasis on southeast asia. *Coral Reefs* 1997:16: S121–S127.
39. De'ath G, Fabricius KE, Sweatman H, Puotinen M. The 27-year decline of coral cover on the great barrier reef and its causes. *Proceedings of the National Academy of Sciences of the United States of America* 2012:109(44): 17995–17999.
40. Jackson J, Donovan M, Cramer K, Lam V. *Status and Trends of Caribbean Coral Reefs: 1970–2012*, Washington, DC: Global Coral Reef Monitoring Network; 2014.
41. Hughes TP. Catastrophes, phase shifts, and larg-scale degradation of a caribbean coral reef. *Science* 1994:265: 1547–1551.
42. Sutherland KP, Porter JW, Torres C. Disease and immunity in caribbean and indo-pacific zooxanthellate corals. *Marine Ecology-Progress Series* 2004:266: 273–302.
43. Harvell CD, Kim K, Burkholder JM, Colwell RR, Epstein PR, Grimes DJ, Hofmann EE, Lipp EK, Osterhaus ADME, Overstreet RM et al. Emerging marine diseases: Climate links and anthropogenic factors. *Science* 1999:285(5433): 1505–1510.
44. Albins MA, Hixon MA. Invasive indo-pacific lionfish pterois volitans reduce recruitment of atlantic coral-reef fishes. *Marine Ecology-Progress Series* 2008:367: 233–238.
45. Douglas AE. Coral bleaching: How and why? *Marine Pollution Bulletin* 2003:46(4): 385–392.
46. van Oppen MJ, Lough JM. *Coral Bleaching*, Cham, Switzerland: Springer; 2018.
47. Heron SF, Maynard JA, Van Hooidonk R, Eakin CM. Warming trends and bleaching stress of the world's coral reefs 1985–2012. *Scientific Reports* 2016:6: 38402.

48. Hughes TP, Kerry JT, Álvarez-Noriega M, Álvarez-Romero JG, Anderson KD, Baird AH, Babcock RC, Beger M, Bellwood DR, Berkelmans R. Global warming and recurrent mass bleaching of corals. *Nature* 2017:543(7645): 373.
49. Hughes TP, Kerry JT, Baird AH, Connolly SR, Dietzel A, Eakin CM, Heron SF, Hoey AS, Hoogenboom MO, Liu G et al. Global warming transforms coral reef assemblages. *Nature* 2018:556(7702): 492–496.
50. Bruno JF, Selig ER, Casey KS, Page CA, Willis BL, Harvell CD, Sweatman H, Melendy AM. Thermal stress and coral cover as drivers of coral disease outbreaks. *Plos Biology* 2007:5(6): 1220–1227.
51. Miller J, Muller E, Rogers C, Waara R, Atkinson A, Whelan KRT, Patterson M, Witcher B. Coral disease following massive bleaching in 2005 causes 60% decline in coral cover on reefs in the us virgin islands. *Coral Reefs* 2009:28(4): 925–937.
52. Kleypas JA, Yates KK. Coral reefs and ocean acidification. *Oceanography* 2009:22(4): 108–117.
53. Hughes TP, Barnes ML, Bellwood DR, Cinner JE, Cumming GS, Jackson JB, Kleypas J, Van De Leemput IA, Lough JM, Morrison TH. Coral reefs in the anthropocene. *Nature* 2017:546(7656): 82.
54. Andersson AJ, Gledhill D. Ocean acidification and coral reefs: Effects on breakdown, dissolution, and net ecosystem calcification. *Annual Review of Marine Science* 2013:5: 321–348.
55. Cote IM, Reynolds JD. *Coral Reef Conservation*, Cambridge [u.a.]: Cambridge University Press.

10

Fisheries: Conservation and Management

Steven X. Cadrin
University of Massachusetts Dartmouth

Introduction

Fisheries involve the utilization of living aquatic resources by individuals, communities, businesses, and societies. Fisheries include recreation, subsistence, commercial harvesting, seafood processing and distribution, and sale for consumption. Advanced fishing technologies have the potential to deplete natural resources and alter aquatic ecosystems, requiring management of fishing activities to achieve societal objectives. Fisheries science is a multidisciplinary field involving limnology, oceanography, ecology, biology, engineering, economics, anthropology, and policy. Effective fisheries management can achieve sustainable fisheries and conserve productive ecosystems.

The goal of this entry is to briefly summarize the field of fisheries science, with a focus on stock assessment and fishery management. Similar to many disciplines, a description of the historical foundations of fisheries science helps to understand the basis of fishery conservation and management. Several recent trends in fisheries science and management are expected to continue in the future.

Sustainable Fisheries

In 1882, Thomas Henry Huxley provoked the scientific community with his confident affirmation that "*... probably all the great sea fisheries are inexhaustible...*".[1] The subsequent debate on the exhaustibility of fishery resources was central to the formation of the International Council for the Exploration of the Seas and its initial task of considering the possibility of overfishing.[2–4] In the early 1900s, overfishing was defined as a magnitude of fishery removals that exceeds the fished population's capacity to replace itself. By contrast, sustainable yield was defined as a harvest that is less than the population's biological production (Figure 10.1).[5] Each component of biological production (somatic growth, reproduction, and survival) is density dependent (i.e., their contribution to biological production depends on

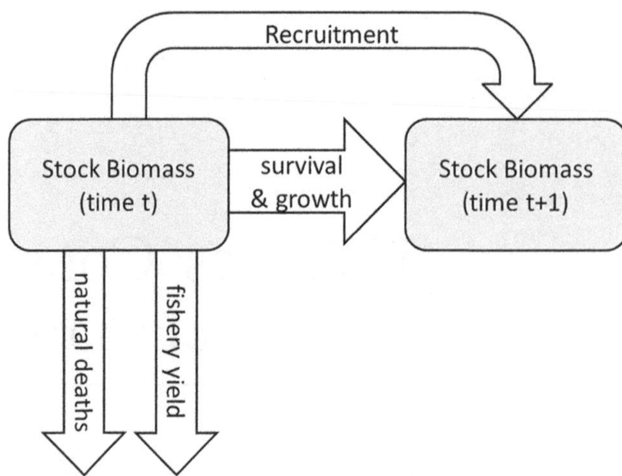

FIGURE 10.1 Schematic representation of the components of biological production (survival, somatic growth, and recruitment) and fishery yield.

how many fish are in the population); so, sustainable yield is also a function of population size, and there is an intermediate population size that can produce maximum sustainable yield (MSY).[6] Given that population size is expected to decrease as harvest rate increases, there is also a harvest rate, or fishing mortality, associated with MSY (Fmsy; Figure 10.2).

Single-Species Stock Assessment and Fishery Management

Stock assessment involves fitting population dynamics models to fisheries information to estimate stock size and fishing mortality. The primary information used in stock assessment is from monitoring of the fishery removals and surveys of the resource. Surveys are statistically designed to represent trends in stock abundance or biomass as well as demographic information (e.g., age, size, gender, or spatial distributions) and life history traits (e.g., size and maturity at age, food habits). Fishery monitoring involves all components of catch, including landings and discarded catch, as well as fishing effort, catch rates, and demographic distributions of catch.

Fishery production can be modeled as aggregate biomass dynamics[7–9] or demographic processes to explicitly account for mortality, growth, and reproduction.[10–12] Most stock assessment models assume that the fishery exploits a unit stock that is more influenced by internal dynamics than interactions with adjacent stocks, but many management units also reflect practical considerations of jurisdiction and fishing grounds.[13]

Models with size and age structure are considered the most informative approach to stock assessment.[14,15] Population dynamics models are fit to observed data from fisheries and fishery resources to estimate parameters for fishery management (e.g., historical and current stock size and fishing mortality). Demographic approaches include length-based models,[16] stage-based models,[17] size and age structure[15], or models with greater complexity to account for sex, spatial, or seasonal structure. Models that sufficiently represent population and fishery dynamics can be used to predict how the fishery resource will respond to alternative management actions.[14] Long-term projection can be used to derive the fishing mortality that produces maximum average yield and associated MSY reference points.[18]

Age-based survival is modeled using exponential mortality rates (Equation 10.1).

$$N_{t+1} = N_t e^{-(F+M)} \tag{10.1}$$

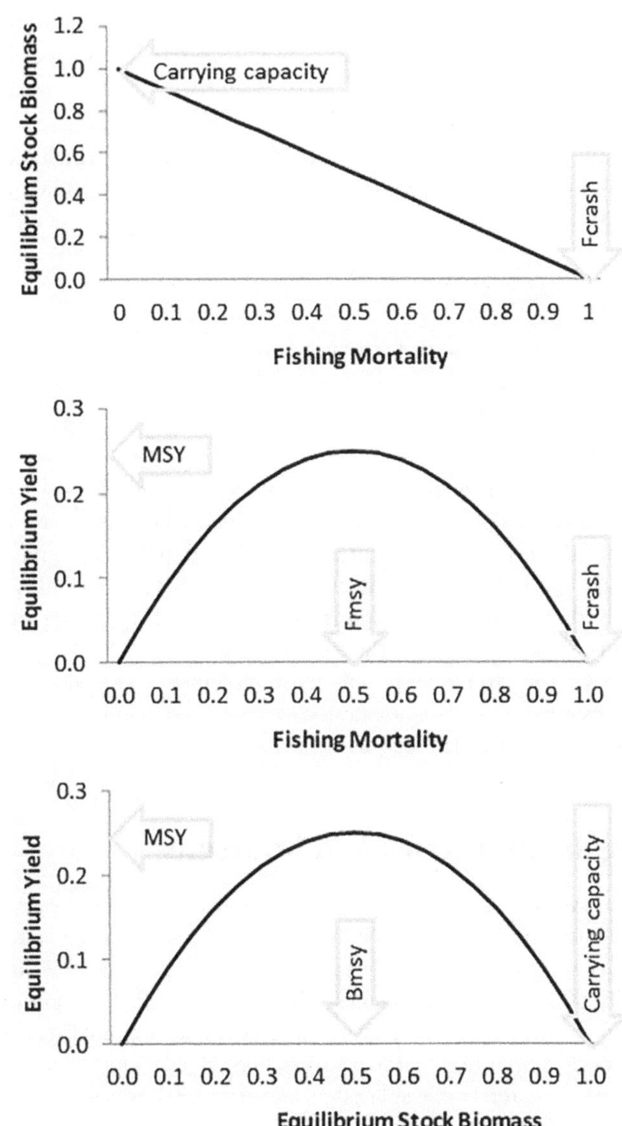

FIGURE 10.2 Equilibrium expectations of fishing, with reduction of stock biomass from the unfished carrying capacity (upper panel), MSY at intermediate fishing mortality rate (Fmsy), and equilibrium stock size (Bmsy) (middle and lower panels, respectively).

N_t: abundance at time t
F: fishing mortality rate
M: natural mortality rate

Somatic growth is either derived empirically, using samples of size at age, or theoretically modeled (Equation 10.2).[19]

$$L_a = L_\infty \left[1 - e^{-k(a-a_0)}\right] \tag{10.2}$$

L_t: *length at age a*
L_∞: *asymptotic size*
k: *growth coefficient*
a_0: *age at length* = 0

Reproduction is typically considered to be a theoretical expectation in which recruitment (the abundance of young fish) is a density-dependent function of reproductive potential or the spawning stock biomass (Equation 10.3).

$$\text{Ricker stock-recruit relationship: } N_{a,\,t+a} = \alpha S_t e^{-\beta S_t}$$
$$\text{Beverton-Holt stock-recruit relationship: } N_{a,t+a} = \frac{aS_t}{b+S_t} \tag{10.3}$$

$N_{a,t}$: *abundance at age of recruitment a*
S_t: *biomass of spawners in time t*

Recruitment can also be derived through more empirical relationships.[20–22]

Each of these components of biological production is influenced by the environment. Recruitment is quite variable for many fisheries, so recruitment is often allowed to deviate from the theoretical expectation, and annual deviations reflect environmental variability. Natural mortality can be modeled as a function of predators or their consumption. Empirical approaches to modeling growth by sampled size at age are responsive to energetic changes in the environment.

Age-based methods are used to model population replacement as a function of fishing mortality, and the properties of sustainable yield demonstrated by biomass dynamics (Figure 10.2) are also found with demographic models, so that MSY and associated fishing mortality and population size can be estimated.[20]

Fishery Management Tactics and Strategies

Governance of fisheries systems varies widely from "top-down" regulation by governments for fishery resources that are under their jurisdiction to "bottom-up" systems that involve cooperative behavior of fishermen to achieve sustainable fisheries and responsible fishing practices.[23] Fisheries can be managed through a large variety of regulatory approaches. "Input controls" include limited entry (e.g., fishing licenses), limiting fishing effort (e.g., limited days fishing, seasonal or annual area closures), and restrictions on fishing gear specifications (e.g., net mesh sizes, hook sizes). "Output controls" involve limiting the total fishery removals (e.g., total allowable catch, annual catch limits, daily catch limits, minimum legal size of fish). In limited-entry fisheries with output controls, individuals can be allocated a portion of the total allowable catch each year, and some individual allocations are transferable to others (i.e., individual transferable quotas). Management tactics are typically tailored to the type of fishery (e.g., small-scale artisanal vs. large-scale industrial), the availability of fishery-monitoring information, and management objectives. Effective fishery management requires enforcement of regulations to avoid illegal and unreported fishing.

Production of wild capture fisheries can be enhanced through protection of essential fish habitat, mitigation of incidental bycatch, as well as culture and stocking of young fish. Habitat conservation extends to regulation of human uses of aquatic ecosystems beyond fisheries (e.g., coastal development, fish passage in rivers, pollution), and marine protected areas are designed to protect habitats from multiple human activities. Bycatch in mixed-species fisheries can be mitigated by using more selective fishing gears or bycatch avoidance through spatiotemporal fishing patterns.[24] Aquaculture programs can also rear fish to marketable sizes and contribute to seafood production.[25] Stocking programs and aquaculture can be productive but may also have the unintended consequences of polluting natural systems or introducing nonnative species, which have caused major perturbations of many aquatic ecosystems.[26]

Many fishery management systems attempt to achieve optimum yield by limiting fisheries to the fishing mortality that produces MSY (Fmsy).[18] Simple economics can be coupled with population dynamics models to incorporate costs and derive revenue and profit as well as maximum economic yield.[27] These biological and economic reference points can form fishery management objectives. Fishery management can also be aimed at other utilities, such as employment and sustaining fishing communities.

As a result of environmental variability and limited sampling of fishery resources, fisheries science is inherently uncertain. Recognition of it as an uncertain science led to the widespread adoption of a precautionary approach to fishery management in the 1990s that involves quantification of uncertainty in estimates (e.g., overfishing limits), derivation of risk-averse targets,[28] and risk-based management decisions.[29,30] Precautionary approaches involve an evaluation of trade-offs between the costs and benefits of conservation and utilization.

Performance of alternative management tactics and strategies for achieving stated objectives can be evaluated through adaptive learning.[31] Best practices or management procedures can also be determined using simulation techniques that allow management strategy evaluation before decisions are implemented.[32] In some situations, simple harvest control rules based on survey or fishery-monitoring information perform better for achieving objectives than more complicated stock assessment and management procedures.[33] A recent trend in fisheries is seafood certification to inform consumers about sustainability of fisheries to promote best practices in fishery management.

Ecosystem Approaches to Fishery Management

The single-species convention for stock assessment and fishery management developed during the industrialization of fishing technology and widespread depletion of fishery resources in the 20th century. Although overfishing persists in many fisheries, the implementation of management strategies to end overfishing has effectively reduced fishing mortality to within-sustainable limits in many regions and rebuilt many depleted stocks. The successes of single-species management strategies have reduced the effect of fishing on many populations and increased the influence of abiotic factors (e.g., climate change)[34] and biological interactions (e.g., predation and competition) with other species, prompting a transition to multispecies and ecosystem models. Ecological approaches to fisheries science and management involve the incremental inclusion of abiotic or biotic factors in single-species population dynamics models[35] or the holistic modeling of energy through aquatic ecosystems.[36] The integration of ecosystem components as well as the consideration of all human uses of aquatic systems promoted the development of integrated ecosystem assessments.[37] Ecosystem approaches to management require confrontation of many trade-offs for multiple ecosystem utilities.[38] Although environmental variability is considered in many fishery stock assessments and management plans, the effects of climate change have been observed in many fisheries through shifting geographic distributions, phenological changes, or trends in productivity, and these changes pose a challenge for stock assessment and estimating MSY.[39]

Conclusion

Overfishing persists in many fisheries, but fishery conservation and management have effectively ended overfishing in many regional management systems and achieved rebuilding toward the production of MSY. Fishery management procedures vary, and management strategy evaluation is emerging as best practice for testing their performance for meeting objectives. Although environmental variability is often considered in fishery stock assessment and management, climate change is a growing challenge for estimating MSY.

Acknowledgments

I thank my mentors, peers, and students for the many lessons reflected in this brief summary of the diverse and challenging field of fisheries science.

References

1. Huxley, T.H.H. Inaugural Address. Fisheries Exhibition, London. 1883; Blinderman, C., Joyce, D. 1998. The Huxley File. http://aleph0.clarku.edu/huxley/SM5/fish.html.
2. Sinclair, M. *Marine Populations: An Essay on Population Regulation and Speciation*; Washington Sea Grant Program: Seattle, 1988.
3. Smith, T.D. Stock assessment methods: the first fifty years. In *Fish Population Dynamics*, 2nd edn.; Gulland, J.A. Ed.; Wiley: New York, 1988; 1–34.
4. Smith, T.D. *Scaling Fisheries, The Science of Measuring the Effects of Fishing, 1855–1955*; Cambridge University Press: Cambridge, England, 1994.
5. Russell, E.S. Some theoretical considerations on the "overfishing" problem. *Conseil Permanent International pour l'Exploration de la Mer* **1931**, *6*, 3–20.
6. Graham, M. Modern theory of exploiting a fishery and application to North Sea trawling. *J. Cons. Int. Explor. Mer.* **1935**, *10*, 264–274.
7. Schaefer, M.B. Some considerations of population dynamics and economics in relation to the management of the commercial marine fisheries. *J. Fish. Res. Bd. Can.* **1957**, *14*, 669–681.
8. Polachek, T.; Hilborn, R.; Punt, A.E. Fitting surplus production models: comparing methods and measuring uncertainty. *Can. J. Fish. Aquat. Sci.* **1993**, *50*, 2597–2607.
9. Prager, M.H. A suite of extensions to a nonequilibrium surplus-production model. *Fish. Bull.* **1994**, *92*, 374–389.
10. Thompson, W.F.; Bell, F.H. Effect of changes in intensity upon total yield and yield per unit of gear. *Rep. Int. Fish. Comm.* **1934**, *8*, 7–49.
11. Ricker, W.E. Stock and recruitment. *J. Fish. Res. Bd. Can.* **1954**, *11*, 559–623.
12. Beverton, R.J.H.; Holt, S.H. On the dynamics of exploited fish populations. *Fish. Invest., Ser. II, Mar. Fish., G. B. Minist. Agric. Fish. Food* **1957**, *19*, 1–533.
13. Cadrin, S.X.; Friedland, K.D.; Waldman, J. *Stock Identification Methods: Applications in Fishery Science*; Elsevier Academic Press: San Diego, 2005.
14. Hilborn, R.; Walters, C.J. *Quantitative Fisheries Stock Assessment: Choice, Dynamics & Uncertainty*; Chapman and Hall: New York, 1992.
15. Quinn, T.J. II; Deriso, R.B. *Quantitative Fish Dynamics*; Oxford University Press: New York, 1999.
16. Fournier, D.A.; Sibert, J.R.; Majkowski, J.; Hampton, J. MULTIFAN: a likelihood-based method for estimating growth parameters and age composition from multiple length frequency data sets illustrated using data for southern bluefin tuna *(Thunnus maccoyii)*. *Can. J. Fish. Aquat. Sci.* **1990**, *47*, 301–317.
17. Collie, J.S.; Sissenwine, M.P. Estimating population size from relative abundance data measured with error. *Can. J. Fish. Aquat. Sci.* **1983**, *40*, 1871–1879.
18. Mace, P.M. A new role for MSY in single-species and ecosystem approaches to fisheries stock assessment and management. *Fish and Fisheries*, **2001**, 2, 2–32.
19. Bertalanffy, L. von A quantitative theory of organic growth (Inquiries on growth laws. II). *Human Biol.* **1938**, *10*, 181–213.
20. Shepherd, J.G. A versatile new stock-recruitment relationship for fisheries, and the construction of sustainable yield curves. *J. Cons. Int. Explor. Mer.* **1982**, *40*, 67–75.
21. Barrowman, N.J.; Myers, R.A. Still more spawner-recruitment curves: the hockey stick and its generalizations. *Can. J. Fish. Aquat. Sci.*, **2000**, *57*, 665–676.

22. Getz, W.M.; Swartzman, G.L. A probability transition matrix model for yield estimation in fisheries with highly variable recruitment. *Can. J. Fish. Aquat. Sci.*, **1981**, *38*, 847–855.
23. Ostrom, E. *Governing the Commons: The Evolution of Institutions for Collective Action*; Cambridge University Press: New York, 1990.
24. Alverson, D.L.; Freeburg, M.H.; Murawski, S.A.; Pope, J.G. A global assessment of fisheries bycatch and discards. *FAO Fish. Tech. Paper* **1994**, *339*, 1–233.
25. Nash, C. *The History of Aquaculture*; John Wiley and Sons: New York, 2011.
26. Naylor, R.L.; Williams, S.L.; Strong, D.R. Aquaculture – A gateway for exotic species. *Science* **2001**, *294*, 1655–1656.
27. Gordon, H.S. An economic approach to the optimum utilization of fishery resources. *J. Fish. Res. Bd. Can.* **1953**, *10*, 442–457.
28. Garcia, S.M. The precautionary approach to fisheries and implications for fishery research, technology and management: an updated review. *FAO Fisheries Tech. Paper* **1996**, *350* (2), 1–76.
29. Smith, S.J.; J; Hunt, J.J.; Rivard, D. Risk evaluation and biological reference points for fisheries management. *Can. Spec. Publ. Fish. Aquat. Sci.* **1993**, *120*, 442.
30. Prager, M.H.; Shertzer, K.W. Deriving acceptable biological catch from the overfishing limit: Implications for assessment models. *N. Am. J. Fish. Manag.* **2010**, *30*, 289–294.
31. Walters, C. *Adaptive Management of Renewable Resources*; Macmillan: New York, 1986.
32. Cooke, J.G. Improvement of fishery-management advice through simulation testing of harvest algorithms. *ICES J. Mar. Sci.* **1999**, *56*, 797–810.
33. Butterworth, D.S.; Punt, A.E. Experiences in the evaluation and implementation of management procedures. *ICES J. Mar. Sci.* **1999**, *56*, 985–998.
34. Cushing, D.H. *Fisheries Biology: A Study in Population Dynamics*; 2nd edn.; University of Wisconsin Press: Madison, 1982.
35. Rothschild, B.J. *Dynamics of Marine Fish Populations*; Harvard University Press: New York, 1986.
36. Steele, J.H.; Ruzicka, J.J. Constructing end-to-end models using ECOPATH data. *J. Mar. Syst.* **2011**, *87*, 227–238.
37. Levin P.S.; Fogarty, M.J.; Murawski, S.A.; Fluharty, D. Integrated ecosystem assessments: developing the scientific basis for ecosystem-based management of the ocean. *Public Libr. Sci. (PLOS) Biol.* **2009**, 7 (1), 23–28.
38. Link, J.S. *Ecosystem-Based Fisheries Management: Confronting Tradeoffs*; Cambridge University Press: New York, 2010.
39. Brander, K. Impacts of climate change on fisheries. *J. Marine Systems* **2010**, *79*, 389–402.

11

Mangrove Forests

Ariel E. Lugo
International Institute of Tropical Forestry

Ernesto Medina
International Institute of Tropical Forestry
Instituto Venezolano de Investigaciones Científicas

Introduction

Mangroves are woody plants that grow in saline soils. Tomlinson[1] considered 20 genera with 54 species as mangrove species of which 9 genera and 34 species constituted "true mangroves." He established five criteria to typify a true mangrove, including the ecophysiological capacity for excluding salinity and tolerating high tissue concentrations of Na and Cl. The other four criteria were the following: complete fidelity to the mangrove environment, a major role in the structure of the community and capacity to form pure stands, morphological specialization that adapts them to their environment, and taxonomic isolation. Tomlinson also listed 46 genera and 60 woody species as mangrove associates and was emphatic that the number of mangrove associates is a potentially large list of species, given the many habitats that converge with the mangrove environments:

> The ecological literature seems incapable of being reduced to a simple set of rules to account for the diversity of vegetation types within the broad generic concept of mangal. Lack of uniformity is… a measure of the plasticity of mangroves and their ability to colonize such an enormous range of habitats.
>
> *Tomlinson, p. 5*[1]

In 2005, mangrove forests covered about 15–17 million ha. worldwide,[2,3] while more recent assessments detected a considerable reduction of mangrove area to 13.75 million ha.[4] It should be clear that those global estimates oscillate widely depending on the resolution of images and criteria for identifying mangroves. They occur on low-latitude coastal zones where waters with different levels of salinity flood forests at different frequencies and depths. The mangrove environment is diverse, and we use the many environmental gradients under which mangroves grow to organize this review. At the extremes of any of the environmental gradients that we discuss, mangroves not only function differently but also may appear to be exceptions to generalities. For example, some mangroves appear to grow in freshwater, whereas others appear never to flood. In both cases, the incursion of seawater or floods occurs but at very low frequencies that require long-term observation. Despite the complexity of gradient space under which mangroves occur, they are all forested and tidal wetlands in estuarine environments. Mangroves have global importance because their carbon sequestration and dynamics are in the same order as the unaccounted global carbon sink.[5] The term "blue carbon" is used to depict the carbon sink function associated with sediment burial in coastal vegetation, particularly mangroves.[6] Mangroves are also important to the functioning of coastal ecosystems and have economic and cultural importance to people.[7] However, between 1980 and 2005, there was a 20%–30% loss in the global mangrove area.[2,3] Fortunately, mangroves can recover from deforestation if socioeconomic and environmental conditions are favorable.[8] We focus only on natural environmental gradients and ignore anthropogenic gradients, which are reviewed in Lugo et al.[9] and Cintrón and Schaeffer Novelli.[10]

The Ecophysiological Challenge of the Mangrove Environment

Growing on saline, periodically flooded, and low oxygen environments impose severe restrictions on plant growth. Mangroves species exhibit a high degree of structural plasticity as a response to different kinds of environmental stress (Figure 11.1). The effects of other environmental variables are exerted through salinity stress and oxygen supply at the root level. Salt exclusion from roots has high energy cost and effects on plant structure and nutrient uptake such that for plants in saline environments, the salt balance is more critical than the water balance. For example, restricted water uptake in halophytes prevents rapid accumulation of salt in their tissues. Thus, in contrast to non-halophytes, high atmospheric evaporative demand (caused by high temperature and irradiation) cannot be compensated in halophytes by high transpiration because of the resulting salt accumulation inside the plant. Restriction of freshwater supply to mangroves primarily affects salt concentration at the root level, thus affecting water and nutrient uptake. In the high-radiation environment where mangroves grow, maintenance of leaf temperature within physiological limits is strongly dependent on evaporative cooling provided by transpiration. However, because salinity restricts water uptake and transpiration, mangroves maintain leaf temperature and optimize water-use efficiency through variations in leaf inclination, leaf area, and

FIGURE 11.1 (See color insert.) Examples of plasticity in structural development of mangrove species determined by fresh water and nutrient availability (**a**) Riverine *R. mangle* forest near the mouth of the San Juan river, Sucre and Monagas state, Venezuela; (**b** and **c**) Fringe and dwarf *R. mangle* populations in Los Machos wetlands eastern Puerto Rico. The site C strongly P deficient. (**d**) Riverine *A. germinans* forest, along the Yaguaraparo river, Paria Gulf, Sucre state, Venezuela. (**e**) Pure *A. germinans* stand along the western coast of the Gulf of Venezuela. The green carpet is formed by the extreme halophyte *Batis maritima*. (**f**) Dwarf shrub of *A. germinans* in a salt flat behind the bay of Patanemo, Carabobo State, Venezuela.

TABLE 11.1 Variation of Leaf Inclination with Degree of Solar Exposure in Five Mangrove Species in Hinchinbrook Island, Queensland

Species	Exposure	Leaf Area (cm^2)	Projected Fraction (g/m^2)	Succulence
Bruguiera gymnorrhiza	LS	67	0.83	236.5
	MS	70	0.79	332.0
	HS	58	0.56	262.5
R. apiculata	BS	78	0.94	262.4
	MS	75	0.60	285.9
	HS	69	0.37	348.4
R. stylosa	BS	61	0.95	258.5
	MS	60	0.57	321.4
	HS	44	0.30	387.9
Ceriops Tagal	BS	49	0.93	310.1
	MS	20	0.63	351.9
	HS	8	0.36	463.2

Source: Data from Ball et al. *Aust. J. Plant Physiol.* 1988, *15*, 262–276.
The leaf area projected = cos α, where α is the leaf blade angle in respect to the horizontal; sun exposure described as low (LS), medium (MS), and high (HS).

succulence (Table 11.1, from Ball et al.[11]). Increase in leaf inclination is very effective in avoiding high leaf temperatures and photoinhibition by reducing the radiation load. High degree of leaf inclination is commonly observed in fully exposed mangroves, most markedly in semiarid environments.

The hypoxia factor in mangrove wetlands is counteracted mainly through structural devices allowing oxygen to reach waterlogged roots. Mangroves do not tolerate deep flooding for prolonged periods although in true mangroves, tolerance to periodic flooding is high. Finally, the gradient analysis that we undertake here is complex, particularly in the case of nutrients where actual gradients in the field are difficult to establish. Although there are certainly nutrient-rich (estuarine, high runoff, high rainfall) and nutrient-poor (calcareous islands in semiarid areas) sites, an actual nutrient availability gradient is observed only in connection with salinity gradients, as salinity interferes with nutrient uptake.

Temperature

Mangroves are considered a lowland tropical ecosystem, where normally tree species do not tolerate frost. However, mangrove forests occur over a wide range of temperature and a wide latitudinal range that extends from the equator to 30° and 38° north and south, respectively, for *Avicennia marina*.[12] Similar latitudinal range is observed for *Avicennia schaueriana* and *Laguncularia racemosa* in the Neotropics.[13,14] Taking advantage of an unusual chilling event, Chen et al.[15] analyzed the effect of low temperature in a variety of native and introduced mangrove species along a latitudinal range from 19°34′ to 26°41′N in China. Less sensitive were species of *Kandelia* and *Bruguiera* (Rhizophorac), *Aegiceras* (Primulac), and *Avicennia* (Verbenac). High sensitivity was detected in *Sonneratia* species (Lythrac) and *Rhizophora apiculata*. These high-latitude mangroves are clearly no longer tropical forests *sensu stricto*, as they may reproduce and regenerate in lowlands with frost, that is, in warm temperate and temperate life zones.

Woody vegetation on coastal systems is completely replaced by herbaceous vegetation above 32° and 40° north and south, respectively.[13,16] In Florida, *Spartina alterniflora* salt marshes compete with mangroves for space at the coastal fringe, a competition that is mediated by freezing events,[17] with mangroves losing ground with increased freezing frequency or intensity or both.[18] *Spartina brasiliensis* and *S. alterniflora* co-occur with mangroves in the Neotropics,[19,20] but their competition outcome depends on a different ecological constraint, possibly depth of flooding and sediment erosion and deposition dynamics.

Mangroves cope with the latitudinal temperature gradient with a change of species. In Florida, e.g., four mangrove species occur in the keys and Everglades, but as frost frequency increases northward, all species but *Avicennia germinans* drop out.[21] In the Brazilian coast, two species of mangroves reach the same latitudinal limit (28°30′S) but only *A. schaueriana* grows as a large tree, *L. racemosa* grows stunted as a shrub.[13] In China, the number of mangrove species is larger and their number decreases with latitude, the species surviving at the northern most site (26°41′N) is *Kandelia obovata.*[15] Among mangrove species, *A. marina, A. schaueriana, A. germinans,* and *L. racemosa* are the most tolerant to lower temperatures. Duke[12] found that as the temperature decreases, *A. marina* undergoes significant phenological changes involving leaf production, flowering, and fruiting, such that with lower temperatures, growth rates and reproductive potential diminish. Distributional limits for this species coincide with trends toward zero reproductive success. For each 10°C increase, growth increased by a factor of two or three. Stuart et al.[16] found that at freezing temperatures, water deficits cause freeze-induced xylem failure (xylem embolism). Species with wider vessels experienced 60%–100% loss of hydraulic conductivity after freezing and thawing under tension, whereas species with narrower vessels lost as little as 13%–40% of hydraulic conductivity. They suggest that freeze-induced embolism may play a role in limiting latitudinal distribution of mangroves either through massive embolism or through constraints on water transport because of vessel size. For canopy leaves of *Rhizophora stylosa* at its northern limit in Japan (26°11′N), light-saturated maximum photosynthesis rate decreased with decreasing temperature reaching minimum values in January and maximum in June.[22] Near the southern limit of mangrove distribution below 28°S in Brazil, photosynthesis measured in the spring was consistently lower in *Rhizophora mangle* compared to *L. racemosa* and *A. schaueriana*. The latter species had consistently higher water-use efficiency measured both by gas exchange and carbon isotope natural abundance.[23]

Our observations of mangroves at high latitudes along the coasts of Florida, Brazil, and Australia show that trees become shrubby and leaves turn yellowish at the extremes. Multiple sprouting increases as well. However, in global warming scenarios for the future, mangrove species are expected to move to higher latitudes in both hemispheres and establish closed-canopy mangrove forests where *Spartina* marshes grow today. Those movements have been precisely measured in Florida,[24] Texas,[25] and Australia[26,27] (Figure 11.2).

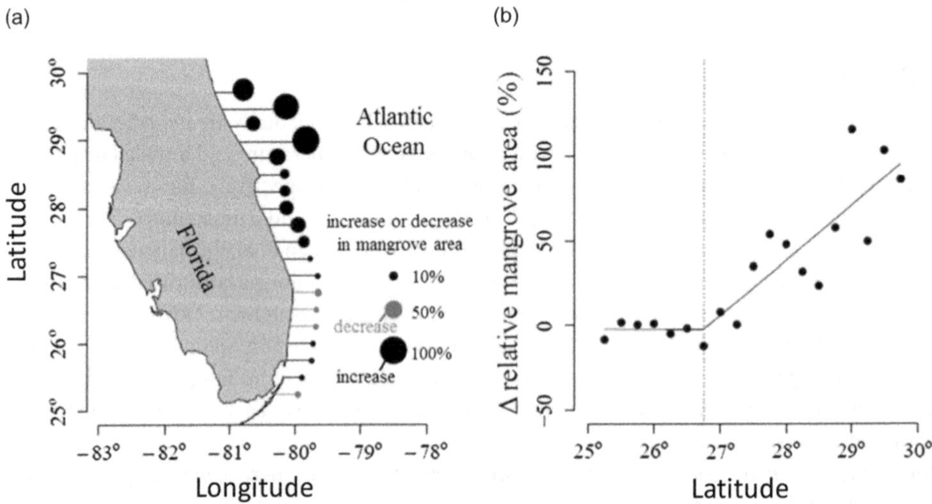

FIGURE 11.2 (**a**) Map of the Florida study area showing the long-term increase or decrease in mangrove area for each 0.25° latitudinal band from 25.25°N to 29.75°N. Long-term change was defined as the relative change in 5-year mean mangrove area from 1984–1988 to 2007–2011. (**b**) Relationship between latitude and relative change in mangrove area. Solid line represents a piecewise regression, and the vertical dotted line gives the breakpoint (26.75°N) of the piecewise regression. (Cavanaugh et al. PNAS et al. 2014, 111, 723–727.)

The effects of high insolation and its influence on photoinhibition confound the determination of high-temperature tolerance of mangroves. However, Biebl[28] found species differences in the temperature range of tolerance of four mangrove species in Puerto Rico: *Conocarpus erectus* −3°C to 54°C, *L. racemosa* −1°C to 51°C, *R. mangle* −3°C to 50°C, and *A. germinans* −3°C to 48°C. Notably, the high-temperature tolerance is higher than temperatures normally encountered by mangroves under natural conditions. We could not find recent experimental assessments of high-temperature tolerance of mangrove leaves.

Rainfall

Mangroves occur from rain forest to very dry forest life zones, which cover a gradient from 500 to about 8,000 mm annual rainfall. This wide rainfall gradient influences soil salinity and forest structure. In Mexico, e.g., Méndez Alonzo et al.[29] found that average tree height, average tree dbh (diameter at breast height), and leaf mass per unit area of *Avicennia* trees increased with both precipitation (500–3,000 mm/year) and minimum annual temperature (-2°C to 14°C). High rainfall areas also increase the likelihood of coupling mangroves to terrestrial nutrient sources through runoff. Those mangroves that receive both high rainfall and runoff are usually the most complex and productive mangroves of all, particularly those growing under riverine conditions.

Many mangroves, however, are isolated from significant terrestrial runoff and, thus, depend on rainfall for sustaining their primary productivity. In these cases, their development is proportional to rainfall. Those that receive terrestrial runoff (including groundwater) are less dependent on rainfall for sustaining primary productivity. Mangroves in dry life zones sometimes receive riverine runoff that originates in upland moist, wet, or rain forest life zones as happens with the mangroves at Tumbes in Peru[10] and those in the strait connecting Lake Maracaibo with the Caribbean Sea in western Venezuela, which receives runoff from Río Limón.[30,31] However, mangroves in dry life zones are generally exposed to drought and high soil salinity. *Avicennia germinans* exhibits reduced photosynthetic rates when exposed to drought, irrespective of salinity,[32] although the reduction was stronger in plants grown at lower salinities. Coincidentally, plants under higher salinity have a lower osmotic potential in their leaves, which allows them to be more effective in water uptake as we discuss in the following text.

Salinity

Mangroves *sensu stricto* are halophytes with elevated salt concentration in their cells. They require a higher level of sodium (Na) than non-halophytes for optimum growth, achieved under controlled conditions at salinity equivalent to 25% of seawater.[33–35] There is a controversy about the concept of mangroves as facultative or obligate halophytes.[36,37] Seedlings of high salt-tolerant *Avicennia* species, cultivated in nutrient solutions with low salt contents (0%–10% seawater) for more than 20 weeks, either do not grow at all[38] or behave similar to seedlings grown with 45% seawater.[39] However, it seems clear that mangroves accumulate more ions, particularly Na, than non-halophytes growing in similar low or nonsaline environments.[40,41] The salinity gradient under which mangroves grow ranges from freshwater to about three times the salinity of seawater.[42] For a given species, such as *R. mangle*, mangrove height[42] and leaf size decrease[43,44] whereas leaf thickness increases[45] with increasing salinity. Osmolality (concentration of osmotically active solutes) of leaf sap in mangroves increases with soil salinity.[46] Leaf area decreases linearly as osmolality increases with *R. mangle* appearing to be more sensitive than *A. germinans* and *L. racemosa*. Also, as salinity increases along a spatial gradient, mangrove species respond by forming monospecific zones according to their salinity tolerance.[47] The changes in species along a salinity gradient are usually accompanied by changes in photosynthetic rates and nutrient-use efficiency,[43] and when exposed to hypersaline conditions (>100‰), quick and massive mangrove mortalities ensue.[48,49]

Tolerance to high levels of salt in the vacuoles of mangroves requires the accumulation of "compatible solutes" in the cytoplasm to prevent dehydration of proteins. Those compatible solutes, which accumulate in the cytoplasm without toxic effects, counteract the osmotic effect of ions in the vacuoles. Among the most common are cyclitols such as mannitol in the species of *Aegiceras* and *Sonneratia,* pinitol in the species of *Bruguiera* and *Aegialitis,* methyl-muco-inositol in *Rhizophora* spp., nitrogen-containing glycinbetaine in *Avicennia* spp., and proline in the species of *Aegialitis* and *Xylocarpus.*[50,51] Mangroves accumulating betaines and proline as compatible solutes also have higher total nitrogen (N) concentration in their leaves than mangroves accumulating cyclitols.[31,43,46,51]

Although under natural conditions mangroves can grow in freshwater, at some point in their life cycle, they must be exposed to seawater, as they are not effective competitors with non-halophytic species. Ball and Pidsley[52] showed that the distribution of *Sonneratia alba* and *S. lanceolata,* two species that grow at low salinities and freshwater, followed the temporal distribution of salt incursion nicely. Neither species grew under conditions where salinity was always absent, reaching just as far into the salinity gradient as was seasonally present. Biomass accumulation in *S. lanceolata* peaked at very low salinities and rapidly decreased with increasing salinity.[53] Mangrove trees grow to the largest sizes and biomass in estuaries with salinities well below seawater. Some species of mangroves such as *A. germinans* behave as euryhaline species with a wide tolerance to salinity. They tend to grow in basins where salinity can range through the whole range of tolerance of mangroves.[54] In contrast, species such as *R. mangle* behave as a stenohaline species with a narrower range of salinity tolerance. They occur mostly on fringes where salinities range narrowly and their range is from almost freshwater to salinities in the order of twice seawater.[55]

The salinity gradient in mangroves develops with distance from the ocean (higher-salinity inland in dry and moist environments and lower-salinity inland in wet and rainy environments), with distance from freshwater runoff (higher salinity toward the ocean), or seasonally depending on rainfall and runoff events or periodic droughts. Salinity gradients are sharper in arid coastlines where differences between seawater and inland hypersaline soils can span the range of tolerance of all mangroves. Interstitial soil water also exhibits a vertical salinity gradient from surface waters (usually lower salinity) to deep soil water (salinity increases with depth).[56] However, freshwater discharges of aquifers can reverse the gradient and make soil water less saline than surface water, if the surface water is fed by tides.

The photosynthetic rate of mangrove species decreases with increasing salinity.[32,54,57–62] In *L. racemosa* and *A. germinans,* net photosynthesis and leaf conductance decrease in parallel as the salt concentration of the nutrient solution increases from 0‰ to 30‰ and 55‰, respectively.[57,59] *Avicennia germinans* grows well and shows higher photosynthetic rates in nutrient solutions without added salt than those grown at salinities above 10‰,[59] in contrast to reports on other *Avicennia* species under natural conditions.[58,61] Potassium (K) is the main ion accumulated in the leaves under those conditions.[59] Measurements of gas exchange during rainy and dry seasons showed that *A. germinans* and *C. erectus* maintain much higher photosynthetic rates and lower osmolalities of leaf sap during the rainy season.[54]

At intermediate to high salinities (30‰ to 55‰), fertilization with N or phosphorus (P) slightly increases photosynthetic rates of *Avicennia* and *Rhizophora* but not over the levels of control trees,[62] that is, fertilization is less effective in overcoming the effects of high salinity.

Mangrove associates are species that are either tolerant to salinity to a certain degree or that are present at those stages in their life cycle that are tolerant to high salt concentrations. *Pterocarpus officinalis* belongs to the first group. It is a mangrove associate throughout the Caribbean and northern South America in areas with high rainfall or surface runoff.[63,64] This species coexists with species of *Rhizophora* and *Laguncularia,* has a high affinity for K, and can restrict Na input from terminal veins in the leaves into parenchyma cells. The fern *Acrostichum aureum* is in the second group of mangrove associates, a species found throughout the tropics and that at times competes with mangrove establishment. The sporophytic phase of this fern has a salt tolerance similar to that of coexisting mangrove species,[65] but the gametophytic phase cannot survive under saline conditions; therefore, the distribution of the mangrove fern is restricted to areas with high rainfall or low salinities that allow the sexual reproduction of the species.

Hydroperiod

The importance of hydroperiod to mangrove structure and species zonation was first recognized by Watson,[66] who estimated the hydroperiod based on the number of tidal events that flooded particular areas of mangroves. Mangroves that grow over patches of coral reef or are located on off-coast overwash islands, or on fringes below low tide, are usually continuously flooded or experience a long hydroperiod. In contrast, some inland mangroves appear to always grow on dry land, except perhaps during the highest tidal events, storm tides, or excessive rainfall or runoff events; their hydroperiod is short. These extreme points in the hydroperiod gradient create numerous ecophysiological challenges to mangroves. Krauss et al.[67] found little effect of hydroperiod on gas exchange of seedlings and saplings of three mangrove species from south Florida, confirming studies that showed that the flooding regime affected mostly the maintenance of leaf area and biomass partitioning.[68]

Oxygen availability to roots can become limiting under long hydroperiods. Mangroves exposed to long hydroperiods or to abnormally high water depth develop adventitious roots on the water surface, which supplement oxygen supply to below-water parts. Water movement by tidal or runoff forces also mitigate low oxygen conditions during chronic inundation. Low root respiration in mangroves[69] is another mitigating effect to low oxygen supply. In contrast to the long hydroperiod, mangroves with a short hydroperiod can experience excessive soil salinity or drought. Between these two extremes grow most of the mangroves with different degrees of soil oxygenation and salinity. Both long and short hydroperiods inhibit understory development and mangrove regeneration and reduce mangrove height.

As sea level rises, some coastal zones flood beyond the tolerance of mangroves. On the other hand, the saline wedge penetrates much further inland into estuaries and provides an opportunity for mangrove expansion. An example of how this phenomenon may proceed has been observed in one of the main tributaries of the Orinoco River, from which a large fraction of its freshwater supply was reduced due to the damming upriver. The consequence of the change in hydrology was that mangroves moved hundreds of kilometers inland.[70] In south Florida, mangroves expanded 3.3 km inland with a sea level rise of 10 cm between 1940 and 1994.[71]

Hydrologic Energy

In general, mangroves occupy low-energy coastlines. In high-energy coastlines, mangroves grow behind sand dunes where waters are calm. However, fringe mangrove forests on different sections of low-energy coastlines face different levels of wave energy. This energy gradient is poorly studied but is now relevant to understanding mangrove responses to increased sea level[72] and tsunamis. Nevertheless, mangrove response to sea level rise will depend not only on the amount of area with saline soils but also on air temperature (greater frost frequency at higher latitudes), competition with salt marshes[17,74–77] and other vegetation,[72] and the balance between the rate of sea level rise, input of allochthonous sediments, and mangrove production of peat.[73,74]

Usually, mangrove peat and root systems are undercut by rising sea level, and it is common to see overturned mangroves in places where this process occurs. Mangrove stems and prop roots resist incoming tides and waves and, in so doing, reduce the hydrologic energy dissipated on coastlines. Under high-energy conditions, prop root and stem density increase, allowing mangroves to perform this ecological service of coastline protection. However, this process occurs within a limited range of tolerance to high wave or tidal energy. In the tropical Atlantic Brazilian coast, *Avicennia* is frequently at the fringe, and this species is less tolerant to the erosive force of tides, coastal currents, and strong winds.[56]

In locations affected by macrotides, where large rivers discharge, as in the Atlantic coast of Brazil north of the mouth of the Amazon river, mangroves face strong effects from erosion and sediment deposition. Batista et al.[74] used remote sensing to show that along the coast of Amapa, Brazil, during the period 1980–2003, large erosion rates caused the disappearance of 1.37 km^2/year of mangroves on one location, whereas in another location, progradation added 56 km^2 of mangrove areas to the shoreline.

Such dramatic shifts in erosive and sedimentary forces maintain mangrove vegetation in a constant state of successional change.

When mangroves are massively killed by drought, hurricanes, or other disturbances, the organic peat on which they grow collapses under the erosive power of tides and waves and because the production of organic matter by the forest ceases. This collapse of the forest floor causes the intrusion of seawater and converts the forest into a lagoon.[42,75] This sets succession back by many decades as it takes time for the mangroves to reverse a lagoon environment back to forest growing at the higher elevation afforded by the accumulation of peat and roots. McKee et al.[76] increased the rate of mangrove root accumulation and, therefore, the rate of peat accretion by fertilizing mangroves in the Caribbean coast of Belize.

Mangroves are overtaken by tsunamis and are exposed to significant structural effects.[77,78] However, behind the mangroves, there is less destruction of property because of the energy dissipation involved in overcoming mangrove resistance to wave energy (search for mangroves and tsunamis in www.fao.org).

Nutrients

The levels of nutrient availability experienced by mangroves range from oligotrophic with extreme P limitation to highly eutrophic. Oligotrophy occurs in carbonate environments[79,80] or over acid peat soils.[81,82] The area covered by nutrient-limited mangroves is large and mostly in the wider Caribbean (including the Everglades of Florida) and many Pacific atolls. Nevertheless, mangroves in general are eutrophic systems and without nutrient limitation. Eutrophy occurs on alluvial floodplains and riverine fringes. Polluted coastlines also provide eutrophic conditions for mangrove growth. Within this generally favorable nutrient availability, conditions may vary with the type of water entering the mangroves (nutrient rich from land, nutrient poor from the ocean). For example, Chen and Twilley[82] found a gradient of mangrove structure and productivity from the mouth of the Shark River to inland sites in South Florida. The biotic gradient was not responding to salinity or sulfide concentration but to N and P concentrations in soil pore water. Fertile sites were dominated by *L. racemosa*, whereas *R. mangle* grew in the less fertile sites. These forests exhibited seasonal changes in photosynthetic rates in response to changes in air temperature and light intensity.[83]

Oligotrophy leads to the formation of dwarf mangroves, which are low-height trees (no more than 1 m) with normal leaf sizes and reduced leaf turnover rates. Fertilization experiments with mangroves in Belize, Florida, and Panama illustrate some of the complexities associated with nutrient limitations in mangroves.[84–87] Dwarf *R. mangle* and *A. germinans* mangroves in carbonate environments always respond to P fertilization, but surrounding fringing *R. mangle* mangroves respond only to N fertilization, and mangroves under intermediate conditions responded to both N and P fertilization.[74] In a disturbed mangrove forest in the Indian River Lagoon, both *R. mangle* and *A. germinans* responded to N fertilization but not to P fertilization along a tree height gradient.[85] Contrasting responses to nutrient fertilization by the same species under different environments differentially affected ecological process of the species.[87] Among the biotic responses affected by fertilization were changes in plant habit from stunted to larger-sized individuals, which were accomplished by increasing wood relative to leaf biomass and changes in leaf-specific area, nutrient uptake, and leaf herbivory.

Resorption efficiency of N and P in mangroves is in general higher than by other angiosperms, suggesting that internal cycling of these elements may account for a significant fraction of their nutrient requirements for growth.[88] Feller et al.[87] observed changes in the cycling of nutrients at a stand level through leaf fall and within stand cycling. They found that in P-limited environments, retranslocation (or resorption) of P by *R. mangle* is much higher (≈70%) than that of N (≈45%). Nitrogen fertilization did not change those percentages, but P fertilization decreased P resorption efficiency (<50%) and increased N resorption (≈70%). Studies of dwarf mangroves on P-deficient peats in Puerto Rico[81] and stunted mangroves under hypersaline conditions[43] confirmed larger resorption values for P than for N. It appears that under conditions of P limitation, resorption of P is greater than that of N and that a sufficient supply of P reduces its resorption levels below those of N. Under natural conditions, resorption of N appears to be higher than that of P in *Avicennia* spp., and the resorption of both N and P increases with interstitial water salinity.[89–92] Comparison of resorption efficiencies of *R. mangle* and *L. racemosa* growing side by side in a coastal lagoon in the Gulf of Venezuela showed that resorption of N by the latter species is much higher than that of P.[93]

In Panama, Lovelock et al.[94] found that fertilized dwarf mangroves responded to N and P fertilization by increasing hydraulic conductance sixfold by P and 2.5-fold by N. The response of the hydraulic conductivity, more than the photosynthetic response per unit leaf area, accounted for the increase in size of the fertilized mangroves.

Lovelock et al.[69] found that *R. mangle* root respiration per unit mass was low in Belize compared to temperate tree species at the same temperature. Root respiration did not differ significantly between zones (fringe vs. dwarf) and fertilization treatments (N or P), although rates were consistently higher after fertilization, particularly in dwarf mangroves. The fine roots fertilized with P responded to fertilization with increased P concentrations.

Wind

Coastal zones are usually windy mostly due to sea breezes, which favor gas exchange through their influence on gaseous gradients across leaf surfaces. Sea breezes ventilate the forest and moderate air temperatures. A significant fraction of mangroves occurs in the hurricane belt, which results in periodic exposures to extreme wind events (velocities >100 km/h are common and can exceed 250 km/h). In the process, hurricanes dissipate high levels of energy over forests, as it happened when Hurricane Hugo dissipated about 210 J/m^2s over the northeastern mangroves of Puerto Rico.[95] Cintrón and Schaeffer Novelli[96] found that for Neotropical mangroves, the maximum canopy height decreased with increasing latitude, where winds (Caribbean) and frost (Florida and Brazil) become significant factors affecting forest structure. The large tree sizes in the Pacific Island of Kosrae vs. Pohnpei[97] or San Juan River mangroves in Venezuela[98] vs. mangroves in nearby Caribbean Islands are examples of wind effects on mangrove stature. Hurricane winds exert selective pressure on forests by periodically trimming or overturning taller trees, which are usually the ones most affected by wind energy.[95] As a result, hurricane-affected forests have lower stature and wind-sculptured canopies.

Redox

Redox gradients occur in mangroves in association with anoxic conditions in mud and the relatively high concentration of sulfate in seawater. Sulfur is the fourth-most abundant element in seawater after chlorine (Cl), Na, and magnesium (Mg). Under reduced conditions (no oxygen), sulfur is present as sulfide, which is toxic for mangroves and reduces photosynthesis and growth.[99] Mangroves mitigate sulfide accumulation around the roots by the transport of oxygen through aerenchymatic tissues.[100,101] The ecological significance of the redox gradient in soils is expressed by its effects on the rates of microbial decomposition processes as well as in the diversity of both microbial communities and metabolic pathways associated with increasingly reducing conditions with soil depth (Figure 5.5 and Table 5.6 in Alongi).[102]

Other Organisms

In addition to mangrove trees, other groups of organisms respond to environmental gradients within the mangroves and, in some cases, affect mangrove trees and ecosystem functioning. For example, gastropods respond to sediment metal concentration.[103] Crabs respond to tidal range, and epibionts increase with light availability but decrease with tidal energy. In the case of crabs, they act as ecosystem engineers, not only facilitating mangrove regeneration but also influencing soil aeration and transport of oxygen to anaerobic soil layers.[104]

Avicennia germinans mangrove forests in Panama growing under different environmental conditions (Caribbean and Pacific coasts with different rainfall, hydroperiod, and soil salinity) differed in phenology and invertebrate and bird composition.[106] Although the coastal zones shared 95% of the bird species, the mangroves only shared 34%, and each forest had a different feeding guild assemblage in the bird community. The results suggest that despite being the same mangrove species, the different environmental conditions along the environmental gradients ripple through the food chain and result in different community composition.

Integration and Spatial Scales of Mangroves

There is no mangrove model that considers all the gradients discussed here and uses the responses of mangrove organisms to these gradients to holistically explain mangrove structure and productivity (but see Lugo[47] for a zonation/succession diagram with many of the gradients included). Such a model would be extremely useful to mangrove conservation actions, including restoration and rehabilitation of mangrove sites. However, there have been several efforts to model or conceptualize mangrove functioning. Odum[105] developed trophic-level models of mangroves for South Florida, and Lugo et al.[106] used energy flow to simulate mangrove productivity and response to hurricanes. Cintron et al.[42] developed a model of the effects of salinity on mangrove functioning, and Twilley et al.[107] modeled mangrove succession and applied it to restoration. Kangas[108] developed an energy theory for the landscape classification of wetlands, which dovetails with the stand classification of mangrove ecosystem types of Lugo and Snedaker.[109]

Thom[110–112] was the first to relate mangrove ecosystem structure and function with the geomorphology of coastlines. Twilley et al.[113] developed a spatial and functional hierarchy for assessing mangrove forests (Figure 13.3 in Twilley et al.[115]). Their concept included the latitudinal or global distribution of mangroves at the top of the hierarchy. Within the latitudes, they included the environmental settings, which were based on Thom's geomorphological types. Geomorphologic conditions expose mangroves to different energy conditions such as direction and force of hydrological fluxes and origin and quality of waters interacting with the mangroves, that is, coastal vs. inland waters. Inside these environmental settings, Twilley et al. included the ecological types of Lugo and Snedaker, which function in relation to topography and hydrology. Mangrove stands occur within any of the ecological types, and within a stand, both aboveground and belowground processes take place. Twilley and Rivera Monroy[114] modified and refined their 1996 model and compiled what they called five ecogeomorphic models of nutrient biogeochemistry for mangrove wetlands. These models represent the state of understanding of the whole mangrove ecosystem functioning, taking into consideration the effects of multiple environmental gradients on these ecosystems.

Climate Change Effects

Drivers of global climate change have been intensifying for more than 50 years without signs of deceleration (Figure 11.3). Steady increments in atmospheric CO_2 concentration and temperature anomalies since 1960 are correlated with changes in patterns of rainfall distribution, oceanic currents, melting of polar ice, and sea level rise.

The complex interactions between atmospheric and terrestrial dynamics were schematically depicted by Ward et al.[115] providing a frame to visualize their effect on ecophysiological properties of mangrove forests (Figure 11.4). The variety of ecosystem and species responses has been analyzed in a series of recent reviews that emphasize the need of developing management procedures to confront the consequences of mangrove ecosystems perturbation. Lovelock et al.[116] emphasize physiological aspects, Jennerjahn et al.[117] deal with biogeochemistry, and Alongi[118] describes the complex pathways altering the nutrient relationships of mangrove forests.

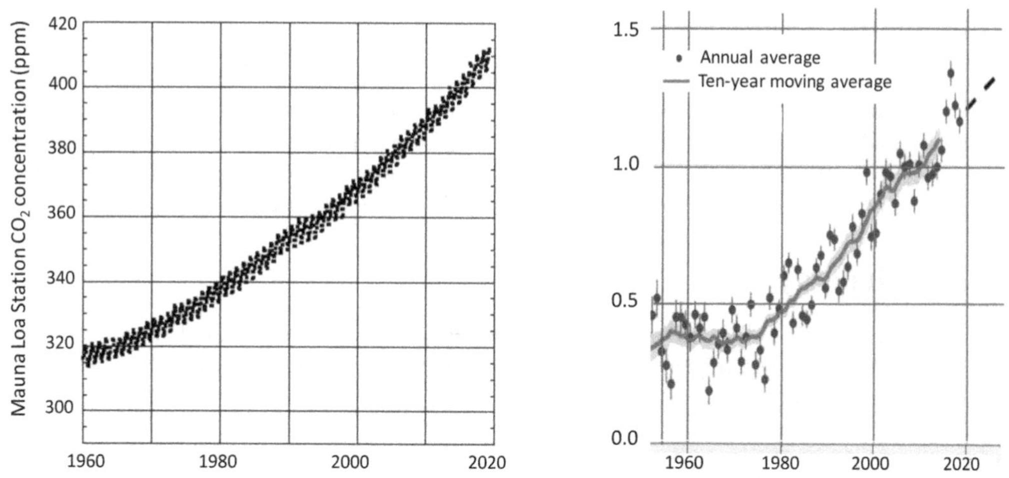

FIGURE 11.3 Steady increments in atmospheric CO_2 concentrations measured at the Mauna Loa Research station in Hawaii. (Scripps CO_2 Program http://scrippsco2.ucsd.edu, and temperature anomalies compiled by Berkeley Earth.org, since 1960 to the month of March 2019.)

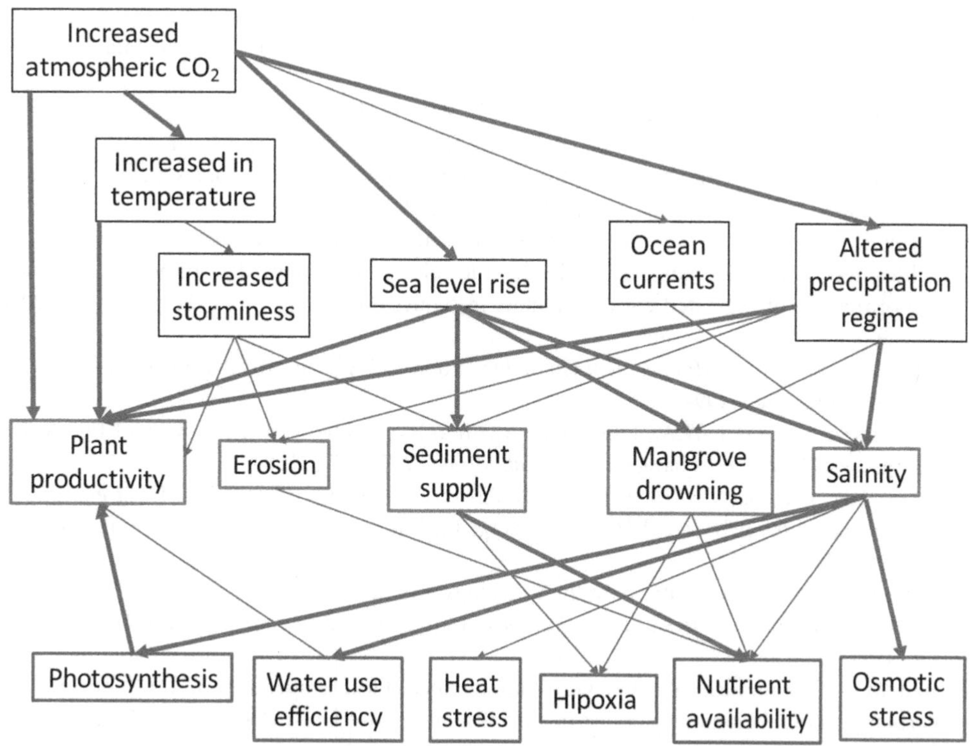

FIGURE 11.4 Scheme depicting the effect of expected changes in climatic parameters on soil water, salinity, and sediment and how they potentially affect physiological properties of mangrove forests constraining plant productivity. (Modified from Ward et al. *Ecosystem Health and Sustainability* 2016, 2(4), e01211. doi:10.1002/ehs2.1211.)

Conclusion

Mangrove ecosystem function results in different rates of ecological processes and structural development depending on whether the mangroves are above or below the frost line, exposed or not to hurricanes, and in dry or rain forest climates. These overarching environmental extremes (temperature, rainfall, and wind disturbances) have controlling effects on how mangroves respond to environmental gradients and which species might predominate at the environmental extremes.

Acknowledgments

This work was done in collaboration with the University of Puerto Rico. Mildred Alayón edited the manuscript. We received useful comments from Gilberto Cintrón, Frank Wadsworth, Blanca Ruiz, and two anonymous reviewers.

References

1. Tomlinson, P.B. *The Botany of Mangroves*; Cambridge University Press: Cambridge, England, 1986.
2. Valiela, I.; Bowen, J.L.; York, J.K. Mangrove forests: one of the world's threatened major tropical environments. *BioSci.* **2001**, *51*, 807–815.
3. FAO. *The World's Mangroves 1980–2005*; The Food and Agriculture Organization of the United Nations: Rome, Italy, 2007.
4. Giri, C.; Ochieng, E.; Tieszen, L.L.; Zhu, Z.; Singh, A.; Loveland, T.; Masek, J.; Duke, N. Status and distribution of mangrove forests of the world using earth observation satellite data. *Global Ecol. Biogeogr.* **2010**, *20*, 154–159.
5. Bouillon, S.; Borges, A.V.; Castañeda-Moya, E.; Diele, K.; Dittmar, T.; Duke, N.C.; Kristensen, E.; Lee, S.Y.; Marchand, C.; Middleburg, J.J.; Rivera-Monroy, V.H.; Smith III, T.J.; Twilley, R.R. Mangrove production and carbon sinks: a revision of global budget estimates. *Global Biogeochem. Cycles* **2008**, *22*, GB2013. doi:2010.1029/2007GB003052.
6. Mcleod, E.L.; Chmura, G.L.; Bouillon, S.; Salm, R.; Bjork, M.; Duarte, C.M.; Lovelock, C.E.; Schlesinger, W.H.; Silliman, B.R. A blueprint for blue carbon: toward an improved understanding of the role of vegetated coastal habitats in sequestering CO_2. *Front. Ecol. Environ.* **2011**, *9*, 552–560.
7. FAO. *Mangrove Forest Management Guidelines*; Food and Agriculture Organization of the United Nations: Rome, Italy, 1994.
8. Martinuzzi, S.; Gould, W.A.; Lugo, A.E.; Medina, E. Conversion and recovery of Puerto Rican mangroves: 200 years of change. *Forest Ecol. Manag.* **2009**, *257*, 75–84.
9. Lugo, A.E.; Cintrón, G.; Goenaga, C. Mangrove ecosystems under stress. In *Stress Effects on Natural Ecosystems*; Barret, W.G.; Rosenberg, R., Eds.; John Wiley and Sons Limited: Sussex, England, 1981; 129–153.
10. Cintrón Molero, G.; Schaeffer Novelli, Y. Ecology and management of new world mangroves. In *Coastal Plant Communities of Latin America*; Seeliger, E., Ed.; Academic Press, Inc.: San Diego, CA, 1992; 233–258.
11. Ball, M.C.; Cowan, I.R.; Farquhar, G.D. Maintenance of leaf temperature and the optimization of carbon gain in relation to water loss in a tropical mangrove forest. *Aust. J. Plant Physiol.* **1988**, *15*, 262–276.
12. Duke, N.C. Phenological trends with latitude in the mangrove tree *Avicennia marina*. *J. Ecol.* **1990**, *78*, 113–133.
13. Schaeffer Novelli, Y.; Cintrón Molero, G.; Rothleder Adaime, R.; de Camargo, T.M. Variability of mangrove ecosystems along the Brazilian coast. *Estuaries* **1990**, *13*, 204–218.

14. Tognella, M.M.P.; Soares, M.L.G.; Cuevas, E.; Medina, E. Heterogeneity of elemental composition and natural abundance of stables isotopes of C and N in soils and leaves of mangroves at their southernmost West Atlantic range. *Braz. J. Biol.* **2016**, *76*, 994–1003.
15. Chen, L.; Wang W.; Li, Q.Q.; Zhang, Y.; Yang, S.; Osland, M.J.; Huang, J.; Peng, C. Mangrove species' responses to winter air temperature extremes in China. *Ecosphere* **2017**, *8*(6), e01865. doi:10.1002/ecs2.1865.
16. Stuart, S.A.; Choat, B.; Martin, K.C.; Holbrook, N.M.; Ball, M.C. The role of freezing in setting the latitudinal limits of mangrove forests. *New Phytol.* **2007**, *173*, 576–583.
17. Kangas, P.C.; Lugo, A.E. The distribution of mangroves and saltmarshes in Florida. *Trop. Ecol.* **1990**, *31*, 32–39.
18. Lugo, A.E.; Patterson Zucca, C. The impact of low temperature stress on mangrove structure and growth. *Trop. Ecol.* **1977**, *18*, 149–161.
19. Braga, C.F.; Beasley, C.R.; Isaac, V.J. Effects of plant cover on the macrofauna of Spartina marshes in northern Brazil. *Braz. Archi. Biol. Technol.* **2009**, *52*, 1409–1420.
20. Colonnello, G. *Spartina alterniflora* Loisel, primer registro para Venezuela. *Memorias Fundación La Salle de Ciencias Naturales* **2001**, *59*, 29–33.
21. Odum, W.E.; McIvor, C.C. Mangroves. In *Ecosystems of Florida*; Myers, R.L.; Ewel, J.J., Eds.; University of Central Florida Press: Orlando, FL, 1990; 517–548.
22. Suwa, R.; Hagihara K.C. Seasonal changes in canopy photosynthesis and foliage respiration in a *Rhizophora stylosa* stand at the northern limit of its natural distribution. *Wetlands Ecol. Manag.* **2008**, *16*, 313–321.
23. Gomes Soares, M.L.; Pereira Tognella, M.M.; Cuevas, E., Medina, E. Photosynthetic capacity and intrinsic water-use efficiency of *Rhizophora mangle* at its southernmost western Atlantic range. *Photosynthetica* **2015**, *53*, 464–470.
24. Cavanaugh, K.C.; Kellner, J.R.; Forde, A.J.; Gruner, D.S.; Parker, J.D.; Rodriguez, W.; Feller, I.C. Poleward expansion of mangroves is a threshold response to decreased frequency of extreme cold events. *PNAS* **2014**, *111*, 723–727.
25. Armitage, A.R.; Highfield, W.E.; Brody, S.D.; Louchouarn, P. The contribution of mangrove expansion to salt marsh loss on the Texas Gulf Coast. *PLoS One* **2015**, *10*(5), e0125404. doi:10.1371/journal.pone.0125404.
26. Saintilan, N.; Nicholas, C.; Wilson, N.C.; Rogers, K.; Rajkaran, A.; Krauss, K.W. Mangrove expansion and salt marsh decline at mangrove poleward limits. *Glob. Change Biol.* **2014**, *20*, 147–157.
27. Saintilan, N.; Rogers, K.; McKee, K.L. Chapter 26 The shifting saltmarsh-mangrove ecotone in Australasia and the Americas. In *Coastal Wetlands (Second Edition), An Integrated Ecosystem Approach*; Perillo, G.; Wolanski, E.; Cahoon, D.; Hopkinson, C., Eds.; Elsevier: Amsterdam, Netherland; Oxford, UK; Cambridge, MA, 2019; 915–945. doi:10.1016/B978-0-444-63893-9.00026-5.
28. Biebl, R. Temperaturresistenz tropischer Pflanzen auf Puerto Rico (Verglichen mit jener von Pflanzen der gemässigten Zone). *Protoplasma* **1965**, *59*, 133–156.
29. Méndez Alonzo, R.; López Portillo, J.; Rivera Monroy, V.H. Latitudinal variation in leaf and tree traits of the mangrove Avicennia germinans (Avicenniaceae) in the central region of the Gulf of Mexico. *Biotropica* **2008**, *40*, 449–456.
30. Medina, E.; Fonseca, H.; Barboza, F.; Francisco, M. Natural and man-induced changes in a tidal channel mangrove system under tropical semiarid climate at the entrance of the Maracaibo lake (Western Venezuela). *Wetlands Ecol. Manag.* **2001**, *9*, 243–253.
31. Barboza, F.; Barreto, M.B.; Figueroa, V.; Francisco, M.; González, A.; Lucena, M.; Mata, K.Y.; Narváez, E.; Ochoa, E.; Parra, L.; Romero, D.; Sánchez, J.; Soto, M.N.; Vera, A.J.; Villarreal, A.L.; Yabroudi, S.C.; Medina, E. Desarrollo estructural y relaciones nutricionales de un manglar ribereño bajo clima semi-árido. *Ecotropicos* **2006**, *19*, 13–29.
32. Sobrado, M.A. Drought effects on photosynthesis of the mangrove, *Avicennia germinans*, under contrasting salinities. *Trees* **1999**, *13*, 125–130.

33. Pannier, F. El efecto de distintas concentraciones salinas sobre el desarrollo de *Rhizophora mangle* L. *Acta Cient. Venez.* **1959,** *10,* 68–78.
34. Downton, W.J.S. Growth and osmotic relations of the mangrove *Avicennia marina*, as influenced by salinity. *Aust. J. Plant Physiol.* **1982,** *9,* 519–528.
35. Clough, B.F. Growth and salt balance of the mangroves *Avicennia marina* (Forsk.) Vierh. and *Rhizophora stylosa* Griff. in relation to salinity. *Aust. J. Plant Physiol.* **1984,** *11,* 419–430.
36. Wang, W.; Yan, Z.; You, S.; Zhang, Y.; Chen, L.; Lin, G. Mangroves: obligate or facultative halophytes? A review. *Trees* **2011,** *25,* 953–963.
37. Krauss, K.W.; Ball, M.C. On the halophytic nature of mangroves. *Trees* **2013,** *27,* 7–11.
38. Nguyen, H.T.; Stanton, D.E.; Schmitz, N.; Farquhar, G.D.; Ball, M.C. Growth responses of the mangrove *Avicennia marina* to salinity: development and function of shoot hydraulic systems require saline conditions. *Ann. Bot.* **2015,** *115,* 397–407.
39. Kodikara, K.A.S.; Jayatissa, L.P.; Huxham, M.; Dahdouh-Guebas, F.; Koedam, N. The effects of salinity on growth and survival of mangrove seedlings changes with age. *Acta Bot. Bras.* **2018,** *32,* 37–46.
40. Medina, E; Lugo, A.E.; Novelo, A. Contenido Mineral del Tejido Foliar de Especies de Manglar de la Laguna de Sontecomapan (Veracruz, Mexico) y su Relacion con la Salinidad. *Biotropica* **1995,** *27,* 317–323.
41. Medina E, Barboza F, Francisco M. Occurrence of red mangrove (Rhizophora mangle L.) in the south-western wetlands of the Maracaibo lake: leaf sap analysis detects halophytic physiology in low salinity environments. In *Ökologische Forschung im Globalen Kontext*; Veste, M.; Wucherer, W.; Homeier, J., Eds.; Cuvillier: Göttingen, 2005; 45–54.
42. Cintrón, G.; Lugo, A.E.; Pool, D.J.; Morris, G. Mangroves of arid environments in Puerto Rico and adjacent islands. *Biotropica* **1978,** *10,* 110–121.
43. Lugo, A.E.; Medina, E.; Cuevas, E.; Cintron, G.; Laboy Nieves, E.N.; Novelli, Y.S. Ecophysiology of a mangrove forest in Jobos Bay, Puerto Rico. *Caribb. J. Sci.* **2007,** *43,* 200–219.
44. Duke, N.C. Morphological variation in the mangrove genus Avicennia in Australasia: systematic and ecological consideration. *Aust. Syst. Bot.* **1990b,** *3,* 221–239.
45. Camilleri, J.C.; Ribi, G. Leaf thickness of mangroves (*Rhizophora mangle*) growing in different salinities. *Biotropica* **1983,** *15,* 139–141.
46. Medina, E.; Francisco, M. Osmolality and $\delta^{13}C$ of leaf tissues of mangrove species from environments of contrasting rainfall and salinity. *Estuar. Coast. Shelf Sci.* **1997,** *45,* 337–344.
47. Lugo, A.E. Mangrove ecosystems: successional or steady state? *Biotropica* **1980,** *12*(supplement 2), 65–72.
48. Botero, L. Massive mangrove mortality on the Caribbean coast of Colombia. *Vida Silvestre Neotrop.* **1990,** *2,* 77–78.
49. Jiménez, J.A.; Lugo, A.E.; Cintrón, G. Tree mortality in mangrove forests. *Biotropica* **1985,** *17,* 177–185.
50. Stewart, G.R.; Popp, M. The ecophysiology of mangroves. In *Plant Life in Aquatic and Amphibious Habitats*; Crawford, R.M.M., Ed.; Blackwell Science Publishers: Oxford, 1987; 333–345.
51. Popp, M.; Polania, J. Compatible solutes in different organs of mangrove trees. *Annales des Sci. Forest.* **1989,** *46*(suppl.), 842–844.
52. Ball, M.C.; Pidsley, S.M. Growth responses to salinity in relation to distribution of two mangrove species, *Sonneratia alba* and *S. lanceolata*, in Northern Australia. *Funct. Ecol.* **1995,** *9,* 77–85.
53. Ball, M.C.; Sobrado, M.A. Ecophysiology of mangroves: challenges in linking physiological processes with patterns in forest structure. In *Physiological Plant Ecology*; Press, M.C.; Scholes, J.D.; Barker, M.G., Eds.; Blackwell Science: Oxford, UK, 1999; 331–346.
54. Smith, J.A.C.; Popp, M.; Lüttge, U.; Cram, W.J.; Diaz, M.; Griffiths, H.; Lee, H.S.J.; Medina, E.; Schäfer, C.; Stimmel, K.-H.; Thonke, B. Water relations and gas exchange of mangroves. *New Phytol.* **1989,** *111,* 293–307.

55. Lugo, A.E. Fringe wetlands. In *Forested Wetlands*; Lugo, A.E.; Brinson, M.M.; Brown, S.; Eds.; Elsevier: Amsterdam, the Netherlands, 1990; 143–169.
56. Mehlig, U.; Menezes, M.P.M.; Reise, A.; Schories, D.; Medina, E. Mangrove vegetation of the Caeté Estuary. In *Mangrove dynamics and management in north Brazil*; Saint-Paul, U.; Schneider, H., Eds.; Springer: Berlin, 2010; 71–108.
57. Sobrado, M.A. Leaf characteristics and gas exchange of the mangrove *Laguncularia racemosa* as affected by salinity. *Photosynthetica* **2005**, *43*, 217–221.
58. Naidoo, G.; Tuffers, A.V.; von Willert, D.J. Changes in gas exchange and chlorophyll fluorescence characteristics of two mangroves and a mangrove associate in response to salinity in the natural environment. *Trees* **2002**, *16*, 140–146.
59. Suárez, N.; Medina, E. Influence of salinity on Na+ and K+ accumulation, and gas exchange in *Avicennia germinans*. *Photosynthetica* **2006**, *44*, 268–274.
60. Nandy (Datta), P.; Sauren Das, S.; Ghose, M.; Spooner- Hart, R. Effects of salinity on photosynthesis, leaf anatomy, ion accumulation and photosynthetic nitrogen use efficiency in five Indian mangroves. *Wetlands Ecol. Manag.* **2007**, *15*, 347–357.
61. Tuffers, A.; Naidoo, G.; von Willert, D.J. Low salinities adversely affect photosynthetic performance of the mangrove, *Avicennia marina*. *Wetlands Ecol. Manag.* **2001**, *9*, 235–242.
62. Lovelock, C.E.; Feller, I.C. Photosynthetic performance and resource utilization of two mangrove species coexisting in a hypersaline scrub forest. *Oecologia* **2003**, *134*, 455–462.
63. Medina, E.; Cuevas, E.; Lugo, A.E. Nutrient and salt relations of *Pterocarpus officinalis* L. in coastal wetlands of the Caribbean: assessment through leaf and soil analyses. *Trees: Struct. Func.* **2007**, *21*, 321–327.
64. Medina, E.; Francisco, M.; Quilice, A. Isotopic signatures and nutrient relations of plants inhabiting brackish wetlands in the northeastern coastal plain of Venezuela. *Wetlands Ecol. Manag.* **2008**, *16*, 51–64.
65. Medina, E.; Cuevas, E.; Popp, M.; Luga, A.E. Soil salinity, sun exposure, and growth of *Acrostichum aureum*, the mangrove fern. *Bot. Gaz.* **1990**, *151*, 41–49.
66. Watson, J.G. *Mangrove Forests of the Malay Peninsula*. Malayan Forest Records, 6; Forest Dept., Federated Malay Peninsula: Kuala Lumpur, 1928, 1–274.
67. Krauss, K.W.; Twilley, R.R.; Doyle, T.W.; Gardiner, E.S. Leaf gas exchange characteristics of three neotropical mangrove species in response to varying hydroperiod. *Tree Physiol.* **2006**, *26*, 959–968.
68. Pezeshki, S.R.; DeLaune, R.D.; Patrick Jr. W.H. Differential response of selected mangroves to soil flooding and salinity: gas exchange and biomass partitioning. *Can. J. Forest Res.* **1990**, *20*, 869–874.
69. Lovelock, C.E.; Ruess, R.W.; Feller, I.C. Fine root respiration in the mangrove Rhizophora mangle over variation in forest stature and nutrient availability. *Tree Physiol.* **2006**, *26*, 1601–1606.
70. Colonnello, G.; Medina, E. Vegetation changes induced by dam construction in a tropical estuary: the case of the Manamo River, Orinoco delta (Venezuela). *Plant Ecol.* **1998**, *139*, 145–154.
71. Ross, M.S.; Meeder, J.F.; Sah, J.P.; Ruiz, P.L.; Telesnicki, G.J. The southeast saline everglades revisited: 50 years of coastal vegetation change. *J. Veg. Sci.* **2000**, *11*, 101–112.
72. Doyle, T.W.; Krauss, K.W.; Conner, W.H.; From, A.S. Predicting the retreat and migration of tidal forests along the northern Gulf of Mexico under sea-level rise. *For. Ecol. Manag.* **2010**, *259*, 770–777.
73. Ellison, J.C.; Stoddart, D.R. Mangrove ecosystem collapse during predicted sea-level rise: Holocene analogues and implications. *J. Coast. Res.* **1991**, *7*, 151–165.
74. Batista, E.D.M.; Souza Filho, P.W.M.; Machado da Silveira, O.F. Avaliação de áreas deposicionais e erosivas em cabos lamosos da zona costeira Amazônica através da análise multitemporal de imagens de sensores remotos. *Rev. Bras. Geof.* **2009**, *27*. doi:10.1590/S0102-261X2009000500007.
75. Cahoon, D.R.; Hensel, P.; Rybczyk, J.; McKee, K.L.; Proffitt, C.E.; Perez, B.C. Mass tree mortality leads to mangrove peat collapse at Bay Islands, Honduras after Hurricane Mitch. *J. Ecol.* **2003**, *91*, 1093–1105.

76. McKee, K.L.; Cahoon, D.R.; Feller, I.C. Caribbean mangroves adjust to rising sea level through biotic controls on change in soil elevation. *Global Ecol. Biogeogr.* **2007**, *16*, 545–556.
77. Baird, A.H.; Kerr, A.M. Landscape analysis and tsunami damage in Aceh: comment on Iverson and Prasad (2007). *Landscape Ecol.* **2008**, *23*, 3–5.
78. Iverson, L.R.; Prasad, A.M. Modeling tsunami damage in Aceh: a reply. *Landscape Ecol.* **2008**, *23*, 7–10.
79. Feller, I.C. Effects of nutrient enrichment on growth and herbivory of dwarf red mangrove (*Rhizophora mangle*). *Ecol. Monogr.* **1995**, *65*, 477–505.
80. Cheeseman, J.M.; Lovelock, C.E. Photosynthetic characteristics of dwarf and fringe *Rhizophora mangle* L. in a Belizean mangrove. *Plant, Cell Environ.* **2004**, *27*, 769–780.
81. Medina, E.; Cuevas, E.; Lugo, A.E. Nutrient relations of dwarf *Rhizophora mangle* L. mangroves on peat in eastern Puerto Rico. *Plant Ecol.* **2010**, *207*, 13–24.
82. Chen, R.; Twilley, R.R. Patterns of mangrove forest structure and soil dynamics along the Shark River estuary, Florida. *Estuaries* **1999**, *22*, 955–970.
83. Barr, J.G.; Jose, D.; Fuentes, J.D.; Engel, V.; Zieman, J.C. Physiological responses of red mangroves to the climate in the Florida Everglades. *J. Geophys. Res.* **2009**, *114*, G02008. doi:10.1029/2008JG000843.
84. Feller, I.C.; McKee, K.L.; Whigham, D.F.; O'Neill, J.P. Nitrogen vs. phosphorus limitation across an ecotonal gradient in a mangrove forest. *Biogeochemistry* **2002**, *62*, 145–175.
85. Feller, I.C.; Whigham, D.F.; McKee, K.L. Nitrogen limitation of growth and nutrient dynamics in a disturbed mangrove forest, Indian River Lagoon, Florida. *Oecologia* **2003**, *134*, 405–414.
86. Feller, I.C.; Lovelock, C.E.; McKee, K.L. Lovelock, C.E. Nutrient addition differentially affects ecological processes of *Avicennia germinans* in nitrogen vs. phosphorus limited mangrove ecosystems. *Ecosystems* **2007**, *10*, 347–359.
87. Feller, I.C.; Whigham, D.F.; O'Neill, J.P.; McKee, K.L. Effects of nutrient enrichment on within-stand cycling in a mangrove forest. *Ecology* **1999**, *80*, 2193–2205.
88. Reef, R.; Feller, I.C.; Lovelock, C.E. Nutrition of mangroves. *Tree Physio.* **2010**, *30*, 1148–1160.
89. Ochieng, C.A.; Erftemeijer, P.L.A. Phenology, litterfall and nutrient resorption in *Avicennia marina* (Forssk.) Vierh in Gazi Bay, Kenya. *Trees* **2002**, *16*, 167–171.
90. Alam, M.R.; Mahmood, H.; Biswas, T.; Rahman, M.M. Physiologically adaptive plasticity in nutrient resorption efficiency of Avicennia officinalis L. under fluctuating saline environments in the Sundarbans of Bangladesh. *Hydrobiologia* **2019**, *828*, 41–56.
91. Almahasheer, H.; Duarte, C.M.; Irigoien, X. Leaf nutrient resorption and export fluxes of Avicennia marina in the Central Red Sea area. *Front. Mar. Sci.* **2018**, *13*. doi:10.3389/fmars.2018.00204.
92. Wei, S.; Liu, X.; Zhang, L.; Chen, H.; Zhang, H.; Zhou, H.; Lin, Y. Seasonal changes of nutrient levels and nutrient resorption in Avicennia marina leaves in Yingluo Bay, China. *J. South. For.* **2015**, *77*, 237–242.
93. Medina, E.; Fernandez, W.; Barboza, F. Element uptake, accumulation, and resorption in leaves of mangrove species with different mechanisms of salt regulation. *Web Ecol.* **2015**, *15*, 1–11. www.web-ecol.net/15/1/2015.
94. Lovelock, C.E.; Feller, I.C.; Mckee, K.L. The effect of nutrient enrichment on growth, photosynthesis and hydraulic conductance of dwarf mangroves in Panamá. *Funct. Ecol.* **2004**, *18*, 25–33.
95. Lugo, A.E. Visible and invisible effects of hurricanes on forest ecosystems: an international review. *Aust. Ecol.* **2008**, *33*, 368–398.
96. Cintrón, G.; Schaeffer Novelli, Y. *Los manglares de la costa brasileña: revisión preliminar de la literatura*. Oficina Regional de Ciencia y Tecnología de UNESCO para América Latina y el Caribe y Universidad Federal de Santa Catalina: Santa Catarina, 1981.
97. Allen, J.A.; Ewel, K.C.; Jack, J. Patterns of natural and anthropogenic disturbance of mangroves on the Pacific Island of Kosrae. *Wetlands Ecol. Manag.* **2001**, *9*, 279–289.
98. Twilley, R.R.; Medina, E. Forest dynamics and soil characteristics of mangrove plantations in the San Juan River estuary, Venezuela. In *Annual Letter*; Lugo, A.E., Ed.; International Institute of Tropical Forestry, USDA Forest Service: Río Piedras, PR, 1997; 87–94.

99. Lin, G.H.; Sternberg; L.D.S.L. Effect of growth form, salinity, nutrient and sulfide on photosynthesis, carbon isotope discrimination and growth of red mangrove (Rhizophora mangle L.). *Aust. J. Plant Physiol.* **1992**, *19*, 509–517.
100. McKee, K.L.; Mendelssohn, I.A.; Hester, M.K. A reexamination of pore water sulfide concentrations and redox potentials near the aerial roots of *Rhizophora mangle* and *Avicennia germinans*. *Am. J. Bot.* **1988**, *75*, 1352–1359.
101. McKee, K.L. Soil physicochemical patterns and mangrove species distribution - reciprocal effects? *J. Ecol.* **1993**, *81*, 477–487.
102. Alongi, D.M. *The Energetics of Mangrove Forests*; Springer: New York, 2009.
103. Amin, B.; Ismail, A.; Arshad, A.; Yap, C.H.; Kamarudin, M.S. Gastropod assemblages as indicators of sediment metal contamination in mangroves. *Water Air Soil Pollut.* **2009**, *201*, 9–18.
104. Lefebvre, G.; Poulin, B. Bird communities in Panamanian black mangroves: potential effects of physical and biotic factors. *J. Trop. Ecol.* **1997**, *13*, 97–113.
105. Odum, W.E. Pathways of energy flow in a south Florida estuary; Sea Grant Technical Bulletin 7; University of Miami: Miami, FL, 1971.
106. Lugo, A.E.; Sell, M.; Snedaker, S.C. Mangrove ecosystem analysis. In *Systems Analysis and Simulation Ecology*; Patten, B.C., Ed.; Academic Press: New York, 1976; 13–145.
107. Twilley, R.R.; Rivera Monroy, V.H.; Chen, R.; Botero, L. Adapting an ecological mangrove model to simulate trajectories in restoration ecology. *Mar. Pollut Bull.* **1999**, *37*, 404–419.
108. Kangas, P.C. An energy theory of landscape for classifying wetlands. In *Forested Wetlands*; Lugo, A.E.; Brinson, M.; Brown, S., Eds.; Elsevier: Amsterdam, 1990; 15–23.
109. Lugo, A.E.; Snedaker, S.C. The ecology of mangroves. *Annu. Rev. Ecol. Syst.* **1974**, *5*, 39–64.
110. Thom, B.G. Mangrove ecology and deltaic geomorphology, Tabasco, Mexico. *J. Ecol.* **1967**, *55*, 301–343.
111. Thom, B.G. Mangrove ecology from a geomorphic viewpoint. In *Proceedings of the International Symposium on Biology and Management of Mangroves*; Walsh, G.; Snedaker, S.; Teas, H., Eds.; Institute of Food and Agricultural Sciences, University of Florida: Gainesville, FL, 1974; 469–481.
112. Thom, B.G. 1982. Mangrove ecology-a geomorhological perspective. In *Mangrove Ecosystems in Australia: Structure, Function and Management*; Clough, B.F., Ed.; Australian Institute of Marine Sciences: Canberra, Australia, 1982; 3–17.
113. Twilley, R.R.; Snedaker, S.C.; Yáñez Arancibia, A.; Medina, E. Biodiversity and ecosystem processes in tropical estuaries: perspectives of mangrove ecosystems. In *Functional Role of Biodiversity: A Global Perspective*; Mooney, H.A.; Cushman, J.H.; Medina, E.; Sala, O.E.; Schulza, E.-D., Eds.; John Wiley & Sons: New York, 1996; 327–370.
114. Twilley, R.R.; Rivera Monroy, V.H. Ecogeomorphic models of nutrient biogeochemistry for mangrove wetlands. In *Coastal Wetlands: An Integrated Approach*; Perillo, G.M.E.; Wolanski, E.; Cahoon, D.R.; Ironson, M.M. Eds.; Elsevier: Amsterdam, 2009; 641–683.
115. Ward, R.D.; Friess, D.A.; Day, R.H.; MacKenzie, R.A. Impacts of climate change on mangrove ecosystems: a region by region overview. *Ecosyst. Health Sustainability* **2016**, *2*(4), e01211. doi:10.1002/ehs2.1211.
116. Lovelock, C.E.; Krauss, K.W.; Osland, M.J.; Reef, R.; Ball, M.C. The physiology of mangrove trees with changing climate. In *Tropical Tree Physiology*, Tree Physiology 6; Goldstein, G.; Santiago, L.S., Eds.; Springer International Publishing: Switzerland, 2016; 149–179. doi:10.1007/978-3-319-27422-5_7.
117. Jennerjahn, T.C.; Gilman, E.; Krauss, K.W.; Lacerda, L.D.; Nordhaus, I.; Wolanski, E. Chapter 7 Mangrove ecosystems under climate change. In *Mangrove Ecosystems: A Global Biogeographic Perspective*; Rivera-Monroy, V.H. et al., Eds.; Springer International Publishing AG: Cham, Switzerland, 2017; 211–244. doi:10.1007/978-3-319-62206-4_7.
118. Alongi, D.M. Impact of global change on nutrient dynamics in mangrove forests. *Forests* **2018**, *9*, 596. doi:10.3390/f9100596.

12

Tourism Management: Marine and Coastal Recreation

C. Michael Hall
University of Canterbury
University of Eastern Finland
University of Oulu

Introduction

The coastal environment is a major focus for tourism and recreational activity in many parts of the world and is often associated with tourism images of "sun, sand and surf."[1] However, the development of beach resorts and the increasing popularity of coastal and marine tourism (e.g., fishing, scuba diving, whale watching, windsurfing, and yachting) have all placed increased pressure on coastal regions, an area in which use is often already highly concentrated in terms of agriculture, human settlements, fishing, and industry.

The concept of coastal tourism includes a range of tourism, leisure, and recreationally oriented activities that occur in the coastal zone and immediate offshore coastal waters. These include tourism-related development (accommodation, restaurants and food services, attractions, and second homes), and the infrastructure supporting coastal and marine tourism development (e.g., retail businesses, transport hubs, marinas, and activity suppliers). Also included are tourism activities such as recreational boating, coast- and marine-based ecotourism, cruises, swimming, recreational fishing, snorkeling, and diving. Marine tourism is closely related to coastal tourism, but it is more geographically expansive and also includes ocean-based tourism activities, such as cruise ships and yacht cruising.[2] The entry first discusses the historical development of coastal and marine tourism and the values associated with it before examining coastal resort morphology and the associated environmental impacts of coastal and marine tourism. The entry then discusses coastal and marine tourism in the specific contexts of coral reefs, cruise tourism, and conservation.

Historical Development

Although coastal areas are now synonymous with tourism and recreation, this has not always been the case. In western landscape traditions, natural coastal areas were often perceived negatively until the late 1700s when the Romantic Revolution transformed perceptions of landscape and seascape. Simultaneously, a belief developed in the healing and recuperative benefits of sea air and seawater. The period 1750–1840 witnessed a fundamental reassessment of coastal areas as leisure place. In that period, the beach developed as an activity space for recreation and tourism, with distinct cultural and social forms emerging in relation to resort fashions, tastes, and innovations. The development of piers, jetties, and promenades for formal recreational activities led to new ways of experiencing and appreciating the sea and development of positive coastal amenity values dates from this time.[3]

From the mid-19th century to the mid-20th century, the development of coastal resorts was closely linked to the expansion of rail and shipping transport networks. For example, for what was arguably the world's first modern seaside resort, Margate in southeast England, the visits by the wealthy gentry and upper middle classes, were replaced by lower middle class and working class tourists from the 1820s as a result of direct steamer access from London. This expansion of tourist numbers but change in their socioeconomic composition received further impetus when Margate was connected to London by railway after the 1840s. This particularly innovation, together with the growth of personal car access in the twentieth century, also meant that visitation could take the form of day excursions rather than overnight stays with consequent effects on the economic returns from tourism at the destination level as well as increased pressures on the transit routes to and from coastal resorts.[4]

Charting the history of seaside resorts in northern Europe and North America also highlights how continued changes in transport technology led to the decline of many traditional resorts as, for a similar cost, it became possible for tourists in major urban centers to travel to the Mediterranean or the Caribbean. This shift in coastal tourism activity has had significant implications for resource management in newer coastal resorts as well as for climate change given growth in emissions from aviation-related travel to coastal resorts. Such historical change also indicates that, like many industries, tourism also moves through development cycles and that resorts can enter into significant decline and even abandonment as tourist destinations.[5] This can therefore create new problems for such cities and towns as they seek to regenerate infrastructure and find new forms of economic development at a time of diminished returns from tourism. In some cases, this may mean focusing on new markets or attractions, for example, casinos in the case of Atlantic City, New Jersey, or cultural tourism in the case of Margate, that has few direct links to the water-based tourism activities. In older coastal resorts, the tourism base may also be replaced by a second-home and retirement home function that still includes a coastal leisure component.[2]

Coastal Resort Morphology and Impacts

Tourism urbanization is spatially and functionally different from other urban places.[6] In many coastal areas, tourism urbanization is highly linear with resultant impacts on the coastal environment. These are particularly pronounced in coastal resort areas in the Mediterranean where tourism and leisure-related urbanization is held to be primarily responsible for coastal urbanization. Of the 220 million people who visit the Mediterranean region each every year, over 100 million visit Mediterranean beaches.[7] In Italy, over 43% of the coastline is completely urbanized, 28% is partly urbanized, and less than 29% is still free of construction. There are only six stretches of coast over 20 km long that are free of construction and only 33 stretches between 10 and 20 km long without any construction.[8] Similarly, in Cyprus, 95% of the tourism industry is located within two kilometers of the coast,[9] placing the coastal environment as well as archeological heritage under extreme pressure. In Tunisia, it is expected that about 150 km of shoreline, over 13% of the total Tunisian coastline, will eventually be occupied by tourism and leisure facilities and infrastructure.[10] In many coastal resorts, the inappropriate siting of tourist infrastructures on foredunes has accelerated beach erosion processes as a result of sediment processes being interrupted,[11] while tourism development has also altered the water dynamics of coastal regions.[10]

Island destinations are a special case of coastal and marine tourism particularly affected by tourism urbanization and the resource demands of tourism. The government of the Caribbean Island of Barbados estimates that 98% of renewable freshwater resources are already being used, and that industrial and commercial uses (including tourism) had increased from 20% in 1996 to 44% by 2007. Tourism operations (hotels, cruise ships, golf courses), which represented approximately one-sixth of total consumption in 1996, is projected to represent one-third of overall demand by 2016.[12]

Impacts of Seasonality

The seasonal nature of coastal tourism can place significant stress on infrastructure with populations frequently more than doubling in resort destinations during high season. Island destinations with already limited resources may be especially affected. For example, in the case of Anguilla in the Caribbean, annual tourist numbers are equivalent to a 30.5% increase in permanent population, while in the Cayman Islands it is equivalent to an 89% increase.[13] Such high levels of tourist visitation place extreme strain on water and sewage infrastructure.[14] In the Mediterranean, only 80% of the effluent of residents and tourists is collected in sewage systems, with the remainder being discharged directly or indirectly into the sea or to septic tanks. However, only 50% of the sewage networks are actually connected to wastewater treatment facilities with the rest being discharged into the sea.[15] The lack of appropriate sewage treatment leads to nutrient enrichment of littoral waters and the growth of algal blooms which can affect both coastal ecology as well as tourist perception of beaches and destinations.[16] The effects of seasonal tourism demands on infrastructural systems such as sewage and water supply, as well as resources means that in a number of coastal destinations local water supplies have to be supplemented by water imports. The Mediterranean is a coastal tourism region identified as being extremely vulnerable to water scarcity,[17] while small islands and regions such as Baja California Sur, Sharm El Sheikh, Egypt, Zanzibar, and Almería province, Spain, also have overused water capacity or are at high-risk of overusing their water capacity.[18]

Beach Environments

Beaches and wetland systems are the coastal environments most under threat from tourism urbanization and leisure activities. Visitor pressure increases dune degradation and vulnerability highlighting the need for close monitoring of impacts and changes in dune morphology as well as the development of appropriate management regimes.[11] Many significant wildlife habitats have been lost through resort development. This includes not only the construction of hotels and tourism infrastructure but also the development of activities, such as golf courses. The Parliamentary Assembly, Council of Europe[19] estimated there has been a reduction in the Mediterranean wetland area of approximately three million hectares by 93% since Roman times. Of which, a third has been lost since 1950. In the Asia-Pacific many coastal wetlands have been drained or substantially modified for resort development, although the loss of mangrove swamps has substantial long-term implications for increased coastal erosion especially as a result of increased storm events and sea-level rise while habitat loss affects local fish stocks given the nursery role of mangroves.[20]

Coral Reefs

Coral reefs are one of the most important marine ecosystems for tourism. However, they are highly susceptible to urban run-off and sedimentation; over-fishing; souveniring of coral and shells; as well as over use by divers and snorkelers. Coral reefs and are also considered one of the most vulnerable to climate change and other anthropogenic environmental change,[21] including several climate change-related impacts: ocean acidification, coral bleaching, and for in-shore coral reefs, greater land runoff as a result of increased storm events.[20]

Coral bleaching can occur for multiple reasons, but temperature change is the primary cause, and acidification can be a strong contributor. Mass coral bleaching transforms large reef areas from a mosaic of color (if healthy) to a stark white. With very high confidence, the IPCC[22] concluded that a warming of 2°C above 1990 levels would result in mass mortality of coral reefs globally. However, the vulnerability of reefs to the impacts of climate change will vary spatially and temporally, with shallow reefs, reefs with species closest to their thermal maximum threshold, and those closest to sources of pollution or other human impacts the most vulnerable and where impacts will be visible to tourists the soonest. Nevertheless, even moderate further warming will consequently affect the attractiveness of coral reefs to tourists in some destinations. Further, coral reefs provide other important ecosystem services to the tourism sector, including as a fishery resource and coastal protection against storms. For destinations where reefs are the key attraction for tourists, the long-term damage arising from bleaching incidents will have important implications for the quality of dive tourist experiences, and therefore for the sustainability of tourism operations.[23] However, previous bleaching events suggest that the effects may be variable among different markets at different destinations.[20]

In the Caribbean, 76% of tourists at the diving destination of Bonaire (where 99% of respondents took at least one dive during their trip) indicated they would be unwilling to return for the same holiday price in the event that corals suffered "severe bleaching and mortality."[24] Andersson's[25] study of tourism perceptions at the African islands of Zanzibar and Mafia found that serious divers who visit Mafia were more aware of bleaching than the recreational divers who visit Zanzibar (62% versus 29%). When asked if they would be willing to dive on a bleached reef, 40% of respondents at Zanzibar and 33% at Mafia indicated that they would. In their study of Mauritius, Gössling et al.[26] found that there are significant differences between dive destinations. They found that the state of coral reefs was largely irrelevant to dive tourists and snorkelers, as long as a threshold level was not exceeded. This level was defined by visibility, abundance and variety of species, and the occurrence of algae or physically damaged corals. The results by Gössling et al.[26] are consistent with the findings of Main and Dearden[27] that 85% of recreational divers failed to perceive any damage to reefs in Phuket, Thailand after the 2004 Indian Ocean tsunami. Dearden and Manopawitr[28] even question whether future generations of divers will perceive environmental changes, such as bleaching and degraded reef conditions, in the same way as contemporary divers if they have no frame of reference of previous, more pristine conditions. Such results highlight how different scuba diving market segments will be differentially affected by the impacts of climate and other forms of environmental change depending on motivations, prior experiences, and budgets.

Cruise Tourism

The cruise sector is one of the fastest-growing areas of tourism with cruise demand in North America alone increasingly at an average annual rate of 7.6% between 1990 and 2010.[29] In 2010 14.89 million people travelled on CLIA member cruise lines. Cruising impacts the marine environment as a result of discharging untreated waters and other waste at sea,[30] the role of shipping as a vector of alien species,[31] and the high-energy intensity of cruise ships on a per-passenger basis.[32] Global ocean-going cruise emissions for 2005 were estimated at 34 $MtCO_2$, less than 5% of global shipping emissions.[33] However, this figure does not include the full range of tourist passenger vessels.[20] Marine transport is a major contributor to the spread of disease and biological invasion. Cruise ships and yachts, as well as site visitation by tourists, have been recognized as particularly significant avenues of species introduction via hull fouling[34] and ballast water.[35]

Tourism and Coastal and Marine Management and Conservation

Although tourism is responsible for a range of impacts on coastal and marine environments, it also provides an economic justification for species and habitat conservation. For example, it is estimated that in 2008 more than 13 million people took whale watching tours in 119 countries worldwide, generating US$2.1 billion in total expenditures.[36] However, while individual species may receive legal protection the conservation of marine habitats lags far behind their terrestrial counterparts. As of 2010, the global total are for marine-protected areas (MPAs) was 4.7 million km^2, or 1.31% of the global ocean surface (3.21% of potential marine jurisdictions/EEZ areas, and 1.06% of off-shelf areas).[7] Only 12 out of 190 states and territories with marine jurisdictions have an MPA coverage of 10% or more in the areas under their jurisdiction. Even in island systems such as the Caribbean and the South Pacific, the proportion of protected marine area in both regions is much lower than the terrestrial area. In the Caribbean, Jamaica has the highest proportion of marine area set aside at 3.56% and in the Pacific, Palau has 8.74% of its marine territory as protected area.[13]

Given the large number of stakeholders in the coastal and marine areas, tourism is often but one component of multiple-objective spatial management and planning strategies. Marine spatial planning often aims to reduce the environmental impacts of tourism, among other marine area users, but simultaneously ensures that the resource base is sufficient to encourage further visitation, especially as tourism, fees, licenses, and income can assist in financing biodiversity conservation. Management and spatial planning strategies aim to reduce conflict between marine users via the concentration of compatible and separation of incompatible uses.[37] However, such integrated management approaches are harder to achieve in the coastal zone because of the larger numbers of stakeholders and population base that is experienced there, especially in urban areas. Nevertheless, in the Integrated Coastal Management Framework, tourism is usually recognized as one of the most important activities in coastal areas as part of their sustainable development[38]

Conclusion

Tourism is a major economic activity in coastal and marine areas. Yet, in historical terms, the association of tourism with the coast and the sea is relatively recent. Nevertheless, in a short period of time, tourism has become an industry critical to many coastal regions and communities as well as generating income for coastal and marine conservation activities. Yet, at the same time, tourism has also been one of the biggest sources of environmental impact of coastal and near-shore areas particularly when seasonal tourist visitation overwhelms water and sewage infrastructure. Tourism is also a significant source of greenhouse gas emissions which may have long-term consequences for significant recreational resources such as coral reefs as well as the threat of sea level rise to coastal tourism resources and infrastructure. Integrated marine area spatial strategies and coastal zone management are the primary planning means to manage coastal and marine tourism although tourism is only one, sometimes conflicting, use of coastal areas. Yet, the income provided by tourism may provide some of the strongest economic justifications to conserve coastal and marine ecosystem services and resources.

Acknowledgment

The support of Academy of Finland (project SA 255424) is gratefully acknowledged.

References

1. Hall, C.M. Trends in coastal and marine tourism: the end of the last frontier? Ocean Coast Manage **2001,** *44*, 9–10, 601–618.
2. Hall, C.M.; Page, S. *The Geography of Tourism and Recreation*, 3rd Ed.; Routledge: London, 2006.

3. *Towner, J. An Historical Geography of Recreation and Tourism in the Western World 1540–1940; Wiley: Chichester, 1996.*
4. Walton, J. The *English Seaside Resort: A Social History 1750–1914*; Leicester University Press: Leicester, 1983.
5. *Walton, J. The British Seaside: Holidays and Resorts in the Twentieth Century; Manchester University Press: Manchester, 2000.*
6. Hall, C.M. Tourism urbanization and global environmental change. In *Tourism and Global Environmental Change: Ecological, Economic, Social and Political Interrelationships*; Gössling, S., Hall, C.M., Eds.; Routledge: London, 2006; 142–156 pp.
7. Toropova, C.; Meliane, I.; Laffoley, D.; Matthews, E.; Spalding, M. *Global Ocean Protection: Present Status and Future Possibilities*; IUCN: Gland, 2010.
8. World Wide Fund for Nature. *Tourism Threats in the Mediterranean*; WWF Mediterranean Programme: Rome, 2001.
9. Loizidou, X. Land use and coastal management in the Eastern Mediterranean: the Cyprus example. In *International Conference on the Sustainable Development of the Mediterranean and Black Sea Environment*; Thessaloniki: Greece; May 28–31, 2003. http://www.iasonnet.gr/abstracts/loizidou.html; 2003.
10. De Stefano, L. *Freshwater and Tourism in the Mediterranean*; WWF Mediterranean Programme: Rome, 2004.
11. Williams, A.; Micallef, A. *Beach Management: Principles and Practice*; Earthscan: London, 2009.
12. Emmanuel, K.; Spence, B. Climate change implications for water resource management in Barbados tourism. Worldwide Hospitality Tourism Themes **2009,** *1*, 252–268.
13. Hall, C.M. An island biogeographical approach to island tourism and biodiversity: An exploratory study of the Caribbean and Pacific Islands. Asia Pac. J. Tourism Res. **2010,** *15*, 383–399.
14. UN Division for Sustainable Development. *Trends in Sustainable Development: Small Island Developing States*; United Nations: New York, 2010.
15. Scoullos, M.J. Impact of anthropogenic activities in the coastal region of the Mediterranean Sea. In *International Conference on the Sustainable Development of the Mediterranean and Black Sea Environment*; Thessaloniki: Greece; May 28–31, 2003. http://www.iasonnet.gr/abstracts/Scoullos.pdf; 2003.
16. Bauer, M.; Hoagland, P.; Leschine, T.M.; Blount, B.G.; Pomeroy, C.M.; Lampl. L.L.; Scherer, C.W.; Ayres, D.l.; Tester, P.A.; Sengco, M.R.; Sellner, K.G.; Schumacker, J. The importance of human dimensions research in managing harmful algal blooms. Front Ecol. Environ. **2010,** *8*, 75–83.
17. Iglesias, A.; Garrote, L.; Flores, F.; Moneo, M. Challenges to manage the risk of water scarcity and climate change in the Mediterranean. Water Resource Manag. **2010,** *21*, 775–788.
18. Gössling, S.; Peeters, P.; Hall, C.M.; Ceron, J.P.; Dubois, G.; Lehmann, L.V.; Scott, D. Tourism and water use: Supply, demand, and security. An international review. Tourism Manage **2012,** *33*, 1–15.
19. Parliamentary Assembly, Council of Europe. *Erosion of the Mediterranean Coastline: Implications for Tourism, Doc. 9981 16 October 2003, Report Committee on Economic Affairs and Development*; Brussels: Council of Europe, 2003.
20. Scott, D.; Gössling, S.; Hall, C.M. *Tourism and Climate Change: Impacts, Adaptation and Mitigation*; Routledge: London, 2012.
21. Hughes, T.P.; Graham, N.A.J.; Jackson, J.B.C.; Mumby, PJ, Steneck, R.S. Rising to the challenge of sustaining coral reef resilience. Trends Ecol. Evol. **2010,** *25*, 633–642.
22. Schneider, S.H.; Semenov, S.; Patwardhan, A.; Burton, I.; Magadza, C.H.D.; Oppenheimer, M.; Pittock, A.B.; Smith, J.B.; Suarez, S.; Yamin, F. Assessing key vulnerabilities and the risk from climate change. In *Climate Change 2007: Impacts, Adaptation and Vulnerability. Contribution of Working Group II to the Fourth Assessment Report of the Intergovernmental Panel on Climate Change*; Parry, M.L.; Canziani, O.F.; Palutikof, J.P.; van der Linden, V.D.; Hanson, C.E., Eds.; Cambridge University Press: Cambridge, 2007; 779–810 pp.

23. Flugman, E., Mozumder, P.; Randhir, T. Facilitating adaptation to global climate change: perspectives from experts and decision makers serving the Florida Keys. Climatic Change **2011,** *112*, 1015–1035.
24. Uyarra, M.C.; Côté, I.; Gill, J.; Tinch, R.R.T.; Viner, D.; Watkinson, A.R. Island specific preferences of tourists for environmental features: implications of climate change for tourism-dependent states. Environ. Conserv. **2005,** *32*, 1, 11–19.
25. Andersson, J. The recreational costs of coral bleaching – a stated and revealed preference study of international tourists. Ecol. Econ. **2007,** *62*, 704–715.
26. Gössling, S.; Lindén, O.; Helmersson, J.; Liljenberg, J.; Quarm, S. Diving and global environmental change: a Mauritius case study. In *New Frontiers in Marine Tourism: Diving Experiences, Management and Sustainability*; Garrod, B., Gössling, S., Eds.; Amsterdam: Elsevier: Amsterdam, 2007; 67–92 pp.
27. Main, M.; Dearden, P. Tsunami impacts on Phuket's diving industry: geographical implications for marine conservation. Coastal Manag. **2007,** 35, 4, 1–15.
28. Dearden, P.; Manopawitr, P. Climate change – coral reefs and dive tourism in South-east Asia. In *Disappearing Destinations*; Jones, A., Phillips, M., Eds.; CABI Publishing: Wallingford, 2011; 144–160 pp.
29. Cruise Lines International Association. 2011 CLIA Cruise Market Overview: Statistical Cruise Industry Data through 2010; Cruise Lines International Association: Fort Lauderdale, 2011.
30. Klein, R. The cruise sector and its environmental impact. In *Tourism and the Implications of Climate Change: Issues and Actions. Bridging Tourism Theory and Practice*; Schott, C., Ed.; Emerald Group Publishing: Bingley, **2011**; 113–130 pp.
31. Hall, C.M.; James, M.; Wilson, S. Biodiversity, biosecurity, and cruising in the Arctic and sub-Arctic. J. Herit. Tourism **2010,** 5, 351–364.
32. Eijgelaar, E.; Thaper, C.; Peeters, P. Antarctic cruise tourism: the paradoxes of ambassadorship, "last chance tourism" and greenhouse gas emissions. J. Sustain Tour **2010,** *18*, 337–354.
33. World Economic Forum. *Towards a Low Carbon Travel & Tourism Sector*; World Economic Forum: Davos, 2009.
34. Drake, J.M.; Lodge, D.M. Hull fouling is a risk factor for intercontinental species exchange in aquatic ecosystems. Aquatic Invasions **2007,** 2, 121–131.
35. Endresen, O.; Behrens, H.L.; Brynestad, S.; Bjørn Andersen, A.; Skjong, R. Challenges in global ballast water management. Mar. Pollut. Bull. **2004,** *48*, 615–623.
36. *O'Connor, S.; Campbell, R.; Cortez, H.; Knowles, T. Whale Watching Worldwide: Tourism Numbers, Expenditures and Expanding Economic Benefits, Prepared by Economists at Large; International Fund for Animal Welfare: Yarmouth, MA, 2009.*
37. Beck, M.W.; Ferdana, Z.; Kachmar, J.; Morrison, K.K.; Taylor, P. *Best Practices for Marine Spatial Planning*; The Nature Conservancy: Arlington, VA, 2009.
38. *UN Environmental Programme. Sustainable Coastal Tourism: An Integrated Planning and Management Approach; UNEP: Milan, 2009.*

Bibliography

Dimmock. K.; Musa, G., Eds.; *Scuba Diving Tourism*; Routledge: London, 2013.

Dowling, R.K. *Cruise Ship Tourism: Issues, Impacts, Cases*; CABI Publishing: Wallingford, 2010.

Gössling, S.; Scott, D.; Hall, C.M., Ceron, J.P.; Dubois, G. Consumer behaviour and demand response of tourists to climate change. Ann. Tourism Res. **2012,** *39*, 36–58.

Hall, C.M.; Lew, A.A. *Understanding and Managing Tourism Impacts: An Integrated Approach*; Routledge: London, 2009.

13

Science Communication for Natural Resource Managers: Techniques and Examples in Marine Systems

Stephanie I. Anderson
University of Rhode Island

Katharine McDuffie
University of Rhode Island

Sunshine Menezes
University of Rhode Island

Science Communication: A Useful Tool

Approximately 40% of the global population or 2.4 billion people live within 100 km of the coast [1]. This coastal populace and the USD 3–6 trillion per year global ocean economy [1] pose challenges, such as overfishing, pollution, and coastal development, to fragile ecosystems. Exacerbated by anthropogenic climate change, these coastal issues have been identified as "wicked problems" [2]. Wicked problems are environmental issues that are ill defined, lack concrete solutions, and rely on political judgment for resolution. Addressing and mitigating each of these challenges requires knowledgeable personnel who can mobilize constituents and advocate for marine policy that protects resources [3]. Through effective communication, more stakeholders could be encouraged to engage in the policy process (e.g., via co-production of important policy questions, public comments, or opinion pieces for local news outlets), enabling improved quality and legitimacy of management decisions [4] and greater stakeholder support [5].

While science communication can greatly contribute to ecosystem-based management (EBM) decision-making [6], facilitating meaningful discussions about scientific research can be challenging. Most people have difficulty staying current on scientific developments [7] and comprehending the statistics [8] and uncertainties [4] necessary to make informed decisions. For example, Takahashi and Tandoc [9] found that only 28% of American adults had the knowledge needed to comprehend science policy issues. This is compounded by the confusing array of information sources and the concomitant uncertainties about which sources to trust [9,10]. At the same time, newsrooms have significantly reduced their reporting staff, leaving many news outlets without dedicated reporting on science and the environment [8].

Increasingly, consumers are turning to online and social media sources for their science news [11]. These sources are seldom subjected to editorial review, limiting the accuracy and comprehensibility of scientific information [12]. In a study by the Pew Research Center, the majority (64%) of Americans felt that science news provided a significant source of confusion [13]. At the same time that newsrooms have faced cutbacks, globalization has helped news to travel faster, creating a gap between the need for scientific information and the ability to provide it. Thus, researchers and natural resource managers are increasingly charged with filling this gap by conveying their results directly to public audiences [14,15].

Research has shown that one's knowledge of environmental impacts, such as those caused by overfishing or coastal erosion, can greatly influence one's level of concern [3]. Resource managers who engage with public audiences can help to clarify the issues by battling the psychological distance, unfamiliarity, and politicization associated with coastal concerns [3]. With effective public engagement, resource managers can build trust while bridging knowledge gaps and facilitating the informed discussions needed to enact change. This chapter will outline skills and techniques used by natural resource managers and science communication experts to effectively communicate scientific findings and engage public audiences. Examples of organizations that have successfully implemented public engagement programs will also be highlighted along with areas that still need improvement.

Methods for Natural Resource Managers to Engage Target Audiences

Natural resource managers may wish to engage audiences for many reasons. Three nonexclusive aims are to inform communities about environmental issues [16], to promote specific behaviors such as the conservation of natural resources [17], and to engage communities in co-creation of knowledge [18]. The first may include notifying constituents about harmful algal blooms that can toxify commercial shellfish stocks, the second may involve promoting a change in land use around a watershed, and the third may engage community members in development of research questions and data collection to evaluate the relationship between land use change and harmful algal bloom occurrence. Each of these examples requires the consideration of multiple stakeholders, including, but not limited to, resource users, local, regional, or national interest groups; affected communities; managers; and public officials [5]. In order to accomplish these goals and reach relevant audiences, natural resource managers should practice rhetorical strategies (see Ref. [19]) and provide opportunities for public audiences and journalists to encounter and discuss digestible scientific information [7]. A growing number of organizations are already actively engaged in these pursuits; examples include the University of Rhode Island's Metcalf Institute, which leads trainings for journalists and scientists on effective science communication, and the United States' National Park Service, which offers resources for designing public outreach materials [20]. Additional organizations and the methodologies that can be learned from each will be explored in the sections that follow.

Professional Development Opportunities

The field of science communication has gained traction in recent years as more people recognize the need to engage a variety of audiences in discussions about complex scientific and environmental topics [21]. Many programs have emerged to provide formal communication trainings with quantifiable results. For example, in Australia, the Science Circus offers science postgraduates the opportunity to take part in a year of field-intensive science communication trainings, increasing their retention in the communication field [22]. The Northwestern Hawaiian Islands Research Partnership (NWHI-RP) conducts a biannual symposium in collaboration with the Hawai'i Institute of Marine Biology and the Office of National Marine Sanctuaries. This meeting invites scientists and natural resource managers to engage in discussions about scientific findings and relate them to potential environmental management concerns [6]. In the United States, science communication trainings have become widely accessible, from ongoing programs offered by COMPASS and the Alan Alda Center for Communicating Science to conference-based programs such as ComSciCon and the Inclusive SciComm Symposium. For a partial list of professional development offerings in science communication, refer to the chapter appendix.

Communicating with Public Audiences

There are several techniques currently practiced in coastal and marine management that can help to strengthen the impact of environmental communication. To begin, experts highlight the importance of identifying communication objectives [5,23]. Many researchers and managers make the mistake of citing increased "awareness" as a primary goal, but the ultimate goal is more commonly to shift stakeholder behavior [23]—e.g., voting for a new marine protected area or implementing green infrastructure to reduce stormwater runoff. Once the communication objective has been clearly identified, the next step is to discern the target audience, which will determine the most effective means of building relationships, engaging in dialogue, or crafting specific messages. Historically, researchers have sought to transmit scientific knowledge through a one-way, facts-only approach referred to as the "deficit" model of communication, in which the communicator aims to fill an audience's knowledge "deficit" with information. This approach has been proven ineffective [24] because it assumes passive, identical, and rational interpretations of scientific information by all audiences [25–27]. Instead, communicators should strive for a dialogue model characterized by engagement rather than transmission, taking into account an audience's background, ideology, and cultural identity, which are more influential than facts in the acceptance and support of a message [4,10,14,28]. For example, the audience's background should inform the communicator's use of language [5] and their values should influence both content and focus of an argument [4,29]. For controversial or polarizing issues, being mindful of an audience's ideologies can also help to avoid a boomerang effect in which a message produces the opposite of the desired outcome (i.e., antilitter messages increasing the propensity to litter) [30]. Strategies to address these will vary based on the delivery mechanism (e.g., in-person dialogue, social media, podcasts), but some examples of successful implementations include incorporating a glossary of key terms at the conclusion of scientific reports, a technique used by the NWHI-RP [6], presenting multimedia information in lay terms, as was done by the California marine protected area planning committee [5], or facilitating collaborations with the target audience directly, as *Science-Links* has done with scientists and policy or natural resource managers [16].

With the audience in mind, the message can be crafted. One commonly used tool is the message box; a worksheet that assists users in identifying key aspects of their argument, such as benefits and implications [31]. Framing information in a particular context can also be used to influence the reach and interpretation of an argument [10,32]. For example, it is sometimes helpful to describe climate change impacts in the context of public health or economic effects if the audience responds negatively to climate change references [3]. Additional strategies include the use of culturally relevant metaphors [3] and a delivery style suited to the presenter's skill set [5]. The important goal to keep in mind is that audience members are more likely to develop positive attitudes towards science and scientists if the presenter is both engaging and perceived as trustworthy, someone who is willing to listen as well as contribute [23]. For more information on strategies for effective and inclusive science communication, see Canfield et al. [33], Pandya [34], Jarreau et al. [35], and Nadkarni et al. [36].

Increasing Engagement to Enact Change

In addition to a well-crafted message, stakeholder engagement is essential to achieve support for specific environmental policies. While many assume a linear model of communication, in which scientists carry out studies and policy makers develop solutions, this model is inaccurate [37]. Instead, inclusion of diverse groups and viewpoints in the decision-making process is necessary to garner support and enact change [38]. Research shows that engagement, exchange, and collaboration are all important to success: engagement, in which scientists and communities convene to share relevant scientific insights; exchange, in which a network of audiences can participate in the learning process; and collaboration, in which organizations can interact, share ideas, and provide valuable feedback to one another [16,34]. This type of forum requires the integration of interdisciplinary and transdisciplinary research and working groups that engage scientists, policy makers, and other stakeholders [37]. While it may be laborious to initiate these multiparty interactions, broad participation often improves understanding of complex issues and leads to participants feeling more confident about taking action [39].

One notable example used to increase participation in environmental management was the creation of an educational tool, CoastRanger MS. This program offers audiences a chance to virtually manage a coast in a gaming environment and learn about management decision-making in the process [40]. Although game construction may not always be feasible, institutions should allocate resources and incentives toward encouraging scientists and resource managers to engage with those outside their disciplines [37]. A more achievable example may be to fund scientific convenings or interdisciplinary research with the goal of educating participants about one another's knowledge, needs, and concerns [16]. In some cases, having a trusted "translator" [37] or "knowledge broker" [38] can also be useful for organizations. This may be someone with good interpersonal skills who has the ability to see and comprehend both sides of an argument [38]. It is important to keep in mind that disagreement may also reflect successful discourse [41].

Evaluating Engagement

Evaluation is an essential, yet frequently overlooked, aspect of effective science communication [42]. It ensures that interactions are producing the desired outcomes and offers an objective mechanism for improving future interactions. When engaging with a new audience, both pre- and post-surveys are recommended methods of evaluation [5]. Assessments should be structured to measure the degree to which an engagement effort achieved its objective; e.g., assessing how messages were received by an audience or tracking what outcomes resulted from the engagement effort [5]. The Great Barrier Reef (GBR) management strategy offers a valuable example of this. Zoning of the reef requires that public audiences offer their inputs prior to the commencement of zoning and again via comments on the drafted plan [43]. This strategy provides valuable feedback to the GBR legislation, ensures that stakeholders feel heard throughout the zoning process, and results in a constituency that is more likely to follow reef guidelines [43]. In the United Kingdom, natural resource managers have also utilized Q methodology, a type of factor analysis, to quantitatively assess the public's subjective viewpoints about coastal flood management issues [44]. This pre-assessment has helped managers discern the values and backgrounds of their constituents, which were then influential in coastal management decisions. Additional resources for implementing these evaluation techniques can be found in Peterman et al. [45] and Nisbet [32].

Conclusion and Future Directions

This chapter summarizes fundamental aspects of science communication and specific approaches that natural resource managers are using around the world to increase public engagement and message impact. By building trusting relationships with diverse audiences and carefully crafting their messages, managers can improve the quality of policy decisions and garner greater support from stakeholders.

The field of science communication is still young; more research is needed to assess the influence of various science communication strategies in natural resource management, especially in coastal and marine settings. Partnerships between natural and social scientists, managers, science communication practitioners, and stakeholders could offer the relevant expertise needed to plan, implement, and evaluate science communication activities. With continued research, more audiences can be engaged in important environmental decision-making and management.

APPENDIX Science Communication Professional Development Course and Workshop Offerings around the World

Country	Institution	Program	Participants	Duration	Focus
Australia	Australian National University	Many	All disciplines	4 days	Varies based on workshop
	Science and Technology	Superstars of STEM	All disciplines	1 day	Multimedia communication techniques
England	University of the West of England, Bristol	Science Communication: Master Class	All disciplines	4 days	Public perceptions and engagement
Scotland	Dundee Science Centre	Create and Inspire	All disciplines	3 months (part time)	Public engagement
United States	Alan Alda Center for Communicating Science	STEM Immersion Program	All disciplines	2 days	Improvement-based approaches to storytelling and communication
	American Association for the Advancement of Science	Leshner Leadership Institute	All disciplines	1 week	Public engagement training and plan development
		Mass Media Science and Engineering Fellowship	STEM students	10 weeks	Science writing and reporting for public audiences
	COMPASS	Many	All disciplines	Varies	Message development and delivery
	Metcalf Institute, University of Rhode Island	Many	All disciplines	Varies	Communicating with different audiences; message development and delivery; using specific communication platforms/tools
	Metcalf Institute, University of Rhode Island	Inclusive SciComm Symposium	All disciplines	3 days	Inclusive, equitable, intersectional approaches for science communication and public engagement
	Science Talk	Science Talk Conference	All disciplines	3 days	Best practices in science communication
	University of Michigan	RELATE	STEM Academics	6–10 weeks	Multimedia communication techniques
United States and Canada	ComSciCon	Flagship Workshop	STEM graduate students	3 days	Written and verbal communication skills; networking

References

1. U. Nations, "The Ocean Conference Factsheet: People and Oceans," 2017. [Online]. Available https://www.un.org/sustainabledevelopment/wp-content/uploads/2017/05/Ocean-fact-sheet-package.pdf.
2. H. Rittel and M. Webber, "Dilemmas in a General Theory of Planning," *Policy Sci.*, vol. 4, pp. 155–169, 1973.
3. J. P. Schuldt, K. A. McComas, and S. E. Byrne, "Communicating about ocean health: Theoretical and practical considerations," *Philos. Trans. R. Soc. B Biol. Sci.*, vol. 371, no. 1689s, 2016.
4. T. Dietz, "Bringing values and deliberation to science communication," *Proc. Natl. Acad. Sci.*, vol. 110, no. Supplement_3, pp. 14081–14087, 2013.
5. K. Grorud-Colvert, E. Neeley, S. Airame, S. E. Lester, and S. D. Gaines, "Communicating marine reserve science to diverse audiences," *Proc. Natl. Acad. Sci.*, vol. 107, no. 43, pp. 18306–18311, 2010.
6. C. S. Wiener et al., "Creating effective partnerships in ecosystem-based management: A culture of science and management," *J. Mar. Biol.*, pp. 1–8, 2011.

7. M. J. Novacek, "Engaging the public in biodiversity issues," *Proc. Natl. Acad. Sci.*, vol. 105, no. Supplement 1, pp. 11571–11578, 2008.
8. S. Menezes, "Science training for journalists: An essential tool in the post-specialist era of journalism," *Front. Commun.*, vol. 3, no. 4, pp. 1–5, 2018.
9. B. Takahashi and E. C. Tandoc, "Media sources, credibility, and perceptions of science: Learning about how people learn about science," *Public Underst. Sci.*, vol. 25, no. 6, pp. 674–690, 2016.
10. D. A. Scheufele, "Communicating science in social settings," *Proc. Natl. Acad. Sci.*, vol. 110, no. Supplement_3, pp. 14040–14047, 2013.
11. H. Smith, S. Menezes, and C. Gilbert, "Science training and environmental journalism today: Effects of science journalism training for midcareer professionals," *Appl. Environ. Educ. Commun.*, vol. 17, no. 2, pp. 161–173, 2018.
12. D. Brossard, "New media landscapes and the science information consumer," *Proc. Natl. Acad. Sci.*, vol. 110, no. Supplement_3, pp. 14096–14101, 2013.
13. Pew Research Center, "The State of the News Media 2013," 2013. Available https://assets.pewresearch.org/files/journalism/State-of-the-News-Media-Report-2013-FINAL.pdf.
14. A. I. Leshner, "Capably communicating science," *Science (80-.)*., vol. 337, no. 6096, p. 777, 2012.
15. S. E. Brownell, J. V. Price, and L. Steinman, "Science communication to the general public: Why we need to teach undergraduate and graduate students this skill as part of their formal scientific training," *J. Undergrad. Neurosci. Educ.*, vol. 12, no. 1, pp. E6–E10, 2013.
16. D. L. Osmond et al., "The role of interface organizations in science communication and understanding," *Front. Ecol. Environ.*, vol. 8, no. 6, pp. 306–313, 2010.
17. C. G. Druschke, "Watershed as common-place: Communicating for conservation at the watershed scale," *Environ. Commun.*, vol. 7, no. 1, pp. 80–96, 2013.
18. L. E. Van Kerkhoff and L. Lebel, "Coproductive capacities: Rethinking science-governance relations in a diverse world," *Ecol. Soc.*, vol. 20, no. 1, p. 14, 2015.
19. C. G. Druschke and B. McGreavy, "Why rhetoric matters for ecology," *Front. Ecol. Environ.*, vol. 14, no. 1, pp. 46–52, 2016.
20. J. L. Tuxill, N. J. Mitchell, and D. Clark, *Stronger Together: A Manual on the Principles and Practices of Civic Engagement*, Woodstock, VT: Conservation Study Institute, 2009.
21. Engineering National Academies of Sciences and Medicine, *Communicating Science Effectively: A Research Agenda*. Washington, DC: The National Academies Press, 2017.
22. M. McKinnon and C. Bryant, "Thirty years of a science communication course in Australia," *Sci. Commun.*, vol. 39, no. 2, pp. 169–194, 2017.
23. A. Dudo and J. C. Besley, "Scientists' prioritization of communication objectives for public engagement," *PLoS One*, vol. 11, no. 2, p. e0148867, 2016.
24. M. C. Nisbet and D. A. Scheufele, "What's next for science communication? Promising directions and lingering distractions," *Am. J. Bot.*, vol. 96, no. 10, pp. 1767–1778, 2009.
25. P. Sturgis and N. Allum, "Science in society: Re-evaluating the deficit model of public attitudes," *Public Underst. Sci.*, vol. 13, no. 1, pp. 55–74, 2004.
26. A. G. Gross, "The roles of rhetoric in the public understanding of science," *Public Underst. Sci.*, vol. 3, no. 1, pp. 3–23, 1994.
27. M. J. Simis, H. Madden, M. A. Cacciatore, and S. K. Yeo, "The lure of rationality: Why does the deficit model persist in science communication?" *Public Underst. Sci.*, vol. 25, no. 4, pp. 400–414, 2016.
28. D. L. Medin and M. Bang, "The cultural side of science communication," *Proc. Natl. Acad. Sci.*, vol. 111, no. Supplement 4, pp. 13621–13626, 2014.
29. S. T. Fiske and C. Dupree, "Gaining trust as well as respect in communicating to motivated audiences about science topics," *Proc. Natl. Acad. Sci.*, vol. 111, no. Supplement_4, pp. 13593–13597, 2014.
30. P. S. Hart and E. C. Nisbet, "Boomerang effects in science communication: How motivated reasoning and identity cues amplify opinion polarization about climate mitigation policies," *Commun. Res.*, vol. 39, no. 6, pp. 701–723, 2012.

31. COMPASS Science Communication, *The Message Box Workbook*. 2017.
32. M. Nisbet, *Scientists in Civic Life : Facilitating Dialogue-Based Communication*, Washington, DC: American Association for the Advancement of Science, 2018.
33. Canfield, K. et al., "Science communication demands a critical approach that centers inclusion, equity and intersectionality," *Frontiers in Communication*, (In Press), 2020.
34. R. E. Pandya, "A framework for engaging diverse communities in citizen science in the US," *Frontiers in Ecology and the Environment*, vol. 10, no. 6, pp. 314–317, 2012.
35. P. B. Jarreau, Z. Altinay, and A. Reynolds, "Best practices in environmental communication: A case study of Louisiana's coastal crisis," *Environmental Communication*, vol. 11, no. 2, pp. 143–165, Mar. 2017.
36. N. M. Nadkarni, C. Q. Weber, S. V. Goldman, D. L. Schatz, S. Allen, and R. Menlove. "Beyond the deficit mode: the ambassador approach to public engagement," *BioScience*, vol. 69, no. 4, pp. 305–313, Apr. 2019.
37. J. C. Young et al., "Improving the science-policy dialogue to meet the challenges of biodiversity conservation: Having conversations rather than talking at one-another," *Biodivers. Conserv.*, vol. 23, no. 2, pp. 387–404, 2014.
38. S. Michaels, J. Holmes, and L. Shaxson, "Science Communication and the Tension between Evidence-Based and Inclusive Features of Policy Making," in *New Trends in Earth-Science Outreach and Engagement*, Cham, Switzerland: Springer International Publishing, vol. 38, pp. 83–92, 2014.
39. P. M. Groffman et al., "Restarting the conversation: Challenges at the interface between ecology and society," *Front. Ecol. Environ.*, vol. 8, no. 6, pp. 284–291, 2010.
40. N. I. Pontee and K. Morris, "CoastRanger MS: A tool for improving public engagement," *J. Coast. Res.*, vol. 27, no. 1, pp. 18–25, 2011.
41. W. De Nooy, "Communication in natural resource management: Agreement between and disagreement within stakeholder groups," *Ecol. Soc.*, vol. 18, no. 2, p. 44, 2013.
42. J. Varner, "Scientific outreach: Toward effective public engagement with biological science," *Bioscience*, vol. 64, no. 4, pp. 333–340, 2014.
43. J. C. Day, "Effective public participation is fundamental for marine conservation—Lessons from a large-scale MPA," *Coast. Manag.*, vol. 45, no. 6, pp. 470–486, 2017.
44. J. Smith and A. Bond, "Delivering more inclusive public participation in coastal flood management: A case study in Suffolk, UK," *Ocean Coast. Manag.*, vol. 161, pp. 147–155, 2018.
45. K. Peterman, J. Robertson Evia, E. Cloyd, and J. C. Besley, "Assessing public engagement outcomes by the use of an outcome expectations scale for scientists," *Sci. Commun.*, vol. 39, no. 6, pp. 782–797, Nov. 2017.

14

Quantifying Reef Ecosystem Services

Joshua Drew and
Kate Henderson
State University of New York College of Environmental Science and Forestry

Emma McKinley
*Cardiff University
Cardiff Wales*

Introduction

Coral reefs contain both structural and biological features that contribute to a mosaic ecosystem, which in turn provides a wide range of ecosystem services and benefits, contributing to their natural capital value (Ruckelshaus et al., 2013; MA, 2005; Moberg and Folke, 1999). These services are many and range from having a role in coastal protection to supporting fish nursery habitat to providing space for coastal tourism and recreation, accruing to a variety of stakeholders, including communities both proximal to those reefs and, perhaps less obviously, to individuals and communities living hundreds of kilometers away.

As global efforts to deliver sustainable marine and coastal management continue, a number of concepts and models (including ecosystem services and natural capital) have been proposed and adopted as a lens to better understand the complex relationship between people and the sea. Increasingly, there is a growing acceptance that in order for management to be effective and well implemented, understanding the diverse values (including both monetary and nonmonetary) of marine and coastal systems (e.g., reefs) and the benefits they provide (including ecological, social, economic, and cultural) is crucial. While still a relatively young concept, ecosystem services are increasingly accepted by both researchers and practitioners as a way of understanding the intricacies and interrelationships between ecological systems and human well-being (McKinley et al., 2019; Costanza et al., 2014; de Groot et al., 2012; MA, 2005). Assigning value can be a useful tool to support development of ocean literacy, raise public awareness, and inform policy development and, increasingly, used to support the design of sustainable finance schemes for marine conservation and management. While there are those who challenge the use of monetary value as a method of quantification (Kallis et al., 2013; Dempsey and Robertson, 2012), it remains the most common approach to valuation of ecosystem services. In the context of coral reefs, numerous authors have set out evaluations and assessments of the value of reef ecosystems, the benefits they provide, as well as the monetary cost of damage and/or deterioration experienced by these environments (see e.g., Moberg and Folke, 1999; Brander et al., 2007).

This chapter presents an overview of the existing evidence base relating to the assessment of values attributed to the spectrum of ecosystem services and benefits provided by reef environments, providing an insight into the methods used to quantify these values, the role of this in management of reef systems, and finally, the challenges associated with quantifying the sometimes unquantifiable services provided by reefs globally.

Valuing Reef Ecosystems

Reef systems provide society with a diverse spectrum of goods and services, which in turn they derive, ecological, economic, social, and cultural benefits from. With this variety in mind, rather than covering every ecosystem service and benefit that could be related to reef environments, this chapter focuses on existing work relating to a subset of the more commonly researched (and perhaps those that are easier to assign quantitative values to) ecosystem services—including coastal protection, provision of fish nursery habitat, resource extraction, and tourism. The challenge of quantifying services and benefits that fall more commonly in the nonmarketable or intangible category of benefits is also addressed. While we recognize that this is by no means a comprehensive list, and does not account for the less tangible services, this chapter provides managers, policy decision makers, and the interested public with an overview of the complex nature of coral reef ecosystems and the services and benefits they provide.

Ecosystem Services Provided by Structural Components of Reefs

To understand the processes through which reefs provide these ecosystem services, it is first necessary to tease apart what we mean by a "coral reef" and to examine those constituent parts through the lens of ecosystem service provisioning. Reefs contain a stereotypical structure (Figure 14.1) wherein one moves from deep reefs through to reef crests and lagoons and finally sheltered backreef habitats that may include iconic seagrass and mangrove habitats. Each of these zones contributes to the wider system and provides an ecological function maintaining the overall health of the physical structure of the reef as well as the existence of the denizens living within it.

The most oceanic of structures, deep reefs and reef walls, provide stabilizing habitat, helping to break up the force of large oceanic waves. Moreover, deep reefs contain an incredible diversity of life including fish and other macroorganisms (Friedlander et al., 2019; Stefanoudis et al., 2019) where they are somewhat sheltered from shallow-water thermal anomalies suggesting that they are important refugia for

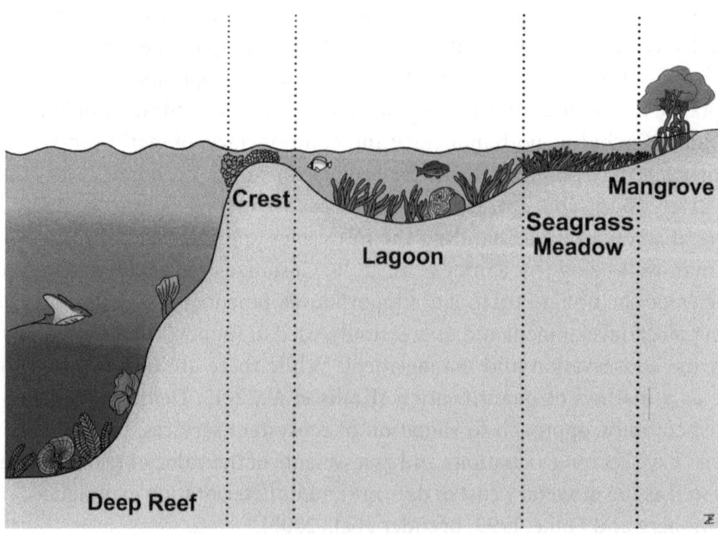

FIGURE 14.1 The major structural features of a reef ecosystem as described in the text. (Artwork by Ilana Zeitzer.)

some heat-intolerant species of coral (Muir et al., 2017). While the coral communities may be similar, there are distinct fish communities that "turn over" at depth often resulting from physical aspects of the reef like reduced light and changes in coral morphology. These different environments often result in distinct shallow water and deep water assemblages of fish, often occurring at a distinct break point below approximately 50 m (Brokovich et al., 2008; Rocha et al., 2018; Semmler et al., 2017). Unfortunately, the differences in fish communities and other parameters between deep and shallow reefs limit the potential for "re-seeding" of shallow reefs from deeper areas. In fact, Rocha et al. (2018) suggest that deep reefs may be as critically threatened as their shallower cousins. In part, this is because deep reefs are literal sinks for anthropogenic waste (Woodall et al., 2015), which is as disruptive toward ecological processes on deep reefs as it is on shallow ones (de Carvalho-Souza et al., 2018).

Reef crests are areas where the reef structure is emergent at times and subsequently faces the brunt of oceanic waves and the associated hydrodynamic stress those waves bring. Many corals living at the crest have evolved morphotypes to deal with this wave stress (Hopley et al., 2007) and, in doing so, have created an environment wherein biological and structural features combine to lower wave energy. In reducing wave energy, reefs provide a highly valuable ecosystem service through supporting coastal protection. Evidence of this is presented by researchers in both Maui (HI, USA) and Grenada who modeled the impact of a loss of reef with hindcasted wave functions. The results suggested that communities in both countries were more at risk to flooding, and, at least in Grenada, loss of coral reefs was a larger driver than sea level rise in explaining historical flooding (Storlazzi et al., 2017; Reguero et al., 2018).

Similarly, modeling suggests that reef structural complexity, not the amount of sea level rise, is a better predictor of coastal habitat protection (Harris et al., 2018). Reef structural complexity is also a strong predictor for reef fish diversity (Richardson 2017; Darling et al., 2017; Ferrari et al., 2016), which is important in sustaining coastal fisheries, as well as the ecosystem services provided by those reef fish. As structural complexity decreases with human disturbance, especially bleaching (Magel et al., 2019), this seems to be an important area to target future conservation interventions.

The wave reducing structures of reefs' fore reef and reef crest zones reduce the need for artificially constructed breakwaters, and coral reef restoration projects are often less expensive than building breakwaters in the tropics (Ferrario et al., 2014). Beck et al. (2018) estimate that coral reefs reduce storm damages by more than US $4 billion each year worldwide; this flood protection was the most notable in the Philippines, Malaysia, Cuba, and the Dominican Republic in terms of both total damage averted and damage relative to the size of the countries' economy. Coral reefs provide risk reduction from storms and flooding to nearly 200 million people worldwide who live within 50 km of a reef (Ferrario et al., 2014). Reefs also reduce coastal erosion by sheltering beaches from waves (Frihy et al., 2004), and their loss can exacerbate erosion (Baldock et al., 2015). For example, in the Seychelles, where warm waters led to a coral bleaching event and die-off in 1998, models predicted increasing wave energy and beach erosion (Sheppard et al., 2005), while in Grenada, the beaches inland from the degraded northern reefs were more eroded than the beaches inland from the more intact southern ones (Reguero et al., 2018).

Behind the wave-dissipating reef crest lies lagoons—shallow areas that tend to be calm and sheltered. In addition to being nurseries for commercially important fisheries (Hamilton et al., 2017; Matich et al., 2017; Lefcheck et al., 2019), these calm waters allow for the development of two ecologically important habitats, seagrass beds, and mangroves. Mangroves are taxonomically diverse coastal forests, which slow wave functions through the presence of prop and buttress roots that act as a flexible filter through which wave energy is dispersed. Similarly, the blades of seagrasses dissipate wave energy as the water passes through them. In both cases, slower water is unable to keep particulate matter suspended and that organic material precipitates out of solution in the form of soil accrual; in doing so, both mangroves and seagrass beds create fertile soil for further propagation of these ecosystems.

Reef lagoons, mangroves, and seagrass beds combine synergistically to provide a number of coastal protection services, Guannel et al. (2016) found that the combined impact of the three systems was greater not only than the sum of the individual ecosystems. However, it should be noted that the impacts of these coupled bio/structural systems depend on the exact location of the organisms. They found that *live corals in the fringing reef profile supply more protection services than seagrasses; seagrasses in the barrier reef profile supply more protection services than live corals; and seagrasses can even compensate for the long-term degradation of the barrier reef.*

In addition, mangroves and seagrasses also serve as vital nursery areas providing fisheries (Liquete et al., 2016; Jackson et al, 2015), and their loss can result in direct and indirect monetary losses (Estoque et al., 2018). It is important to note that in many least developed countries, the fisheries' benefits of mangroves can be social as well as economic, with women often experiencing direct benefits of mangrove fisheries, suggesting that beyond raw dollars, the economic distribution of wealth within communities and the influence of this on community dynamics and gender roles should also be considered as an economic benefit of these ecosystem services (Lau and Scales, 2016; Carney, 2017; Pearson et al., 2019).

Tourism-Based Ecosystem Services

Reefs and their associated ecosystems are an important tourism destination, with tourists demonstrating a strong preference for, and a commensurate desire to pay more to visit, healthy reefs (Grafeld et al., 2016; Gill et al., 2015; Peng and Oleson, 2017). While the methodologies for calculating ecotourism differ, a recent meta-analysis proposed a global value for reef and reef adjacent ecotourism of $USD 36 billion/year (Spalding et al., 2017). In addition, reef-associated habitats also garner significant economic value. In the Galapagos, mangroves were found to contribute $USD 62 million/ha/year (Tanner et al., 2019). Often considered to be one of the dominant examples of nature-based tourism, the Great Barrier Reef was estimated to bring in up to $USD 1.6 billion/year in 2003 (Carr and Mendelsohn, 2003); recent reports of its deterioration seem to have actually spawned an increase in a new class of tourism. This so-called "last chance tourism" which centers around ecosystems and species threatened by climate change may also be a motivating factor to spur conservation (Curnock et al., 2019; Piggot-McKellar and McNamara, 2017).

In addition to the reefs themselves, the associated biodiversity, including charismatic megafauna, also provide an additional tourism-based ecosystem service. These ecotourism opportunities can range in their foci from general reef related observation of animals in their natural habitat to immersive experiences with marine megafauna being attracted through food provisioning (Murphy et al., 2018; Macdonald et al., 2017). A special case of the latter, elasmobranch tourism (sharks and rays) is a growing market with multiple bioeconomic studies suggesting that sharks in these areas may be worth much more alive than dead. For example, Manta ray watching generated US $8.1 million annually in the Maldives from 2006 to 2008, shark-diving tourism generated US $18 million/year in Palau in 2011, and whale shark tourists spent more than US $4 million in Australia's Ningaloo Coast region in 2006 (Anderson et al., 2011; Vianna et al., 2012; Jones et al., 2009; Zimmerhackel et al., 2019).

Quantifying the Unquantifiable: A Challenge for Reef Management

Reefs and their associated environments also provide less quantifiable, but no less valuable, ecosystem services to the people who live within, and have connections with, those environments. These are often considered so-called "cultural services" and tend to represent the value of the place that is manifested through a variety of mechanisms including spiritual connection, cultural heritage, sense of place and identity, religious, and familial history (Kittinger et al., 2016; MA, 2005). In a recent review, Rodrigues et al. (2017) highlighted that the most frequently mentioned cultural ecosystem services in marine systems were recreation, aesthetics, cultural heritage, and religious significance. Of these, recreation is generally quite readily quantifiable, with numerous authors providing assessments of this in different locations (e.g., Brander et al., 2007). Other aspects of cultural services, such as sense of place, identity, perceptions, and cultural values, are often much harder to elucidate and measure. Despite this challenge, in recent years, there has been a growing recognition of the need to be cognizant of these nonmonetary values and the need to understand these intangible aspects of marine and coastal systems to support conservation and management (Chan et al., 2012a, 2012b).

Understanding these less quantifiable values has a significant role in coastal management more broadly as the restoration of degraded ecosystems can improve cultural ecosystem services as well as more easily quantified services. For example, in Hawaii, the removal of 1.32 million kg of invasive algae led to new opportunities to revitalize intergenerational Hawaiian knowledge transfer and to restore a traditional Hawaiian temple (Kittinger et al., 2016) practices. In Milne Bay, Papua New Guinea facilitated conversations with local communities leading to reimagining traditional patterns of trade (the Kula ring) as avenues for sharing information and best practices as one way to build more resilient coastal communities (Bohensky et al., 2011). Finally, Oles (2007) elegantly speaks to the ways that an atoll provides a sense of place for both residents and absentee members. Writing of the construction of a landing strip on Mwoakilloa atoll (Federated States of Micronesia):

> During the initial construction of the runway, residents were paid by the weight of the coral that they harvested. A number of coral heads were destroyed, despite their cultural, historical, and biological significance (T. Nagaoka, personal communication). Virtually every reef feature on Mwoakilloa has a name and associated story that refer to historical occurrences associated with the place, legendary events featuring the exploits of mythological ancestors and creatures, or biophysical aspects of the area. The symbolic sacrifice of a subsistence lifestyle to market trade is poignantly captured in the very real act of building, for wages, a portal to the western world on the bones of living coral.

Conclusions

Coral reefs and their associated ecosystems are the result of a combination of biotic and abiotic factors resulting in a rich mosaic of ecosystems. Each component within this mosaic ecosystem provides numerous ecosystem services ranging from dissipating the impact of storms, to mitigating the rise in sea level, to providing blue space and access to nature, which can be beneficial to health and well-being. Moreover, the rich life inhabiting this mosaic ecosystem sustains fisheries, livelihoods, and tourism opportunities—diversifying and supporting the economies of the communities built on and around them. Finally, coral reefs provide a sense of place, an aesthetic that resonates with people, including both local communities and visitors. While harder to monetize, these cultural ecosystem services may help dissipate the impact of an increasingly stressful and urbanized world.

Our review highlights some of the mechanisms through which corals generate ecosystem services; however, as pressures on the natural environment continue to increase, coral reefs remain one of the most threatened ecosystems in the ocean. A combination of a warming world with its associated changes in sea levels and ocean acidification, with an increasing demand for marine resources, present numerous threats to these ecosystems. The cost of degradation has been widely documented (see e.g., Burke et al., 2004 and Becken and Hay, 2007). Should reefs continue to become degraded, we will potentially lose the species that help make them recognizable and a phase shift toward an algal dominated system, with an associated loss of ecosystem services (Ainsworth and Mumby, 2015); however, by clearly quantifying the numerous ecosystem services healthy reefs can provide, it is our hope that we can help decision makers to envision a world with vibrant reefs and the rich communities, both ecological and social, that they nurture.

References

Ainsworth, Cameron H., and Peter J. Mumby. "Coral–algal phase shifts alter fish communities and reduce fisheries production." *Global Change Biology* 21.1 (2015): 165–172.

Anderson, R. Charles, et al. "Extent and economic value of manta ray watching in Maldives." *Tourism in Marine Environments* 7.1 (2011): 15–27.

Baldock, T. E., et al. "Impact of sea-level rise on cross-shore sediment transport on fetch-limited barrier reef island beaches under modal and cyclonic conditions." *Marine Pollution Bulletin* 97.1–2 (2015): 188–198.

Beck, Michael W., et al. "The global flood protection savings provided by coral reefs." *Nature Communications* 9.1 (2018): 2186.

Becken, Susanne, and John E. Hay. *Tourism and Climate Change: Risks and Opportunities*. Vol. 1. Multilingual Matters: Clevedon Buffalo, NY and Toronto: Channel View Publications, 2007.

Bohensky, Erin L., James R. A. Butler, and David Mitchell. "Scenarios for knowledge integration: exploring ecotourism futures in Milne Bay, Papua New Guinea." *Journal of Marine Biology* 2011(2011): 504651.

Brander, Luke M., Pieter Van Beukering, and Herman S. J. Cesar. "The recreational value of coral reefs: a meta-analysis." *Ecological Economics* 63.1 (2007): 209–218.

Brokovich, Eran, et al. "Descending to the twilight-zone: changes in coral reef fish assemblages along a depth gradient down to 65 m." *Marine Ecology Progress Series* 371(2008): 253–262

Burke, Lauretta Marie, et al. *Reefs at Risk in the Caribbean*: Smithsonian Libraries, Washington, D.C, 2004.

Carney, Judith A. "Shellfish collection in Senegambian mangroves: a female knowledge system in a priority conservation region." *Journal of Ethnobiology* 37.3 (2017): 440–458.

Carr, Liam, and Robert Mendelsohn. "Valuing coral reefs: a travel cost analysis of the Great Barrier Reef." *AMBIO: A Journal of the Human Environment* 32.5 (2003): 353–358.

Chan, Kai M. A., et al. "Where are cultural and social in ecosystem services? A framework for constructive engagement." *BioScience* 62.8 (2012a): 744–756.

Chan, Kai M. A., Terre Satterfield, and Joshua Goldstein. "Rethinking ecosystem services to better address and navigate cultural values." *Ecological Economics* 74(2012b): 8–18.

Costanza, Robert, et al. "Changes in the global value of ecosystem services." *Global Environmental Change* 26(2014): 152–158.

Curnock, Matthew I., et al. "Shifts in tourists' sentiments and climate risk perceptions following mass coral bleaching of the Great Barrier Reef." *Nature Climate Change* 9.7 (2019): 535.

Darling, Emily S., et al. "Relationships between structural complexity, coral traits, and reef fish assemblages." *Coral Reefs* 36.2 (2017): 561–575.

de Carvalho-Souza, Gustavo F., et al. "Marine litter disrupts ecological processes in reef systems." *Marine Pollution Bulletin* 133(2018): 464–471.

De Groot, Rudolf, et al. "Global estimates of the value of ecosystems and their services in monetary units." *Ecosystem Services* 1.1 (2012): 50–61.

Dempsey, Jessica, and Morgan M. Robertson. "Ecosystem services: tensions, impurities, and points of engagement within neoliberalism." *Progress in Human Geography* 36.6 (2012): 758–779.

Estoque, Ronald C., et al. "Assessing environmental impacts and change in 'Myanmar's mangrove ecosystem service value due to deforestation (2000–2014)." *Global Change Biology* 24.11 (2018): 5391–5410.

Ferrari, Renata, et al. "Quantifying the response of structural complexity and community composition to environmental change in marine communities." *Global Change Biology* 22.5 (2016): 1965–1975.

Ferrario, Filippo, et al. "The effectiveness of coral reefs for coastal hazard risk reduction and adaptation." *Nature Communications* 5(2014): 3794.

Friedlander, Alan M., et al. "Marine biodiversity from zero to a thousand meters at Clipperton Atoll (Île de La Passion), Tropical Eastern Pacific." *PeerJ* 7 (2019): e7279.

Frihy, Omran E., et al. "The role of fringing coral reef in beach protection of Hurghada, Gulf of Suez, Red Sea of Egypt." *Ecological Engineering* 22.1 (2004): 17–25.

Grafeld, Shanna, et al. "Divers' willingness to pay for improved coral reef conditions in Guam: an untapped source of funding for management and conservation?" *Ecological Economics* 128(2016): 202–213.

Guannel, Greg, et al. "The power of three: coral reefs, seagrasses and mangroves protect coastal regions and increase their resilience." *PloS one* 11.7 (2016): e0158094.

Gill, David A., Peter W. Schuhmann, and Hazel A. Oxenford. "Recreational diver preferences for reef fish attributes: economic implications of future change." *Ecological Economics* 111(2015): 48–57.

Hamilton, Richard J., et al. "Logging degrades nursery habitat for an iconic coral reef fish." *Biological Conservation* 210(2017): 273–280.

Harris, Daniel L., et al. "Coral reef structural complexity provides important coastal protection from waves under rising sea levels." *Science Advances* 4.2 (2018): eaao4350.

Hopley, David, Scott G. Smithers, and Kevin Parnell. *The Geomorphology of the Great Barrier Reef: Development, Diversity and Change*. Cambridge University Press, 2007.

Jackson, Emma L., et al. "Use of a seagrass residency index to apportion commercial fishery landing values and recreation fisheries expenditure to seagrass habitat service." *Conservation Biology* 29.3 (2015): 899–909.

Jones, Tod, et al. "Expenditure and ecotourism: predictors of expenditure for whale shark tour participants." *Journal of Ecotourism* 8.1 (2009): 32–50.

Kallis, Giorgos, Erik Gómez-Baggethun, and Christos Zografos. "To value or not to value? That is not the question." *Ecological Economics* 94(2013): 97–105.

Kittinger, John N., et al. "Restoring ecosystems, restoring community: socioeconomic and cultural dimensions of a community-based coral reef restoration project." *Regional Environmental Change* 16.2 (2016): 301–313.

Lau, Jacqueline D., and Ivan R. Scales. "Identity, subjectivity and natural resource use: how ethnicity, gender and class intersect to influence mangrove oyster harvesting in The Gambia." *Geoforum* 69 (2016): 136–146.

Lefcheck, Jonathan S., et al. "Are coastal habitats important nurseries? A meta-analysis." *Conservation Letters* (2019): 12(4).

Liquete, Camino, et al. "Perspectives on the link between ecosystem services and biodiversity: the assessment of the nursery function." *Ecological Indicators* 63(2016): 249–257.

Macdonald, Catherine, et al. "Conservation potential of apex predator tourism." *Biological Conservation* 215(2017): 132–141.

Magel, Jennifer M. T., et al. "Effects of bleaching-associated mass coral mortality on reef structural complexity across a gradient of local disturbance." *Scientific Reports* 9.1 (2019): 2512.

Matich, Philip, et al. "Species co-occurrence affects the trophic interactions of two juvenile reef shark species in tropical lagoon nurseries in Moorea (French Polynesia)." *Marine Environmental Research* 127(2017): 84–91.

McKinley, Emma, et al. "Ecosystem services: a bridge or barrier for UK marine stakeholders?" *Ecosystem Services* 37(2019): 100922.

Millennium Ecosystem Assessment. *Ecosystems and Human Well-Being: Synthesis*. Island Press, Washington, DC, 2005.

Moberg, Fredrik, and Carl Folke. "Ecological goods and services of coral reef ecosystems." *Ecological Economics* 29.2 (1999): 215–233.

Muir, Paul R., et al. "Species identity and depth predict bleaching severity in reef-building corals: shall the deep inherit the reef?" *Proceedings of the Royal Society B: Biological Sciences* 284.1864 (2017): 20171551.

Murphy, Shannon E., Ian Campbell, and Joshua A. Drew. "Examination of tourists' willingness to pay under different conservation scenarios; Evidence from reef manta ray snorkeling in Fiji." *PloS One* 13.8 (2018): e0198279.

Oles, B. "Transformations in the sociocultural values and meanings of reefs and resources on Mwoakilloa." *Coral Reefs* 26.4 (2007): 971–981.

Pearson, Jasmine, Karen E. McNamara, and Patrick D. Nunn. "Gender-specific perspectives of mangrove ecosystem services: case study from Bua Province, Fiji Islands." *Ecosystem Services* 38 (2019): 100970.

Peng, Marcus, and Kirsten LL Oleson. "Beach recreationalists' willingness to pay and economic implications of coastal water quality problems in Hawaii." *Ecological Economics* 136(2017): 41–52.

Piggott-McKellar, Annah E., and Karen E. McNamara. "Last chance tourism and the Great Barrier Reef." *Journal of Sustainable Tourism* 25.3 (2017): 397–415.

Reguero, Borja G., et al. "Coral reefs for coastal protection: a new methodological approach and engineering case study in Grenada." *Journal of Environmental Management* 210(2018): 146–161.

Richardson, Laura E., et al. "Structural complexity mediates functional structure of reef fish assemblages among coral habitats." *Environmental Biology of Fishes* 100.3 (2017): 193–207.

Rocha, Luiz A., et al. "Mesophotic coral ecosystems are threatened and ecologically distinct from shallow water reefs." *Science* 361.6399 (2018): 281–284.

Rodrigues, João Garcia, et al. "Marine and coastal cultural ecosystem services: knowledge gaps and research priorities." *One Ecosystem* 2: e12290. doi: 10.3897/oneeco.2.e12290 (2017).

Ruckelshaus, M., et al. "Securing ocean benefits for society in the face of climate change." *Marine Policy* 40(2013): 154–159.

Semmler, Robert F., Whitney C. Hoot, and Marjorie L. Reaka. "Are mesophotic coral ecosystems distinct communities and can they serve as refugia for shallow reefs?" *Coral Reefs* 36.2 (2017): 433–444.

Sheppard, Charles, et al. "Coral mortality increases wave energy reaching shores protected by reef flats: examples from the Seychelles." *Estuarine, Coastal and Shelf Science* 64.2–3 (2005): 223–234.

Spalding, Mark, et al. "Mapping the global value and distribution of coral reef tourism." *Marine Policy* 82(2017): 104–113.

Stefanoudis, Paris V., et al. "Depth-dependent structuring of reef fish assemblages from the shallows to the rariphotic zone." *Frontiers in Marine Science* 6:307. doi: 10.3389/fmars.2019.00307 (2019).

Storlazzi, Curt, et al. "Rigorously valuing the role of coral reefs in coastal protection: an example from Maui, Hawaii, USA." *Coastal Dynamics 2017* (2017): 665–674.

Tanner, Michael K., et al. "Mangroves in the Galapagos: ecosystem services and their valuation." *Ecological Economics* 160(2019): 12–24.

Vianna, Gabriel M. S., et al. "Socio-economic value and community benefits from shark-diving tourism in Palau: a sustainable use of reef shark populations." *Biological Conservation* 145.1 (2012): 267–277.

Woodall, Lucy C., et al. "Deep-sea litter: a comparison of seamounts, banks and a ridge in the Atlantic and Indian Oceans reveals both environmental and anthropogenic factors impact accumulation and composition." *Frontiers in Marine Science* 2 (2015): 3.

Zimmerhackel, Johanna S., et al. "Evidence of increased economic benefits from shark-diving tourism in the Maldives." *Marine Policy* 100(2019): 21–26.

II

Marine Environment

15

Archaeological Oceanography

Michael L. Brennan and Robert D. Ballard
University of Rhode Island

Introduction

Modern deep submergence technology has advanced in recent decades, increasing our ability to find archaeological sites in the deep ocean.[1,2] As the depths of exploration have expanded below those accessible by human divers, so too has our reliance on oceanographic methods and data for the investigation and interpretation of these sites. Archaeological oceanography is a multidisciplinary field that combines archaeology, oceanography, and ocean engineering to develop ways to locate and document archaeological sites in deep water. The methods with which we approach, investigate, and conserve these sites require an understanding of the environments in which these wrecks lie, and how these alien materials have impacted and have been impacted by the marine environment. Over the past few decades, archaeology has been increasingly reliant on the application of scientific techniques for the analysis of cultural materials. As the capabilities of archaeological science continue to expand, so too does our responsibility to survey, accurately document, and conserve these sites. The objective of archaeological oceanography as a multidisciplinary field is the location, documentation, characterization, and finally in situ preservation of underwater cultural heritage. These sites are then used as platforms from which long-term monitoring can be conducted to better understand the surrounding environment and how each site has come into equilibrium with it as part of the modern landscape.[3]

Why the title "Archaeological Oceanography"? George Bass, one of the founders of underwater archaeology, wrote that it is easier to teach an archaeologist to work underwater than it is to teach archaeology to an accomplished diver.[4] Similarly, Ballard writes, "An archaeological oceanographer is an archaeologist working in the ocean."[1] However, the compounding factor for conducting archaeology underwater is depth. The development of maritime archaeology included the adaptation of terrestrial archaeological methods to underwater sites: grids were set up over a wreck, the site was drawn, and then excavated by divers in its entirety?[4] This was when technological access was limited to shallow coastal zones. The discovery of ancient shipwrecks in deep water has typically indicated a better degree of preservation than shallow water–wreck sites due to their deposition in low-energy environments, where they are removed from dynamic coastal processes and human activity, and where sedimentation rates are low.[1] However, the cost, time, and technology required to work at sites at these depths vastly limit any attempt to fully excavate and recover a shipwreck site. Therefore, in the case of deepwater sites, the goal must shift from full-scale excavation to high-resolution imaging, environmental characterization, and in situ preservation. The inclusion of oceanography and ocean engineering then becomes essential, not only for the technology and methodology required to access deepwater sites, but also to develop and refine ways to approach and document them.

Location

The exploration of our deep oceans was greatly furthered by President Clinton's Panel on Ocean Exploration, which recommended that the United States develop a national program in ocean exploration that includes both natural and social sciences.[5] The first objective listed in the panel's report is "Mapping the physical, geological, biological, chemical, and archaeological aspects of our ocean, such that the U.S. knowledge base is capable of supporting the large demand for this information from policy makers, regulators, commercial ventures, researchers, and educators."[5] Archaeological oceanography parallels this exploration initiative by aiming to document the locations of previously unknown archaeological sites in deep water.

Archaeological survey techniques in deep water require both acoustic and visual imaging including side-scan sonar, sub-bottom profiling, and video and still-camera photography from remote or autonomous vehicle platforms.[2,6] Locating archaeological sites involves not only historical parameters such as trade routes, but also oceanographic parameters such as currents, sedimentation rates, seafloor morphology, and modern trawling patterns.[7] Systematic surveys are important for documenting the submarine landscapes surrounding archaeological sites to interpret processes that have affected the formation of the modern site. These processes can include the ship's impact on the seabed, such as the landslide created by the German battleship, *Bismarck*;[7,8] chemical and physical impacts of shipwrecks on ocean ecology;[9,10] damage to shipwreck sites by bottom-trawling activities;[11,12] and environmental parameters affecting a site's preservation on the seabed.[13] Surveys should also include areas where archaeological sites are not expected to be found, as well as where they are; an absence of sites helps support interpretation of the archaeology of a region.[6]

Documentation

Excavation of shipwrecks in deep water is difficult because of the cost and limitations of accessing the sites. Therefore, the focus of archaeological oceanography shifts to high-resolution imaging and documentation of the wrecks. Imaging in deep water faces unique challenges that include underwater navigation, light attenuation, and the physics of mapping three-dimensional, non-planar sites.[14] Such challenges, along with the need for precise mapping for archaeological interpretations of sites, have helped drive the development of deepwater imaging systems.[2] Current remotely operated vehicle (ROV)-mounted mapping systems are able to produce subcentimeter-level precision bathymetric maps of archaeological sites over hundreds of square meters, which meet the standards of maps drawn by divers at shallow sites. A combination of high-frequency multibeam sonar, stereo cameras, and structured light laser profile imaging have created some of the highest-resolution images of shipwreck sites to date (Figure 15.1).[15,16]

Photomosaic and bathymetric maps of shipwrecks are important products for the investigation of deepwater sites because they can provide high-resolution images of a site's plan, orientation, and artifact provenience that excavations at these depths cannot (Figures 15.2 and 15.3). These surveys with a single ROV take mere hours, and are therefore practical, given the challenges of working in deep water. Such imagery of shipwrecks can also be used for oceanographic studies, including the analysis of the changing conditions of wreck sites to document damage to the seabed by bottom trawling.[16] These high resolution imaging systems have broad applications beyond archaeological sites, such as the detailed

FIGURE 15.1 High-definition video capture of remotely operated vehicle (ROV) *Hercules* surveying an ancient shipwreck with structured light laser profile-imaging system.
Source: Copyright the Institute for Exploration (IFE)/the Ocean Exploration Trust (OET).

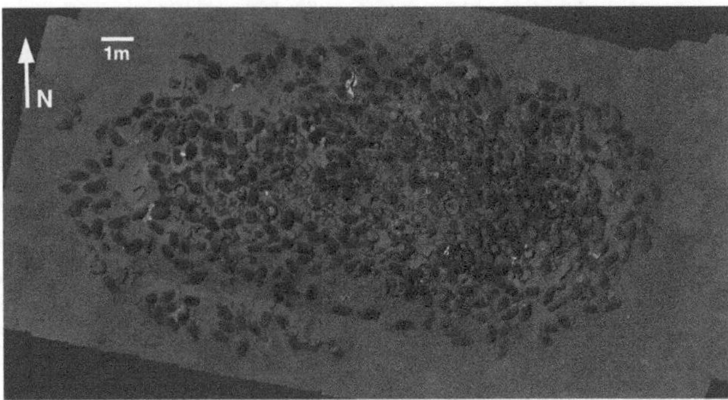

FIGURE 15.2 Photomosaic of *Byzantine C* shipwreck site, Knidos C, located south of Datcha, Turkey.
Source: Image by Chris Roman.

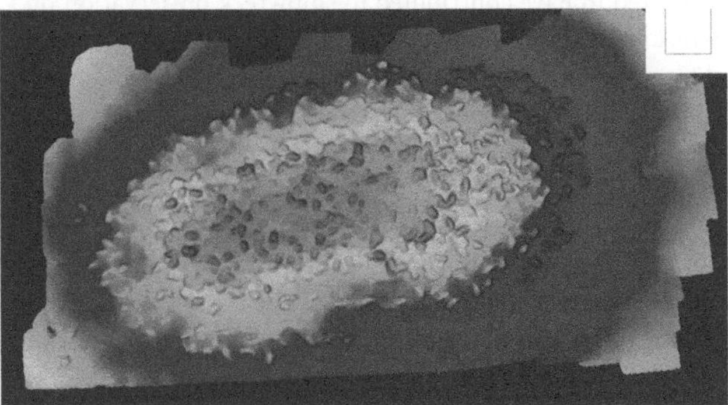

FIGURE 15.3 **(See color insert.)** Multibeam bathymetric map of *Byzantine C* shipwreck site, Knidos C, located south of Datcha, Turkey.
Source: Image by Chris Roman.

bathymetric imaging of hydrothermal vents and other geological features.[16,17] However, the mapping of shipwreck sites has been important for the development of imaging techniques because artifacts such as amphoras are of well-known shapes and sizes, which greatly aid the fine tuning of the imaging calibrations for use on unknown objects, such as those of a geological nature.[17]

Characterization

Following the inception of underwater archaeology, George Bass wrote[4] "No archaeologist specializes in the environment in which he works." Over the past two decades, most archaeologists have realized the importance of collaborating with scientists to gain a more rounded picture of the site and the processes that went into forming it. The birth of the field of geoarchaeology, for example, is testament to the need to understand a site's complete history. Likewise, in marine archaeology, there is much that can be learned about a shipwreck by studying the modern site with respect to the environment in which it is deposited.[7] Archaeological oceanography not only investigates cultural sites, but also the history of those sites from the time of their deposition underwater to the present day, documenting the modern landscapes as part of the environment.[13] Understanding the history of a shipwreck site from its sinking to the present day is as important as learning about the history of the ship from its construction to its sinking. Characterizing the physical and chemical actions that have continuously affected a site is necessary to define ongoing site formation processes. This is a principle objective behind archaeological oceanography.

A prime example of this application can be found in the Black Sea. The stratification of this body of water into three layers—oxic, suboxic, and anoxic—provides an environment with a high preservation potential for cultural materials.[11,13] Additionally, investigations of archaeological sites in these waters, particularly those located close to the interface between these layers, can help provide a greater understanding of the physical and chemical dynamics of the Black Sea.[11,13] Over the past decade, expeditions by the Institute for Exploration and the Ocean Exploration Trust have located over a dozen shipwrecks off the southern Black Sea coast, near Sinop, Turkey, all in various states of preservation. These wrecks, located between 100 and 300 m depth, span the shelf break and the oxic/anoxic interface. In addition to documenting the wreck sites with high-resolution imagery, sediment and water samples were collected in the area of these wrecks to begin to characterize the oceanographic environment in which they sit, including microbiological, geochemical, meiofaunal, and mineralogical analyses.[11] Most of these wrecks lie between 100 and 115 m depth, which is likely due to a combination of physical parameters rather than historical. Heavy bottom trawling along the shallower, coastal shelf areas, and slumping along the steep slopes at the shelf break may have destroyed or concealed some shipwreck sites.[11] At the same time, anoxic waters have likely moved up along the shelf during episodes of internal wave action, which has been hypothesized to have contributed to shipwreck preservation off Crimea, Ukraine,[18–20] helping to preserve wrecks in this specific depth range. This research in the Black Sea illustrates the importance of including oceanographic research in deepwater archaeological investigations, and the need for environmental specialists during such research.

In Situ Preservation

The scientific excavation of underwater cultural heritage sites versus their salvage by commercial companies has been a topic of rigorous legal and ethical debates.[3,21–23] The 2001 UNESCO Convention on the Protection of the Underwater Cultural Heritage is the result of an international response to the illegal looting of shipwreck sites. The recent technological ability to access sites in deep water, in both academic and private sectors, opens new discussions about the responsibility of investigating and preserving these sites.[23] The UNESCO Convention calls for the in situ preservation of underwater archaeological sites as a first option, followed by the careful recovery of materials for either scientific or protective purposes.[24] Recent research has shown that the expansion and better enforcement of marine protected areas can have a positive impact on the protection of sites from bottom trawlers, and that such steps toward their in situ preservation may be more effective than their salvage.[15]

Technological developments in underwater imaging and remote sensing, both surficial bathymetry and sub-bottom profiles, require us to consider seriously when excavation in deep water is in fact appropriate because sites can be documented in great detail remotely. The UNESCO Convention also states,

"Responsible non-intrusive access to observe or document in situ underwater cultural heritage shall be encouraged to create public awareness, appreciation, and protection of the heritage except where such access is incompatible with its protection and management."[24] Such goals of the Convention are in line with the objectives of archaeological oceanography, where the nonintrusive documentation and characterization of sites are the primary goals, as discussed earlier. Additionally, recent expeditions by the *E/V Nautilus* broadcast live video from the seafloor for educational outreach, as public access to the deep sea is severely limited, helping to increase knowledge and excitement of ocean exploration, as well as awareness of the processes, human and environmental, that threaten both marine environments and underwater cultural sites.[25]

Conclusion

Archaeological oceanography does not only apply to the study of sites in deep water, but can be applied to any submerged cultural site. Underwater archaeology has been a successful discipline for decades; the application of archaeological oceanography emphasizes the inclusion of oceanographic methods and technology as a collaborative study between these disciplines. Although the study of archaeological sites in deep water requires oceanographic technology and resources for access and interpretation, the investigation of these sites can aid ocean exploration as well. Once shipwreck sites are dated, based on artifact type or by other means, they can serve as important references for sediment processes, such as for detecting differences in sedimentation observed at two wrecks off Knidos, Turkey from the Archaic Greek and Byzantine periods.[15] Additionally, the fixed locations of shipwreck sites can be used as platforms for long-term monitoring of oceanographic environments, such as in the Black Sea, where these sites are helping to pinpoint the depths and locations of internal wave activity and the dynamics of the Black Sea water column.[11,13]

Many archaeologists working underwater today utilize a variety of scientific methods to great effect. Archaeological oceanography does not modify the methodology of marine archaeology, but rather reinforces the importance of oceanographic research principles and practices in such studies. Beyond the technology required to access cultural sites in deep water, oceanography should also be employed to gather simultaneous, related data sets from the shipwrecks in their in situ environments. These data can broaden our understanding of both the ocean environment and the condition of the wreck site. These sites have come into equilibrium with the ocean over the course of hundreds to thousands of years underwater, and much can be learned about both the ships and the environment through the investigations of the modern sites. The application of oceanography to archaeological studies aims to maximize the potential information that can be gathered from the investigation of these sites and provides critical insights into site dynamics and management decisions. Given the time and cost associated with accessing sites in deep water, multidisciplinary collaboration is essential for the documentation and preservation of underwater cultural heritage.

Acknowledgments

The authors wish to thank Katy Croff Bell, Alexis Catsambis, Dwight Coleman, Dan Davis, Jim Delgado, Carter DuVal, Gabrielle Inglis, Chris Roman, Art Trembanis, and Sandra Witten for their help in preparing this manuscript.

References

1. Ballard, R.D. Ed. Introduction. In *Archaeological Oceanography*; Princeton University Press: New Jersey, 2008; ix-x pp.
2. Mindell, D.A.; Croff, K.L. Deep water, archaeology and technology development. MTS J. **2002**, *36* (3), 13–20.

3. Brennan, M.L. Quantification of trawl damage to premodern shipwreck sites: Case studies in the Aegean and Black Sea. UNESCO Scientific Colloquium on Factors Impacting the Underwater Cultural Heritage. 2011. http://www.unesco.org/new/fileadmin/MULTIMEDIA/HQ/CLT/pdf/UCH_S2_Brennan.pdf.
4. George B. F. *Archaeology under Water*; Praeger, New York, 1966.
5. U.S. Dept of Commerce/NOAA. Discovering Earth's Final Frontier: A U.S. Strategy for Ocean Exploration. The Report of the President's Panel on Ocean Exploration. U.S. Department of Commerce/National Oceanic and Atmospheric Administration. 2000. explore.noaa.gov/media/http/pubs/pres_panel_rpt.pdf.
6. Coleman, D.F.; Ballard, R.D. Oceanographic methods for underwater archaeological surveys. In *Archaeological Oceanography*; Ballard, R.D., Ed.; Princeton University Press: New Jersey, 2008; 3–14.
7. Ballard, R.D. The search for contemporary shipwrecks in the deep sea: lessons learned. In *Archaeological Oceanography*; Ballard, R.D., Ed.; Princeton University Press: New Jersey, 2008; 95–127.
8. Ballard, R.D. *The Discovery of the Bismarck*; Scholastic/ Madison Press: New York, 1990.
9. Delgado, J.P. Recovering the past of the *USS Arizona*: symbolism, myth, and reality. His. Archaeol. **1992**, *26* (4), 69–80.
10. Kelly, L.W.; Barott, K.L.; Dinsdale, E.; Friedlander, A.M.; Nosrat, B.; Obura, D.; Sala, E.; Sandin, S.A.; Smith, J.E.; Vermeij, M.J.; Williams, G.J.; Willner, D.; Rohwer, F. Black reefs: iron-induced phase shifts on coral reefs. Int. Soc. Microbial. Ecol. **2011**, *6*, 638–649.
11. Brennan, M.L.; Davis, D.; Roman, C.; Buynevich, I.; Catsambis, A.; Kofahl, M.; Ürkmez, D.; Vaughn, J.I.; Merrigan, M.; Duman, M. Ocean dynamics and anthropogenic impacts along the southern Black Sea shelf examined by the preservation of pre-modern shipwrecks. *Continental Shelf Res.* **2013**, *53*, 89–101.
12. Foley, B. Archaeology in deep water: Impact of fishing on shipwrecks. Woods Hole Oceanographic Institution. 2008. http://www.whoi.edu/sbl/liteSite.do?litesiteid=2740&articleId=4965. (accessed December 2009).
13. Brennan, M.L.; Ballard, R.D.; Croff Bell, K.L.; Piechota, D. Archaeological oceanography and environmental characterization of shipwrecks in the Black Sea. In *Geology and Geoarchaeology of the Black Sea Region: Beyond the Flood Hypothesis*·, Buynevich, I., Yanko-Hombach, V., Gilbert, A., Martin, R.E., Eds.; Geological Society of America Special Paper, **2011**; *473*, 179–188.
14. Singh, H.; Roman, C.; Pizarro, O.; Foley, B.; Eustice, R.; Can, A. High-resolution optical imaging for deep-water archaeology. In *Archaeological Oceanography*; Ballard, R.D., Ed.; Princeton University Press: New Jersey, 2008; 30–40.
15. Brennan, M.L.; Ballard, R.D.; Roman, C.;. K.L.C.; Buxton, B.; Coleman, D.F.; Inglis, G.; Koyagasioglu, O.; Turanli, T. Evaluation of the modern submarine landscape off southwestern Turkey through the documentation of ancient shipwreck sites. *Continental Shelf Res.* **2012**, *43*, 55–70.
16. Roman, C.; Inglis, G.; Rutter, J. *Application of Structured Light Imaging for High Resolution Mapping of Underwater Archaeological Sites.* IEEE OCEANS: Sydney, 2010.
17. Roman, C.N.; Inglis, G.; Vaughn, J.I.; Williams, S.; Pizarro, O.; Friedman, A.; Steinberg, D. Development of high-resolution underwater mapping techniques. In *New Frontiers in Ocean Exploration*; Bell, K.L.C., Fuller, S.A., Eds.; The *E/V Nautilus* 2010 Field Season. Oceanography **2011,** *24* (1), supplement, 14–17.
18. Murray, J.W.; Stewart, K.; Kassakian, S.; Krynytzky, M.; DiJulio, D. Oxic, suboxic, and anoxic conditions in the Black Sea. In *The Black Sea Flood Question: Changes in Coastline, Climate and Human Settlement*; Yanko-Hombach, V., Gilbert, A.S., Panin, N., Dolukhanov, P.M., Eds.; *Dordrecht*; Springer: 2007; 1–22.
19. Ward, C.; Ballard, R. Black Sea shipwreck survey 2000. Int. J. Nautical Archaeol. **2004,** *33*, 2–13.

20. Trembanis, A.; Skarke, A.; Nebel, S.; Coleman, D.F.; Ballard, R.D.; Fuller, S.A.; Buynevich, I.V.; Voronov, S. Bedforms, hydrodynamics, and scour process observations from the continental shelf of the northern Black Sea. In *Geology and Geoarchaeology of the Black Sea Region: Beyond the Flood Hypothesis*; Buynevich, I., Yanko-Hombach, V., Gilbert, A., Martin, R.E., Eds.; Geolo. Soc. Am. Special Paper, **2011**; *473*, 165–178.
21. Aznar-Gomez, M.J. Treasure hunters, sunken state vessels and the 2001 UNESCO Convention on the Protection of the Underwater Cultural Heritage. Int. J. Marine Coastal Law **2010**, *25*, 209–236.
22. Delgado, J.P. Underwater archaeology at the dawn of the 21st century. His. Archaeol. **2000**, *34* (4), 9–31.
23. Greene, E.S.; Leidwanger, J.; Leventhal, R.M.; Daniels, B.I. Mare nostrum? Ethics and archaeology in Mediterranean waters. Am. J. Archaeol. **2011**, *115*, 311–319.
24. UNESCO. Convention on the Protection of the Underwater Cultural Heritage 2001. *UNESCO.* 2001. http://portal.unesco.org/en/ev.php-URL_ID=13520&URL_DO=DO_TOPIC&URL_SECTION=201.html.
25. Witten, A; O'Neal, A. Education and outreach activities enabled by telepresence technology. In *New Frontiers in Ocean Exploration: The E/V Nautilus 2010 Field Season*; Bell, K.L.C, Fuller, S.A., Eds.; Oceanography **2011**, *24* (1), supplement, 12–13.

16

Arctic Hydrology

Bretton Somers and
H. Jesse Walker
Louisiana State University

Introduction

After a discussion about the boundaries of the Arctic, its hydrological elements are analyzed. It is noted that, in the Arctic, the solid phase of water dominates the landscape during most of the year. In the solid phase, water occurs as snow, river and lake ice, sea ice, glacial ice, and ground ice. Ground ice is integral to permafrost, which is nearly continuous. Although all of the hydrologic elements can be considered separately as though each is in storage, they are in fact interconnected and should be considered as an integral part of a cycle that progresses through all three phases among the atmosphere, lithosphere, and cryosphere. Recent research has shown that many of the Arctic's hydrologic elements are changing rapidly.

The Arctic

Although no one denies the existence of the region known as the Arctic, few actually agree on its southern boundary. Astronomers usually use the Arctic Circle, oceanographers use the distribution of sea ice, biologists use the tree line, and cryosphere specialists use a permafrost boundary.[1] In one way or another, all of these boundaries are of significance to hydrologists, because they must contend with distributions that often spread well beyond the Arctic. For example, many of the rivers that drain into the Arctic Ocean originate in temperate latitudes (Figure 16.1), the Arctic Ocean is impacted by flow from both the Atlantic and Pacific Oceans, and the atmosphere can be affected by non-arctic weather patterns. Nonetheless, the main hydrologic factor that distinguishes the Arctic from other climatic zones is the long period of time during which snow and ice dominate the landscape. From a regional standpoint, the Arctic may best be considered a moderately sized ocean almost completely surrounded by a fringe of land, i.e., a configuration that is the opposite of its counterpart in the Southern Hemisphere, the Antarctic.

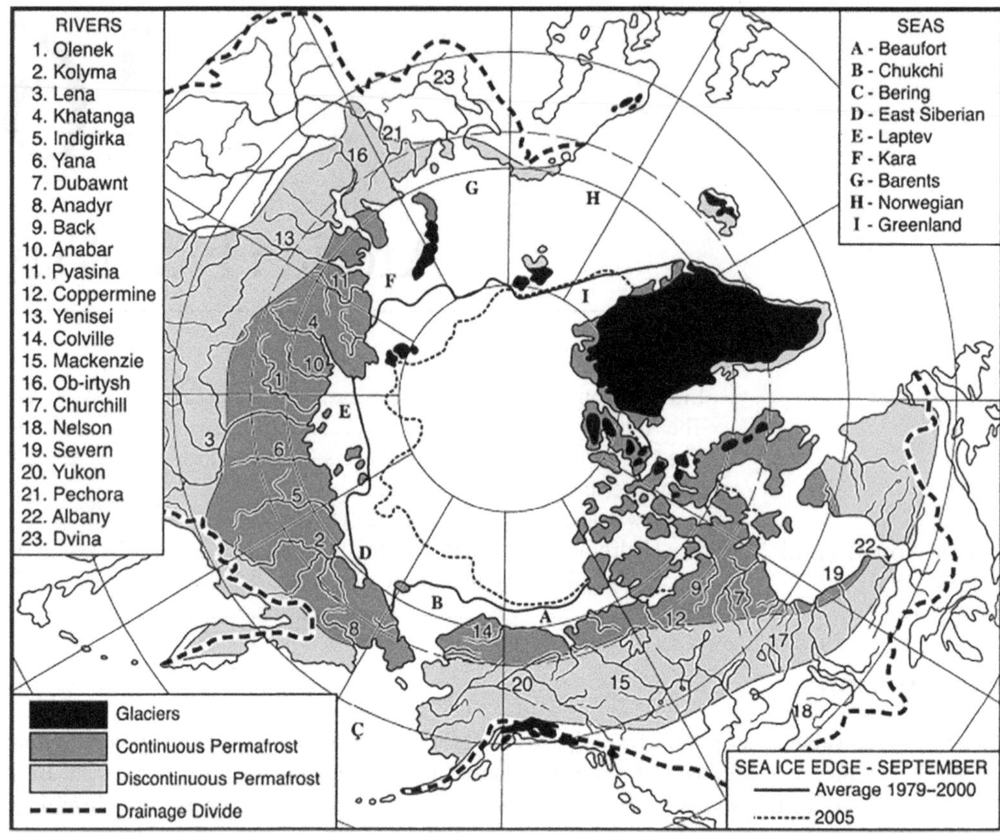

FIGURE 16.1 Map of the Arctic showing the major rivers, permafrost distribution, September sea-ice locations, and bordering seas in the Arctic Ocean.
Source: Adapted from Walker & Hudson.[9]

The Ocean, Rivers, Lakes, and Glaciers

The major features of the Arctic are the Arctic Ocean with its numerous bordering seas, the rivers that drain into it, its lakes and ponds, and the numerous islands (many with glaciers) that occur north of the continents (Figure 16.1). Some scientists contend that the Arctic Ocean should be classified as a Mediterranean sea rather than an ocean because in addition to being nearly surrounded by land, it occupies less than 4% of the Earth's ocean area and only about 1% of the Earth's ocean volume.

The Arctic Ocean is also unique in that it is only 40% as large as the continental area that drains into it. The rivers that enter the ocean vary greatly in size and discharge. Included among them are four of the Earth's 12 longest—the Yenisey (5870 km), the Ob (5400 km), the Lena (4400 km), and the Mackenzie (4180 km). They contribute nearly 60% of the fresh water that drains from the continents.[2] This total equals some 11% of the Earth's runoff from the continents. Because of its unique relationship between land and sea, the Arctic Ocean is impacted more by water draining into it than any other ocean.[3]

In addition to its seas, rivers, and islands, the Arctic also possesses numerous lakes and ponds. Most Arctic lakes are small. One of the most studied lake types is the oriented lake. Generally elliptical in shape, oriented lakes originate as thaw lakes and range in length up to several kilometers. However, by far, the most common bodies of water dotting the land surface are small ponds whose existence is partly the result of the poor drainage and limited infiltration that characterizes permafrost environments.

Glaciers in the Arctic, except for the Greenland ice cap, are generally limited to the Canadian and Russian arctic islands. Arctic glaciers are major contributors of water and ice bergs to the ocean.

Snow, Ice, and Permafrost

One of the unique characteristics of water is that it is the only chemical compound that occurs on Earth in three phases, i.e., as a gas (water vapor), a liquid, and a solid. In its solid phase, water appears as snow, surface ice (river, lake, and sea ice), and ground ice.

Snow

Although snow is not limited to high latitudes, it is there that it is nearly ubiquitous. Over most of the Arctic, it covers the surface for nine or more months. Snow fall is related to the volume of water vapor in the air, a volume that is temperature dependent. Therefore, the actual amount of snow that falls is limited in quantity. Once it falls, it tends to retain its solid form until melt season. However, because most of the Arctic has low-lying vegetation (tundra), snow drifting is extensive. On flat surfaces, such as lake ice, the snow-pack is thin or even missing. Uneven surfaces, such as river banks, trap snow into sizable drifts, some of which may last through most of the summer.

Snow is important in several major ways in the Arctic. Because it is highly reflective, much of the sun's energy that reaches it is returned to space. The reflected energy, or albedo, of fresh snow is more than 75%, whereas that from water is usually less than 30% and from tundra less than 20%.[4] Further, snow, because of its low thermal conductivity, is a good insulator—affecting the occurrence and maintenance of permafrost and in helping keep vegetation and fauna from freezing. Snow, when it melts, provides the bulk of the water that feeds the rivers and streams that drain into the Arctic Ocean. Because the melt season is relatively short and the melt rate is rapid, most rivers in the Arctic have peaks of discharge that correlate closely with river breakup (Figure 16.2).

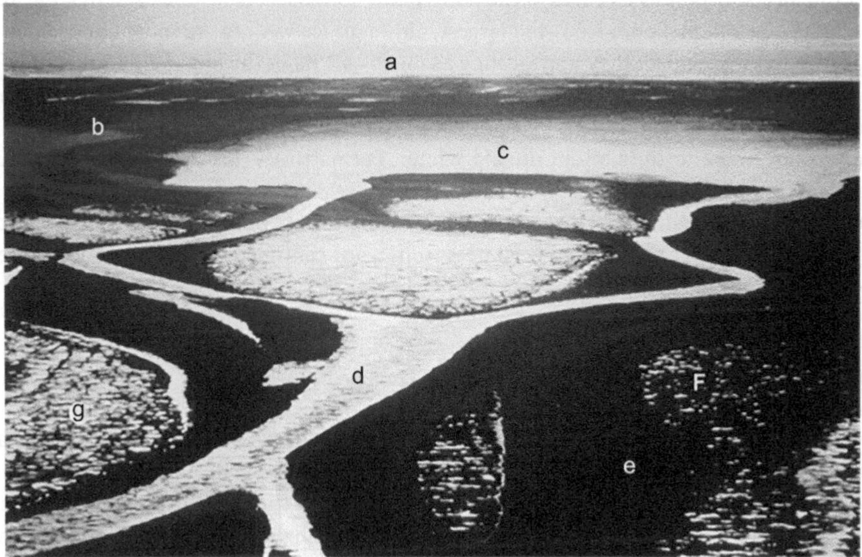

FIGURE 16.2 The mouth of the Colville River, Alaska showing the distribution of snow, water, river ice and sea ice during breakup. (**a**) sea ice, (**b**) floodwater on top of sea ice, (**c**) sea ice floating on top of river floodwater, (**d**) river ice floating on top of floodwater, (**e**) flooded mudflats, (**f**) flooded island with remnant snow patches, (**g**) snow and ice covering ice-wedge polygons. Photograph by Donald Nemeth.

River and Lake Ice

In the Arctic, the temperature regime is such that surface ice, whether on rivers, lakes, or the sea, is present for most of the year. On fresh-water bodies, the ice reaches thicknesses of about 2 meters. Because most of the ponds and many of the streams have depths less than that, water freezes to the bottom. In deeper lakes and rivers, water will remain in the liquid state beneath the ice. In some rivers, especially those originating in lower latitudes (Figure 16.1), flow continues throughout the winter beneath the ice. Ice in the deeper lakes lasts longer than ice on rivers, where breakup follows after snow-melt water enters the river channels.

Sea Ice

Some authors maintain that sea ice is the most dramatic feature of the Arctic. Formed by the freezing of sea water, sea ice covers nearly all of the Arctic Ocean during winter when it is attached to much of the continental coastline of Siberia and North America. Except for a narrow band of fast ice, the bulk of the sea ice is in motion, steered by ocean currents and winds. Its predominant direction of flow is clockwise. Because its motion is erratic, pressure ridges many meters high and open bodies of water, known as leads, develop. First year sea ice averages about 2 meters in thickness whereas ice that survives through the summer becomes thicker during the following winter. The minimum extent of sea ice in the Arctic is in September (Figure 16.1). Sea ice suppresses the energy exchange between the ocean and the atmosphere, affects the circulation of ocean currents, and dampens wave action.[5]

Permafrost and Ground Ice

Permafrost is defined as earth material in which the temperature has been below 0°C for two or more years. Because it is defined only by its temperature, water is not necessary for its existence. However, most permafrost, which underlies more than 20% of the Earth's land area, does contain ice in various amounts and forms. Ground ice occurs in the pores of the soil as lenses or veins and in large forms such as ice wedges.[6] By volume, pore ice is the largest, although ice wedges are more conspicuous. Where ice wedges are well developed, they may occupy as much as 30% of the upper 2 or 3 meters of the land surface. Their surface expression is distinctive and takes the form of ice-wedge polygons. In the Arctic, permafrost is continuous (Figure 16.1) except beneath those water bodies that are more than 2 meters deep and do not freeze to the bottom during winter. Permafrost is also present beneath near-shore waters off Siberia and North America.

Associated with permafrost is the active layer—the portion of terrain that thaws and freezes seasonally. The active layer can vary in thickness from a few centimeters to several meters, depending mainly on vegetation cover and soil texture.

The Hydrologic Cycle in the Arctic

In the above discussion, each hydrologic element was considered individually, as if it was a pool of water in storage as a gas, a liquid, or a solid. However, in reality, the hydrologic elements are interconnected and mobile, moving from one phase to another through time. The examination of water from this standpoint is best done through the concept of the hydrologic or water cycle, a cycle that is considered the most fundamental principle of hydrology.[7] The conceptual model (Figure 16.3) illustrates hydrologic links among the atmosphere, the land, and the ocean.[3] As a concept, it can be applied to small units within the hydrosphere or large units such as the Arctic. In the Arctic, the movement of water from one phase or location is, to a large extent, dependent on freeze-thaw cycles of snow, ice, and permafrost.

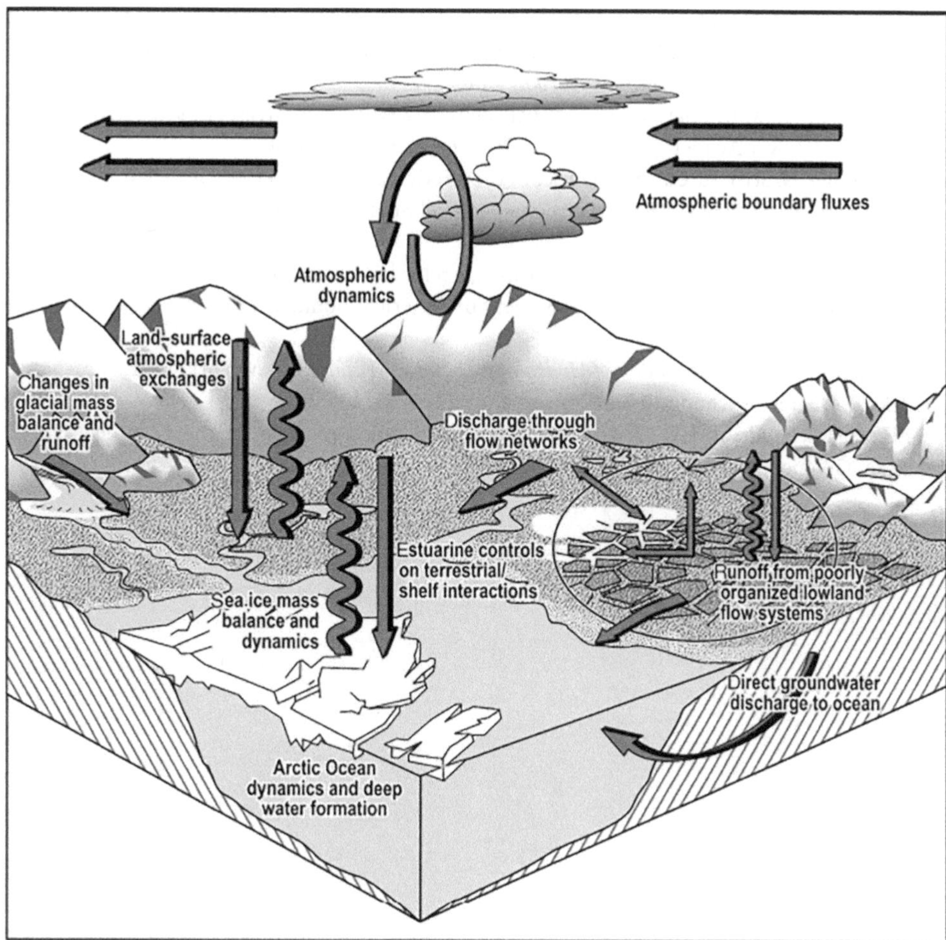

FIGURE 16.3 Conceptual model of the arctic hydrologic cycle.
Source: Adapted from Vörösmarty, et al.[3]

Arctic Hydrology and the Future

In recent years, research on climate change has increased dramatically. Much of it has been conducted in and about the Arctic and especially about arctic hydrology. Numerous changes in the hydrologic elements of the Arctic have been recently documented. Included are a shortening of the snow-cover season, later freeze-up and earlier breakup of river, lake, and sea ice, increased fresh-water runoff, melting glaciers, the degradation of permafrost and an increase in active-layer thickness, increased groundwater flow, decreased sea-ice extent (Figure 16.1), and decreased albedo, among others. In 2005, Dan Endres of the National Oceanic and Atmospheric Administration, in a discussion on climate change, stated, "Whatever is going to happen is going to happen first in the Arctic and at the fastest rate."[8] This is a prophecy that is especially applicable to virtually every hydrologic and ecological element in the Arctic.

References

1. Walker, H.J. Global change and the arctic. Curr. Top. Wetland Biogeochem. **1998**, *3*, 44–60.
2. http://www.awi-bremerhaven.dc/GEO/APARD/Backgr.html (accessed November 2005).

3. *Vörösmarty, C.J.; Hinzman, L.D.; Peterson, PJ. et al. The Hydrologic Cycle and Its Role in Arctic and Global Environmental Change: A Rationale and Strategy for Synthesis Study; Arctic Research Consortium of the U.S.: Fairbanks, Alaska, 2001.*
4. Collier, C.G. Snow. In *Encyclopedia of Hydrology and Water Resources*; Herschy, R.W., Fairbridge, R.W., Eds.; Kluwer Academic Publishers: Dordrecht, 1998; 621–624.
5. Untersteiner, N. Structure and dynamics of the Arctic ocean ice cover. In *The Arctic Region*; Grantz, A., Johnson, L., Sweeney, J.F., Eds.; The Geological Society of America, Inc.: Boulder, Colorado, 1990; 37–51.
6. Carter, L.D.; Heginbottom, J.A.; Woo, M.-K. Arctic lowlands. In *Geomorphic Systems of North America*; Graf, W.L., Ed.; Geological Society of America: Boulder, CO, 1987; Centennial Special; Vol. 2, 583–628.
7. Maidment, D.R.; Ed. *Handbook of Hydrology*; McGrawHill, Inc.: New York, 1992.
8. Glick, D. Degrees of change. Nat. Conservancy **2005**, *55*, 40–50.
9. Walker, H.J.; Hudson, P.F. Hydrologic and geomorphic processes in the Colville River Delta, Alaska. Geomorphology **2002**, *56*, 291–303.

17

Bathymetry: Assessment

Heidi M. Dierssen
Norwegian University of Science and Technology
University of Connecticut

Albert E. Theberge, Jr.
Central Library, National Oceanic and Atmospheric Administration (NOAA)

Introduction

The term "bathymetry" most simply refers to the depth of the seafloor relative to sea level, but the concepts involved in measuring bathymetry are far from commonplace. The mountains, shelves, canyons, and trenches of the seafloor have been mapped with varying degrees of accuracy since the mid-nineteenth century. Today, the depth of the seafloor can be measured from kilometer-to centimeter-scale using techniques as diverse as multi beam sonar from ships, optical remote sensing from aircraft and satellite, and satellite radar altimetry. Various types of bathymetry data have been compiled and modeled into gridded matrices spanning the seafloor at 1 arc-minute spatial resolution per pixel (<2 km) or even finer resolution for portions of the globe (e.g., <100 m resolution for U.S. coastal waters).[1] Even so, errors are prevalent in these modeled datasets and remote regions of the global ocean have yet to be accurately mapped. Indeed, the topography of Mars and Venus may be better known than our own seafloor.

Bathymetry is measured using "remote sensing" methods that investigate the seafloor indirectly without making physical contact. The majority of methods for estimating are based on the concept of using time to infer distance. Specifically, sensors emit a beam of sound, light, or radio waves and measure the round-trip travel time it takes for the beam to be reflected from a surface and return back to the sensor. The time elapsed is then related to the distance the beam traveled and used to infer bathymetry. The longer the elapsed time for the beam to return, the greater the distance traveled. However, limitations are inherent with all of these methods and no single technique is ideal for measuring the diversity and complexity of the underwater landscape and coastline. In addition to collecting additional datasets across the global ocean, new techniques are required to measure bathymetry and to interpret and process the datasets that have already been collected.

Methods for Assessing Bathymetry

This section outlines the primary modern techniques for estimating bathymetry from acoustics to the use of both the visible and radio wave portions of the electromagnetic spectrum. Measurements are made through the water medium in the case of acoustics, through the water and air for visible light measurements, and through the air medium alone for radar altimetry remote sensing to obtain an estimate of the seafloor depth. As will be outlined further, each method has its own advantages

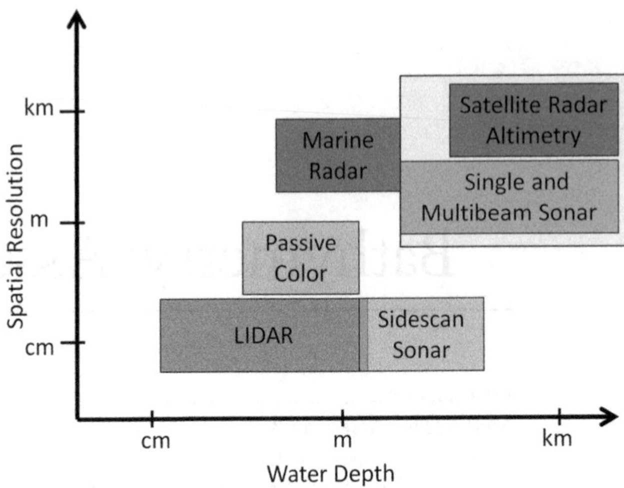

FIGURE 17.1 Schematic showing the applicability of different techniques for estimating bathymetry in terms of spatial resolution of measurement and the range in water depths that can be sampled. The yellow box indicates the datasets that are integrated and interpolated to form the global gridded bathymetry datasets at 1 arc-minute resolution.

and disadvantages in terms of scale of feature observed, accuracy, and water depth that can be sensed (Figure 17.1). Acoustics is applicable for determining centimeter- to kilometer-scale features and is the prevailing method used to validate optical and radar bathymetry. Because of limits to penetration of visible light in a water medium, measurements using active and passive light are only applicable in shallow water (generally <60 m) and are the primary means for delineating coastlines. Radar altimetry measurements from satellites provide large-scale bathymetry across the deep ocean and are particularly valuable in remote regions of the world ocean with little ship traffic. Marine radars from ships and shore have also been used to map bathymetry of shallow seas. The platforms for measuring bathymetry are also variable and include satellites orbiting high above the Earth, suborbital aircraft, ships on the sea surface, and remotely operated vehicles hovering along the seafloor.

Bathymetry measurements are not static. The rise and fall of the tides can change bathymetry up to several meters in height, depending on when and where measurements are taken. For bathymetric maps seaward of the continental shelf, usually no tidal corrections are made to ship soundings. Most international navigational charts used over shallow areas such as the Great Bahama Bank commonly adjust soundings to the lowest astronomical tide while mean lower low water is the datum for United States charts. However, tidal models have not yet been made perfect for shallow water mapping and tidal forecasting is an active area of research.[2] Having accurate tidal models are required for delineating coastlines, planning for storm surges and flooding,[3] and assessing bathymetry from satellite altimetry.

Acoustics

Acoustic bathymetry or echo sounding came into play in the 1920s and was instrumental in determining the configuration of the seafloor as we know it today.[4] The initial technique was to emit a single pulse of sound and measure the elapsed time for it to travel to the seafloor and be reflected and be detected by a shipboard hydrophone. One-half of the round-trip travel time multiplied by the velocity of sound in sea water equals the depth at a given point. As noted by Leonardo da Vinci in 1490 and later by Benjamin Franklin in 1762, sound travels far in water with little attenuation relative to air.[5] Sound also travels significantly faster in water than in air and great ocean depths can be probed acoustically without substantial degradation of the signal. The velocity of sound in seawater is ~1500 m/sec, but the precise

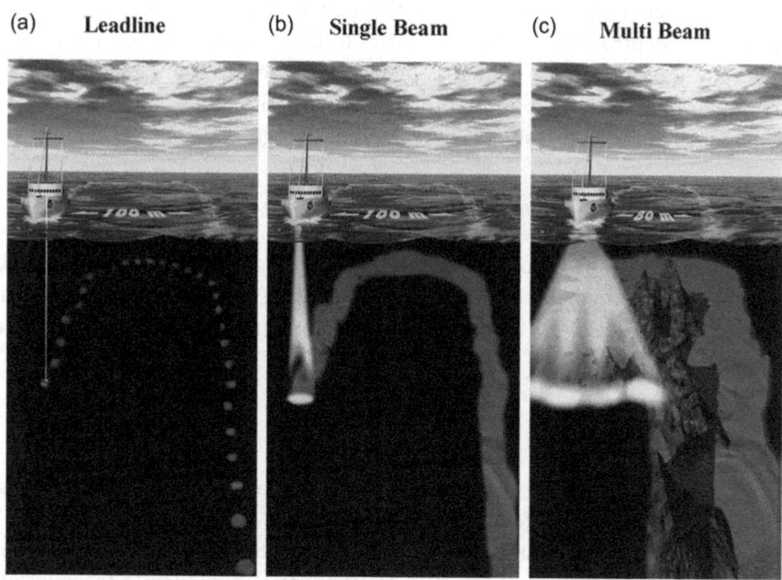

FIGURE 17.2 **(See color insert.)** Illustration of different methods for measuring bathymetry from **(a)** line and sinker technology; **(b)** echo sounding using single-beam acoustics; and **(c)** multibeam acoustics providing the highest spatial resolution of seafloor features.

velocity depends on ocean temperature, salinity, and pressure. In addition to the velocity of sound, the character of the seabed, vegetative cover, biota, and other particles in the water column, can affect the accuracy of the measured depth.

Today, high-resolution measurements of ocean depth and bottom reflectivity are produced by multi beam sonar. Each ping of a multibeam sonar emits a single wide swath of sound (up to 153°) that reflects off the seafloor (Figure 17.2). The return echo is received by an array of transducers and then separated electronically into a number of individual beams for each of which a depth is calculated. Very high resolution is attainable in shallow water, but the swath width is decreased. Conversely, the efficiency of ship operation is increased in deep water as the swath width expands geometrically but resolution is decreased. A swath of the seafloor is acoustically imaged with each pass of a survey ship as it follows a pattern similar to "mowing the grass." A series of overlapping swaths produce a bathymetric map of the area being surveyed.

Various sound frequencies (e.g., 12–400 kHz) are employed for different depth ranges; the lower the frequency, the deeper the depth measurement possible, while higher frequency is associated with higher resolution but shallower depth. Bathymetry is now being estimated at levels of resolution and accuracy that were previously unattainable. For example, shallow-water multibeam systems can measure bathymetry roughly at a 10 cm scale in 10 m of water and have been used to effectively map coastal waters of the United States.[6,7] Concentrated mapping programs also incorporate "sidescan" sonar for a qualitative view of seafloor reflectivity characteristics.[7] Acoustic sensors can also be placed on tethered remotely operated vehicles or autonomous underwater vehicles, such as gliders, that can maintain a constant position relative to the seafloor and provide high-resolution bathymetry.[8]

The primary disadvantages of acoustic measurements are the time and cost associated with making measurements from a ship in deep waters or a small vessel in shallow waters. In order to build up coherent images at a high resolution, many survey lines with overlapping tracks must be run. Because the swath width decreases in shallow water, many more ship or glider tracks are required in coastal estuaries and bays with shallower water. Hence, detailed surveys in coastal regimes require considerable time and effort to cover relatively small portions of the sea bed. In general, acoustic methods can be used throughout all oceanic depths from shallow estuaries to the deepest trenches. However, ship time is costly even in deep water, and because of increasing time and effort to operate in shallow waters, acoustic systems are not ideal for such tasks as monitoring bathymetric changes and shoreline configuration changes caused by such phenomena as tidal currents, storm surge, and sea-level change.

Optics

Remote sensing by visible light has also been used to estimate shallow water bathymetry, where acoustic methods become limited. Bathymetry has been quantified using passive methods, which measure only the natural light reflecting from the seafloor, and active methods, which use lasers to measure the distance to the seafloor.

Passive Ocean Color Remote Sensing

Of the sunlight that hits the ocean surface, only a small percentage is scattered back out again and can be remotely detected by aircraft or satellite. Passive ocean color sensors measure this small amount of solar radiation that has entered the water column and been scattered back out.[9] In waters that are shallow and clear enough for light to reach the bottom (called "optically shallow"), the color of the seafloor also contributes to the water leaving radiance in a way that depends on the bathymetry, substances in the water column, and bottom composition. In the clearest natural waters, bright sandy bottoms can be detected in 30 m or more. In most coastal areas, water is clarity is reduced by algae, sediment and other colored substances and generally depths less than 10 m can be measured. As the water becomes more turbid with sediment or dense phytoplankton, much of the light entering the water column is absorbed in the top layer of the water and the sea becomes "optically deep" at only a few meters in depth.

In sufficiently shallow and clear water, the magnitude and spectral quality of light reflected from the seafloor can be interpreted from remotely sensed ocean color. Methods to retrieve shallow water bathymetry from passive ocean color reflectance measurements have been either empirical[10] or radio metrically based using iterative modeling or look-up tables.[11] Many of the empirical approaches use far red or near-infrared wavelengths where water is highly absorbing and the signal is less influenced by phytoplankton, sediments, and other absorbing and scattering material in the water column.[12] Radio metrically based approaches generally use most of the visible spectrum and require many wavelengths of data (i.e., hyper spectral) to retrieve bathymetry. In addition, the radiometric methods also solve for the color of the seafloor (i.e., benthic reflectance), as this is a key component of the water-leaving radiance signal. Using passive ocean color, bathymetry has been mapped with high spatial resolution from aircraft[12] to regional scales from satellites[13] (Figure 17.3).

Because ocean color sensors are already up in space for oceanographic research with data freely available,[14] the use of satellites to derive shallow water bathymetry is quite effective and cost-efficient compared to acoustic methods. However, cloud cover masks the ocean color and only clear sky imagery can be analyzed. In addition, only a low percentage of the incident light that reaches the ocean is reflected back to a satellite. When viewed remotely, the ocean color is observed through a thick atmosphere containing gases and aerosols which also reflect sunlight back to the sensor. The atmosphere contributes more photons to a satellite sensor than the ocean surface itself and accurate "atmospheric correction" is one of the many challenges to obtaining an accurate water-leaving radiance required for estimating shallow water bathymetry. If too much signal is removed from the atmospheric correction, the ocean will appear too dark and bathymetry can be overestimated and vice versa.[15] Most methods also require some a priori knowledge of the benthic reflectance in the region and such measurements are difficult to make and integrate over the appropriate scales required for remote sensing. Bathymetry estimated with passive ocean color methods are generally not accurate enough for navigational purposes,

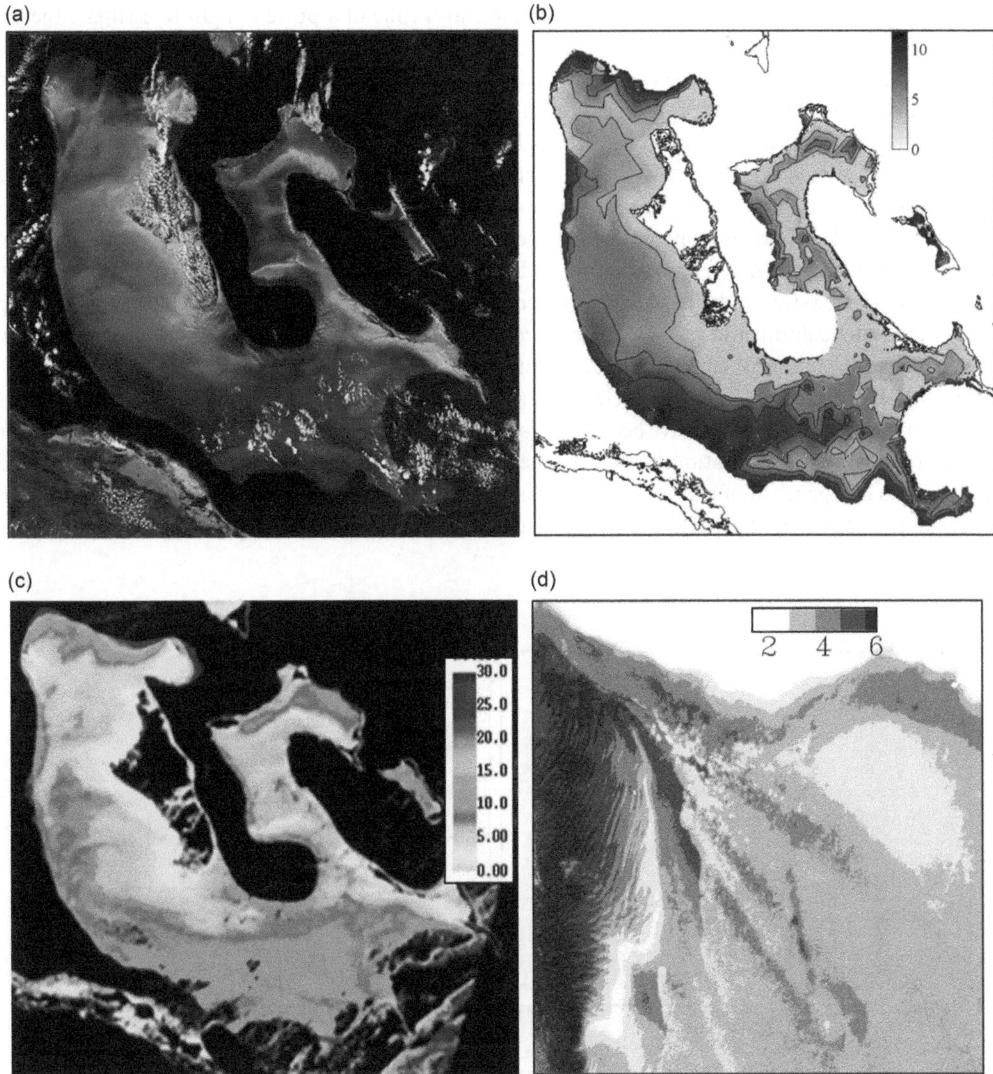

FIGURE 17.3 The Great Bahama Bank is a large optically shallow bank that can be mapped with passive ocean color techniques. **(a)** A pseudo-true-color image from the MODIS Aqua sensor at 250 m resolution from 6 March 2004 shows the bright reflected color off of the shallow Bank; **(b)** bathymetry gridded using selected soundings from navigational chart;[26] **(c)** bathymetry modeled from the MERIS ocean color sensor;[13] **(d)** shallow water bathymetry mapped from a hyperspectral sensor on an aircraft for a small region south of Lee Stocking Island, Bahamas.[27] All bathymetry is in units of meters (m).

but they provide much better results than gravimetric measurements (see further) over shallow basins (Figure 17.3c) and the spectral information can also provide an estimate of the seafloor composition.

Active Lidar

At present, the most effective means to map shallow water bathymetry is with active lidar systems. Active sensors produce and sense their own stream of light (e.g., light detection and ranging, or LIDAR) and are generally flown on aircraft, although several space-based lidars have been launched. Similar to acoustic measurements, LIDAR uses the round-trip travel time of a pulse of light to estimate the distance to the seafloor. Aircraft-mounted lasers pulse a narrow, high-frequency laser beam toward the Earth and are capable of recording elevation measurements at rates of hundreds to thousands of pulses per second. Because they do not rely on sunlight, LIDAR systems can be operated at night.

Generally, two lasers are employed to estimate bathymetry: 1) an infrared laser (1064 nm), which does not penetrate water, is used to detect the sea surface and 2) a green laser (532 nm) is used to penetrate into the water column and provide a return signal from the seafloor.[16] In coastal waters, green light is the least absorbed and generally penetrates the deepest into the water column. Similar to passive ocean color measurements, the green laser is attenuated by the absorbing and scattering particles in the water column, and the maximum depth of measurement is determined by a combination of the optical properties of the water column and seafloor. In ideal conditions, depths up to 60 m have been measured with LIDAR systems,[17] but most applications are limited to depths <40 m.

Current LIDAR measurements with elevation accuracy of 10–30 cm can provide point measurements from 0.1 to 8 pixels per m^2. They have been successfully used to map bathymetric features, beach erosion, coral reefs, and coastal vegetation. Unlike acoustic sensors on ships, the imagery is provided in fixed-width swaths independent of water depth and can provide high area coverage rates of up to $70 km^2 hr^{-1}$ (Figure 17.4a). The method is particularly effective in shallow water where larger ships are excluded and

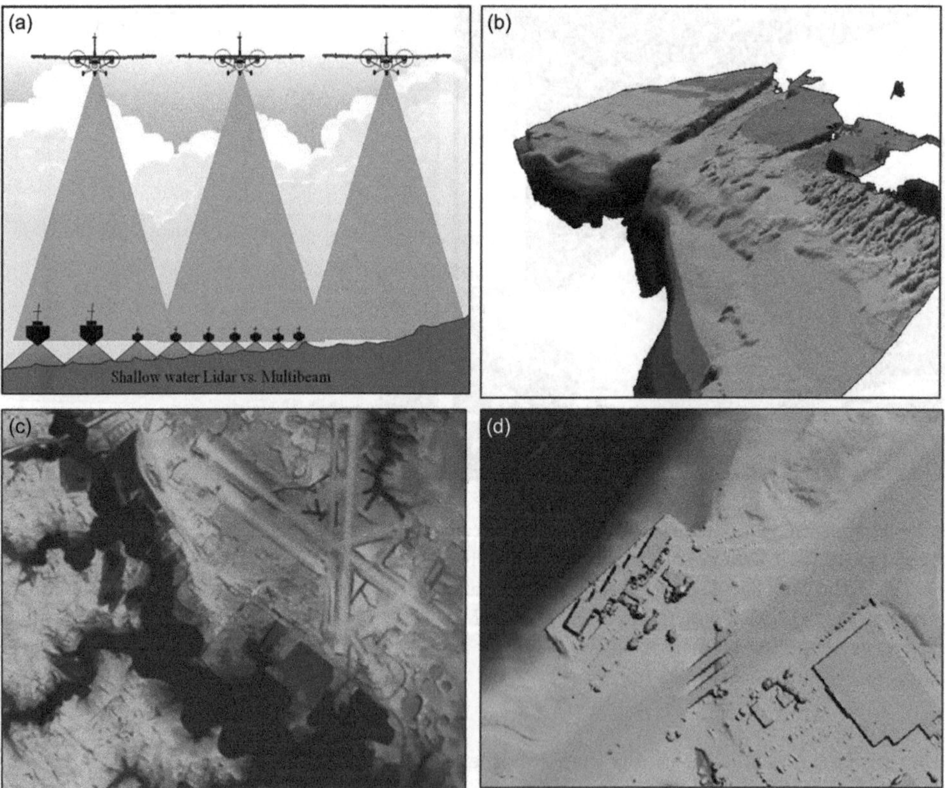

FIGURE 17.4 (See color insert.) (a) Airborne LIDAR provides a large sampling swatch for collecting bathymetry data compared to ship-based sonar systems and **(b)** can seamlessly blend water and land elevations in one image for precise coastline delineation. LIDAR data collected in **(c)** coastal Delaware can be used for mapping projected sea-level rise; and **(d)** Monterey Bay for evaluating erosion patterns along a seawall jutting into the coastal zone. Images in Panel (a) and Panel (b) provided courtesy of Optech Industries. Images from Panel C and Panel D are available from the NOAA Coastal Services Center Digital Coast project.

acoustic surveying is least efficient. Another major advantage of LIDAR is that it can map land and water in the same mission and provide continuity of shoreline mapping. Such combined land and water measurements have been used by environmental managers to map coastlines and conduct city planning for storms and potential changes in sea level (Figure 17.4b–d). Airborne systems are also useful in shallow waters with reefs or in regions where tidal flows make ship-based measurements difficult if not impossible. Often LIDAR systems are coupled with passive hyper spectral imagers to assess both the bathymetry and seafloor composition simultaneously.

Radar Systems

Marine radars can be used from satellite to detect large-scale changes in bathymetry that are related to sea-surface height. Imaging radar systems from ships and shore have been used to measure wave fields and deduce shallow water bathymetry from wave theory.

Satellite Altimetry

The presence of bathymetric features, such as ridges and troughs, create changes in Earth's gravity field that produce small fluctuations in the height of the sea surface. The sea surface bulges slightly upward in response to a seamount, for example, as water is attracted to the greater mass (Figure 17.5). Satellite-mounted radar altimeters orbiting the Earth can measure these slight variations in the sea surface-height by sending out radio wave pulses at high frequencies, usually in the range of 13 GHz, for determining sea-surface height variations. The radar pulse scatters off of the sea surface and the round-trip travel time of the signal is measured. If the satellite's position in orbit is well-characterized, then the round-trip travel time of the pulse can be related quite precisely to the height of the sea surface relative to the satellite sensor and can be estimated to within a few centimeters (Figure 17.5). At wavelengths

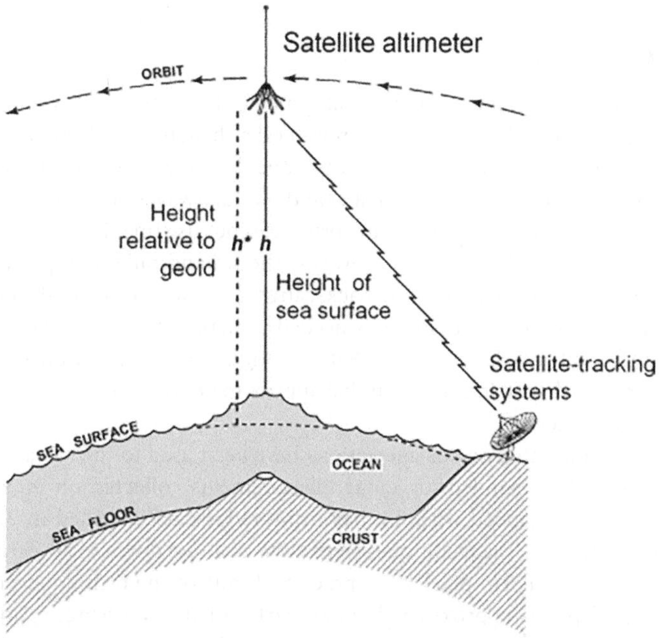

FIGURE 17.5 Illustration of the use of radar altimetry to estimate large-scale bathymetry of the deep ocean floor.[18] The radar pulse provides the height of the sea surface to centimeter-scale, h, which is combined with the height relative to Earth's geoid, h^*, and acoustic soundings in order to estimate bathymetry.

from 1–200 km, gravity anomaly variations are highly correlated with seafloor topography and have been used to gravimetrically map bathymetry of the world oceans with a spatial resolution of ~10 km resolving features ~20 km in scale.[18]

Sea-surface heights are not impacted by bathymetry alone, but also by oceanographic factors such as waves, currents, and tides, as well as the underlying density of the seafloor. To effectively use altimetry for bathymetric purposes, the sea-surface height that would exist without any perturbing factors or "geoid" must be modeled (Figure 17.5). Determining the geoid over the oceans is an active area of research. The footprint of the pulse must be large enough to average out the local irregularities in the surface due to ocean waves. Tidal corrections must be applied. Corrections to the travel time of the pulse are also made for water vapor and electrons in the atmosphere. Profiles from many satellites, collected over many years, are combined to make high resolution images of the geoid showing major mass and density differences on the seafloor.

To produce reasonable estimates of bathymetry, the gravimetric anomaly measurements are combined with acoustically determined soundings to construct uniform grids of seafloor topography.[19,20] The methods work best in the deep-ocean basins, where sediments are thin and geological features can be mapped at coarser resolution. For example, gravimetric techniques are the primary method to locate seamounts >2000 m in height across the global ocean floor. However, errors arise in regions with sparse acoustic soundings and where thick sediments can bury the underlying "basement" topography. Gravimetric methods for estimating depth are of limited value over continental shelves, where sediments are thick and conventional bathymetric soundings are plentiful.

Comprehensive high-resolution bathymetry grids have been constructed from quality-controlled ship depth soundings with interpolation between sounding points guided by satellite-derived gravity data.[1,21] These data, freely available online, are used in a variety of applications, including tsunami path forecasting and ocean current models. However, such gridded datasets are not accurate enough to assess hazards to navigation, and in shallow water where other remote sensing techniques, such as LIDAR, would produce significantly better results.

Imaging Radar Systems

Wave theory has been used to deduce shallow bathymetry since the 1940s when sequences of aerial photographs were used by the Allied forces to map water depth along the Normandy coast of France.[22] As waves move into shallow water, their speed decreases and the wavelengths get shorter. Wave dispersion theory can be used to infer the bottom depth that would produce waves of a given speed (or celerity) and wavelength. Modern techniques rely on the same principles but use imaging radar systems to get accurate measurements of wavelength and speed of waves. The X-band radars, in particular, interact with the small sea-surface capillary waves and the backscattered signal can be used to resolve wave fields. A minimum degree of sea-surface roughness is needed to make such measurements with significant wave heights >1 m and wind speeds >3 ms^{-1}.[23] Water depth is calculated both from linear wave theory and from a nonlinear wave dispersion equation that approximates the effects of amplitude dispersion of the waves in the shallow water.[24]

Marine radar systems mounted on coastal stations have been used to infer nearshore bathymetry.[25] The technique was recently expanded for radar measurements collected on moving vessel.[23] Over 64 km^2 of coastal seas were mapped within a bay using radar data collected in 2 h using equipment already available on the ship.[23] The techniques were quite accurate down to 40–50 m water depth with a horizontal resolution of 50–100 m pixels. The practical limits of the technique vary by location and occur when the water depth is approximately one-quarter of the wavelength. Imaging radars measure a wide swath of sea in 360° around the vessel and, unlike acoustic systems, only a single pass is needed to estimate bathymetry over a wide area. Because ships already have radars for navigation, only a recorder is necessary to collect the data and the method is quite cost effective. The method does not resolve bathymetry at high enough resolution for most navigational purposes, but is especially useful for reconnaissance of coastal regions for military and scientific purposes and surveying migrating features such as interisland channels.

Conclusion

Knowledge of ocean bathymetry has progressed rapidly in the past century due to the advancement of techniques using acoustics, optics, and radar. The ocean has been mapped at a variety of spatial resolutions from kilometers to centimeters, but considerable work has yet to be done to accurately map the vast underwater landscape. More acoustic soundings are required to validate gravimetric bathymetry in remote regions of the world (e.g., the Southern Ocean), better characterization of the geoid is necessary for refining gravimetric techniques, more accurate tidal models are needed to assess nearshore bathymetry, and higher spatial resolution data are vital to assess shallow water bathymetry to delineate coastlines for storm surges and potential changes to sea level. The seafloor may be considered one of the last largely unexplored and most dynamic landscapes on Earth.

Acknowledgments

Acknowledgments to Optic Industries, Dave Sandwell, and the U.S. National Oceanic and Atmospheric Association (NOAA) who provided figures for this entry. Thanks to Rick Stumpf, Jerry Mills, and Walter Smith for reviewing this entry. Funding was provided from the U.S. Office of Naval Research and U.S. National Aeronautics and Space Administration (Dierssen) and National Oceanographic and Atmospheric Administration (Theberge).

References

1. Amante, C.; Eakins, B.W. ETOPO1 1 Arc-Minute Global Relief Model: Procedures, Data Sources and Analysis. *NOAA Technical Memorandum NESDIS NGDC-24*, 2009, 19.
2. Wang, Y.; Fang, G.; Wei, Z.; Wang, Y.; Wang, X. Accuracy assessment of global ocean tide models base on satellite altimetry. Adv. Earth Sci. **2010**, *25*, 353–359.
3. Stoker, J.M.; Tyler, D.J.; Turnipseed, D.P.; Van Wilson Jr, K.; Oimoen, M.J. Integrating disparate lidar datasets for a regional storm tide inundation analysis of Hurricane Katrina; J. Coast. Res. **2009**, *53*, 66–72.
4. Dierssen, H.M.; Theberge, A.E. Bathymetry: History of Seafloor Mapping. In *Encyclopedia of Ocean Sciences*; Taylor and Francis: New York, 2012.
5. Theberg, A. Appendix: The history of seafloor mapping. In *Ocean Globe*; Breman, J. Ed.; ESRI Press: 2010; pp 237–274.
6. Kvitek, R; Iampietro, P. In *Ocean Globe*; ESRI Press; Redlands, CA, J. Coast. Res. **2010**, *53*, 66–72.
7. Poppe, L; Polloni, C. Long Island Sound Environmental Studies. U.S. Geolog. Surv. Open File Rep. **1998**, *98*, 1.
8. Moline, M.A.; Woodruff, D.L.; Evans, N.R. Optical delineation of benthic habitat using an autonomous underwater vehicle. J. Field Robot. **2007**, *24*, 461–471.
9. Dierssen, H.M. Perspectives on empirical approaches for ocean color remote sensing of chlorophyll in a changing climate. Proc. Nat. Acad. Sci. **2010**, *107*, 17073.
10. Lyzenga, D.R. Passive remote sensing techniques for mapping water depth and bottom features. Appl. Opt. **1978**, *17*, 379–383.
11. Dekker, A.G. Intercomparison of shallow water bathymetry, hydro-optics, and benthos mapping techniques in Australian and Caribbean coastal environments. Limnol. Oceanogr. Methods **2011**, *9*, 396–425.

12. Dierssen, H.M.; Zimmerman, R.C.; Leathers, R.A.; Downes, T.V.; Davis, C.O. Ocean color remote sensing of seagrass and bathymetry in the Bahamas Banks by high-resolution airborne imagery. Limnol. Oceanogr. **2003**, *48*, 444–455.
13. Lee, Z.P.; Hu, C.; Casey, B.; Shang, S.; Dierssen, H.; Arnone, R. Global shallow-water high resolution bathymetry from ocean color satellites. EOS Trans. Amer. Geophy. Union **2010**, *91*, 429–430.
14. McClain, C.R. A decade of satellite ocean color observations. Annual Rev. Mar. Sci. **2009**, *1*, 19–42.
15. Dierssen, H.M.; Randolph, K. In *Encyclopedia of Sustainability Science and Technology*; Springer: Verlag Berlin Heidelberg 2013, http://www.springerreference.com/index/chapterdbid/310809.
16. Quadros, N.D.; Collier, P.A. A New Approach to Delineating the Littoral Zone for an Australian Marine Cadastre, 2010.
17. Wozencraft, J.; Millar, D. Airborne lidar and integrated technologies for coastal mapping and nautical charting. Mar. Techno. Soc. J. **2005**, *39*, 27–35.
18. Sandwell, D.T.; Smith, W.H.F. Marine gravity anomaly from Geosat and ERS 1 satellite altimetry. J. Geophys. Res. **1997**, *102*, 10039–10054.
19. Smith, W.H.F.; Sandwell, D.T. Bathymetric prediction from dense satellite altimetry and sparse shipboard bathymetry. J. Geophys. Res. All Ser. **1994**, *99*, 21.
20. Dixon, T.; Naraghi, M.; McNutt, M.; Smith, S. Bathymetric prediction from Seasat altimeter data. J. Geophys. Res. **1983**, *88*, 1563–1571.
21. GEBCO, *GEBCO_08 Gridded Bathymetry Data*. General Bathymetric Chart of the Oceans, 2008.
22. Hart, C.; Miskin, E. Developments in the method of determination of beach gradients by wave velocities. *Air Survey Research Paper No. 15, Directorate of Military Survey*, UK War Office, 1945.
23. Bell, P.S.; Osler, J.C. Mapping bathymetry using X-band marine radar data recorded from a moving vessel. Ocean Dyn. **2011**, *61*, 1–16.
24. Bell, P.; Williams, J.; Clark, S.; Morris, B.; Vila-Concej, A. Nested radar systems for remote coastal observations. J. Coast. Res. **2006**, *39*, 483–487.
25. Bell, P.S. Shallow water bathymetry derived from an analysis of X-band marine radar images of waves. Coast. Eng. **1999**, *37*, 513–527.
26. Dierssen, H.M. Benthic ecology from space: optics and net primary production in seagrass and benthic algae across the Great Bahama Bank. Mar. Ecol. Progress Ser. **2010**, *411*, 1–15.
27. Dierssen, H.M.; Zimmerman, R.C.; Leathers, R.A.; Downes, T.V.; Davis, C.O. Ocean color remote sensing of seagrass and bathymetry in the Bahamas Banks by high resolution airborne imagery. Limnol. Oceanogr. **2003**, *48*, 444–455.

18

Bathymetry: Features and Hypsography

Heidi M. Dierssen
University of Connecticut
Norwegian University of Science and Technology

Albert E. Theberge, Jr.
NOAA Central Library

Introduction

Up until the mid-20th century, the seafloor was thought to be featureless consisting of bits of rock and sediment eroded from the continents and flushed into the ocean reservoirs.[1] However, we now know that ocean bathymetry encompasses a varied seascape including vast mountain ranges, deep trenches, fracture zones extending for thousands of miles, the flattest plains on Earth, and a plethora of lesser meso- and microscale features ranging from individual seamounts to individual ripples and tidal channels. Comprehension of this landscape is a key component of the theory of plate tectonics, an intr insic part of marine ecology, and a significant component of ocean circulation on global scales to individual estuaries. Indeed bathymetry and ocean science are intricately intertwined and one cannot be understood without the other.

Hypsography is the study of the distribution of bathymetric features across the Earth's surface and provides the percentage of the seafloor covered by large-scale features, such as continental shelves and abyssal plains. Fine-scale oceanic features can also be mapped at high resolution using new technology.[2] Assessing the geomorphology or shape of underwater landforms is now an important component in developing marine reserves to protect sensitive marine ecosystems. Many organisms aggregate at discontinuities in the bathymetry that can also affect local eddies and current features that shape aspects of local water properties. Bathymetry of coastal estuaries and bays must also be evaluated in terms of tidal currents driven by the gravitational influence of the moon that can change bathymetry on hourly time scales. In a little over a century, the vast underwater landscapes from large- to small-scale features have been revealed, and this knowledge has contributed to the fundamental understanding of the Earth's processes.

Bathymetric Features and Hypsography

The large-scale primary and secondary features of the deep seafloor and fine-scale features on the marine and coastal seafloor are discussed further.

Large-Scale Features

Covering 3.6×10^8 km^2, or 71% of the Earth's surface, the oceanic basins are dominated by three major physiographic features: 1) deep basins covered with abyssal plains and hills covering ~53% of the seafloor; 2) the world-girdling oceanic ridge system covering ~31% of the seafloor; and 3) continental margins and shallow seas comprised of continental shelf, slope and rise covering over 16% of the seafloor (Figure 18.1). Undulating abyssal hills and gently sloping to flat abyssal plains lying between 4 and 6.5 km spread across the vast ocean floor. Very large features of significant lineal extent include the great *trenches* and *fracture zones* of the world ocean that can extend for thousands of miles either along the margins of ocean basins as in the case of trenches or across ocean basins as in the case of fracture zones. The ridges and rises associated with fractures zones predominantly occur at depths from 2 to 4 km across the world ocean (Figure 18.1). Oceanic trenches from 6.5 to ~11 km in depth cover 0.2% of the world ocean. Continental margins are made up of continental shelves, continental slopes, and continental rises. *Continental shelves* border the continents and vary in size and shape across the world ocean. Comprised of continental crust, they are relatively flat containing thick layers of terrigenous sediments from riverine input. The widest continental shelves tend to occur along

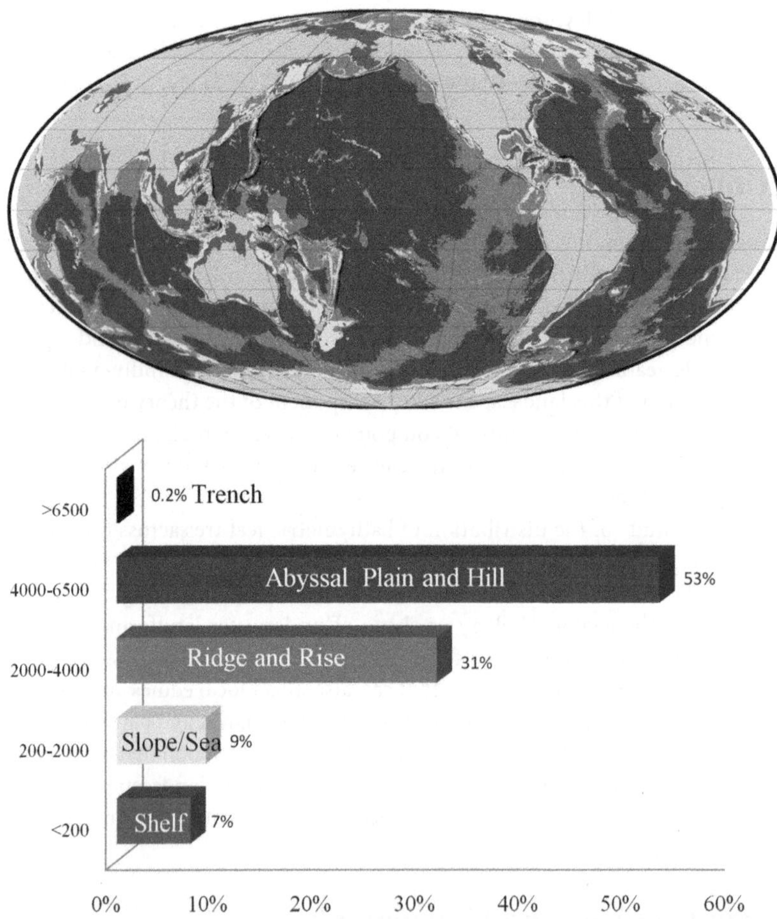

FIGURE 18.1 Depth distribution of seafloor in the global ocean translating roughly to trenches (>6500 m), abyssal plains and hills (4000–6500 m), ridges and rises (2000–4000 m), continental slope and shallow seas (200–2000 m), and continental shelf (<200 m).

Source: Data from ETOPO1.[14]

passive margins where considerable sediment can build up over time. In active continental margins with converging plates, there can be little to no shelf at all. The depth limit describing the continental shelf is often considered to be <200 m, but deeper shelves extending to 500 m commonly occur in polar regions north and south of 60° latitude including Antarctica and Greenland. The *shelf break* indicates the point where the continental shelf abruptly ends and the slope steepens dramatically in what is referred to as the *continental slope.* At the base of the slope, the *continental rise* is characterized by a gentle slope of accumulated sediment that merges into the abyssal plain on passive continental margins.

Superimposed on the primary features of the seafloor landscape are secondary features such as the *median valley* associated with the mid-ocean ridge system, mountain ranges, and ridges not associated with the mid-ocean ridge system. The abyssal hills and plains are interrupted by a spectrum of larger hills, knolls, and cones with varying elevations culminating with the higher underwater mountains having relief >1000 m called *seamounts.* With over 100,000 estimated to occur on the seafloor, seamounts are typically formed from volcanic activity in association with diverging plates, hot spots, or converging plates. Tablemounts or *guyots* are flat-topped seamounts formed by wave erosion which has been transported to deeper water and submerged on a moving plate. Volcanic activity forming mountains (either undersea or continental) commonly occurs on the back side of the trench feature where the downthrust subducting plate heats and melts. *Volcanic island arcs* are formed in this manner and run parallel to a trench at a distance of ~200 km from the trench axis. *Submarine canyons,* generally extensions of large rivers or formed by turbidity flows, periodically cut through the continental slope and transport considerable sediments from the shelf to the deep sea. Other secondary features include individual basins, troughs, deeps, holes, escarpments, benches, terraces, regional and local plains, great canyons, and lesser valleys found on the continental slopes of the world.

Virtually all of the primary and many of the secondary features of the world ocean were discovered by either line and sinker technology or by single-beam echo sounding.[1] However, because the introduction of multibeam sounding systems coupled with the global positioning system,[2] many additional deep-sea features have been discovered including hydrothermal vents and vent fields, mud volcanoes, mud wave and dune fields, individual lava flows, landslide scars, and diapiric structures including commercially valuable salt domes.

Comprehension of the significance and inter-relationships of the primary and secondary features of the seafloor has come about only since the formulation of the theory of plate tectonics. Plate tectonics describes the surface of the Earth in terms of numerous plates that either move away from each other at divergent plate boundaries, collide at convergent plate boundaries, or slide past each other along great faults known as transform faults. The median valleys of mid-oceanic ridges are the primary location of divergent boundaries, also known as seafloor spreading centers, where new seafloor is being produced from upwelling magma. Sites where plates collide with one plate being thrust under another consuming the seafloor, known as subduction zones, are marked by the great oceanic trenches, while colliding plates with no subduction form the great terrestrial mountain ranges of the world such as the Alpine-Himalayan belt of Europe and Asia. Sites where plates slide past one another are marked primarily by the numerous offsets on the mid-oceanic ridge system and are known as transform faults or fracture zones.

The presence of the abyssal plains and hills can also be explained from tectonic processes in combination with sedimentary processes. As new seafloor is formed and then pushed away from the median valley, parallel trains of abyssal hills are formed. With the passage of time, these hills are draped with sediment. When completely covered, the resulting surfaces are termed abyssal plains. Lineal chains of islands and seamounts can also be formed over long-lived stationary "hot spots" that provide conduits for magma to come to the surface. As the seafloor moves over these hot spots, lineal chains of islands and seamounts are formed trending in the direction of seafloor motion. The Hawaiian Islands and Emperor Seamount chains are examples of this phenomenon.

Fine-Scale Features

Marine geomorphology—the study of underwater landforms—is becoming an important component of management of marine ecosystems. Increasingly, the fine-scale ridges and troughs of the underwater landscape are being used to assess and define biological habitats for fish and other marine organisms. In Long Island Sound (Figure 18.2a), the largest urban estuary in the United States, a concentrated mapping program has characterized the physical character of the seafloor using multibeam sonar for depth and "sidescan" sonar for a qualitative view of seafloor reflectivity characteristics.[3] Such high-resolution images showing sand ripples, uplifted ridges, and small depressions in bathymetry (Figure 18.2b) are valuable for assessing benthic habitats, current modeling, and evaluating sedimentary processes.

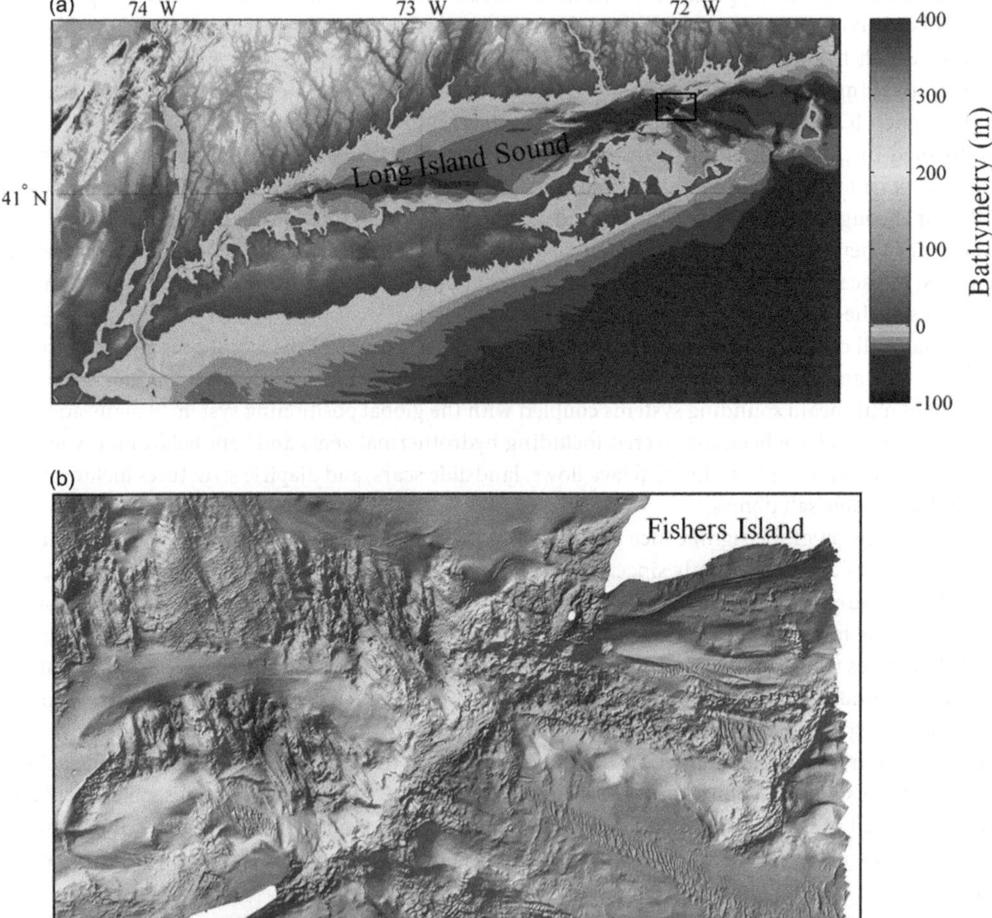

FIGURE 18.2 **(See color insert.)** (a) Bathymetry of Long Island Sound, the United States as mapped from NOAA digital high-resolution coastal gridded data (Amante[14]). (**b**) High-resolution bathymetry from multibeam sonar of the area around Fishers Island, outlined in black in Panel A, highlights a variety of detailed seafloor features (Poppe and Polloni[3]).

Source: Image courtesy of Long Island Sound Resource Center.

Many species aggregate at locations with particular geomorphic features, such as reef promontories, uplifted ridges, and shelf edges, that are associated with abrupt discontinuities in surrounding structure.[4] When related to the seafloor, the terms "habitat heterogeneity" and "keystone structures" are applied to seafloor features that may enhance biodiversity by providing refugia from predation, enhanced or alternative food resources, spawning areas, stress gradients and substratum diversity.[5,6] Not only do undersea landforms provide three-dimensional structure, but they can also influence eddies and currents and create oceanographic conditions important for the growth and survival of marine communities.[7] Submarine canyons, for example, are sites of organic enrichment and benthic productivity due to sinking of particulate materials from the shores.[5]

Modern techniques allow for mapping of the seafloor with higher precision than ever before and assessment of benthic structures at meter scales or less. Because of linkages between bathymetric features, biodiversity, and biogeochemistry, new methods for classifying the seafloor have been developed. A benthic location can be classified according to a terrain ruggedness index[8] or more specifically the bathymetric position index (BPI), for example, which identifies the sloping nature of a region or "neighborhood." A positive BPI denotes ridge features; zero values indicate flat terrain or region with a constant slope; and negative values correspond to valleys. Bathymetric data tends to be spatially autocorrelated where locations closer together are more related than those further apart, and thereby the value of the BPI is related to the scale of data under investigation and the type of topography within the selected region.[9] Often BPIs are estimated at different scales to define different scales of features from small sand ripples to large-seascape features such as channels. Fine-scale benthic features are also characterized by their "rugosity," which is a measure of the bottom complexity or "bumpiness." It is often parameterized as the surface area in relations to the planar area. A rugosity of 1 indicates a smooth, flat surface and values >1 indicate higher relief. Regions with higher rugosity have been linked to higher biodiversity.[10]

Ongoing efforts are underway to organize and define coastal and marine environments in meaningful units that can be compared across different temporal and spatial scales. The Coastal and Marine Ecological Classification Standard (CMECS) is one such effort that defines a habitat using five different components: water column, benthic biotic, surface geology, sub-benthic, and geoform component[11] (Figure 18.3a). The geoform component classifies the major geomorphic or structural characteristics of the seafloor and is often assessed with bathymetric data. The classifiers tend from large to small scale and begin with coastal region, physiographic setting, geoform (natural), and subform, with an additional category called geoform (anthropogenic). The physiographic setting is defined from many of the primary oceanic features described earlier (e.g., mid-ocean ridge, abyssal plain, etc.) and also includes the "coast" defined as the land-water interface. Geoform classifies features from centimeter to kilometer scale and has been separated into structures that are natural and anthropogenic in origin. Anthropogenic geoforms cover jetties, piers, artificial reefs, and other man-made structures and seafloor areas where human activities have changed the landscape (e.g., trawling and prop scars). Within the natural geoforms are definitions to describe marine features (e.g., seamounts, guyots, banks, boulder fields, and megaripples) and those structures in coastal regimes (e.g., fjord, beach, delta, coral reef, and lagoon). Finally, the subform category describes smaller features within the geoform and can include walls, escarpments, and also such factors as substrate orientation which can impact the availability of light and wave energy. Hence, many layers of complexity are required to describe the bottom topography itself, not to mention the additional components describing the type of substrate and biota associated with these features.

In a reef system, these techniques can be used to define zones corresponding to the geomorphology of the coral reef community. A recent analysis mapped the reef structures according to the following categories: shoreline intertidal, vertical wall, lagoon, back reef, reef flat, reef crest, fore reef, bank/shelf, bank/shelf escarpment, channel, and dredged area[12] (Figure 18.3b). These geomorphologic zones can be comprised of different benthic substrate or cover types. The "lagoon," for example, may be covered by patch reefs, sand, and seagrass beds. Integrating and assessing the biota and geomorphology is important for managing these sensitive coastal resources.

(a)

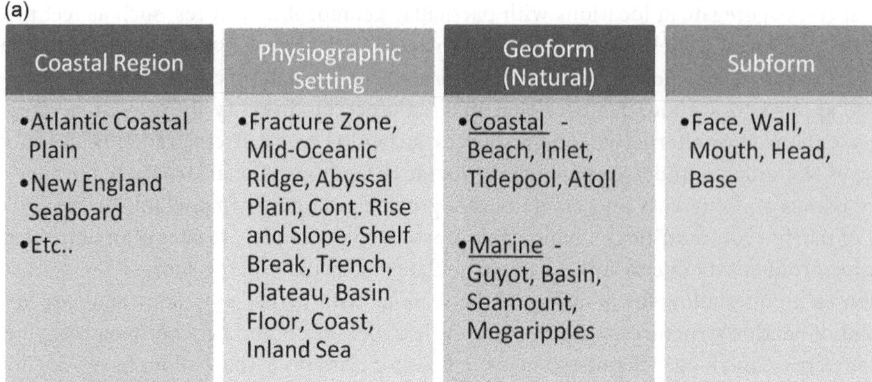

(b)

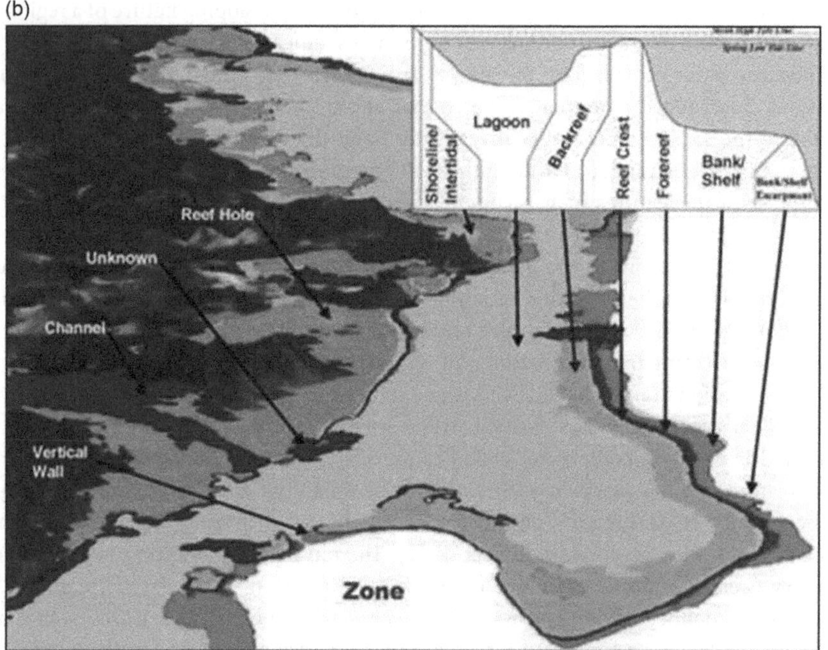

FIGURE 18.3 (**a**) Geomorphologic classification scheme proposed by the Coastal and Marine Ecological Classification Standard (CMECS) categorize features from large (coastal region) to small (subform) spatial scales. (**b**) A classification of a coral reef demonstrating the different bathymetric features mapped from a bathymetric grid (Battista[12]).
Source: Image from NOAA, NOS, NCCOS Biogeography Branch.[12]

Hypsography of the Ocean

The bathymetry of each of the five ocean basins varies considerably (Figure 18.4). Ridges and rises associated with seafloor spreading can be found in all five of the ocean basins. As mapped in the 1800s,[1] the Mid-Atlantic Ridge extends from the north of Iceland to the Southern Ocean boundary. Wide continental shelves (<200 m depth) are visible along much of the western Atlantic Ocean. The vast Pacific Ocean contains mostly broad expanses of deep ocean floor including abyssal hills, chains of seamounts, and lineal scars of great fracture zones. Prominent bathymetric features also occur

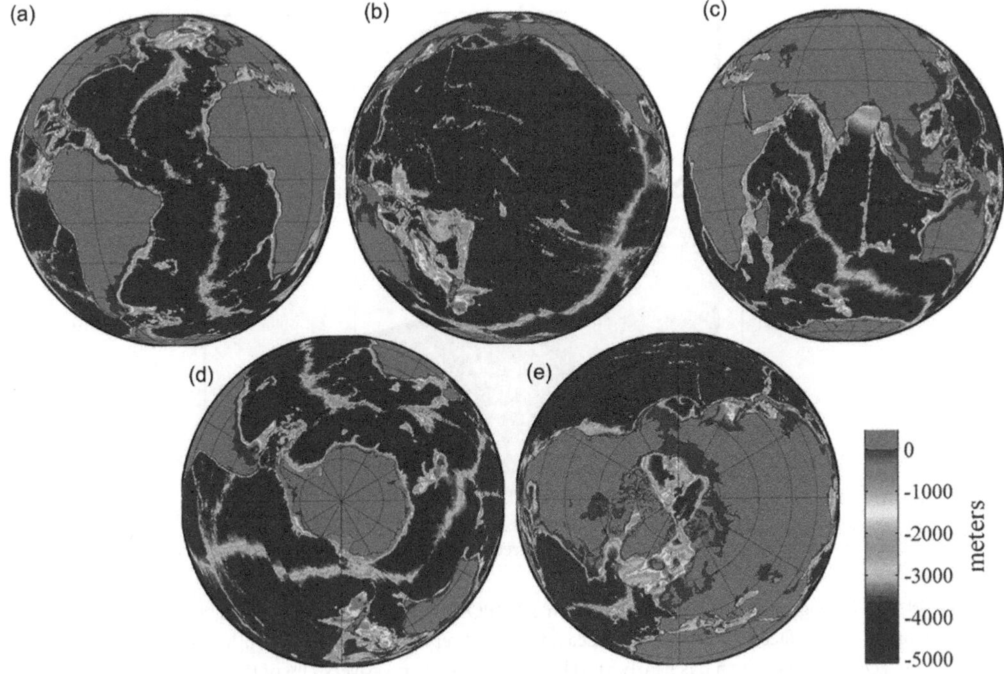

FIGURE 18.4 (See color insert.) Gridded bathymetric data are a compilation of satellite gravimetric data and sonar data from throughout the global ocean.[14] These plots show bathymetry (m) for each ocean basin in orthographic projection: (**a**) Atlantic, (**b**) Pacific, (**c**) Indian, (**d**) Southern, and (**e**) Arctic. Land surfaces are shown in gray.

along the margins of the Pacific Ocean. The Pacific Ocean is bounded by the "Ring of Fire" consisting primarily of subducting plates on the western, northern, and eastern margins with deep trenches and considerable earthquakes and volcanic activity. The southern boundary is defined by the Pacific–Antarctic Ridge and the southern portion of the East Pacific Rise. The Indian Ocean is crossed by ridges and volcanic island structures where the African, Indian, Australian, and Antarctic crustal plates all converge. With the exception of Western Australia, the Indian Ocean has limited continental shelf area. In contrast, the Arctic Ocean represents the shallowest of ocean basins and is characterized by broad expansive deep (500 m) shelves and shallow seas. The Southern Ocean also contains deep continental shelves (500 m), large expanses of abyssal plains, and prominent ridge features bordering the Pacific, Atlantic, and Indian Oceans.

Taken from the Greek word "hypsos," meaning height, hypsography is the study of the distribution of elevations on Earth and is used to relate bathymetry to the projected two-dimensional surface area of a region or feature. The "hypsographic curve" of Earth shows the cumulative percentage of the Earth surface area at different elevations from above and below sea level (Figure 18.5). As shown, the ocean represents ~71% of the Earth's surface and most of this surface area is quite deeply submerged at close to 4 km below sea level. Based on our current and still limited knowledge of bathymetry, the average depth of the oceans is calculated to be 3.7 km. The prevalence of deep seafloor is related to the heavier elemental composition of oceanic crust that makes it much denser than terrestrial crust. Along the perimeters of the ocean basins, the ocean slopes rapidly and very little of the Earth's surface occurs between the continental shelves (<500 m) and the deep sea (4 km). Metaphorically, the ocean basins can be considered like deep bowls with thin rims and steep sides. However, the bowl bottoms are not smooth, but littered with abyssal hills and seamounts, as well as spreading ridges that appear as jagged seams laced across a baseball.

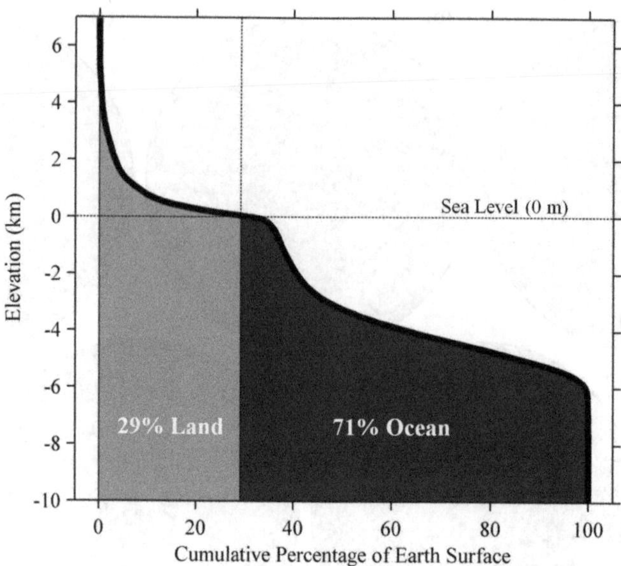

FIGURE 18.5 Hypsographic curve modeled from ETOPO1 gridded bathymetric data[14] shows the cumulative percentage of Earth's surfaces above and below sea level. Land represents a smaller depth range and comprises 29% of Earth's surface; whereas, ocean has nearly an 11 km depth range and covers 71% of the Earth's surface. Over 50% of Earth falls between 4 and 6 km below sea level comprised predominantly of abyssal plains and hills.

Conclusion

Representing less than a third of the Earth's surface, the land is dwarfed by the enormity of ocean. The ocean has more relief with deep 11 km trenches compared to maximum peaks of 7 km on land. The large-scale features of the underwater landscape span the five ocean basins from the Southern Ocean northward to the Arctic Ocean. Describing and classifying coastal and marine geomorphology at fine scales is still an area of active research that requires more data and enhanced computational methods. Bathymetry is not static and the percentage of ocean is likely to increase in the next century as sea-level rises due to thermal expansion of ocean water and melting of land- based glaciers.[13] Ocean bathymetry will need to be revised particularly in the near-shore environment as the oceans get deeper and coastal regions become submerged. While islands and seamounts are continually being formed by volcanic activity, many islands will become submerged by tectonic activity and projected sea-level rise.

Acknowledgments

Funding was provided from the U.S. Office of Naval Research, U.S. National Aeronautics and Space Administration (Dierssen), and U.S. National Oceanic and Atmospheric Administration (Theberge).

References

1. Dierssen, H.M.; Theberge, A.E. Bathymetry: History of seafloor mapping. In *Encyclopedia of Ocean Sciences;* Taylor and Francis: New York, NY, 2012; 360 pp.
2. Dierssen, H.M.; Theberge, A.E. Bathymetry: Assessing methods. In *Encyclopedia of Ocean Sciences*; Taylor and Francis: New York, NY, 2012; 360 pp.
3. Poppe, L.; Polloni, C. Long Island Sound Environmental Studies. U.S. Geol. Surv. Open File Rep. **1998**, *98*, 1.

4. Heyman, W.D.; Wright, D.J. Marine Geomorphology in the Design of Marine Reserve Networks. Prof. Geogr. **2011**, *63*, 429 –442.
5. Vetter, E.W.; Smith, C.R.; De Leo, F.C. Hawaiian hotspots: enhanced megafaunal abundance and diversity in submarine canyons on the oceanic islands of Hawaii. Mar. Ecol. **2010**, *31*, 183–199.
6. Drazen, J.C.; Goffredi, S.K.; Schlining, B.; Stakes, D.S. Aggregations of egg-brooding deep-sea fish and cephalopods on the Gorda Escarpment: a reproductive hot spot. Biol. Bul. **2003**, *205*, 1.
7. Ezer, T.; Heyman, W.D.; Houser, C.; Kjerfve, B. Modeling and observations of high-frequency flow variability and internal waves at a Caribbean reef spawning aggregation site. Ocean Dyn. **2010**, *61*(5), 1 –18.
8. Riley, S.J.; DeGloria, S.D.; Elliot, R. A terrain ruggedness index that quantifies topographic heterogeneity. Intermt. J. sci. **1999**, *5*, 23 –27.
9. Lundblad, E.; Miller, J.; Rooney, J.; Moews, M.; Chojnacki, J.; Weiss, J. Mapping Pacific Island Coral Reef Ecosystems with Multibeam and Optical Surveys. Coastal GeoTools. **2005**.
10. Lundblad, E.; Wright, D.J.; Naar, D.F.; Donahue, B.T.; Miller, J.; Larkin, E.M.; Rinehart, R.W. Classifying benthic terrains with multibeam bathymetry, bathymetric position and rugosity: Tutuila, American Samoa. Mar. Geod. **2006**, *29*, 89 –111.
11. Federal Geographic Data Committee, Standards Working Group *Coastal and Marine Ecological Classification Standard Version 3.1 (Working Draft)* NOAA Coastal Services Center, 2010.
12. *Battista, T.A.; Costa, B.M.; Anderson, S.M. in NOAA Technical Memorandum NOS NCCOS 59, NOAA Center for Coastal Monitoring and Assessment, Biogeography Branch; Silver Spring: MD, 2007.*
13. *IPCC, Synthesis Report. Contribution of Working Groups I, II and III to the Fourth Assessment Report of the Intergovernmental Panel on Climate Change Geneva, Switzerland, 2007.*
14. Amante, C.; Eakins, B.W. ETOPO1 1 Arc-Minute Global Relief Model: Procedures, Data Sources and Analysis. *NOAA Technical Memorandum NESDIS NGDC-24*, 2009, 19 pp.

19

Bathymetry: Seafloor Mapping History

Heidi M. Dierssen
University of Connecticut
Norwegian University of Science and Technology

Albert E. Theberge, Jr.
NOAA Central Library

Early Efforts

The beginning of modern seafloor mapping coincided with the advent of systematic oceanographic observations (i.e., modern oceanography), deep-sea scientific dredging, and the commercial desire to lay deep-sea telegraph cables. Within a century, the concept of a featureless and static seafloor was shattered and the findings of a detailed ocean bathymetry were revealed (Figure 19.1). Some of the first recorded measurements of bathymetry were made by the British explorer Sir James Clark Ross in 1840, by the U.S. Coast Survey beginning in 1845 with systematic studies of the Gulf Stream, and by the U.S. Navy, under the guidance of Matthew Fontaine Maury, beginning in 1849.[1] A weighted hemp or flax rope was dropped over the side of a vessel "lying to" (drifting) and the length of the line recorded once the sinker or lead weight reached the bottom. From a few such measurements, the first bathymetric map was produced and published by Maury in the 1853 Wind and Current Charts of the North Atlantic Ocean (Figure 19.1a). Although this particular map was not very accurate, many important seafloor features were discovered from such measurements. For example, the first map showing the full extent of the Mid-Atlantic Ridge was produced in 1877 by Wyville Thomson from measurements made on HMS *Challenger* supplemented by additional soundings made by British vessels and those of the United States and other nations[1] (Figure 19.1b). In 1875, HMS *Challenger* also discovered the first indication of the Mariana Trench with a measured depth of 4475 fathoms although it was another 30 years before the true configuration of the trench was understood.[2] Just two years later in 1877, the German geographer, Augustus Petermann, produced the first bathymetric chart of the Pacific Ocean with many features including the "Challenger Tiefe" or "Challenger Deep," and the then deepest known spot in the ocean, the Tuscarora Deep, named after the *USS Tuscarora*, which had sounded in the Japan Trench in 1874.

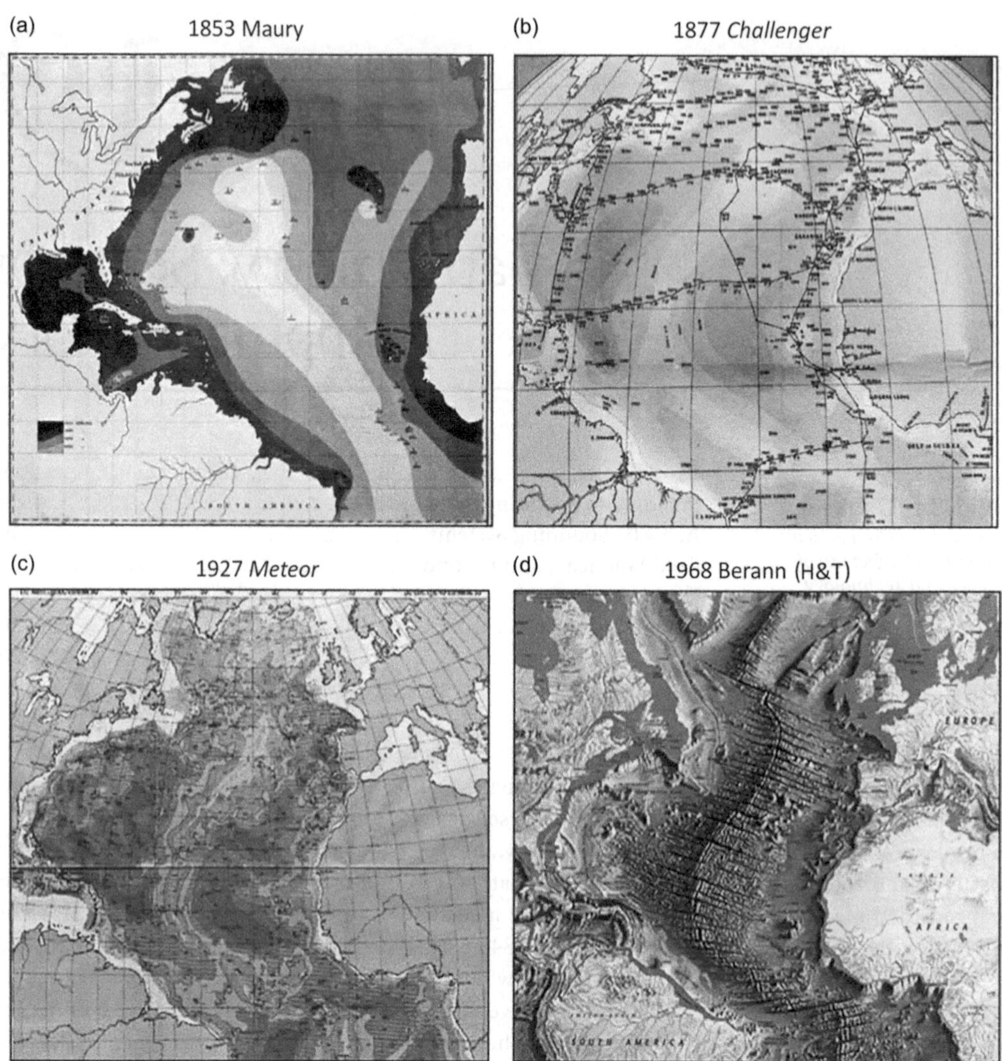

FIGURE 19.1 **(See color insert.)** In roughly one century, the mapping and conceptual understanding of ocean bathymetry were revolutionized as shown in the sequential maps of the North Atlantic. (**a**) First recorded bathymetric map created in 1853 by Maury in collaboration with the U.S. Navy only hints at the mid-ocean ridge. (**b**) Excerpt from the 1877 Thomson map based on the HMS *Challenger* measurements with line-and- sinker techniques shows the first continuous mapping of the mid-ocean ridge. (**c**) Echo sounding techniques allowed for increased frequency and a higher definition of the mid-ocean ridge, as shown in Theodor Stocks map from the *Meteor* cruises in 1927. (**d**) This 1968 Berann illustration, based on the Heezen and Tharp physiographic maps, outlines the ridge system in the North Atlantic (National Geographic Stock).

Piano-Wire Sounding System

The problems and inaccuracies inherent to making "line and sinker" measurements (e.g., angled line due to currents and vessel drifting, determining precisely when the sinker reached the bottom, etc.) led to the development of the piano-wire or Thomson sounding system in the 1870s by Sir William Thomson (later Lord Kelvin) (Figure 19.2a). The piano-wire sounding system was a line-and-sinker technique but approximately three times faster than the old hemp rope system. Because of smaller cross-section of wire

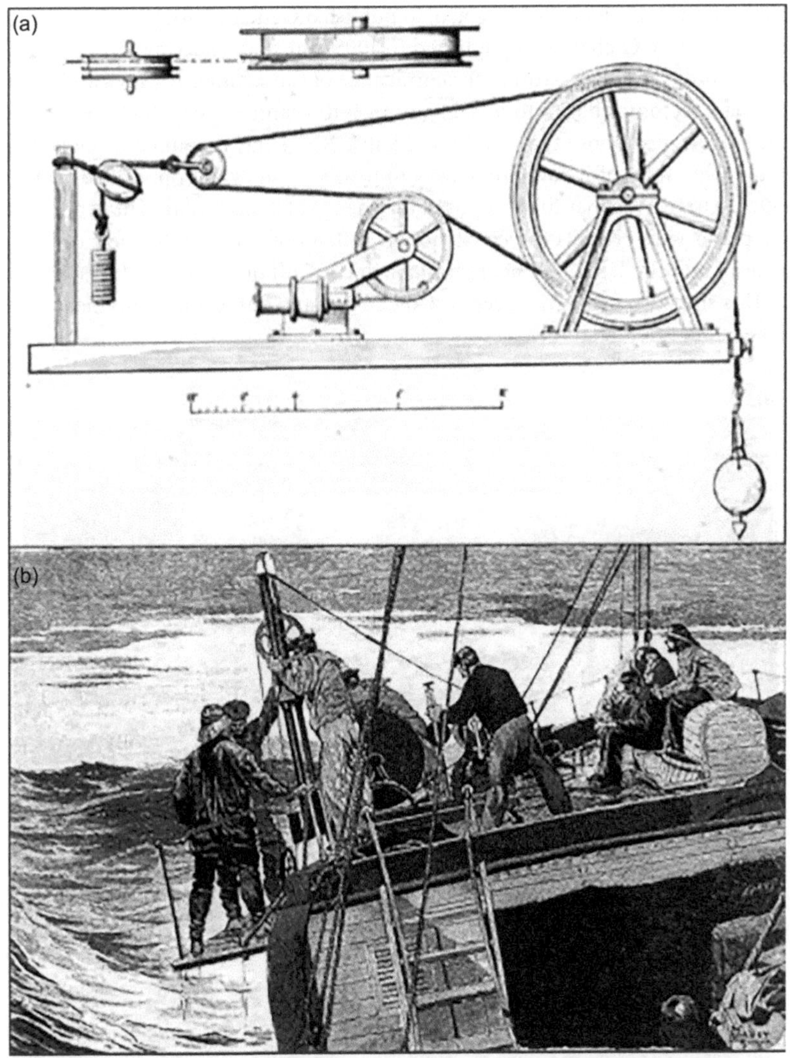

FIGURE 19.2 (**a**) Diagram of the original version of the piano-wire sounding machine or Thompson Sounding Machine invented by Sir William Thomson (later Lord Kelvin) in 1872. (**b**) "The United States Fish Commission" by Richard Rathbun. Century Magazine, Vol. 43, issue 5. 1892. "Sounding the abyss with piano-wire." This image is among the most realistic representations of sounding with the Sigsbee Sounding Machine.

(vs. hemp rope) to be affected by surface and subsurface currents, less time to observe sounding, and exact indication of when the weight hit bottom, this method was approximately an order of magnitude more accurate than hemp rope sounding (10–20 m in 2000–3000 fathom vs. 100 m or more error with hemp rope sounding). It was the Thomson sounding machine and its variants, the Sigsbee Sounding Machine developed in the U.S. Coast and Geodetic Survey (USC&GS) (Figure 19.2b) and the Lucas Sounding Machine developed by British surveyors, which outlined the great features of the world ocean prior to the introduction of acoustic sounding systems following the First World War.

The continental shelves and slopes, mid-ocean ridges, enclosed basins, and major trenches were discovered as a result of the nearly 20,000 soundings made in the deep ocean by the early 20th century. As a result of these discoveries, the 7th International Geographic Congress held in Berlin in 1899 appointed a committee on the nomenclature of undersea features and also formed a commission, under the chairmanship of Prince Albert I of Monaco, to publish a General Bathymetric Chart of the Oceans (GEBCO). This first GEBCO chart was published in 1905 and such charting has continued to modern day (Figure 19.3).[3] From this early work, the significance of the seafloor features and their relationship to modern day plate tectonics began to unfold and in 1910 Frank Bursley Taylor wrote: "It is probably much nearer the truth to suppose that the Mid-Atlantic Ridge has remained unmoved, while the two continents on opposite sides of it have crept away in nearly parallel and opposite directions."[4] Shortly thereafter in 1912, Alfred Wegener first proposed the theory of continental drift. Another notable landmark from the piano-wire era was the construction in 1884 of the first 3D view of the seafloor from soundings made aboard the USC&GS steamer *Blake* in the Gulf of Mexico and western Atlantic Ocean (Figure 19.3a). This ship also pioneered deep-sea anchoring and current measurement techniques while engaged in classic Gulf Stream studies.[1]

(a)

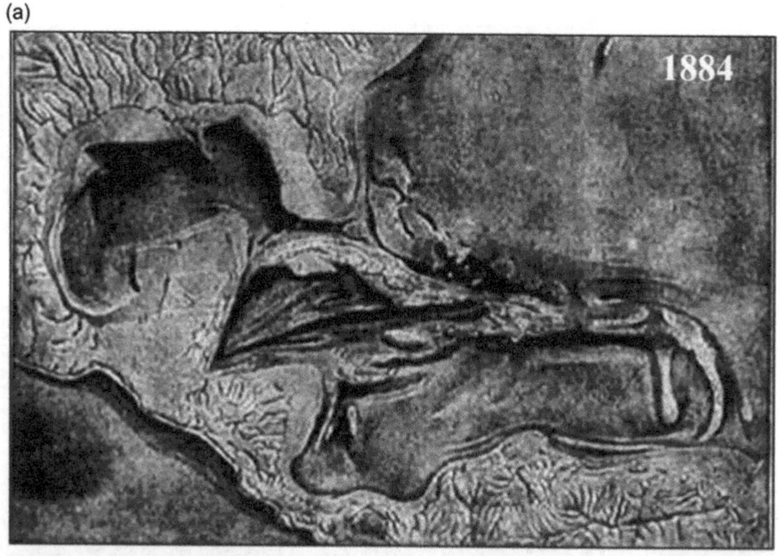

(b)

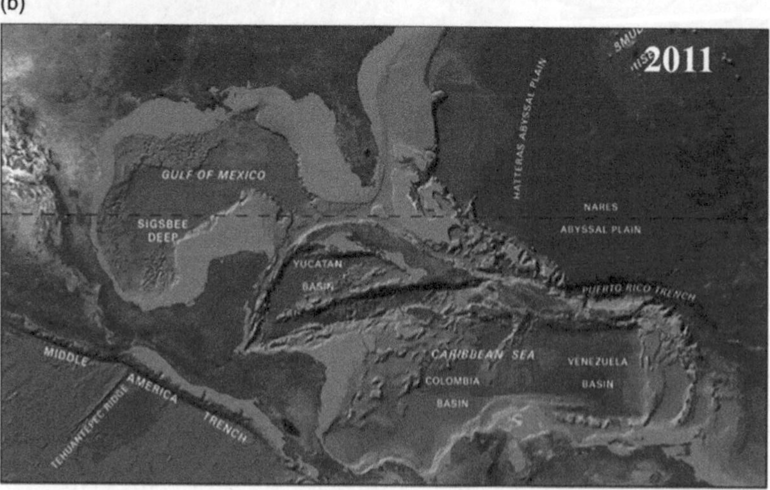

FIGURE 19.3 Early mapping efforts along coastal margins were quite accurate, particularly as shown in (**a**) the first 3D rendition of ocean bathymetry of the Gulf of Mexico and Caribbean Sea produced in 1884 from the Blake soundings. (**b**) Same region highlighted from current bathymetric maps has many of the same features.
Source: Image reproduced from the GEBCO gridded bathymetric data.[3]

Acoustic Sounding Systems

In the early 1900s, Submarine Signal Company, a forerunner of Raytheon Corporation, developed an underwater acoustic navigation system that was deployed from buoys and lightships for helping ships equipped with hydrophones to safely navigate to port during periods of reduced visibility. A similar system was also developed for ship-to-ship communication. Following the *Titanic* disaster, Reginald Fessenden of Submarine Signal Company developed an acoustic transducer that could both transmit and receive sound for the purpose of detecting objects in the water. During tests on the U.S. Revenue Cutter *Miami* in March 1914, reflections were obtained from an iceberg and, unexpectedly, from the bottom. Echo sounding was born. German, French, and American investigators modified and improved this technology for use in both outward looking antisubmarine warfare systems and downward-looking depth finding systems during the First World War. By 1922, the first truly functional acoustic depth measuring devices were in use making piano-wire sounding systems obsolete overnight.

The first issue of the *International Hydrographic Review*, the publication of the newly formed International Hydrographic Organization, contained a profile of the Atlantic Ocean seafloor from Boston to Gibraltar obtained by a U.S. Navy-developed Hayes Sonic Depth Finder mounted on the U.S.S. *Stewart* in 1922 (Figure 19.4a). This profile was derived from over 900 soundings taken during the transit proving both the efficacy of acoustic sounding (the word sounding does not refer to sound; it is derived from the Old French word *sonder* meaning, "to measure") and its ease of use and accuracy.

Overnight, echo sounding became the standard technique for observing bathymetry. Echo sounding determines bottom depth from measuring the time required for a sound pulse to be emitted from a transmitter, travel to the bottom and be reflected, and then travel back to a receiver unit. Dividing this time by 2 and multiplying times the velocity of sound in sea water gives the measure of the depth.[5] Contemporaneous with the *Stewart* work, the U.S. Navy also equipped U.S.S. *Corry* and U.S.S. *Hull* with echo sounders for conducting a survey of the California coast. This survey resulted in the first bathymetric map produced solely by acoustic technology (Figure 19.4b). These successes were followed by the famous German Meteor Expedition (1925–1927), which resulted in over 67,000 soundings of the Atlantic Ocean. In addition to mapping the axis of the Mid-Atlantic Ridge (Figure 19.1c), this expedition delineated for the first time the abyssal hills extending outward from the ridge axis. With the invention of the radio-acoustic ranging navigation system by the USC&GS in 1924, the position of a ship could be determined accurately and literally, millions of soundings were made on the continental shelf and slope of the U.S. prior to World War II.[1] This navigation system was the first ever devised of sufficient accuracy for use in offshore bathymetric surveying operations. The combination of acoustic sounding and precise navigation led to the discovery of many mesoscale features that otherwise would have been impossible solely with the use of celestial navigation.

The 1930s saw a rapid advance of knowledge of the seafloor. The British Egyptian John Murray Mabahiss Expedition to the Indian Ocean discovered the first indications of the median valley of the mid-ocean ridge system; the German *Meteor* expeditions continued and, in 1937, the German oceanographer Gunter Dietrich discovered the median valley of the Mid-Atlantic Ridge; the USC&GS, in addition to its continental shelf and slope surveys, made a number of transects across the Gulf of Alaska discovering lineal chains of seamounts and also flat-topped seamounts, later termed guyots. The USC&GS also made detailed surveys of the bathymetry of the Aleutian Trench. As a result of its continental shelf and slope surveys, the C&GS discovered many large canyons on the east coast of the United States, the Mendocino Escarpment off the California coast which was the first indication of the great lineal features now known as fracture zones, and, of great commercial significance, salt domes on the Texas–Louisiana continental shelf and slope.

The exploration and mapping of the seafloor was interrupted by World War II except for isolated efforts such as the serendipitous discovery and mapping of seamounts and other features by Dr. Harry Hess (serving as a naval officer during the war) in the western Pacific Ocean. He discovered many flat-topped seamounts and called them "guyots." He coined the term "guyot" after Arnold Henry Guyot

(a)

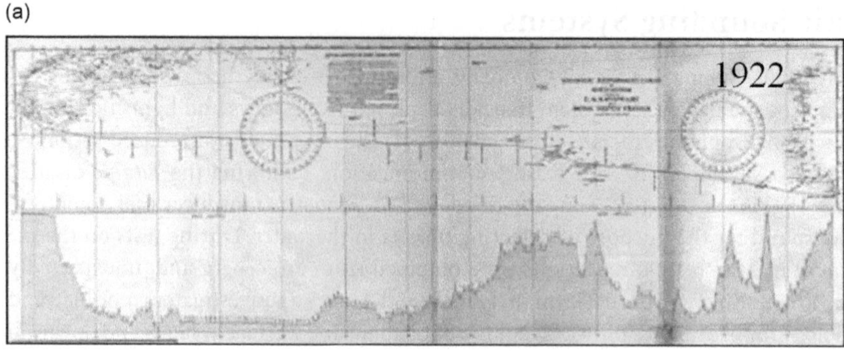

(b)

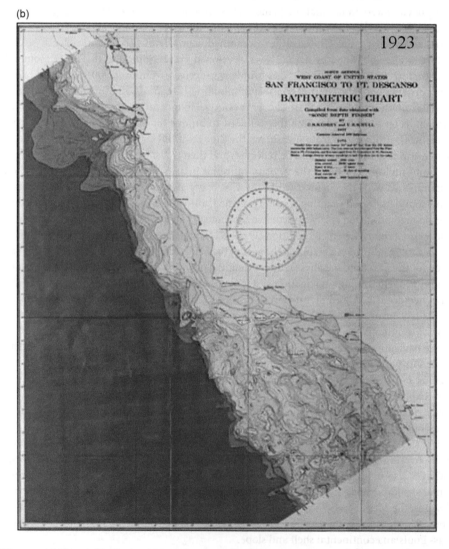

FIGURE 19.4 (**a**) This profile illustrates the first acoustic line of soundings across the Atlantic Ocean obtained by the USS *Stewart* in 1922. (**b**) Published in 1923, Hydrographic Office Miscellaneous Chart 5194 showing the detailed coastal bathymetry of California from San Francisco Bay southward to San Diego, U.S.A. was the first bathymetric chart to be produced solely from acoustic soundings. NOAA/Department of Commerce.

and in honor of the flat-topped building at Princeton University, which also gets its name from Arnold Henry Guyot.

Cold War Years and Beyond

Because of the rise of strategic submarine defense needs during the Cold War years, knowledge of bathymetry and other oceanographic parameters became critical. As a consequence, the world view we have of the seafloor today primarily stems from efforts of U.S. Navy-supported marine geologists and geophysicists, from government agencies, and major U.S. oceanographic institutions following World War II. Beginning in the late 1940s, ships from Columbia University's Lamont Geological Observatory, the Woods Hole Oceanographic Institution, and the Scripps Institution of Oceanography fanned out over the world ocean collecting bathymetry, geophysical data, and other oceanographic information. For the next quarter century, these institutions delineated the abyssal plains, the flattest surfaces on Earth; mapped the full extent of the world-girdling ridge system; defined the great fracture zones and their accompanying scars; surveyed hundreds of seamounts; and established that abyssal hills were the most abundant landform on Earth, particularly because of their widespread occurrence in the Pacific Ocean, where they covered approximately 85% of the seafloor.[6] Other nations and institutions took part in this endeavor. Notably the British *Challenger* II expedition, which established the Mariana Trench as the deepest spot in the ocean; the Swedish *Albatross* expedition; the Danish *Galathea* expedition; Japanese surveys of the northwest Pacific Ocean; and Seamap surveys by the USC&GS in the North Pacific Ocean all contributed to furthering knowledge of the seafloor and ocean in general. Both U.S. Navy and Russian surveys of most of the world ocean were also made, primarily in support of submarine warfare requirements, throughout most of the Cold War era but even today, most of those data remain classified.

Among the most widely known maps of the ocean basins are the iconic physiographic maps produced by Bruce Heezen and Marie Tharp, researchers at Columbia University's Lamont Geological Observatory, beginning in the early 1950s and subsequent collaborative illustrations of H.C. Berann in the 1960s and 1970s (Figure 19.1d). These physiographic illustrations were based on data collected primarily from academic and military surveys conducted during the early Cold War era during which, due to U.S. national security restrictions, new bathymetric maps of many areas were "classified" and not available to the public.[7] The physiographic approach used by Heezen and Tharp portrayed physical features from an oblique perspective and an exaggerated vertical scale, and made detailed extrapolations between the soundings following from their burgeoning knowledge of geomorphology. Marie Tharp wrote: "I had a blank canvas to fill with extraordinary possibilities, a fascinating jigsaw puzzle to piece together: mapping the world's vast hidden seafloor. It was a once-in-a-lifetime—a once-in-the-history-of-the-world—opportunity for anyone, but especially for a woman in the 1940s."[8] During this same era, investigators from the Scripps Institution of Oceanography, such as H. W. Menard and T. E. Chase, were producing lesser-known atlases and physiographic diagrams of the Pacific Ocean basin.

Conclusion

The world view and regional views of bathymetry produced by these investigators were instrumental in helping form concepts of seafloor spreading, continental drift, and the theory of plate tectonics as we know them today. The advent of multibeam sonar in the civil mapping community, access to the remarkably accurate global positioning system, satellite altimetry remote sensing technology, and computer-aided analysis and interpolation has led to further advancements in bathymetric mapping in the last quarter of the 20th century. These advances have had implications in nearly every field of oceanography. The foundation for today's conceptual understanding of seafloor bathymetry, plate tectonics, and related bathymetric effects on many aspects of oceanography and marine ecology was laid down within one century by intensive effort from many talented individuals and institutions.

Acknowledgments

Funding was received from the U.S. Office of Naval Research, U.S. National Aeronautics and Space Administration (Dierssen), and U.S. National Oceanic and Atmospheric Administration (Theberge).

References

1. Theberge, A. Appendix: The History of Seafloor Mapping. In *Ocean Globe*; ESRI Press: Redlands, CA, Breman, J., Ed.; 2010; (ESRI Press): 237–274.
2. Theberge, A. Thirty Years of Discovering the Mariana Trench. Hydro, Int. **2008**, *12*, 38–41.
3. *GEBCO_08 Gridded Bathymetry Data,* General Bathymetric Chart of the Oceans, 2008.
4. Taylor, F.B. Bearing of the tertiary mountain belt on the origin of the Earth's plan. Bull. Geol. Soc. Am. **1910**, *21*, 179–226.
5. Dierssen, H.M.; Theberge, A.E. Bathymetry: Assessing Methods. *Encyclopedia of Ocean Sciences*; Taylor and Francis: New York, NY, 2012.
6. Dierssen, H.M.; Theberge, A.E. Bathymetry: Features and Hypsography. *Encyclopedia of Ocean Sciences*; Taylor and Francis: New York, NY, 2012.
7. Doel, R.E.; Levin, T.J.; Marker, M.K. Extending modern cartography to the ocean depths: military patronage, Cold War priorities, and the Heezen-Tharp mapping project, J. Hist. Geogr. **2006**, *32*, 605–626.
8. Tharp, M. Connect the dots: mapping the sea floor and discovering the Mid-Ocean Ridge. *Lamont-Doherty Earth Observatory of Columbia Twelve Perspectives on the First Fifty Years* 1999, 31–37.

FIGURE 1.3 Cultured oysters (**a**) almost grown to market size in bags; (**b**) juvenile oyster spat, sometimes called oyster seed.
Source: Figure 3B by Trisha Towanda, 2008.

FIGURE 6.3 Hurricane Irene generated significant erosion and overwash on the North Carolina Outer Banks, north of Cape Hatteras. Storm surge completely breached the island here and destroyed a section of Highway 12. Note that the highway makes a broad landward curve at this location, due to relocation of the highway following damage from previous storms.
Source: Adapted from United States Geological Survey Coastal and Marine Geology, August 31, 2011, http://coastal.er.usgs.gov/hurricanes/irene/post-storm-photos/20110830/20110831_150835d.jpg, accessed March 2012.

FIGURE 7.16 Ruined building in Minami-Sanriku town, Miyagi Prefecture. This building was the Disaster Prevention Office in town. Forty-two persons were lost in the 2011 tsunami, which reached up to 2 m higher than the roof top of the building. Many people requested to preserve the ruin as a memorial for the disaster.
Source: Photo by the author.

FIGURE 9.1 Illustration of the loss of coral on Caribbean reefs in the past 30 years. (**a**) A Caribbean reef with roughly 40% of its surface covered with live coral (typical in the 1970s). (**b**) A similar Caribbean reef with less than 4% of its surface covered with live coral (typical in since 2000). (Photographs by the author.)

FIGURE 11.1 Examples of plasticity in structural development of mangrove species determined by fresh water and nutrient availability (a) Riverine *R. mangle* forest near the mouth of the San Juan river, Sucre and Monagas state, Venezuela; (b and c) Fringe and dwarf *R. mangle* populations in Los Machos wetlands eastern Puerto Rico. The site C strongly P deficient. (d) Riverine *A. germinans* forest, along the Yaguaraparo river, Paria Gulf, Sucre state, Venezuela. (e) Pure *A. germinans* stand along the western coast of the Gulf of Venezuela. The green carpet is formed by the extreme halophyte *Batis maritima*. (f) Dwarf shrub of *A. germinans* in a salt flat behind the bay of Patanemo, Carabobo State, Venezuela.

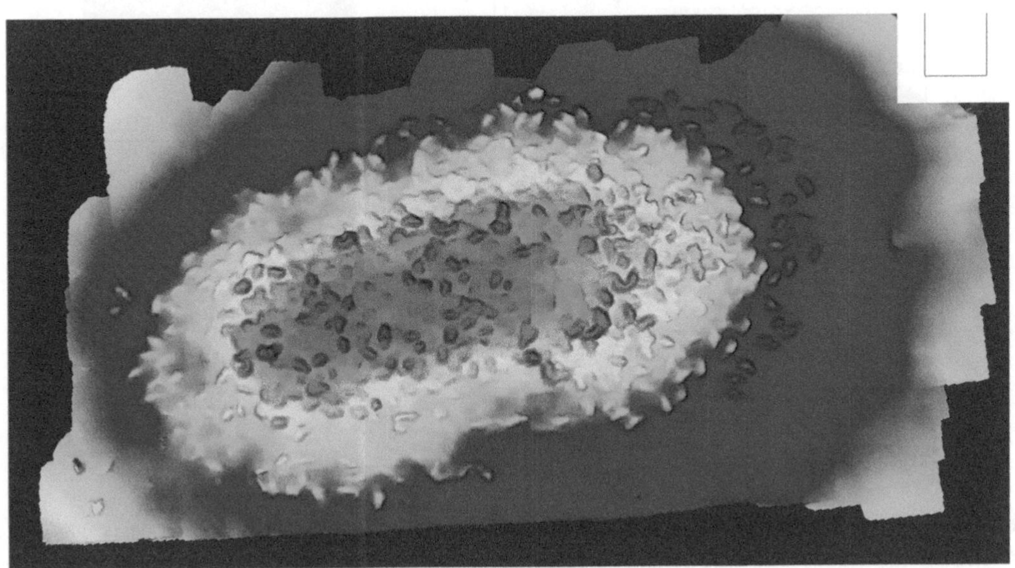

FIGURE 15.3 Multibeam bathymetric map of *Byzantine C* shipwreck site, Knidos C, located south of Datcha, Turkey.

Source: Image by Chris Roman.

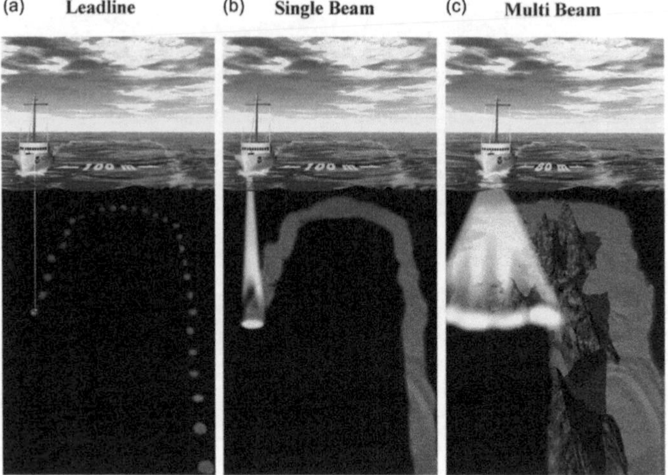

FIGURE 17.2 Illustration of different methods for measuring bathymetry from **(a)** line and sinker technology; **(b)** echo sounding using single-beam acoustics; and **(c)** multibeam acoustics providing the highest spatial resolution of seafloor features.

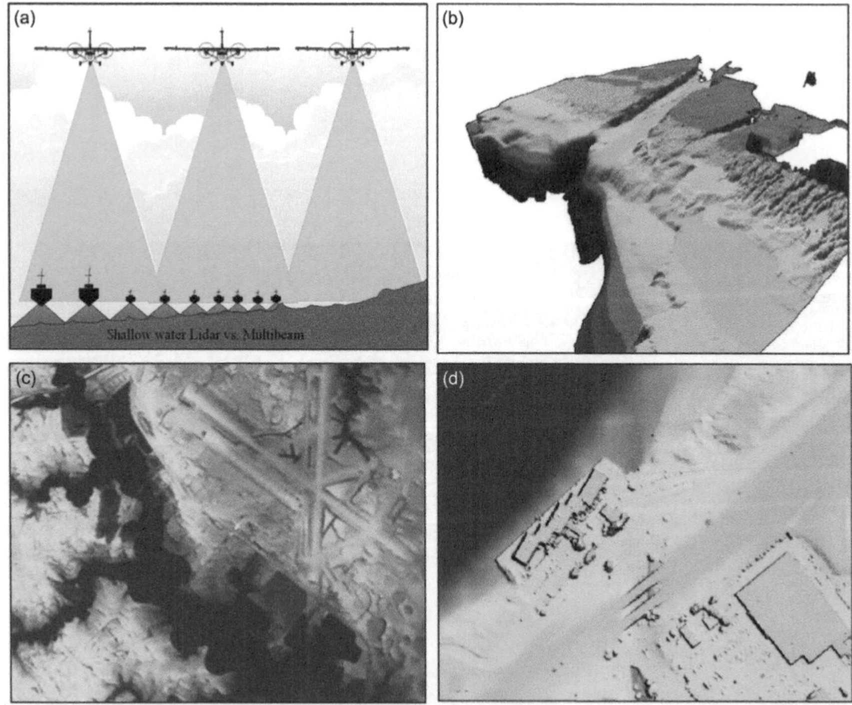

FIGURE 17.4 **(a)** Airborne LIDAR provides a large sampling swatch for collecting bathymetry data compared to ship-based sonar systems and **(b)** can seamlessly blend water and land elevations in one image for precise coastline delineation. LIDAR data collected in **(c)** coastal Delaware can be used for mapping projected sea-level rise; and **(d)** Monterey Bay for evaluating erosion patterns along a seawall jutting into the coastal zone. Images in Panel (a) and Panel (b) provided courtesy of Optech Industries. Images from Panel C and Panel D are available from the NOAA Coastal Services Center Digital Coast project.

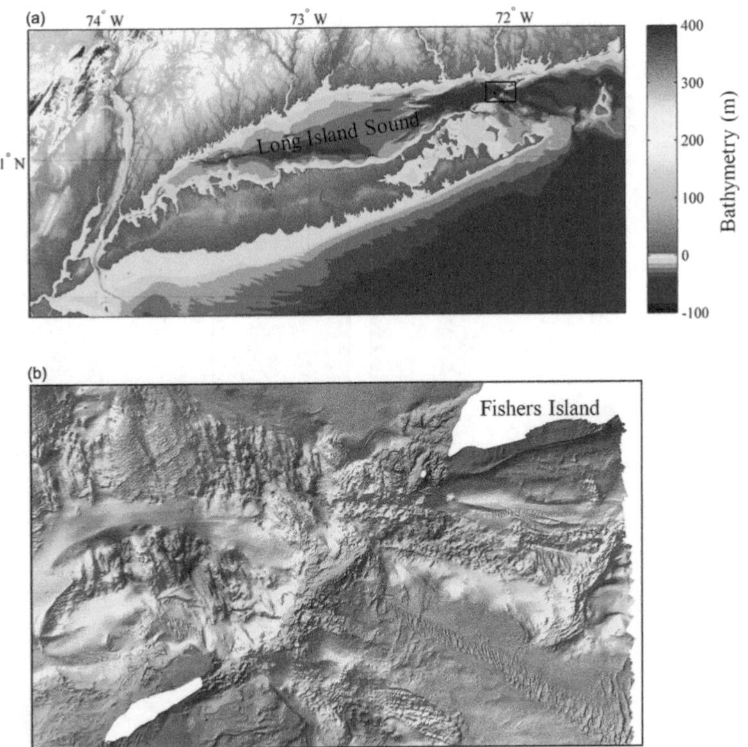

FIGURE 18.2 (**a**) Bathymetry of Long Island Sound, the United States as mapped from NOAA digital high-resolution coastal gridded data (Amante[14]). (**b**) High-resolution bathymetry from multibeam sonar of the area around Fishers Island, outlined in black in Panel A, highlights a variety of detailed seafloor features (Poppe and Polloni[3]). **Source:** Image courtesy of Long Island Sound Resource Center.

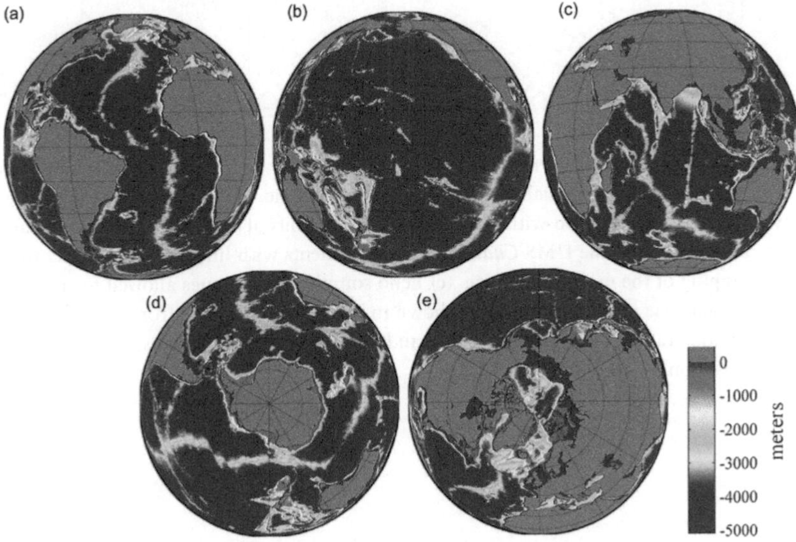

FIGURE 18.4 Gridded bathymetric data are a compilation of satellite gravimetric data and sonar data from throughout the global ocean.[14] These plots show bathymetry (m) for each ocean basin in orthographic projection: (**a**) Atlantic, (**b**) Pacific, (**c**) Indian, (**d**) Southern, and (**e**) Arctic. Land surfaces are shown in gray.

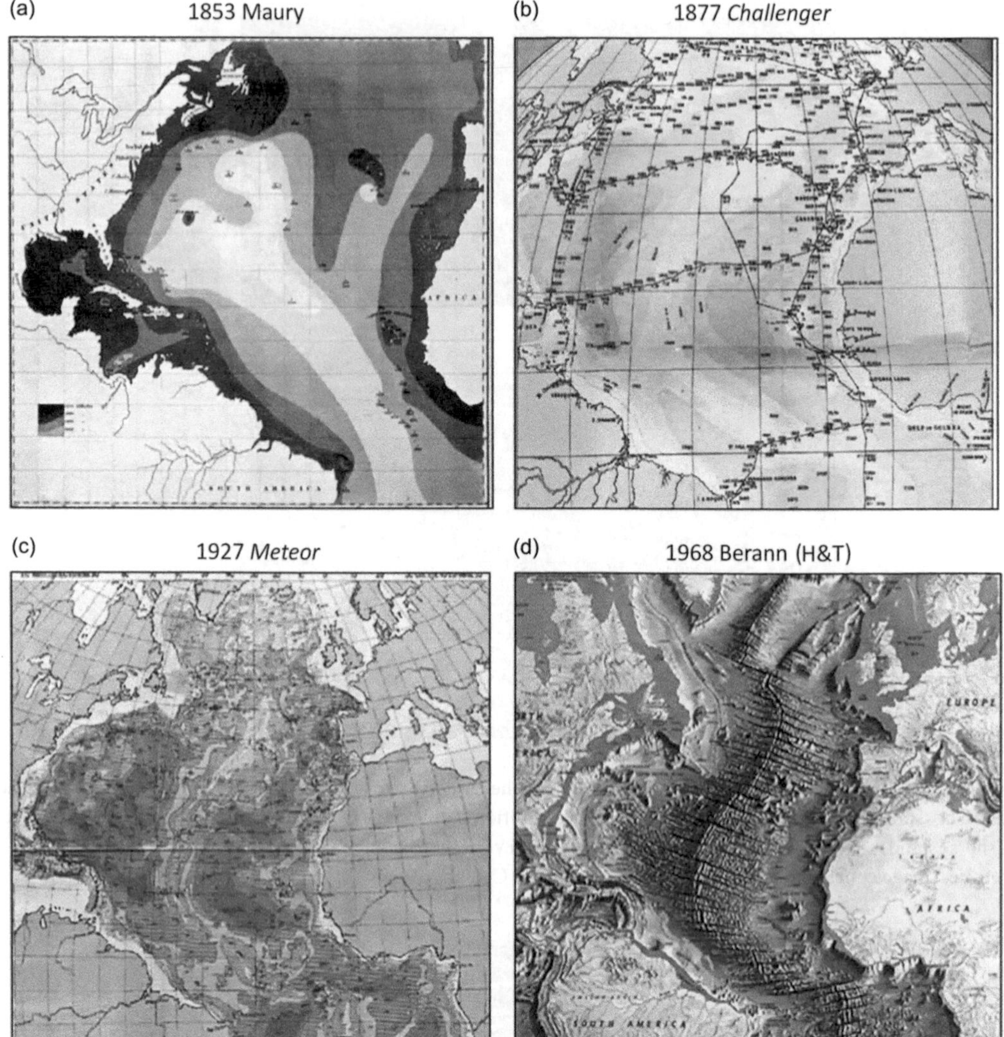

FIGURE 19.1 In roughly one century, the mapping and conceptual understanding of ocean bathymetry were revolutionized as shown in the sequential maps of the North Atlantic. (**a**) First recorded bathymetric map created in 1853 by Maury in collaboration with the U.S. Navy only hints at the mid-ocean ridge. (**b**) Excerpt from the 1877 Thomson map based on the HMS *Challenger* measurements with line-and- sinker techniques shows the first continuous mapping of the mid-ocean ridge. (**c**) Echo sounding techniques allowed for increased frequency and a higher definition of the mid-ocean ridge, as shown in Theodor Stocks map from the *Meteor* cruises in 1927. (**d**) This 1968 Berann illustration, based on the Heezen and Tharp physiographic maps, outlines the ridge system in the North Atlantic (National Geographic Stock).

FIGURE 20.1 Photos of a healthy kelp-dominated marine ecosystem (**a**) and a phase-shifted system dominated by algal turfs (**b**). In this case, the driver of the phase-shift was increased availability of carbon from subtidal CO_2 seeps that have similar CO_2 concentrations to those that are predicted globally by the year 2100.
Source: Photos by Bayden Russell.

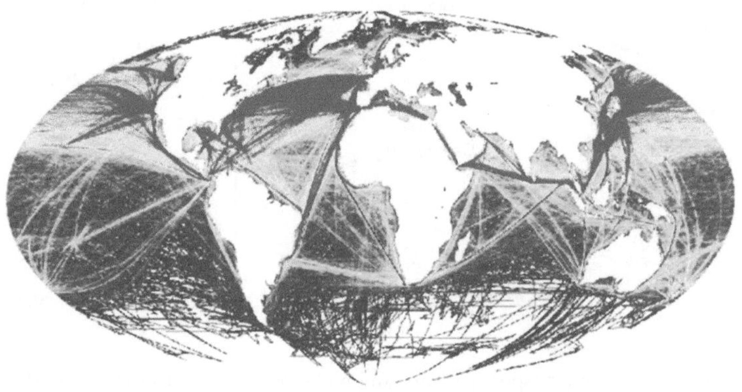

FIGURE 24.1 Global maritime shipping routes.
Source: http://www.nceas.ucsb.edu/GlobalMarine/.

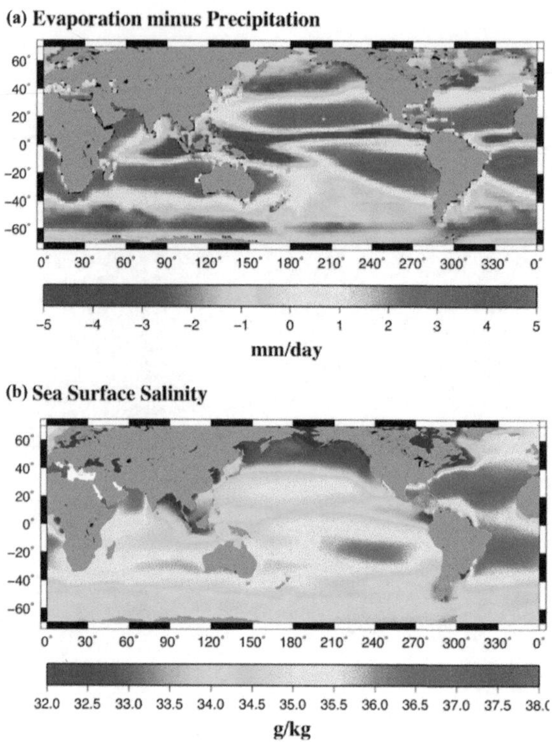

FIGURE 26.1 (**a**) Mean Evaporation - Precipitation in mm/day and (**b**) SSS. Evaporation data are from OAflux (http://oaflux.whoi.edu/),[6] averaged 1979–2009. Precipitation data from Global Precipitation Climatology Project (GPCP) (http://precip.gsfc.nasa.gov/),[7] averaged 1979–2009. Sea surface salinity data from World Ocean Atlas 2009 (http://www.nodc.noaa.gov/OC5/WOA09/pr_woa09.html).[8]

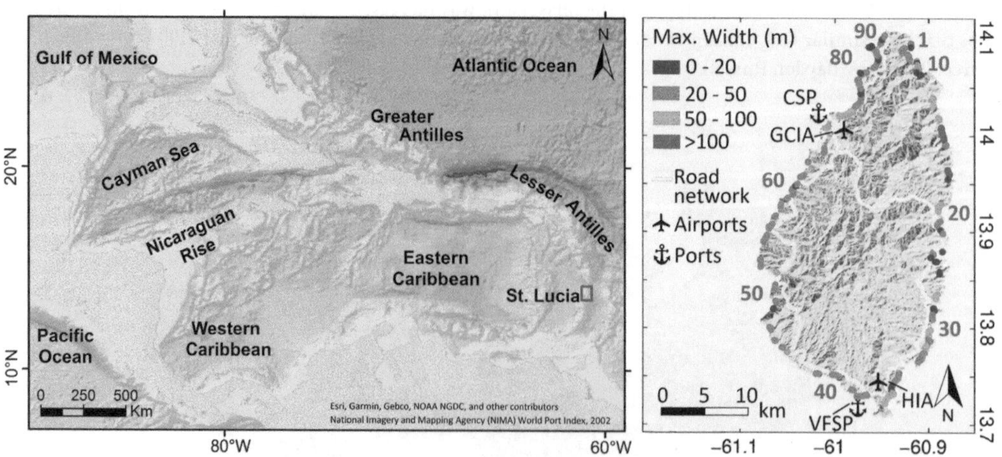

FIGURE 27.1 Saint Lucia: Main transport network and nodes and recorded sandy shore (beach) locations and maximum widths (BMWs). DEM data from Shuttle Radar Topography Mission (SRTM) DTM and bathymetric data from GEBCO_2014 Grid. Key: HIA, Hewanorra International Airport; VFSP, Vieux Fort Seaport; GCIA, George Charles International Airport; and CSP, Castries Seaport.

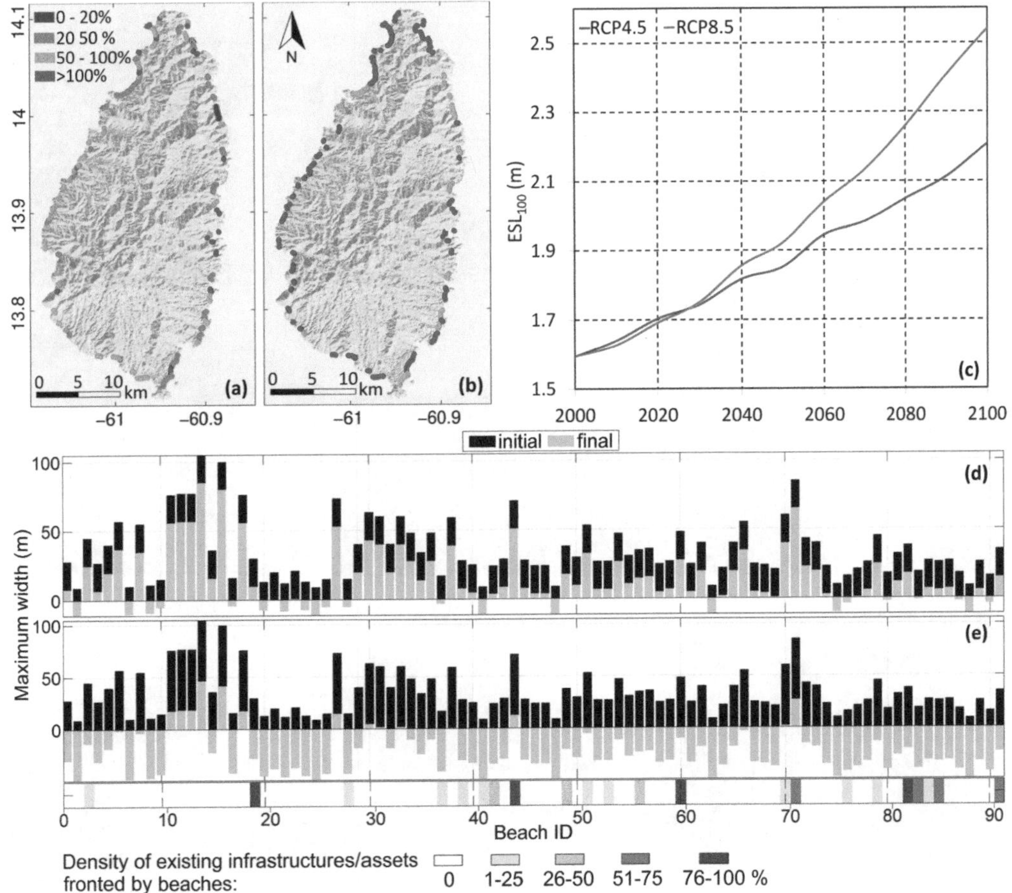

FIGURE 27.3 (**a**) The 10th and (**b**) 90th percentiles of beach retreat projections under the ESL_{100} (for the year 2050 and RCP4.5), showing beaches projected to retreat (erode) by distances equal to different percentages of their initial maximum widths (BMWs). (**c**) Projected time evolution of ESL_{100} in the 21st century under RCP4.5 and RCP8.5. The current (initial) BMWs (black bars) are compared with those resulting from (**d**) the 10th and (**e**) 90th percentiles of beach erosion projections (blue bars); negative values indicate total beach erosion. The recorded density of the frontline backshore assets (as a percentage of the beach length) is also shown.

FIGURE 28.1 The field photos illustrate the characteristics of coastal landscape. The photos show shorebirds in Cape May National Wildlife Refuge, New Jersey (**a**); a simple wooden bridge that crosses over an inlet in a coastal village of Ghana (*photographed in September 2011*) (**b**); a mangrove site along the Pangani River mouth in Tanzania coast (*photographed in January 2002*) (**c**); a tidewater cypress swamp close to Jean Lafitte National Historical Park and Preserve out of New Orleans (**d**); a *S. patens* salt marsh site in Jamaica Bay of the New York City (**e**); a barrier island sand dune on Fire Island National Seashore, New York (**f**); a damaged site by the East Japan earthquake and tsunami that devastated the Sendai and Sanriku Coast on March 11, 2011 (*photographed in August 2011*) (**g**); an overwash site on Fire Island National Seashore caused by Hurricane Sandy on October 24, 2012 (*photographed in May 2016*) (**h**); wild horses graze on salt marsh along the Chincoteague National Wildlife Refuge in Virginia (**i**); a salt marsh restoration site in Jamaica Bay (*photographed in October 2014*), New York (**j**). (Photos: Yeqiao Wang.)

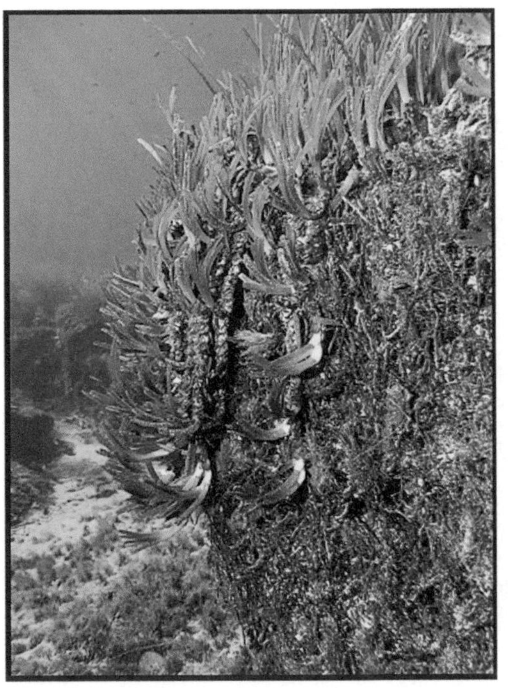

FIGURE 29.1 An erosional escarpment in a *Posidonia oceanica* meadow with organic-rich soils in Calvi Bay, Corsica Island, France. The exposed face of the matte has a thickness of 2 m. (Arnaud Abadie.)

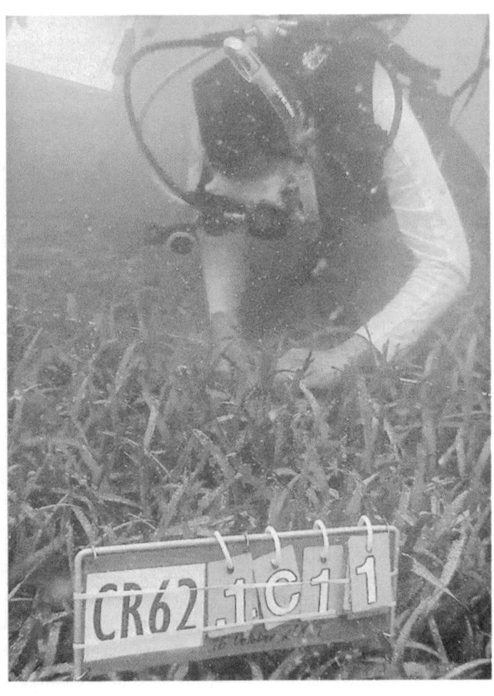

FIGURE 29.3 SeagrassNet sampling in *Thalassia testudinum* meadows at Caulker Marine Reserve, Belize. (Fred Short.)

FIGURE 29.4 A diver transplanting eelgrass in Nantucket Harbor, MA, USA. (Eric Savetsky.)

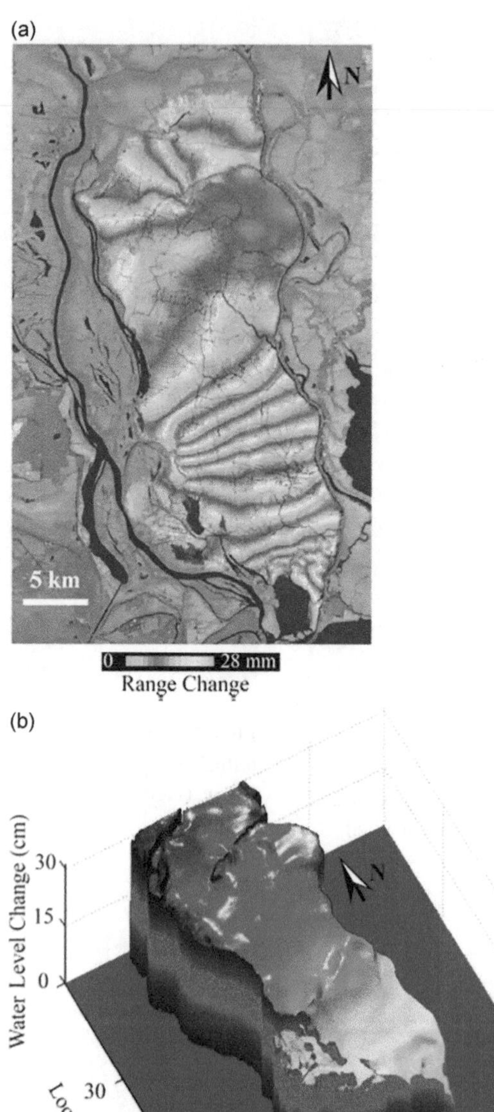

FIGURE 30.2 **(a)** C-band (wavelength of ~5.7 cm) interferometric synthetic aperture radar (InSAR) image produced from RADAR-SAT-1 images, showing heterogeneous water level changes in swamp forests in southeastern Louisiana, U.S.A between May 22 and June 15, 2003. The interferometric phase image is draped over the radar intensity image. Each fringe (full-color cycle) represents a 2.8 cm LOS range change or 3.1 cm vertical change in water level. **(b)** Three-dimensional view of water volume changes derived from the InSAR image combined with radar altimeter and water level gauge measurements. Presumably, the uneven distribution of water level changes results from dynamic hydrologic effects in a shallow wetland with variable vegetation cover.

FIGURE 31.1 (**a**) Landsat OLI pseudo color image display (Bands 5, 6, 4 in RGB) acquired in July 2018 shows the locations of Jamaica Bay and Great South Bay along the Long Island shoreline, New York; (**b**) a field site of saltmeadow cordgrass (*S. patens*) in Jamaica Bay (photographed October 2016); (**c**) a field site of smooth cordgrass (*S. alterniflora*) in Jamaica Bay (photographed August 2015); (**d**) a breached site by Hurricane Sandy in Fire Island National Seashore in October 2012 (photographed May 2016).

FIGURE 34.2 Some examples of mangrove forests around the world. (**a**) Mangroves of Ao Phang Nga National Park in Thailand; (**b**) mangroves of the Sundarbans National Park, a vast forest in the coastal region of the Bay of Bengal; (**c**) mangrove forest in the Esmeraldas-Pacific ecoregion located along the pacific coast of Colombia and Ecuador; (**d**) mangrove tunnels in Florida, United States. (Illustration created by author.)

FIGURE 34.3 Mangrove species are numerous, but red, black, and white mangroves are the most common species found around the globe. Inset shows the salinity gradient they occupy in the intertidal zone. (**a**) Red mangrove (*Rhizophora mangle L.*) forests grow aerial roots that exude salt crystals and oxygenate the tree; (**b**) black mangrove (*Avicennia germinans*) in the salt marsh of northern Florida, United States. *Avicennia* leaves exude salt crystals and also have pencil size aerial roots (Pneumatophores) that help oxygenate the tree; (**c**) white mangrove (*Laguncularia racemosa*) grows to 12–18 m (39–59 ft.) tall. It typically grows inland of black and red mangroves, well above the high tide line. (Illustration created by author.)

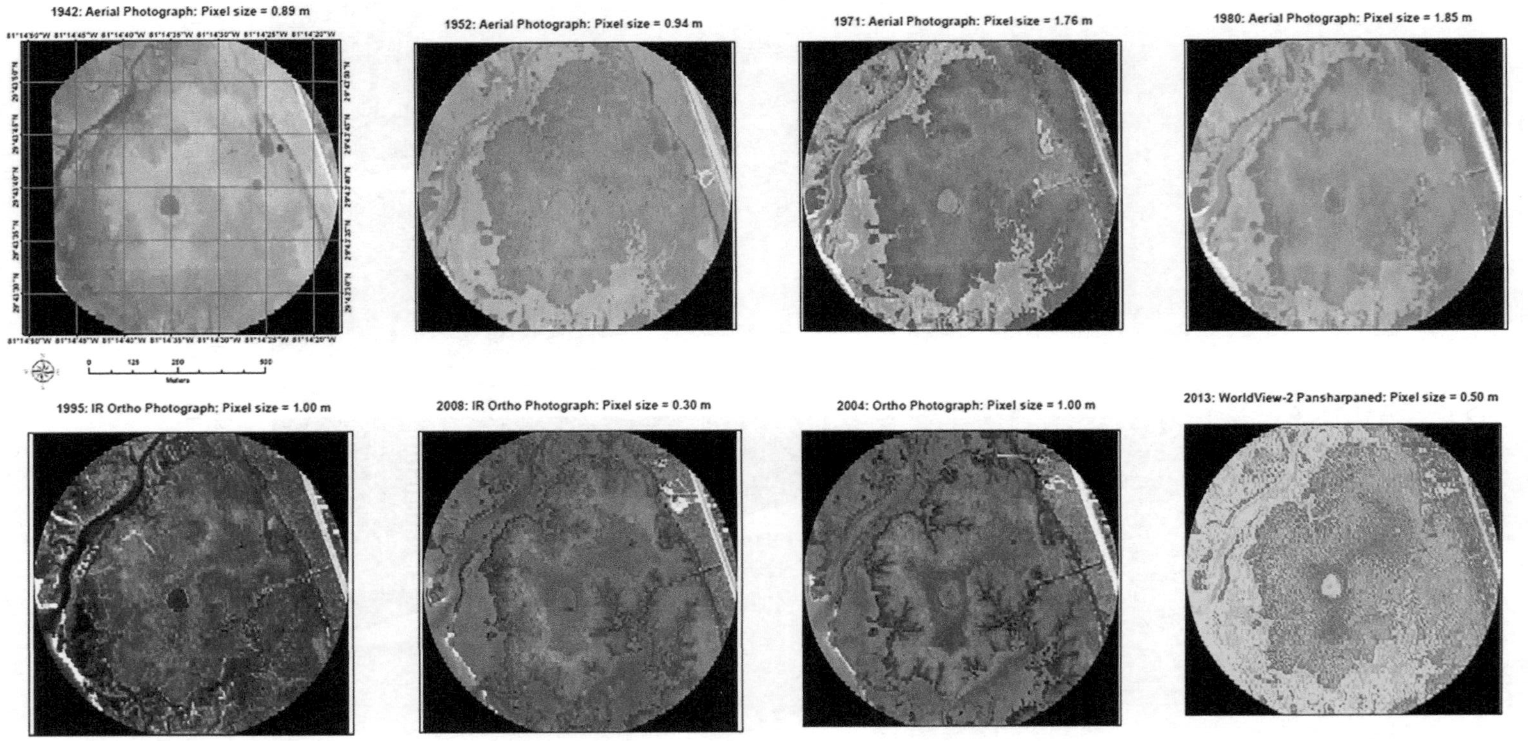

FIGURE 34.7 Use of aerial black and white photography and satellite imagery to quantify changes of mangrove forests over time in northeastern Florida. Satellite data included orthophotos, IKONOS, QuickBird, and WorldView sensors. (Unpublished data, Rodriguez and Feller, 2004.)

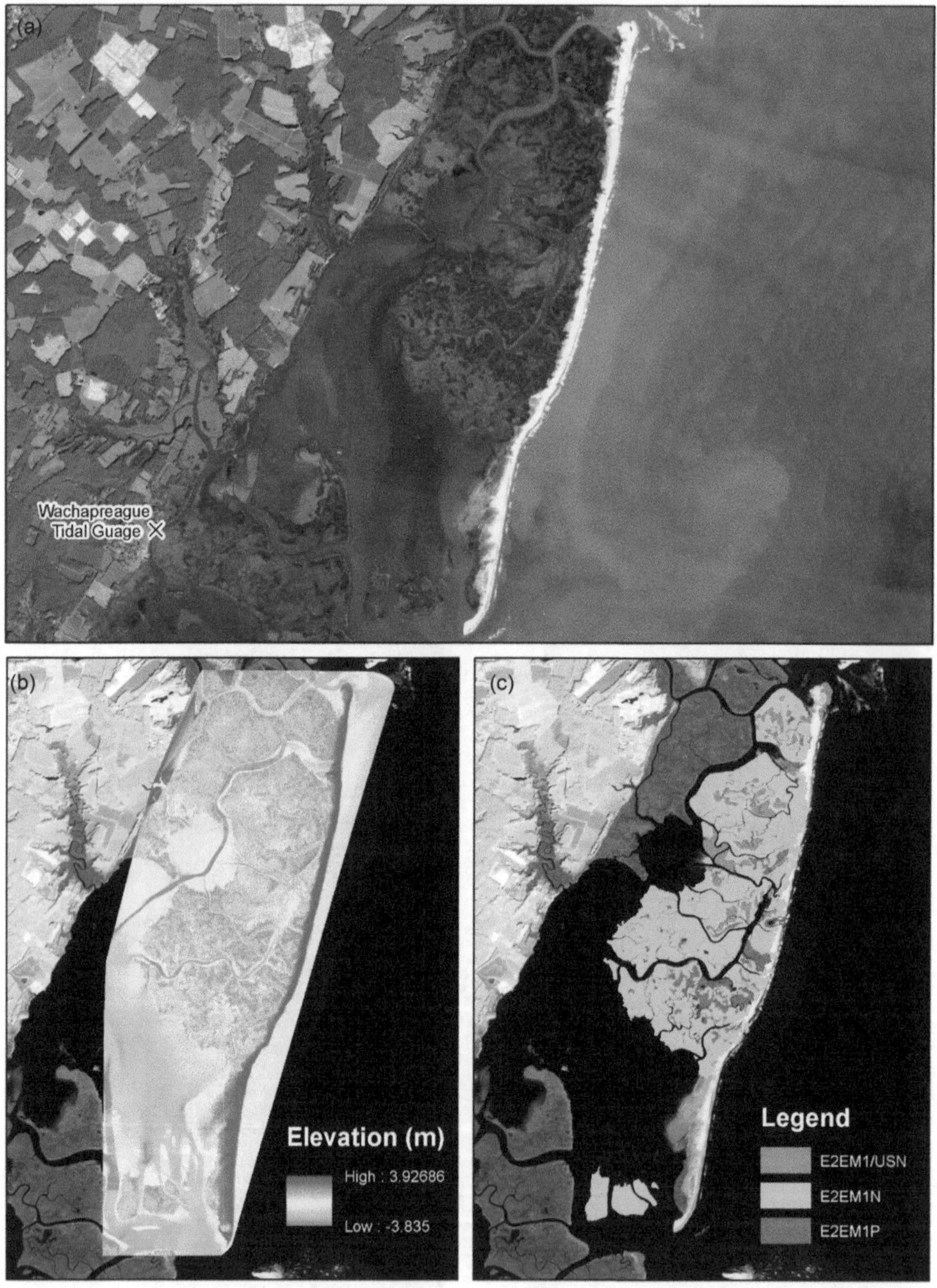

FIGURE 36.1 (**a**) The Planetscope image of Cedar Island acquired on October 28, 2017 and surrounding area including the Wachapreague tidal gauge location. (**b**) The elevation range of the DEM. (**c**) NWI data for Cedar Island which were utilized for the analysis.

20

Marine Benthic Productivity

Bayden D. Russell
and Sean D. Connell
University of Adelaide

Introduction

Primary productivity is a term that refers to the production of energy through the process of photosynthesis. In marine ecosystems, the major primary producers are algae and plants, from phytoplankton in the open oceans to algal forests and seagrass meadows in shallow coastal waters. As a biological process, primary production underpins all marine ecosystems, because the growth of primary producers supplies the energy that supports entire food webs. Yet, the amount of biomass that is produced, and therefore the quantity that enters the ecosystem, is largely determined by abiotic conditions such as light, resource availability [mostly carbon dioxide (CO_2) and nutrients], and temperature. Therefore, a proper understanding of the role of primary productivity in the functioning of marine systems requires an understanding of how changing abiotic conditions, whether natural or anthropogenic in origin, will alter primary productivity.

This entry focuses on the primary productivity of benthic marine systems, in particular macroalgae that inhabit the rocky coastlines of the world. There are other excellent sources that discuss the processes and mechanisms of photosynthesis,[1] and primary productivity in phytoplankton[2] and seagrasses.[3]

Benthic Primary Productivity

Macroalgae

Macroalgae, also known as seaweeds, are the predominant primary producers in benthic marine ecosystems. As light is one of the key requirements for photosynthesis, macroalgae are present from the intertidal to any depth that light can penetrate. All three of the major groups of algae, green (Chlorophyta), red (Rhodophyta), and brown (Phaeophyta), are represented in benthic habitats but as a group the brown algae are the most visually dominant, forming "forests" on many temperate rocky coastlines of the world. Globally, macroalgae produce more than 1 Pg of carbon per year that, relative to habitat available for them to inhabit, is equivalent to phytoplankton, which are generally considered to be more productive. This productivity is transmitted throughout coastal food webs and is responsible for many of the ecosystem services that human populations have come to rely on. However, primary productivity in these systems is strongly controlled by abiotic conditions such as light, temperature, and nutrient availability (both N and C), all of which are changing because of human activities.

Conditions Controlling Benthic Productivity

As with all photosynthetic organisms, light availability is the largest determinant of primary productivity; without light there can be no photosynthesis. This is not a simple linear relationship, however, as many algae have adapted to increase photosynthetic efficiency under low light conditions (e.g., red algae in deep waters). Much of this relationship is determined by the type and quantity of photosynthetic pigments, which varies greatly among species and within species under different environmental conditions.[4]

One of the factors that limits photosynthesis in terrestrial plants is the concentration or the availability of CO_2 In general, the productivity of terrestrial plants increases with increasing CO_2, especially in plants that utilize C3 photosynthesis.[5] This relationship does not necessarily hold for algae, however, because of the number of different mechanisms by which different algae acquire and assimilate inorganic carbon (C_i) as a substrate for photosynthesis. Most notably, CO_2 that is dissolved from the atmosphere into marine waters undergoes a series of chemical reactions, which means that it is not the most abundant source of C_i in marine waters. As such, many marine algae do not use CO_2 directly, but rather have carbon concentrating mechanisms (CCMs) that allow them to use bicarbonate as the dominant source of C for photosynthesis.[4,6] As bicarbonate is abundant in seawater, it is generally thought that photosynthesis of algae that use CCMs is carbon saturated at current CO_2 concentrations.[6,7] However, recent experiments have so far been inconclusive, with some species showing carbon saturation at current CO_2,[8,9] others demonstrating increased photosynthetic production with increasing CO_2,[10] and yet others switching between sources of carbon with greater CO_2 availability.[11,12] Regardless, the availability of resources is one of the largest factors controlling productivity in macroalgae (besides light availability) and as human activities change the availability of these resources (both C and N) the productivity of algae is likely to change.

Abiotic Conditions and Primary Productivity

Resource Availability

The availability of resources has a fundamental role in regulating the productivity of individuals, species, and, ultimately, community structure and function.[13] In the context of primary productivity, the main resources of concern are light, the driver of photosynthesis, and nutrients such as carbon and nitrogen. The availability of these resources varies both spatially and temporally in most ecosystems, meaning that it is rare for primary producers to exist under their ideal conditions; that is, they are likely to be resource-limited.[14] Such resource limitation can be recognized as a change in the rate of processes in response to one resource, such as an increase in photosynthesis with the provision of CO_2. Yet, resource limitation is also partly determined by the ability of organisms to access available resources, meaning that the level of resource limitation will vary among organisms with different physiologies. For example, slower growing species of algae, such as kelp, tend to be able to store nutrients and so are less limited in periods of low nutrient availability. In contrast, fast-growing, ephemeral species of algae, such as filamentous turf-forming algae, cannot store nutrients but are well adapted to quickly acquire them from the environment. Biological communities are, therefore, generally comprised of functional groups experience diverse limitations, a condition that can determine the relative abundance of different primary producers in the community.[15]

The species-specific responses of marine algae to the enrichment of particular resources will manifest not only as changes in primary productivity, but also in the stoichiometry, or the ratio of different nutrients in their tissues.[16] The ratio of carbon to nitrogen in the tissues of primary producers, or C:N ratios, are often used to provide an index of the relative amounts of C and N available to algae.[17,18] High C:N ratios, or relatively more carbon than nitrogen, often indicates strong nitrogen limitation, while lower ratios often seen under nutrient enriched, or eutrophic, conditions generally indicate lower nitrogen limitation because there is a greater availability of N in the tissues. Therefore, the analysis of the tissue of different algae can be used as an indication of differential resource limitations experienced by different groups of algae under a variety of environmental conditions.

In addition to C:N ratio responses, the absolute nutrient content of tissues, primarily the % C and % N, provides insight into the availability of resources in the surrounding environment and the physiological processes by which resources are acquired by the primary producers. For example, the change in C:N ratio of a particular species of algae may indicate that it experiences nutrient limitation under oligotrophic (low nutrient) conditions but not under eutrophic (high nutrient) conditions. A response where the % N and not % C increases under enriched nutrient conditions would indicate that nutrient enrichment enables this alga to access and store more nitrogen but not carbon. Then, if under-elevated CO_2 conditions neither % C or % N change, this would be indicative of a species that is not carbon limited under natural CO_2 concentrations. By understanding these measures of resource limitation, and the mechanisms that underlay them, it is then possible to make informed predictions about changes that may occur when environmental conditions change and resource limitations are removed.

Primary Productivity under Changing Abiotic Conditions

It is becoming increasingly important to develop an understanding of the resource limitations experienced by primary producers as human activities are altering the availability of resources. Of particular concern is the potential for synergistic responses to increased resource availability, where the effects of increased carbon availability, through CO_2 emissions, may be amplified in places where human activities also increase nutrient loads (such as on urban and agricultural coastlines). As discussed above, the responses to increases in the availability of these resources will reflect the extent to which primary producers are carbon-limited as a consequence of the physiological mechanisms by which carbon is acquired for use in photosynthesis.[19–21] There are two key strategies of carbon uptake in marine algae, passive diffusion of CO_2 and active uptake via a carbon concentrating mechanism (CCM). While the majority of marine algae have CCMs that facilitate the active influx of CO_2 and/or HCO_3^- and elevate concentrations at the site of carbon fixation (i.e., Rubisco), some use dissolved CO_2 entering by diffusion and others are able to switch between these mechanisms.[22] Algae with CCMs are predicted to gain little benefit from elevated CO_2 in the world's oceans,[20] and as such will probably continue to grow at rates that are comparable to those seen currently. In contrast, algae that rely on diffusion of CO_2 are possibly carbon limited under current conditions and are likely to show increased primary productivity under elevated CO_2 conditions[19] Species that can switch between these mechanisms are an interesting case because they are likely to increase productivity and growth under elevated CO_2 conditions even though they are not carbon limited at current CO_2 conditions; their increase in growth will be a result of the reduced energetic costs of using direct diffusion of CO_2 compared to CCMs, so their productivity can increase without an increase in photosynthesis per se. Unfortunately, recent work suggests that the taxa of algae that are more likely to benefit from increasing CO_2 availability, either because they do not possess CCMs or because they can switch to passive uptake of CO_2, are those taxa that also show rapid increases in growth in response to elevated nutrients,[23] possibly paving the way for a shift in relative abundance of algae along many coasts of the world.

Loss of Benthic Habitats

In marine systems, the coastal zone is an area in which high productivity and species diversity coincide with human activities that alter abiotic conditions. Historically, this zone has been influenced by large-scale eutrophication from the run-off of nutrients from the land.[15,24,25] Increasingly, the oceans, and this area in particular, are set to be further influenced by the effects of a changing climatic condition,[26] with these waters absorbing approximately 30% of CO_2 emissions produced by humans globally. Therefore, primary productivity in marine systems is likely to increase into the future because of the increased availability of two of the essential resources, nitrogen and carbon.

This increased productivity is unlikely to be realized by all species, however. As discussed above, the faster growing "weedy" species, such as filamentous algal turfs, are likely to benefit from these altered conditions to a greater extent than the slower-growing habitat formers such as kelps. Indeed, evidence to date suggests moderate increases of both nutrients[27] and CO_2 facilitate greater covers and biomass of turfs, potentially turning them from ephemeral to persistent habitats.[28–30] "Turfs" are a recurring component in the reporting of phase-shifts on coral reefs, seagrass meadows, and in kelp forests and are likely to attract attention in climate studies. Turfs seem to become the dominant habitat by taking advantage of a pulse event (e.g., storms) into an ecological phase-shift under altered environmental conditions (e.g., greater nutrient availability). This "shifted" state will then often persist (e.g., see Figure 20.1)

FIGURE 20.1 (See color insert.) Photos of a healthy kelp-dominated marine ecosystem (**a**) and a phase-shifted system dominated by algal turfs (**b**). In this case, the driver of the phase-shift was increased availability of carbon from subtidal CO_2 seeps that have similar CO_2 concentrations to those that are predicted globally by the year 2100. **Source:** Photos by Bayden Russell.

because there is a positive feedback where turfs inhibit the recruitment of habitat formers (e.g., kelps), maintaining their dominance. However, turfs, like many fast-growing and ephemeral algae, lack the ability to store nutrients and have a growth strategy that allows them to exploit increases in resource availability, but collapse when resources are depleted.[17] Therefore, altered environmental conditions allow them to persist but the restoration of "natural" conditions by removing excess resources in the environment, such as by redirecting treated wastewater inland instead of into the sea, may aid in the recovery of these ecosystems.[31]

Conclusion

In conclusion, primary productivity in marine ecosystems is largely controlled by abiotic conditions such as resource availability. Human activities increase the availability of resources, changing the conditions for productivity and growth. The resulting increase in productivity does not affect different taxa uniformly, however, with fast-growing "weedy" species often benefiting more than the slower- growing species. The ensuing change in relative dominance, from slower-growing, habitat-forming species to simpler and fast-growing species, causes a loss of habitat and, subsequently, species diversity. In some cases, this ecosystem shift may be able to be reversed, such as by limiting the input of nutrient-rich wastewater into the sea. It remains to be seen, however, whether this will be possible as global CO_2 emissions continue to increase, providing an almost unlimited source of carbon for enhanced primary productivity in the species of algae that maintain phase-shifts.

References

1. Taiz, L.; Zeiger, E. *Plant Physiology;* Sinauer Associates: Sunderland, Mass, 2002.
2. Reynolds, C.S. Photosynthesis and carbon acquisition in phytoplankton; In *The Ecology of Phytoplankton;* Reynolds, C.S., Ed.; Cambridge University Press: Cambridge, 2006, 93–144.
3. Zimmerman, R.C. Light and photosynthesis in seagrass meadows; In *Seagrasses: Biology, Ecology and Conservation;* Larkum, A.W.D., Orth, R.J., Duarte, C., Eds.; Springer: Netherlands, 2006, 303–321.
4. Raven, J.A.; Hurd, C.L. Ecophysiology of photosynthesis in macroalgae. Photosynth. Res. **2012,** *113* (1–3), 105–125.
5. Ainsworth, E.A.; Long, S.P. What have we learned from 15 years of free-air CO_2 enrichment (FACE)? A metaanalytic review of the responses of photosynthesis, canopy properties and plant production to rising CO_2. New Phytol. **2005,** *165* (2), 351–371.
6. Beardall, J.; Beer, S.; Raven, J. A. Biodiversity of marine plants in an era of climate change: Some predictions based on physiological performance. Bot. Mar. **1998,** *41* (1–6), 113–123.
7. Gao, K.; McKinley, K.R. Use of macroalgae for marine biomass production and CO_2 remediation: A review. J. Appl. Phycol. **1994,** *6* (1), 45–60.
8. Beer, S.; Koch, E. Photosynthesis of marine macroalgae and seagrasses in globally changing CO_2 environments. Mar. Ecol.-Prog. Ser. **1996,** *141* (1996), 199–204.
9. Israel, A.; Hophy, M. Growth, photosynthetic properties and Rubisco activities and amounts of marine macroalgae grown under current and elevated seawater CO_2 concentrations. Glob. Change Biol. **2002,** *8* (9), 831–840.
10. Holbrook, G.P.; Beer, S.; Spencer, W.E.; Reiskind, J.B.; Davis, J.S.; Bowes, G. Photosynthesis in marine macroalgae: Evidence for carbon limitation. Can. J. Bot. **1988,** *66* (3), 577–582.
11. Johnston, A.M.; Raven, J.A. Effects of culture in high CO_2 on the photosynthetic physiology of *Fucus serratus.* Br. Phycol. J. **1990,** *25* (1), 75–82.
12. Schmid, R.; Forster, R.; Dring, M.J. Circadian rhythm and fast responses to blue-light of photosynthesis in *Ectocarpus* (Phaeophyta, Ectocarpales) II. Light and CO_2 dependence of photosynthesis. Planta. **1992,** *187* (1), 60–66.

13. Harpole, W.S.; Ngai, J.T.; Cleland, E.E.; Seabloom, E.W.; Borer, E.T.; Bracken, M.E.S.; Elser, J.J.; Gruner, D.S.; Hillebrand, H.; Shurin, J.B.; Smith, J.E. Nutrient colimitation of primary producer communities. Ecol. Lett. **2011,** *14* (9), 852–862.
14. Andersen, T.; Pedersen, O. Interactions between light and CO_2 enhance the growth of *Riccia fluitans*. Hydrobiologia **2002,** *477* (1–3), 163–170.
15. Connell, S.D.; Russell, B.D.; Turner, D.J.; Shepherd, S.A.; Kildea, T.; Miller, D.; Airoldi, L.; Cheshire, A. Recovering a lost baseline: Missing kelp forests from a metropolitan coast. Mar. Ecol. Prog. Ser. **2008,** *360,* 63–72.
16. Elser, J.J.; Bracken, M.E.S.; Cleland, E.E.; Gruner, D.S.; Harpole, W.S.; Hillebrand, H.; Ngai, J.T.; Seabloom, E.W.; Shurin, J.B.; Smith, J.E. Global analysis of nitrogen and phosphorus limitation of primary producers in freshwater, marine and terrestrial ecosystems. Ecol. Lett. **2007,** *10* (12), 1135–1142.
17. Pedersen, M.F.; Borum, J. Nutrient control of algal growth in estuarine waters. Nutrient limitation and the importance of nitrogen requirements and nitrogen storage among phytoplankton and species of macroalgae. Mar. Eco. Prog. Ser. **1996,** *142* (1), 261–272.
18. Pedersen, M.F.; Borum, J. Nutrient control of estuarine macroalgae: Growth strategy and the balance between nitrogen requirements and uptake. Mar. Ecol.-Prog. Ser. **1997,** *161,* 155–163.
19. Kubler, J.E.; Johnston, A.M.; Raven, J.A. The effects of reduced and elevated CO_2 and O_2 on the seaweed *Lomentaria articulata.* Plant Cell Environ. **1999,** *22* (10), 1303–1310.
20. Hurd, C. L.; Hepburn, C.D.; Currie, K.I.; Raven, J.A.; Hunter, K.A. Testing the effects of ocean acidification on algal metabolism: Considerations for experimental designs. J. Phycol. **2009,** *45* (6), 1236–1251.
21. Hepburn, C.D.; Pritchard, D.W.; Cornwall, C.E.; McLeod, R.J.; Beardall, J.; Raven, J.A.; Hurd, C.L. Diversity of carbon use strategies in a kelp forest community: Implications for a high CO_2 ocean. Glob. Change Biol. **2011,** *17* (7), 2488–2497.
22. Raven, J.A.; Johnston, A.M.; Kubler, J.E.; Korb, R.; McInroy, S.G.; Handley, L.L.; Scrimgeour, C.M.; Walker, D.I.; Beardall, J.; Vanderklift, M.; Fredriksen, S.; Dunton, K.H. Mechanistic interpretation of carbon isotope discrimination by marine macroalgae and seagrasses. Funct. Plant Biol. **2002,** *29* (3), 355–378.
23. Falkenberg, L.J.; Russell, B.D.; Connell, S.D. Contrasting resource limitations between competing marine primary producers: Implications for associated communities under enriched CO_2 and nutrient regimes. Oecologia **2013b,** *172* (2), 575–583.
24. Eriksson, B.K.; Johansson, G.; Snoeijs, P. Long-term changes in the macroalgal vegetation of the inner Gullmar Fjord, Swedish Skagerrak coast. J. Phycol. **2002,** *38* (2), 284–296.
25. Gorman, D.; Connell, S.D. Recovering subtidal forests in human-dominated landscapes. J. Appl. Ecol. **2009,** *46* (6), 1258–1265.
26. Harley, C.D.G. Direct and indirect effects of ecosystem engineering and herbivory on intertidal community structure. Mar. Ecol. Prog. Ser. **2006,** *317,* 29–39.
27. Russell, B.D.; Connell, S.D. A novel interaction between nutrients and grazers alters relative dominance of marine habitats. Mar. Ecol. Prog. Ser. **2005,** *289,* 5–11.
28. Russell, B.D.; Thompson, J.I.; Falkenberg, L.J.; Connell, S.D. Synergistic effects of climate change and local stressors: CO_2 and nutrient driven change in subtidal rocky habitats. Glob. Change Biol. **2009,** *15* (9), 2153–2162.
29. Connell, S.D.; Russell, B.D. The direct effects of increasing CO_2 and temperature on non-calcifying organisms: Increasing the potential for phase shifts in kelp forests. Proc. R. Soc. B **2010,** *277* (1686), 1409–1415.
30. Russell, B.D.; Passarelli, C.A.; Connell, S.D. Forecasted CO_2 modifies the influence of light in shaping subtidal habitat. J. Phycol. **2011,** *47* (4), 744–752.

31. Falkenberg, L.J.; Connell, S. D.; Russell, B.D. Disrupting the effects of synergies among stressors: Improved water quality dampens the effects of future CO_2 on a marine habitat. J. Appl. Ecol. **2013a**, *50* (1), 51–58.

Bibliography

Larkum, A.W.D.; Orth, R.J.; Duarte, C., Eds. *Seagrasses: Biology, Ecology and Conservation;* Springer: Netherlands, 2006.

Reynolds, C.S. Ed. *The Ecology of Phytoplankton*; Cambridge University Press: Cambridge, 2006.

21

Marine Mammals

Kathleen J. Vigness-Raposa
Marine Acoustics, Inc.

Introduction

Marine mammals have captured the attention and imagination of humans for eons. They are fascinating creatures that have encouraged scientists to explore many aspects of their existence, from the evolutionary developments that led mammals into the marine environment to their complex social structures and remarkable acoustic abilities that allow them to survive underwater.

A marine mammal is defined as any mammal that is tied to the oceanic environment for all or part of its life cycle. This includes whales, dolphins, porpoises, seals, sea lions, sea otters, walruses, polar bears, manatees, and dugongs. These species occur in the mammalian order Cetacea (whales, dolphins, porpoises) and the order Sirenia (dugongs and manatees) and include many members of the order Carnivora (polar bear, sea otter, pinnipeds [seals and sea lions], and walruses). The total number of species is continually changing, as species relations are elucidated with modern genetic techniques, and hard-to-detect, elusive species are identified with new research methodologies or serendipitous observations.

This entry provides a brief overview of marine mammals, starting with the evolutionary history of marine mammals and recent phylogenetic discoveries, and concluding with recent advances in understanding these species with new research techniques. The more curious reader is referred to the References and Bibliography for more detailed works that expand on these topics, which are merely jumping-off points into the world of marine mammals.

Evolutionary Biology

Land mammals began to live in the ocean over 50 million years ago, and the results of millions of years of adaptation can be seen in four marine mammal groups: cetaceans, sirenians, pinnipeds, and sea otters. Each group is believed to descend from a different land mammal ancestor and experienced unique evolutionary development of marine adaptations. In addition, in the case of cetaceans and pinnipeds, there is continued controversy about the relationship among existing species and the identification of new species.

Cetaceans: Whales and Dolphins

The earliest marine mammals were predecessors of cetaceans, the artiodactyls (even-toed ungulates), most closely related to the hippopotamus,[1–3] which entered the marine environment approximately 53 million years ago. As the longest living marine mammals, cetaceans show the most pronounced adaptation to an exclusively oceanic existence. In order to breathe air most efficiently, their skulls have elongated, called "telescoping," resulting in the nostrils migrating to the top of the head. The long mobile neck, functional hind limbs, and most of the pelvic girdle of their terrestrial ancestors were lost as the body became more streamlined and torpedo-shaped, and horizontal tail fins, called "flukes," developed.

Another significant adaptation is the profound reliance on sound to acoustically sense their surroundings, communicate, locate food, and protect themselves underwater. Sound travels far greater distances than light underwater and it allows marine animals to gather information and communicate at great distances and from all directions. The reader is referred to the Discovery of Sound in the Sea website (http://www.dosits.org) for an extensive account of the ways in which marine mammals utilize sound.[4]

One unique adaptation of the suborder Mysticeti, comprised of the mysticetes or baleen whales, is the shift from heterodont dentition to rows of baleen plates.[5] Cetacean ancestors had teeth that were differentiated into incisors, canines, and grinding teeth (this is called heterodont dentition), but mysticetes evolved baleen plates, which are keratin structures made of the same tough protein as fingernails, hooves, and horns. The baleen plates are used to sieve great gulps of seawater in order to filter and consume plankton that could not be eaten by toothed whales.

Sirenians: Manatees and Dugongs

Sirenians are believed to have the same 50-million-yearold ancestor as elephants and hydraxes, forming an evolutionary lineage called Paenungulata.[6] Sirenians are the only extant herbivorous marine mammals and have experienced range declines and species extinctions due to oceanographic changes and human disturbance. All four extant species are considered at a high risk of extinction in the wild by the International Union for Conservation of Nature (http://www.iucnredlist.org). Similar to evolutionary adaptations of cetaceans, sirenians have lost their hind limbs and their tail has modified into a paddle used for propulsion; however, they have retained the use of their foreflippers for maneuvering underwater. They also utilize underwater sound for communication, and scientists are trying to use the vocalizations of manatees to warn boaters that animals are in the area. Fatal collisions with boats have contributed to the dwindling numbers of this endangered species.

Pinnipeds: Walruses, True Seals, Fur Seals, and Sea Lions

The earliest pinnipeds, from "pinna" meaning fin and "ped" meaning foot, were aquatic carnivores from 25 to 27 million years ago,[7] but there is continued debate about the exact origins. The traditional, biphyletic view of their evolution proposes that odobenids (walruses) and otariids (fur seals and sea lions) descended from an ursid (bear) ancestor and phocids (earless or true seals) descended from mustelids (weasels, skunks, otters).[8] However, recent morphological evidence and genetic data support the hypothesis that pinnipeds are monophyletic, having a single origin with ursids being the closest relatives.[7]

Pinnipeds are unique in that they are equally able to live in the ocean and on land. They have developed thick blubber and dense fur for marine survival, and the ability to see, hear, and move well both underwater and in the air, though in varying degrees based on the time they spend in each environment. Phocids, accounting for about 90% of pinnipeds, spend little time on land, with brief lactation periods lasting 4 days to several weeks. They have no external ears and their hindflippers cannot be turned forward under the body for mobility on land, resulting in an undulating movement to propel themselves across land. In water, they swim by moving their hindflippers and lower body in a side-to-side motion.[8] Otariids have visible external ears, pronounced sexual dimorphism, and hindflippers that can be turned under the body for moving on land. In water, they use their long foreflippers to swim. They have relatively long lactation periods, from several months to 2 years, and highly polygynous breeding systems in which males defend territories. Odobenids are very similar to otariids, displaying pronounced sexual dimorphism, mobile hindflippers, and long lactation periods lasting 2 or more years; however, they have long tusks and no external ears. They "walk" like otariids when on land, but use their foreflippers to swim like phocids underwater.

Sea Otters

At least 6 of the 13 extant sea otter species are tied to the marine environment for all or part of their existence.[9] The sea otter (*Enhydra lutris*), the only fully marine otter, diverged recently from other mustelids, whereas the marine otter or chungungo (*Lontra felina*) separated by the late Miocene, about 5 to 11 million years ago. All marine species of otters are at high latitudes, with varying degrees of activity in the ocean. The sea otter conducts all activities at sea, including resting, breeding, birthing, and feeding. The chungungo only briefly enters the sea to forage, then returns to land to eat and rest. The remaining marine species exhibit similarly limited excursions into the oceanic environment.

Phylogenetic Discoveries

The ocean is a vastly unknown place and it is remarkable that animals as large as marine mammals have gone undiscovered, but new species continue to be discovered. In several cases, the new species exhibit cryptic behavior in unpopulated portions of the world where direct, at-sea observations are limited. For example, Shepherd's beaked whale (*Tasmacetus shepherdi*), assumed to have a circumpolar distribution in deep, cold temperate waters of the Southern Ocean, was only first reliably identified at sea in 2006.[10] It had been known earlier from stranded individuals, but a definitive description of its distinctive color pattern and external morphology were not completed until recently. Similarly, the external appearance of the Longman's beaked whale (*Indopacetus pacificus*), a deep, warm water species, was only described in 1999.[11,12] The Australian snubfin dolphin (*Orcaella heinsohni*), the first new dolphin species to be described in 56 years, is found off the northern coasts of Australia.[13] It was quickly followed by the discovery of the Burrunan dolphin (*Tursiops australis*), consisting of about 150 individuals in two locations in Victoria, Australia.[14]

The most common method for identifying new species is through DNA sequence data, a method called DNA taxonomy.[15,16] By comparing DNA samples to annotated and curated sets of reference nucleotide sequences from a comprehensive and representative range of cetaceans, insights into genetic diversity can be made. For example, the discovery of greater genetic diversity between minke whales from the North Atlantic and the Antarctic than among all species within the genus *Balaenoptera* led to full species status for the Antarctic minke whale (*Balaenoptera bonaerensis*).[17] The genus *Tursiops* (bottlenose dolphin) has been particularly troublesome, with much variation in color patterns, body dimensions, and cranial structure that as many as 20 different species have been recognized over the years.[18] However, most recently, only the common bottlenose dolphin (*Tursiops truncatus*) was recognized until genetic analyses supported the separate classification of the Indo-Pacific bottlenose dolphin (*T aduncus*), found mainly in the Indian Ocean and around Australia, China, and South Africa. Similarly, the genus *Delphinus* (common dolphin) had many defined species in the past, although they were subsequently considered variations of the single species *Delphinus delphis*.[19] At present, however, two different species have been defined with genetic analyses: a long-beaked form (*D. capensis*) and a short-beaked form (*D. delphis*). As genetic samples continue to be collected and archived, further elucidation of species differentiation will be made.

Advances in Research Techniques

Because of the difficulties in studying marine mammals, new research techniques are continually being developed and refined to extend our understanding of marine mammals. Since marine mammals spend a majority of their time underwater and out of visible view, technologies are being developed to capitalize on the underwater environment. One of the most significant advances has been the design of tags that attach to individuals. Traditional satellite tags include GPS sensors that transmit the location of an animal on a predefined schedule (e.g., one time per day). However, to document a location, the tag must be at the ocean surface to communicate with satellites in space, limiting their reliability with marine mammals. The next step in complexity is time-depth recorder tags that archive the animal's depth in the water column at fine-scale time steps. These tags must be retrieved to obtain the recorded data; however, they provided the first insight into animal behavior below the sea surface. Most recently, tags have been developed that include accelerometers that can measure heading, pitch, and roll, as well as animal depth and location.[20,21] In addition, they include hydrophones that can measure the sounds produced by the animals as well as those from the environment to which the animal is exposed. This type of tag has greatly extended the understanding of marine mammal behavior, providing insight into the rate at which marine mammals vocalize, their behaviors as they vocalize, and the multiple stressors they may encounter during a given time period.

Some of the advances in archival tags have allowed for advances in passive acoustics, in which underwater listening and recording devices are used to monitor the distribution and abundance of marine mammals.[22] Marine mammals have traditionally been surveyed with visual methods; however, visual surveys cannot be conducted at night or during poor weather conditions, and animals are not available for visual detection when they are underwater, which can occur up to 80% of the time. In addition, passive acoustic sensors can be towed behind a ship or mounted on an autonomous glider or other mobile platform to cover large areas, or placed at fixed locations to record in a specific area for potentially long periods of time. In fact, the use of gliders and other autonomous vehicles is increasing rapidly as cost-effective platforms for long-term oceanography studies. With information from archival tags on vocalization rates, data from fixed passive acoustic sensors can be used to estimate marine mammal densities.[23]

Conclusion

Since marine mammals utilize the ocean environment for all or part of their life cycle, they are an important component of ecosystem-based management systems as top-level trophic predators. Further understanding of species relations will inform managers as they attempt to understand the roles that marine mammals play within specific ecosystems. Recent advances in research methodologies have also helped scientists observe greater portions of the marine mammal life cycle and more accurately estimate the role they play in the greater ecosystem and their role as indicator species for the overall health of an ocean basin.

References

1. Shimamura, M.; Yasue, H.; Ohshima, K.; Abe, H.; Kato, H.; Kishiro, T.; Goto, M.; Munechika, I.; Okada, N. Molecular evidence from retroposons that whales form a clade within even-toed ungulates. Nature **1997**, *388* (6643), 666–670.
2. Thewissen, J.G.M.; Williams, E.M.; Roe, L.J.; Hussain, S.T. Skeletons of terrestrial cetaceans and the relationship of whales to artiodactyls. Nature **2001**, *413* (6853), 277–281.
3. Price, S.A.; Bininda-Emonds, O.R.; Gittleman, J.L. A complete phylogeny of the whales, dolphins and even-toed hoofed mammals (Cetartiodactyla). Biol. Rev. Camb. Philos. Soc. **2005**, *80* (3), 445–473.

4. Scowcroft, G.; Vigness-Raposa, K.J.; Knowlton, C.; Morin, H. Discovery of Sound in the Sea, http://www.dosits.org (accessed July 2013).
5. Evans, P.G.H. *The Natural History of Whales and Dolphins;* Facts on File, Inc.: New York, 1987.
6. Domning, D.P. Sirenian evolution. In *Encyclopedia of Marine Mammals,* 2nd Ed.; Perrin, W.F., Wursig, B., Thewissen, J.G., Eds.; Academic Press: San Diego, 2009; 1016–1019.
7. Berta, A. Pinniped evolution. In *Encyclopedia of Marine Mammals,* 2nd Ed.; Perrin, W.F., Wursig, B., Thewissen, J.G., Eds.; Academic Press: San Diego, 2009; 861–868.
8. Riedman, M. *The Pinnipeds: Seals, Sea Lions, and Walruses;* University of California Press: Los Angeles, 1990.
9. Estes, J.A.; Bodkin, J.L.; Ben-David, M. Marine otters. In *Encyclopedia of Marine Mammals,* 2nd Ed.; Perrin, W.F., Wursig, B., Thewissen, J.G., Eds.; Academic Press: San Diego, 2009; 807–816.
10. Pitman, R.L.; van Helden, A.L.; Best, P.B.; Pym, A. Shepherd's Beaked Whale (*Tasmacetus shepherdi*): Information on appearance and biology based on standings and atsea observations. Mar. Mamm. Sci. **2006,** *22* (3), 744–755.
11. Pitman, R.L.; Palacios, D.M.; Brennan, P.L.R.; Brennan, B.J.; Balcomb, K.C.; Miyashita, T. Sightings and possible identity of a bottlenose whale in the tropical Indo-Pacific: *Indopacetus pacificus*? Mar. Mamm. Sci. **1999,** *15* (2), 531–549.
12. Dalebout, M.L.; Ross, G.J.B.; Baker, C.S.; Anderson, R.C.; Best, P.B.; Cockcroft, V.G.; Hinsz, H.L.; Peddemors, V.; Pitman, R.L. Appearance, distribution, and genetic distinctiveness of Longman's beaked whale, *Indopacetus pacificus.* Mar. Mamm. Sci. **2003,** *19* (3), 421–461.
13. Beasley, I.; Robertson, K.M.; Arnold, P. Description of a new dolphin, the Australian snubfin dolphin *Orcaella heinsohni* sp. N. (Cetacea, Delphinidae). Mar. Mamm. Sci. **2005,** *21* (3), 365–400.
14. Charlton-Robb, K.; Gershwin, L.-a.; Thompson, R.; Austin, J.; Owen, K.; McKechnie, S. A new dolphin species, the Burrunan dolphin *Tursiops australis* sp. nov., endemic to southern Australian coastal waters. PLoS ONE **2011,** *6* (9), e24047, doi:10.1371/journal.pone.0024047.
15. Ross, H.A.; Lento, G.M.; Dalebout, M.L.; Goode, M.; Ewing, G.; McLaren, P.; Rodrigo, A.G.; Lavery, S.; Baker, C.S. DNA surveillance: Web-based molecular identification of whales, dolphins and porpoises. J. Hered. **2003,** *94* (2), 111–114.
16. Dalebout, M.L.; Baker, C.S.; Steel, D.; Robertson, K.M.; Chivers, S.J.; Perrin, W.F.; Mead, J.G.; Grace, R.V.; Schofield, T.D. A divergent mtDNA lineage among *Mesoplodon* beaked whales: Molecular evidence for a new species in the tropical Pacific? Mar. Mamm. Sci. **2007,** 23 (4), 954–966.
17. Pastene, L.A.; Goto, M.; Kanda, N.; Zerbini, A.N.; Kerem, D.; Watanabe, K.; Bessho, Y.; Hasegawa, M.; Nielsen, R.; Larsen, F.; Palsboell, PJ. Radiation and speciation of pelagic organisms during periods of global warming: The case of the common minke whale, *Balaenoptera acutorostrata.* Mol. Ecol. **2007,** *16* (7), 1481–1495.
18. Natoli, A.; Peddemors, V.M.; Rus-Hoelzel, A.; Natoli, A. Population structure and speciation in the genus *Tursiops* based on microsatellite and mitochondrial DNA analyses. J. Evol. Biol. **2004,** *17* (2), 363–375.
19. Natoli, A.; Canadas, A.; Peddemors, V.M.; Aguilar, A.; Vaquero, C.; Fernandez-Piqueras, P.; Hoelzel, A.R. Phylogeography and alpha taxonomy of the common dolphin (*Delphinus* sp.). J. Evol. Biol. **2006,** *19* (3), 943–954.
20. Burgess, W.C.; Tyack, P.L.; Le Boeuf, B.J.; Costa, D.P. A programmable acoustic recording tag and first results from free-ranging northern elephant seals. Deep-Sea Res. II **1998,** *45* (7), 1327–1351.
21. Johnson, M.; Tyack, P.L. A digital acoustic recording tag for measuring the response of wild marine mammals to sound. IEEE J. Oceanic Eng. **2003,** *28* (1), 3–12.
22. Mellinger, D.K.; Stafford, K.M.; Moore, S.E.; Dziak, R.P.; Matsumoto, H. An overview of fixed passive acoustic observation methods for cetaceans. Oceanography **2007,** *20* (4), 36–45.
23. Ward, J.A.; Thomas, L.; Jarvis, S.; DiMarzio, N.; Moretti, D.; Marques, T.A.; Dunn, C.; Claridge, D.; Hartvig, E.; Tyack, P. Passive acoustic density estimation of sperm whales in the Tongue of the Ocean, Bahamas. Mar. Mamm. Sci. **2012,** *28* (4), E444–E455.

22

Marine Protected Areas

Jennifer Caselle
University of California

Definition and Description

Marine protected areas (MPAs) are defined as any intertidal or sub tidal areas of the ocean, including all associated fauna, flora, and historical and cultural resources, that have been set aside for protection and management through legal or other means.[1] The general term "marine protected area" encompasses a range of human activities that might be allowed inside each MPA. Areas of the ocean that are completely protected from extractive activities and major human uses are referred to as marine reserves or no-take marine reserves. MPAs throughout the world may be termed nature reserves, marine parks and heritage sites and can be established to protect coral reefs, temperate kelp forests, sea grass beds, mangroves, bays and estuaries, as well as areas of the deep sea and Open Ocean. Data show that MPAs can provide a range of benefits for fisheries, local economies and the marine environment, leading to their increasingly popular use in marine management.

Why Do We Need MPAs?

Globally, oceans are facing a large number of threats including overfishing, pollution, sedimentation, and climate change. Currently, there are no areas left in the oceans that are unimpacted by humans.[2] These changes are impairing the ocean's capacity to provide food, protect livelihoods, maintain water quality, and recover from environmental stress. These and other benefits, collectively called "ecosystem services," depend on healthy ocean ecosystems. The scale of most impacts to ocean ecosystems exceeds single habitats or species and as such, requires a more holistic, ecosystem-based approach to management. As a result, MPAs are increasingly being implemented as part of an ecosystem-based management approach for conserving biodiversity and managing marine resources. By protecting populations, habitats, and ecosystems within their borders, MPAs can provide a spatial refuge for the entire ecological system they contain. MPAs can provide a powerful buffer against a naturally fluctuating environment, catastrophes such as hurricanes, and the uncertainty inherent in establishing marine management schemes.

While the number of protected areas in the ocean is still far fewer than on land, the total ocean area protected has risen by >150% since 2003.[3] The total number of MPAs globally, as of 2010, stands at ~5880, covering >4.2 million km^2 of ocean. This figure equates to just over 1% of the marine area of the world.

Although it is not possible to develop an exact account, fully protected, no-take areas cover only a small portion of MPAs worldwide.[3] Recently, very large protected areas have been implemented in locations such as the northwest Hawaiian Islands, the Phoenix Islands, the Marianas Trench, and the Chagos Archipelago. These large MPAs, together with Australia's Great Barrier Reef, contribute disproportionately to the total area of the ocean that is protected.

MPAs are one tool for managing ocean ecosystems, but they cannot protect the oceans from all human influences. MPAs alone may not address such pervasive problems as pollution and climate change, and they will most directly benefit fishes and invertebrates that do not move long distances. However, they can provide a range of benefits not seen with other management and conservation strategies.

Benefits of Protected Areas

Considerable scientific research shows that MPAs and marine reserves increase the biomass, abundance, diversity, and size of marine species living within their borders (Figure 22.1),[4] with stronger effects in fully protected marine reserves.[5] Generally, species that are subject to fishing pressure outside MPAs show the greatest responses to protection,[6] while other species may show no response or even decline. Such declines generally reflect interactions among species, where larger numbers of predators inside MPAs (i.e., generally those species most prone to overfishing) have cascading effects on lower trophic levels.[7] For example, in the California Channel Islands, the buildup of two urchin predators (CA sheephead and CA spiny lobster) inside a long-standing, fully protected marine reserve resulted in a decline in sea urchin abundance and a subsequent increase in kelp. Urchins are known to graze down giant kelp forests, often resulting in the formation of urchin barrens, which have lower productivity and biodiversity than healthy kelp forests. Indirect effects of the removal of fishing on sea urchin predators resulted in higher abundance of giant kelp and the absence of urchin barrens in the reserve.[8]

MPAs not only affect populations living within their borders but can also influence populations in adjacent areas. Increases in growth, reproduction, and biodiversity in an MPA can replenish fished areas when larvae and adults move outside of the protected area. For example, larger individuals produce exponentially more, and healthier offspring, and higher population densities improve the likelihood of

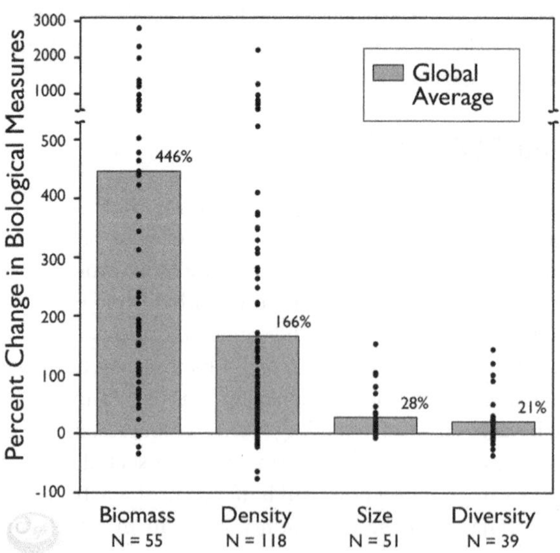

FIGURE 22.1 Average changes in fishes, invertebrates, and algae within marine reserves around the world. Although changes varied among reserves (black dots), most reserves had positive changes.
Source: Adapted from Lester et al.[4] PISCO Science of Marine Reserves.

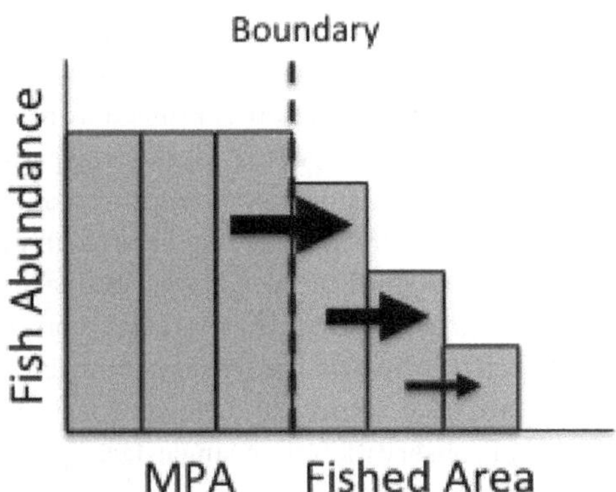

FIGURE 22.2 Spillover from MPAs to fished areas may result from net emigration of adult and juvenile individuals from MPAs due to increased competition for resources or other density-dependent mechanisms (black arrows). Shading represents reduction in abundance outside of MPA due to fishing.
Source: Adapted from Abesamis et al.[10]

reproductive success. By increasing reproductive output, MPAs can serve as a source for larvae that can restock the protected area itself as well as export larvae to adjacent areas open to fishing. Scientists are using genetic data, life-cycle information, computer models, and advanced tagging techniques to document the patterns of larval export from MPAs. As populations increase in MPAs, adult and juvenile organisms may also move from MPAs to open areas in a process called "spillover" (Figure 22.2). Net emigration of adult and juvenile individuals from MPAs can result from density-dependent (e.g., competition for food or shelter, increased predation), or density-independent (e.g., shifts in home ranges, ontogenetic migrations) processes.[9] The rate of adult and juvenile spillover is hypothesized to increase with time after MPA establishment as populations become increasingly dense in the protected area.

Fisheries and Social Benefits

MPAs provide a number benefits to fisheries beyond the expected larval and adult spillover. These include (1) protecting genetic diversity, size distributions, sex ratios, or other stock characteristics; (2) reducing by catch impacts on vulnerable species in multispecies assemblages; (3) rebuilding overfished stocks; (4) protecting critical habitats necessary for productive populations; (5) maintaining ecosystem processes necessary for productive populations; (6) providing a reference point to guide future management decisions; and (7) ensuring future catches even if management mistakes occur. MPAs also have many nonfisheries benefits, such as protecting biodiversity and ecosystem structure, serving as biological reference areas, providing non consumptive recreational activities such as ecotourism, and maintaining other ecosystem services such as shoreline protection, nutrient cycling, and climate control.

MPA Networks

By themselves, small MPAs do not tend to support populations that are large enough to sustain themselves. To ensure that young are available to replenish and sustain populations within MPAs, the protected area must be fairly large. However, in some regions, economic constraints may make it impractical to create a single large reserve or other type of MPA that can support viable populations. Establishing networks of several smaller MPAs can help reduce economic impacts without compromising conservation and fisheries benefits.

A network generally includes a set of multiple MPAs of different sizes, located in critical habitats, and connected by the dispersal of larvae and/or movement of juveniles and adults. The general objective of such networks is to maximize conservation and/or fisheries benefits from MPAs. A network can contain critical components of a particular habitat type, or portions of different kinds of habitats, that are critical to different life stages of organisms. In an effective network, organisms must be able to travel beyond the boundaries of a single protected area into other protected areas. To facilitate this movement, a network should be designed and implemented explicitly considering patterns of connectivity. By using different sizes and spacing of protected areas, a network can protect species with different characteristics. For example, a network of MPAs may include feeding habitats in open waters and breeding and nursery grounds in shallow bays. If MPAs protect these critical habitats, organisms that rely on them throughout their lifespans are likely to grow larger and have greater reproductive success. Networks that allow for redundancy in their protection of particular species and habitats also provide better fishery outcomes by protecting areas that are both sources and sinks of larvae. Networks can offer a better compromise between human use and conservation than single, large protected areas.

Acknowledgments

Thanks to Kirsten Grorud-Colvert and the PISCO science of Marine Reserves team for thoughtful discussion and use of Figure 22.1. This entry also benefitted from discussions with Alan Friedlander and Sarah Lester. This is contribution number 414 from PISCO, the Partnership for Interdisciplinary Studies of Coastal Oceans, funded primarily by the David and Lucile Packard Foundation and the Gordon and Betty Moore Foundation.

References

1. Kelleher, G.E. *Guidelines for Marine Protected Areas*; IUCN: Gland, Switzerland and Cambridge, UK, 1999.
2. Halpern, B.S.; Walbridge, S.; Selkoe, K.A.; Kappel, C.V.; Micheli, F.; D'Agrosa, C.; Brumo, J.F.; Casey, K.S.; Ebert, C.; Fox, H.E.; Fujita, R.; Heinemann, D.; Lenihan, H.S.; madin, E.M.P.; Perry, M.T.; Selig, E.R.; Spalding, M.; Steneck, R.; Watson, R. A Global Map of Human Impact on Marine Ecosystems. Science **2008**, *319*, 948–952.
3. Toropova, C.; Meliane, I.; Laffoley, D.; Matthews, E.; Spalding, M. Eds.; *Global Ocean Protection: Present Status and Future Possibilities;* Brest, France: Agence des aires marines protégées, Gland, Switzerland, Washington, DC and New York, USA: IUCN WCPA, Cambridge, UK: UNEP-WCMC, Arlington, USA: TNC, Tokyo, Japan: UNU, New York, USA: WCS. 2010; 96 pp.
4. Lester, S.E.; Halpern, B.S.; Grorud-Colvert, K.; Lubchenco, J. Ruttenburg, B.I.; Gaines, S.D.; Airamé, S.; Warner, R.R. Biological effects within no-take marine reserves: a global synthesis. Mar. Ecol. Prog. Ser. **2009**, *384*, 33–46
5. Lester, S.E.; Halpern, B.S. Biological responses in marine no-take reserves versus partially protected areas. Mar. Ecol. Prog. Ser. **2008**, *367*, 49–56.
6. Hamilton, S.L.; Caselle, J.E.; Malone, D.P.; Carr, M.H. Incorporating biogeography into evaluations of the Channel Islands marine reserve network. Proc. Nat. Acad. Sci. **2010**, *107*, 18272–18277, http://www.pnas.org/cgi/doi/10.1073/pnas.0908091107.
7. Ling, S.D.; Johnson, C.R.; Frusher, S.D.; Ridgway, K.R. Overfishing reduces resilience of kelp beds to climate-driven catastrophic phase shift. Proc. Nat. Acad. Sci. **2009**, *106*, 22341–22345, http://www.pnas.org/cgi/doi/10.1073/pnas.0907529106.
8. Behrens, M.D.; Lafferty, K.D. Effects of marine reserves and urchin disease on southern Californian rocky reef communities. Mar. Ecol. Prog. Ser. **2004**, *279*, 129–139.
9. Goni, R.; Hilborn, R.; Diaz, D.; Mallol, S.; Adlerstein, S. Net contribution of spillover from a marine reserve to fishery catches. Mar. Ecol. Prog. Ser. **2010**, *400*, 233–243.
10. Abesamis, R.A.; Russ, G.R.; Alcala, A.C. Gradients of abundance of fish across no-take marine reserve boundaries: evidence from Philippine coral reefs. Aquat. Conserv. Mar. Freshw. Ecosyst. **2006**, *16*, 349–371.

23

Marine Resource Management

Richard Burroughs
University of Rhode Island

Introduction

Marine resource management encompasses the uses and properties of the sea that reflect human needs and values. Primary elements include identifying the types of resources, specifying jurisdictions and goals, and selecting policy instruments to influence human behavior. The sections that follow explain each process.

Marine Resources and Values

Marine resources consist of the non-material and material properties of marine environments. The interrelationship of natural systems and human systems starts with the values that people apply to the properties of marine environments (Table 23.1). Consumptive and nonconsumptive uses that involve physical interaction are referred to as direct use values.[1] Consumptive direct uses consist of oil drilling, mineral mining, fishing, and other extractive activities. Nonconsumptive direct uses include transportation, recreation, military, and research activities that depend on the ocean environment, but most often do not directly consume it. Measurable benefits grow from many properties of the marine environment. Nursery habitat supports fisheries. Salt marshes attenuate storm surges from hurricanes. Nutrient cycling can assimilate organic wastes in limited amounts. Each of these properties or processes has indirect benefits that help meet people's needs. Nonuse benefits include knowing about the existence of marine resources, enjoying the view, and reflecting on the culture that surrounds certain marine activities.

How people value benefits from the marine environment shapes decisions. Direct exploitation (Table 23.1) predominated in the past.[2,3] But now, the extents to which the physical beauty of nature inspires (attraction) or the power of the ocean instills fear (aversion) determine emotions that shape human involvement. In addition, individuals may seek to physically control the ocean (dominion), or they may have an emotional attachment to the sea that shapes sharing and cooperation (affection).

TABLE 23.1 Direct, Indirect, and Nonuse of Marine Resources with Examples of Related Values

Use/Property	Values
Direct, Consumptive	
Fisheries	Exploitation, reason, affection
Aquaculture	Exploitation, reason
Oil and gas	Exploitation, dominion
Metallic minerals	Exploitation, dominion
Nonmetallic minerals	Exploitation, dominion
Water	Exploitation, symbolism
Ocean thermal energy conversion	Exploitation, reason
Hydrokinetic, wind	Exploitation, reason
Waste assimilation, accumulation	Exploitation, dominion
Direct, Nonconsumptive	
Transportation	Exploitation
Military	Exploitation, dominion
Recreation	Attraction, affection
Research	Reason, symbolism
Education	Reason, spirituality
Tourism	Attraction, affection, dominion
Indirect	
Nutrient cycling	Aesthetic, reason
Habitat, nursery	Reason, attraction
Sediment stability	Reason, symbolism
Flood control	Exploitation, reason
Non-Use	
Cultural	Affection, aesthetic
Existence	Symbolism, spirituality
Viewshed	Aesthetic
Biodiversity	Affection, attraction, reason

Source: Modified from Skinner,[27] Kellert,[2] Kellert,[3] Barbier.[1]

And subtler values such as study of the ocean (reason), ethical concern and altruism (spirituality), and the use of nature in thought (symbolism) influence how people reach decisions concerning marine resources. All of these values can be reflected in decisions even though they rarely trade in markets.

Direct consumptive uses such as fisheries, oil and gas extraction, waste disposal, and mining dominate the entries in Table 23.1, and unsurprisingly have been the focus for most management initiatives. Historically, each triggered a sector-based management arrangement, if they were managed at all. Direct nonconsumptive uses such as transportation, recreation, and military activities require additional management activities often with separate authorizations.

While the history of marine resource management comprises the direct uses, the indirect uses and nonuse properties will shape the future. Without exception, they require management regimes that respect new and diverse values, and that operate in a holistic manner. Management of nutrient cycles in coastal waters, for example, requires the consideration of multiple sources (ocean dumping, sewage effluent, river contributions), some of which extend to land (agriculture) and air (fossil fuel combustion). Lack of careful attention to nutrient cycles renders the success of fisheries, recreation, aesthetics and other activities in doubt. Adapting current management institutions, which are largely based on single extractive uses, to effectively address indirect and nonuse issues remains a central challenge. More holistic structures that consider all uses in a geographic area will produce more successful results.

Growing support for revised institutional structures is fueled by conflicts among some individual uses and the values that underlie them. Simultaneously using a single area, such as the Gulf of Mexico, for oil extraction, salt marsh restoration, and fishing leads to major conflicts when an oil spill such as Deepwater Horizon cripples, for a time, the biological systems. Promotion of one use at a specific geographic location may diminish or exclude other uses with single use primacy resulting. Decades-long use of urban waters for waste disposal, a trend that is just now being reversed, precluded their use for recreation. To the extent that adverse effects and/or incompatibility arise, institutional change is mandated. The need for change becomes acute as the possibility for resolving conflict through technological innovation and spatial separation is reduced.

Jurisdictions

Resources mentioned in Table 23.1 extend over various geographic areas—international, national, regional, and local. However, management occurs in defined jurisdictions. Therefore, specifying the geographic extent of a regime and identifying the resources covered determine a basis for management. The gradual division of the ocean into jurisdictions has occurred over many decades.

The ocean enclosure movement[4] expands national claims for area usually out to 200 miles and increases the number of resources under national control. In 1945 President Truman asserted that the United States had exclusive rights to exploit resources on and under its continental shelf;[5] and by 1983, the United States had claimed rights to an exclusive economic zone (EEZ) that extended to 200 miles from the shore.[6] This process, repeated by many other nations, transferred 39% of ocean space from high seas where open access predominated to jurisdictions where individual nations control most activities.[7]

The United Nations defined the area seaward of the EEZs as the common heritage of mankind[8] and directs the management of manganese nodules and polymetalic sulfides in that region. On a broad scale, these events signaled increasing jurisdictional clarity for marine resources, and over 150 nations participate in the Law of the Sea Convention.[9]

Where a nation's EEZ delimitation remains uncertain, much more remains to be done. For example, maritime boundaries among exclusive economic zones in the South China Sea remain in dispute, as do many aspects of delimiting control of and managing Arctic seas. Urgency for resolution arises because large quantities of oil and gas may be found in the disputed regions.

As freedom of the seas has given way to various forms of enclosure, the areal extent and the number of resources or uses encompassed in new management regimes has grown (Figure 23.1). If the trend continues, coverage of more resources over greater areas seems likely. At the global scale, individual resources such as whales, biodiversity, straddling stocks, and ocean dumping are treated through separate international agreements and plot in the upper left hand area of Figure 23.1. Regions such as the Mediterranean, the Baltic, Antarctica, and others have arrangements that cover multiple resources for part of the ocean and plot in the middle of the figure. To the lower right are local areas where some claim comprehensive management of all resources at least in theory. The resource management frontier runs from the upper left to the lower right. Most current debate exists along with this frontier and relates to the creation of new institutions to address natural and social issues that lie there.

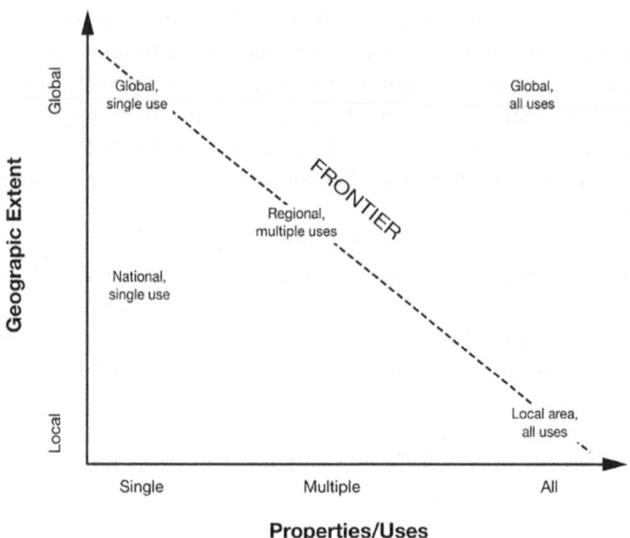

FIGURE 23.1 Scales of management.
Source: Modified from Juda.[26]

Decision Processes

Decision-makers respond to a context and select policy instruments to shape human behavior utilizing a structured means for arriving at a solution. Context specifies the breadth of considerations that go into the decisions. Means for arriving at decisions, whether comprehensive and rational or circumscribed in numerous ways, portray the way that individuals process information. Stakeholders play an increasingly important role in making decisions.[10,11] Well-chosen instruments incorporated in resource management decisions create practical outcomes. Ultimately, environmental decisions can demonstrate success through monitored changes in natural systems that may reasonably be attributed to the program at hand.

Context

Initial efforts at resource management were directed toward individual resources or properties. For example, as offshore oil and gas extraction technologies improved in the 1950s,[12] U.S. law responded to establish decision processes. Later in the 1970s, recognition that oil and gas development affected other resources resulted in legally mandated projections of impacts through environmental statements. Over time, managing one sector, oil and gas, explicitly included other sectors such as fisheries and habitats. Soon thereafter, multiple use management in various forms developed protocols to deal with the interaction of several resources in one geographic area.

By the late 20th century, the decision-context embraced many of the elements of ecosystem-based management (EBM) as evidenced by consideration of interrelationships among resources as well as environmental impacts, and social values related to each. EBM has broadened the discussion about resources by adding properties beyond single consumptive uses and introduced the notion that natural systems have limits to the goods and services they can provide. For example, the U.S. Commission on Ocean Policy (USCOP) declared the need to manage across all ecosystem components and include humans as part of coastal ecosystems.[13] So, direct uses that produce revenue constitute an important but not sole consideration for management that ultimately must include indirect and nonuse values. Ecosystem services, a subsequent extension of EBM, define naturally functioning systems that provision (fish for example), regulate (climate), culturally enrich (recreation), and support (nutrient cycling) activities of great benefit to humans.[14]

Means

The means for reaching a resource management decision take many forms. When operating at the frontier portrayed in Figure 23.1, decision-makers must create new resource management systems. Since there are few pat procedures, the selection of instruments for the first time requires extra creativity and diligence. This process has been identified as constitutive decision-making, which entails determining how policy should be made.[15,16] By indicating what considerations, which instruments, and who should be involved, decision-makers are setting the parameters for new law or guidance about how and when natural resources will be linked to society. Once completed a constitutive decision at the resource management frontier becomes the template for ordinary decisions which proceed within a known and more predictable system of authority and control. Ordinary decision processes can be found below and to the left of the frontier in Figure 23.1 where specific procedures reduce but by no means eliminate controversy.

Constitutive decision-making takes three forms. First, a rational and comprehensive process assumes that the problem has been clearly defined and a goal established.[17] In this context, one need only select evaluation criteria, apply them to the policy alternatives, and recommend the solution or solutions that best meet the objectives.[18,19] By predicting the effects of alternative solutions and comparing across specific criteria, one identifies which policy shows the greatest promise and selects it.

Often however, selection does not present itself so cleanly, and this leads to a second approach to describe how decision makers act. Actual decision processes seldom match the ideal of rationality as described earlier.[20,21] Goals may not be clear, or they may be conflicting. This is often the case for marine resources. Information may not be readily available, or it may be so diverse that establishing an effective way to compare and utilize it is lacking. Complexities in goals and information cause decision makers to utilize an incremental or bounded approach. Organizations have to simplify complex decisions to make them tractable.[21] March and Simon observe that organizations establish boundaries of rationality by seeking only satisfactory not optimal results for a given criterion, and that they do this through actions that fit within a restricted range of familiar situations. In this setting, actions are goal-oriented and adaptive. In addition, solutions are adopted through a series of small changes where each step is scrutinized for its relative effectiveness before the next step is taken.[20] The concept is that if one does not have all the information necessary, but action is mandated, then small actions rather than large ones are preferred. "Muddling through" in this manner copes with information inadequacies and goal deficiencies by limiting the decision-maker's liability should an initial solution prove inadequate.

Others identify a third path. They find that the problem and the preferences have not been clearly specified, that technological solutions remain unclear, and that participants' involvement and effort varies.[22] They envision participants dumping problems and solutions into a garbage can, and, after matching up a problem with a solution, removing both at the same time. At the moment when the organization must act, the decision-maker selects among problems, solutions, participants, and choice opportunities that are in the garbage can. One could envision harmful algal blooms or oil spills forcing decision-makers to hurriedly match existing technical solutions with the problem while the pressure for a resource management decision is high.

Instruments

Policy instruments (Table 23.2) enable society to shape human behavior consistent with common interests. Resource managers select instruments when they specify how to make operational decisions. Instruments define the types of actions that may be directed toward resolving marine resource problems. They vary in the extent to which government acts as opposed to other forces in society such

TABLE 23.2 Policy Instruments and Sample Actions for Marine Resources

Instrument	Sample Actions	Uses/Properties Examples
International treaties	Establish jurisdictions, codify norms	Deep sea mining, biodiversity, ocean dumping
Regulation, sanctions	Change standards for permits, create performance-based approaches, fine	Mining, oil spills, water quality
Taxes, fees, charges	Establish, increase, decrease	Energy production, fishing licenses, beach access
Zoning	Exclusion, single use, multiple use	Marine protected areas, military areas, fish harvest exclusion zones, oil extraction areas
Subsidies, grants	Eligibility, adjust level	Aquaculture, agriculture
Economic activity framework	Competition/concentration	Oil, fisheries
Information, education, capacity building	Warn of hazards, exhort, train	Ecolabeling, management skills
Voluntary agreements	Coercion, norms, negotiation	Waste discharge, fisheries
Property rights, markets	Reallocate and/or redefine rights, establish tradable quotas, co-manage	Fisheries

Source: Modified from[9] Clark,[16] Sterner.[28]

as community values or markets. The instruments at the top of Table 23.2 depend almost exclusively on government action. In the middle of the table subsidies, economic frameworks, and information work with a government action supplying the impetus for nongovernmental entities to act. Voluntary agreements extend arrangements beyond the direct reach of government.

The last entry in Table 23.2, restructuring property rights, could significantly change the role of government. Over half a century ago Gordon[23] asserted that the nature of fisheries and their exploitation meant that successful fishermen must be lucky or in a fishery where social controls convert the open resource into property rights. If economic reasoning predominates, then success is equated to economic efficiency and that objective alone. But as shown here, many other policy instruments besides revising property rights may be used, and a great diversity of values beyond economic efficiency applies to the fisheries. Selective assessments in resource economics and law led proponents of the efficiency approach to define the problem in a manner that demanded changes in property rights as the solution.[24] However, that approach has substantial weaknesses of its own. Who holds the rights and how they handle them will determine success from the perspective of the society as a whole, yet those factors are omitted.[25] Optimizing fisheries decisions to favor one instrument and one value reflects neither the multiplicity of values within society nor the opportunity to reach them through various instruments.

Conclusions

The management of marine resources evolves with associated human needs and values. Technological developments enhance possibilities, but ultimately the limits inherent in the natural world control the extent to which society can intensify uses of marine resources. Initial controls began with a focus on single extractive uses and ultimately expanded to cover more uses over larger areas. Over the past half century, the understanding of interrelationships among properties of the marine environment and anthropogenic impacts have expanded at a time when indirect and nonuse activities gained in importance. Ironically, the society's reliance on marine resources grows at a time when decision processes lag in incorporating shifts in values and limits of the resources themselves.

In the future, the number of uses, properties, and values that influence decisions will likely increase, making resource management even more complicated. As uses and values proliferate, potential for conflict will follow. Calls for comprehensive management to reduce disputes will further focus innovation at the marine resource management frontier.

References

1. Barbier, E. Progress and challenges in valuing coastal and marine ecosystem services. Rev. Environ. Econ. Policy 2012, 6 (1), 1–19.
2. Kellert, S. *The Value of Life: Biological Diversity and Human Society*; Island Press: Washington, DC, 1996.
3. Kellert, S. Birthright: People and Nature in the Modern World; Yale University Press: New Haven, Connecticut, 2012.
4. Eckert, R. The Enclosure of Ocean Resources: Economics and the Law of the Sea; Hoover Institution Press: Stanford, California, 1979.
5. Truman, H. Policy of the United States with respect to the Natural Resources of the Subsoil and Sea Bed of the Continental Shelf: Proclamation 2667. Fed. Regist. 1945, 10, 12305.
6. Reagan, R. Exclusive Economic Zone of the United States: Proclamation 5030. Fed. Regist. 1983, 48 (50), 10605–10606.
7. http://www.NationMaster.com/graph/geo_mar_cla_exc_eco_zon-maritime-claims-exclusive-economic-zone last visited May 26, 2012. This total, compiled from NationMaster, includes double counting where two or more nations have claimed the same space, and national boundaries of their respective EEZs have not been established.
8. http://www.un.org/Depts/los/convention_agreements/texts/unclos.unclos_e.pdf (accessed May 2012).
9. http://www.un.org/Depts/los/reference_files/status2010.pdf (accessed June 2012).
10. Burroughs, R. When stakeholders choose: Process, knowledge, and motivation in water quality decisions. Soc. Nat. Resour. 1999, 12, 797–809.
11. Dalton, T.; Forrester, G.; Pollnac, R. Participation, process quality, and performance of marine protected areas in the wider Caribbean. Environ. Manag. 2012, 49, 1224–1237.
12. Burroughs, R. Coastal Governance; Island Press: Washington, DC, 2011.
13. U. S. Commission on Ocean Policy. An Ocean Blueprint for the 21st Century: Final Report; U. S. Commission on Ocean Policy: Washington, DC, 2004.
14. United Nations Environment Program (UNEP). *Marine and Coastal Ecosystems and Human Well-Being: A Synthesis Report Based on Findings of the Millennium Ecosystem Assessment;* UNEP: Nairobi, Kenya, 2006.
15. Lasswell, H.A Pre-View of Policy Sciences; American Elsevier Publishing Co. Inc.: New York, 1971.
16. Clark, S. *The Policy Process: A Practical Guide for Natural Resource Professionals*; Yale University Press: New Haven, Connecticut, 2011.
17. Birkland, T. *An Introduction to the Policy Process: Theories, Concepts, and Models of Public Policy Making;* M.E. Sharpe: Armonk, New York, 2011.
18. Weimer, D.; Vining, A. Policy Analysis: Concepts and practice; Prentice Hall: Englewood Cliffs, New Jersey, 1992.
19. Bardach, E.A *Practical Guide for Policy Analysis: The Eightfold Path to More Effective Problem Solving;* CQ Press: Washington, DC, 2009.
20. Lindblom, C. The science of "muddling through." Public Adm. Rev. 1959, 19 (2), 79–88.
21. March, J.; Simon, H. Organizations; John Wiley and Sons: New York, 1958.
22. Cohen, M.; March, J.; Olsen, J. A garbage can model of organizational choice. Adm. Sci. Q. 1972, 17, 1–25.
23. Gordon, H.S. The Economic theory of a common-property resource: The fishery. J. Political Econ. 1954, 62, 124–142.
24. Macinko, S.; Bromley, D. Property and fisheries for the twenty-first century: Seeking coherence from legal and economic doctrine. Vermont Law Rev. 2004, 28, 623–661.
25. Charles, A. Human rights and fishery rights in small-scale fisheries management. In Small Scale Fisheries Management; Pomeroy, R., Andrew, N., Eds.; CAB International: Wallingford, Oxfordshire, UK, 2011; 59–74.

26. Juda, L.; Burroughs, R. The prospects for comprehensive ocean management. Mar. Policy 1990, 14 (1), 23–35.
27. Skinner, B.; Turekian, K. Man and the Ocean; Prentice-Hall Inc.: Englewood Cliffs, New Jersey, 1973.
28. Sterner, T. *Policy Instruments for Environmental and Natural Resource Management*; Resources for the Future: Washington, DC, 2003.

24

Maritime Transportation and Ports

Austin Becker
Stanford University

Introduction

Ports and maritime transportation form the backbone of the global economy by facilitating trade that provides civilization with the energy, goods, and materials upon which it depends. However, ports and shipping also bear responsibility for many negative environmental impacts, including oil spills, the introduction of invasive species, and air pollution. National and international policies have been created to address some of these issues, but problems do remain. In addition, ports and shipping contribute to global warming through CO_2 emissions. Climate change presents new challenges to this sector, due to both new emissions requirements and to the new environmental conditions that global warming will present. Ports and shipping will be on the frontline of impacts such as sea-level rise and increases in storminess. The industry may also benefit from certain changes such as the opening of the Northwest and Northeast passages due to melting sea ice.

The Role of Ports and Shipping

A global fleet of 50,000 commercial vessels transports 90% of the world's freight by volume (See Figure 24.1).[1] Ports and maritime transportation play a major role on every scale of the economy. Today, over 3,000 ports around the world serve as transfer points for energy products (coal, oil, and gas), manufactured goods, and raw materials. Ports and shipping fulfill a wide variety of functions for the local, regional, and global economy. They provide jobs, facilitate trade, and serve as critical

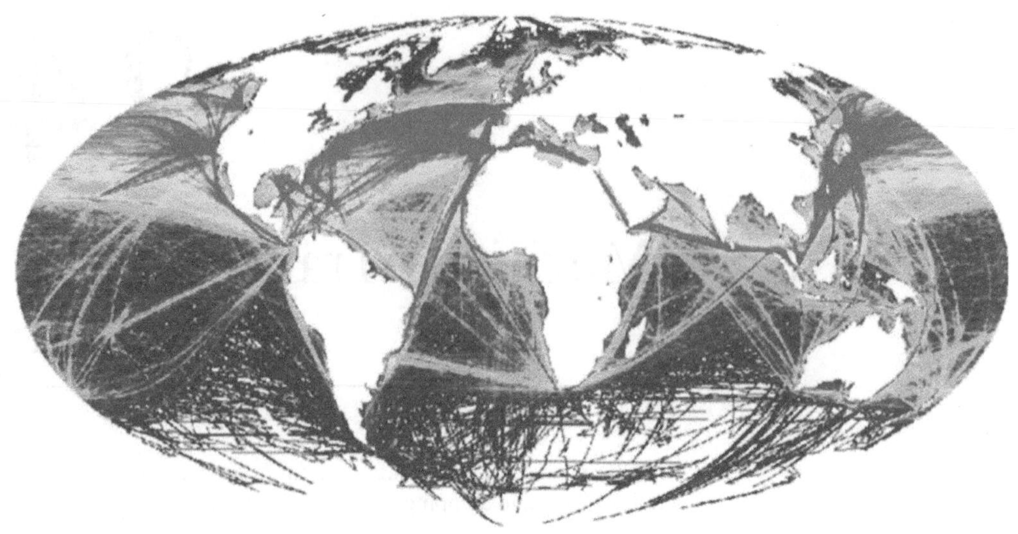

FIGURE 24.1 **(See color insert.)** Global maritime shipping routes.
Source: http://www.nceas.ucsb.edu/GlobalMarine/.

links between the hinterlands (region from which goods come from) and the forelands (the region to which goods are destined). Ports range in specialization from massive container ports (e.g., Los Angeles/Long Beach), to small niche ports that serve one type of freight (e.g., petroleum, coal, grain, or fishing).[2]

The Port's Location

Historically, many of the world's cities sprung from ports. Geographically, areas that connect oceans with rivers allow for economic development that results from the transshipment of products between ocean-going and riverine craft. Before the advent of rail and highway, goods would be transferred between larger ocean-going vessels and smaller river craft so that areas far inland could participate in global markets. During the Industrial Revolution, rivers also provided the power necessary for manufacturing and moving finished products and raw materials along those rivers to a port. This created an efficient means of connecting markets. Although most ports are located along the coast, many inland ports are strategically situated on lakes or along rivers. As cities grew around the trade center of the port, the port infrastructure itself often migrated away from the newly formed city center.[3]

Types of Ports

Ports can be categorized in numerous ways, but they are ultimately difficult to compare. Size may be measured by throughput, cargo value, land footprint, number of vessel calls, or other measures. Similarly, operation and ownership vary widely from port to port, with some being fully privatized and others being entirely public entities. Ports generally fall into one of four categories in terms of operations and management. "Service ports" are predominantly public. A public "port authority" owns the land and all assets and manages all cargo-handling operations. The "tool port" divides responsibility between the public port authority, which owns and maintains the infrastructure, and private firms, which handle the cargo. In a "landlord port," the public port authority owns the land and infrastructure, but leases it to private operating companies. Finally, the "private service port" is entirely owned and operated within the private sector.[4]

Types of Transport

Maritime transportation generally falls into five main categories: bulk, break-bulk, ro–ro, containers, and passenger. Bulk shipping refers to freight carried directly in the hold of a ship without packaging. Examples include grain, coal, liquid petroleum products, and chemicals. Purpose-built ships carry these products between ports equipped to handle loading and offloading of raw materials. Break-bulk consists of cargo that has been "unitized" onto pallets or in barrels. Examples of break-bulk cargo include fruit and lumber. Ro–ro stands for "roll on/roll off" and includes vehicles and other equipment that usually operates under its own propulsion and is transferred on and off of ships via large ramps. Container shipping, technically a form of break bulk, became a global standard in the 1960s. Today, the international standard shipping container measures 20, 40, 45, 48, or 53 feet long. Regardless of the container size, throughput and capacity is usually expressed in twenty- foot equivalent units (TEUs). Containers revolutionized shipping because goods could be loaded from a source into a container, moved intermodally (on rail, truck, and ship), and unloaded at their destination. Finally, passenger ships are used to move people. This category of ships includes ferries, cruise ships, fishing vessels, and other commercial craft that carry people for pleasure or simply as a means of transport.

Ports, Shipping, and the Environment

The strategically advantageous location of the coastal port from an economic standpoint often puts it in an estuarine area of critical importance from an ecological perspective. An "estuary" is defined as the part of a river's mouth where the river current meets the tide. Due to the abundance of nutrients and the mixing of salt and fresh water, these highly productive areas serve as breeding grounds for much of the world's marine life. The nature of a port's business puts it at odds with these ecological functions in a number of ways. Increasingly larger ships require ever- deeper channels so that they may access the port. The deepening of the channel, called dredging, displaces vast amounts of sediment, disturbing habitats, and stirring up contaminants that have settled to the bottom. Ships themselves can introduce new invasive species, which travel from one place to another on the ship's hull or in the ship's ballast water. Runoff from the port lands where industrial activity takes place and accidental spills also might degrade the water quality in an estuary. In the case of an extreme event such as a cyclone, or an earthquake, or a tsunami, the materials stored at the port also could be released into the environment. Port connections to other transportation modes often create a concentration of air and acoustic pollution, which combined with the buildings and other port operations and energy demands creates important implications for global climate change. Ports also contribute to other secondary environmental impacts, in that traffic congestion from trucks naturally increases around port operations. Trucks increase wear and tear on highways and also contribute to lower air quality and noise pollution.

Dredging

The dredging of navigation channels facilitates the movement of deep-draft vessels. Natural siltation and sedimentation slowly fills into these deep thoroughfares. The largest of container vessels today requires a channel that is around 52′ deep, though many ports have channels that are substantially shallower. As vessels grow in size and ports expand to keep up with ever-increasing demand, port planners work to have their channels maintained at their authorized depths or deepened to meet the needs of larger ships. Dredging, however, creates environmental concerns. Dredging disturbs fragile habitats and, in many industrialized areas, stirs up contaminants that lay dormant under a layer of sediment. Disposing of contaminated sediment presents additional challenges. Often, sediment must be hauled by barge far offshore or disposed in special underwater cells and then capped with clean sediment. Dredging also facilitates the passage of larger ships, which can bring their own dangers to local ecosystems if an accident occurs.

Invasive Species and Ballast Water

Throughout history, ships have impacted the world's biodiversity through moving living organisms around the globe. Earthworms, for example, came to North America in ship's ballast in the 17th century.[5] Today, ballast water used to stabilize and adjust the ship's trim may be taken aboard in one ecosystem (fresh, brackish, or salt), transported thousands of miles, and discharged in another ecosystem. Many species survive the voyage and have no natural predators in the new environment. The introduction of zebra mussels (*Dreissena polymorpha*) from Europe to the Great Lakes in the mid-1980s continues to cause millions of dollars of damages annually. The resilient mussel colonizes and clogs power and water piping, as well as displacing other native species. The issues with mussels led to the adoption of ballast water management standards in the United States. The International Maritime Organization (IMO) Ballast Water Convention was adopted in 2004 requiring both ships and ports to properly handle ballast water and minimize environmental impacts.[6]

Oil Spills

Spills of oil and other hazardous materials represent another negative impact of shipping. Although spill frequency has been vastly reduced over the past few decades (see Figure 24.2), spills nevertheless represent a significant threat to coastal and ocean habitats. Oil spills may be caused by a breach in a ship's hull or by the discharging of oily bilge water. Some infamous spills include the Amoco Cadiz in 1978 in which 687 million gallons of oil polluted 200 miles of coastline. In 1991, 9 billion barrels of oil were spilled at the start of the Gulf War in Kuwait. In 1989, the Exxon Valdez spilled 10.8 million gallons of oil into Prince William Sound in Alaska. In 2002, the oil tanker Prestige broke apart spilling oil that washed up on thousands of beaches in France, Spain, and Portugal. Spills such as these kill huge numbers of fish and seabirds and effects can last for decades, despite advances in spill remediation technology and

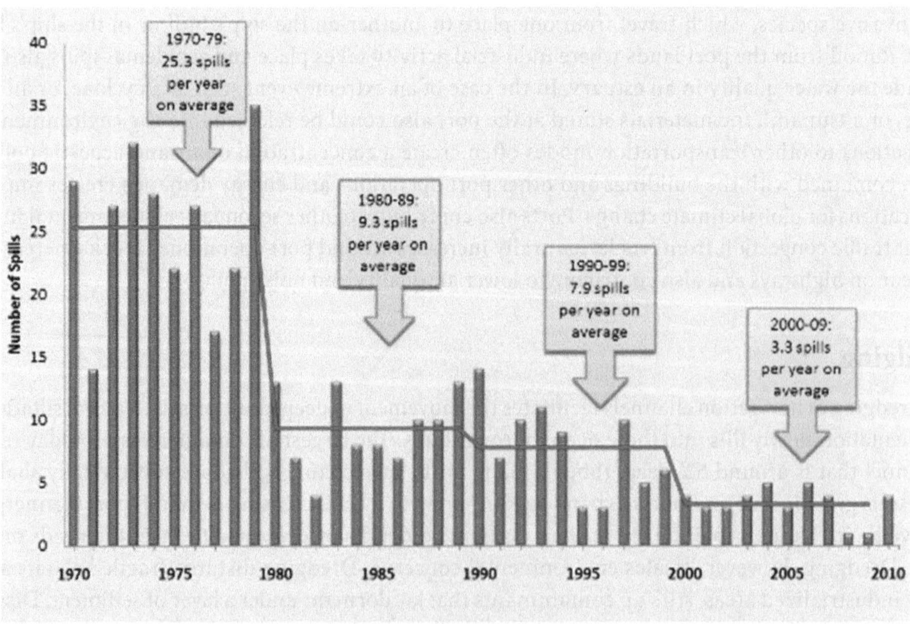

FIGURE 24.2 Number of large (>700 tons) ship spills from 1970 to 2011.
Source: International Tanker Owners Pollution Federation Limited: Information Services - Data & Statistics-Stastics. http://www.kopf.com/information%2Dservices/data%2Dand%2Dstatistics/statistics/.

increased disaster preparedness planning. The International Convention for the Prevention of Pollution from Ships (MARPOL 73/78) was adopted as a measure to help prevent pollution of the marine environment by ships from operational or accidental causes.[7] Annex I of the convention addresses oil spills in particular and led to new requirements for double-hulled tankers. As seen in Figure 24.2, the number of spills has sharply declined since the adoption of the MARPOL convention.

Other Sources of Pollution

Sewage, garbage, noxious liquids, and caustics have also presented environmental problems in the past. Through either accidental or operational discharges, these materials contributed to the degradation of marine habitats in both coastal and deep-water ocean areas. Deliberate ocean dumping of nonbiodegradable materials has led to the accumulation of plastics and other materials accumulating in great oceanic gyres. Seabirds, fish, marine mammals, and other marine life ingest or become ensnared in this waste. Toxic materials then bioaccumulate, working their way up the food chain. Annexes III - V of MARPOL 73/38 address this problem, though noncompliance remains a significant issue.[8]

Emissions (Port and Ship)

Historically, ships have burned the dirtiest type of fuel, bunker fuel. Pollutants from this source include sulfur oxides, nitrogen oxides carbonaceous aerosols, and ozone.[9] Particulate-matter emissions for this fuel source have been associated with asthma, heart attacks, lung cancer, and other illnesses. Terminal operations, too, emit pollutants and new regulations are requiring ports to upgrade their equipment. "Cold ironing," for example, allows ships to utilize shore power rather than relying on their own shipboard power plants. This results in lower emissions at the port and provides the opportunity to utilize cleaner energy from the power utility.

Ports, Shipping, and Climate Change

The ports and shipping industries face new challenges from climate change. On the one hand, the emissions from ships and ports contribute significantly to global warming. Global marine shipping bears responsibility for 1.5–3% of total global CO_2 emissions. This is expected to double by 2050 if emissions-cutting technologies are not adopted.[10] On the other hand, maritime shipping produces less GHG emissions per unit shipped than any other transport mode (Figure 24.3).[11] Both the vessel operators and the port facilities have initiated programs aimed to curb their contributions of greenhouse gasses to the atmosphere. The World Ports Climate Initiative formed in 2008 as a mechanism to assist ports combat climate change by initiating projects to reduce greenhouse gas emissions and improve air quality.[12] For example, the WPCI created a new Environmental Ship Index (ESI) scheme. The ESI creates an incentive for shipping companies to reduce the impacts of their vessels and earn the right to claim a high standard for environmental responsibility and to fly a "clean ship" flag.

In addition, the maritime sectors will certainly be impacted by the many changes the planet will experience in the coming century.[13] Rising sea levels could inundate low-lying ports and will result in much higher storm surge levels than have ever been experience before. Sea level rise also will reduce effective under-bridge clearances (known as "air draft") and could impact other infrastructure that a port relies on (e.g., rail, highway, pipelines). Changes in cyclone behavior could mean more intense storms or more frequent storms, as well as shifting tracks that could result in future catastrophic events in areas that historically have not experienced cyclones. Shipping, too, could be impacted by these changes in storm patterns and have to adjust their routes accordingly to avoid storms.

Impacts on seaports are expected to be significant in many areas. Most ports, however, have not yet taken direct action to reduce their vulnerability.[14] Protecting infrastructure against a 1–2 meter rise in sea levels and more intense storms will require significant investment. Climate change will

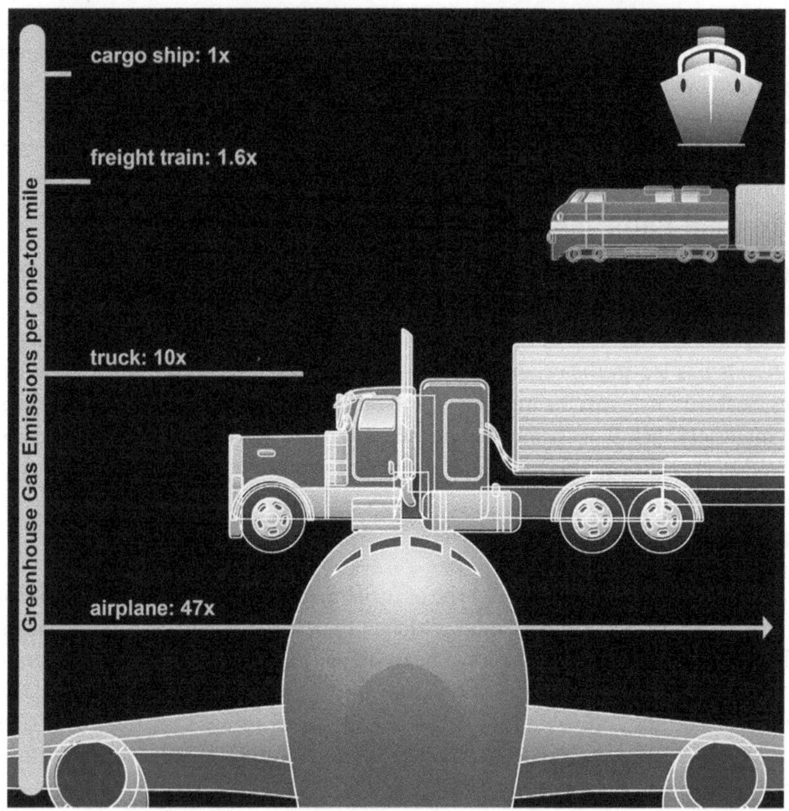

FIGURE 24.3 Comparison of CO_2 emissions between different transport types.
Source: Christine Daniloff Creative Design Director, MIT News. http://web.mit.edu/newsoffice/2010/corporate-greenhouse-gas-1108.html.

disproportionately affect ports and port-based economies, depending on their geographic location and the adaptive capacities of the ports themselves and the communities in which they are located. For example, ports in low-lying areas in a hurricane belt will face different physical challenges than those on emergent arctic coastlines where melting ice lowers the level of protection against storms. Ports in developing nations will have a different suite of options available to them than those in developed nations.[15,16] Ports located in estuaries that provide nursery environments for marine life have an even greater responsibility to protect coastal waters. The complexity and potential risks require the scientific community policy makers· and the port authorities themselves to take an active role to better understand when and how to implement proactive adaptation strategies. Proper stakeholder engagement is also essential to ensure that solutions are sustainable not only from a financial point of view but also in consideration of environmental and social concerns.

To become more resilient to the impacts of climate change and to play a role in mitigating the acceleration of climate change· port decision makers will need to implement new strategies that range from policies (i.e., changing building codes), to design (i.e., creating new protective structures), to practices (i.e., emergency drills). The Port of Gulfport (MS) is currently in the midst of implementing a major resilience-building strategy that serves to illustrate the kind of actions that will inevitably be needed in other ports. After Hurricane Katrina, decision makers chose to elevate the entire port's footprint to 25' above sea level in order to raise it out of the floodplain. Elevating, diking, or moving entire ports are some of the more drastic measures that will need to be considered.

Port Industry and Environmental Goals

The nature of the port industry and global supply chains makes environmental management more complicated, even with improved economic efficiencies. Ports have regionalized and grown in order to accommodate ever- increasing demands. While many of the negative externalities are concentrated in the immediate vicinity of the port itself, the region around the port is not immune to the environmental ramifications that result from the movement of such massive quantities of freight. The growth of ports has in turn spurned the need for larger highways, the development of warehouses, higher bridges to accommodate stacked containers on railways, and a whole host of support services for the port and ships. Termed, "logistics sprawl," the movement of these ancillary facilities away from the port itself and into the suburbs also spreads many of the environmental impacts such as truck emissions and runoff from industrial land use.[17] Finally, due to the global nature of the industry and competition within each link of the supply chain, improving environmental performance through collective actions on the local or regional level can be especially difficult.[18] Some firms, however, are beginning to improve environmental performance. Maersk's shipping, for example, recently began implementing a "fuel switch" program in which its ships will switch to low-sulfur diesel before entering near-shore waters.

The Future of Shipping

As global population increases, concentrating especially in cities and coastal areas, and nations strive to improve their citizens' quality of life, international shipping will grow to meet new demands. Current forecasts project a doubling of cargo movement by 2040. Given its energy efficiency, maritime transport is likely to remain the predominant means for the shipment of freight. A number of new developments in shipping could produce some changes in routes. Short-sea shipping, already widely practiced in Europe, uses smaller ships to transport freight between areas normally served by road or rail. Though transit times may be longer, this reduces the number of trucks and trains and cuts back on overall emissions, traffic congestion, and wear-and-tear on infrastructure. As global warming continues to melt sea ice on the Arctic, new shipping lanes through the Northwest and Northeast Passages may open up, allowing ships to cut transit times down and avoid traveling through the heavily trafficked Panama and Suez Canals, and the Strait of Malacca.[19] However, it is important to note that increased traffic in the Arctic would also bring significant new environmental risks. Fragile Arctic ecosystems could be significantly damaged by contamination resulting from ship emissions and spills. The effectiveness of clean-up technologies for these environments remains uncertain and the infrastructure to support shipping does not yet exist.

Conclusion

Ports and maritime transportation fulfill a critical role in our global economy. Indeed, without ships and seaports, the world would be a very different place. From an environmental perspective, shipping remains one of the most efficient and clean methods of moving products on a unit weight per- distance basis. Shipping does, however, present numerous environmental concerns. Reducing particulate pollution, controlling CO_2 emissions, decreasing reliance on fossil fuels, improving design and operational standards, mitigating climate risk and managing the impacts of dredging will continue to challenge the industry for decades to come.

Acknowledgments

Many thanks to Prof. Robert Desrosiers (Texas A&M), Nathan Chase (Arup), and the anonymous reviewers for insightful comments and suggestions.

References

1. Shipping and World Trade: Key Facts, http://www.marisec.org/shippingfacts/worldtrade/
2. Hoyle, B.; Knowles, R. *Modern Transport Geography*; Belhaven Press: 1992.
3. Charmaz, K. Qualitative interviewing and grounded theory analysis. *Inside interviewing: New lenses, new concerns;* SAGE: 2003; 311–330.
4. Alderton, PM: *Port Management and Operations;* LLP: 2005; 255.
5. Mann, C.C. *Uncovering the New World Columbus Created*; *Alfred a Knopf Inc: 2011.*
6. GloBallastPartnerships-TheNewConvention,http://globallast.imo.org/index.asp?page=mepc.htm.
7. IMO International Convention for the Prevention of Pollution from Ships (MARPOL), http://www.imo.org/about/conventions/listofconventions/pages/international-convention-for-the-prevention-of-pollution-from-ships-(marpol).aspx.
8. Crist, P. Cost Savings Stemming from Non-compliance with International Environenmental Regulations in the Maritime Sector. In *Organization for Economic Cooperation and Development (OECD)*; Edited by Committee MT: 2003.
9. Patton, M.Q. *Qualitative Research and Evaluation Methods;* Sage Publications, Inc: 2002; 392 p.
10. Theys, C.; Notteboom, T.E.; Pallis, A.A.; De Langen, P.W. The economics behind the awarding of terminals in seaports: towards a research agenda. *International Handbook of Maritime Business;* Cheltenham: 2010; 232.
11. Simchi-Levi, D. *Operations Rules: Delivering Customer Value through Flexible Operations;* The MIT Press: 2010; 208 p.
12. World Ports Climate Initiative, http://www.wpci.nl/projects/projects_in_progress.php.
13. Hall, P.V. Seaports, urban sustainability, and paradigm shift. J. Urban Tech. **2007**, *14* (2), 87.
14. Becker, A.; Inoue, S.; Fischer, M.; Schwegler, B. Climate change impacts on international seaports: Knowledge, perceptions, and planning efforts among port adminstrators. J. Clim. Change **2011**.
15. Dasgupta, S.; Laplante, B.; Meisner, C.; Wheeler, D.; Yan, J. The impact of sea level rise on developing countries: a comparative analysis. Clim. Change **2008**, *93* (3–4), 379–388.
16. Nicholls, R.; Hanson, S.; Herweijer, C.; Patmore, N.; Hallegatte, S.; Corfee-Morlot, J.; Chateau, J.; Muir-Wood, R. Ranking Port Cities with High Exposure and Vulnerability to Climate Extremes: Exposure Estimates. In *OECD Environment Working Paper 1; ENV/WKP(2007)1.* Paris, France, OECD: 2007.
17. Dablanc, L.; Rakotonarivo, D. The impacts of logistics sprawl: How does the location of parcel transport terminals affect the energy efficiency of goods' movements in Paris and what can we do about it? Procedia-Social and Behavioral Sciences **2010**, 2 (3), 6087–6096.
18. Hall, P.V.; Jacobs, W. Shifting proximities: The maritime ports sector in an era of global supply chains. Reg. Stud. **2010**, *44* (9), 1103–1115.
19. Arup, *Climate Change and Ports*; Adaptation and Resileince of Assets and Operations. Internal White Paper. In.: 2011.

25

Markets, Trade, and Seafood

Frank Asche
University of Florida

Cathy A. Roheim
University of Idaho

Martin D. Smith
Duke University

Introduction

Oceans, lakes, rivers, and other waterways provide about 16% of animal protein consumed globally. The fishing and aquaculture operations that produce this protein also support livelihoods for 8% of the world's population. Fully 36% of total seafood production (in live weight) enters into international trade, valued at US$141 billion per year in 2016.[1] Seafood from developing countries, in particular, constitutes the most valuable category of agricultural exports, totaling US$75 billion per year. By comparison, coffee exports are approximately US$6 billion per year, and rubber, cocoa, bananas, meat, and tea are each less than US$5 billion per year.[2]

The amount of seafood traded internationally has grown significantly during the past decades. The technologies and logistics associated with globalization of food markets in general have influenced the seafood trade by creating access to new markets, lessening the importance of distance, and ultimately allowing producers to exploit new comparative advantages. These trends have greatly expanded aquaculture production, which accounts for much of the growth in the international seafood trade. Large wholesalers and retailers can now source fish from all over the world, providing new opportunities for some producers and new challenges for others. For instance, substitutes for Atlantic cod are no longer limited to regional substitutes from capture fisheries such as haddock and saithe but now include distant capture fisheries such as Alaskan pollock and New Zealand hoki as well as farmed species such as pangasius and tilapia.

Seafood is distinguished from most other food industries because much of seafood production involves a close connection to an otherwise uncontrolled environment, and often institutions are insufficient to allow markets to organize production efficiently.[3] As the world's last significant hunting-based industry, fisheries are limited by the supporting ecosystem and subject to the tragedy of the commons when poorly managed. There is little doubt that many of the world's fish stocks are degraded[1] However, overfishing is just one of many issues that influence the size of fish stocks and the amount of seafood that enters international seafood trade. Habitat destruction from some types of fishing gear, bycatch (incidental catches of nontarget species), and fishing that affects the size distributions of fish populations all influence the balance of ecosystems and indirectly the future availability of fishery resources. Similarly, most aquaculture production occurs in the open ocean or in ponds, and may lead to surrounding ecosystem effects. But aquaculture overall is more similar to livestock production in agriculture than fishing; it relies on the ecosystem for inputs but not for the product itself.[4] Aquaculture also is less exposed to commons problems. Hence, aquaculture faces weaker environmental and institutional barriers to further growth. Still, the continued growth of aquaculture production has implications for environmental health in producing countries if institutional safeguards are not embedded into the production systems. Aquaculture may have fewer common problems than fisheries, but it must still resolve environmental externalities such that products are priced to reflect the true costs of production.[3]

Interactions among domestic fisheries management, the international retail market, and a wide array of environmental concerns about aquaculture and fisheries could all affect the international seafood trade. In this entry, we first provide background on the growth in seafood production and trade and the main factors causing these developments. We then briefly review the economic literature on the international seafood trade and markets. Because space limitations preclude a complete review, we focus on state-of-the-art analysis of seafood markets with implications for management of fisheries and aquaculture that aim to reduce market inefficiencies.

Production and Trade

The total supply of seafood increased from 68.9 million tonnes in 1976 to 172.7 million tonnes in 2017.[5] Seafood comes from two main modes of production—harvest from capture fisheries and aquaculture. Until the 1970s, aquaculture was unimportant. However, since then, a virtual revolution has taken place. Figure 25.1 shows the changing production from wild fisheries and aquaculture. Aquaculture production has grown from ~3.5 million tonnes in 1970, or 5.1% of total seafood supply, to 80.1 million tonnes, or 46% of total seafood supply in 2017.[5] The combined effect of increased productivity, increased market access, and demand growth has made aquaculture the world's fastest growing animal-based food sector.[1] Capture fisheries production, on the other hand, has fluctuated between 90 and 100 million

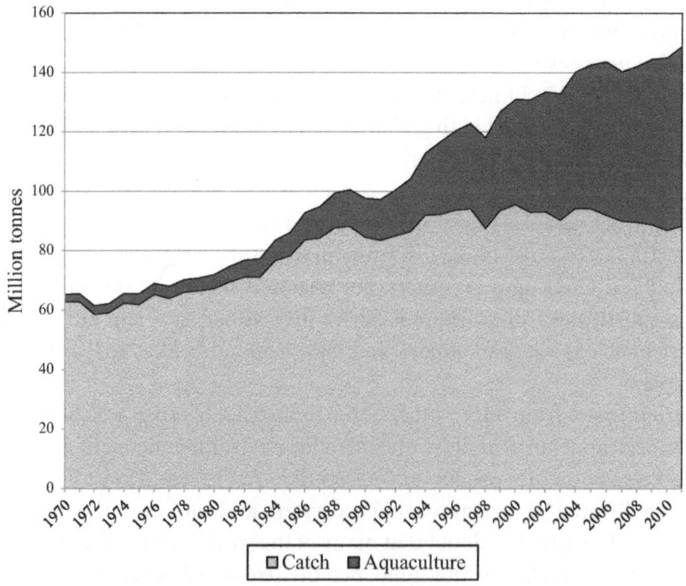

FIGURE 25.1 Global production from capture fisheries and aquaculture: 1970–2017 (FAO FishStat).[5]

tonnes in annual landings with no particular trend. Fisheries supply is not expected to increase in the future, as the majority of fish stocks are either fully or overexploited.[1]

Seafood has long been traded internationally, but trade in seafood has increased dramatically in recent decades. The increase in seafood trade is largely attributable to growth in aquaculture productivity and increased exports from developing countries. International trade has increased much faster than total seafood production. From 1976 to 2016, the export volume of seafood increased from 7.9 million tonnes to 36.5 million tonnes, or more than fourfold. Adjusted for inflation, the export value during this period increased at almost the same rate. It is important to note that export quantities are not directly comparable to the production quantities because exports are measured in product weight. This can lead to dramatic differences. The fillet weight of tilapia, for instance, is only between 30% and 40% of the harvest weight.

A number of factors have caused the increased trade in seafood. Transportation and logistics have improved significantly. Lower transportation costs have also given new producers access to the global market. Coastal nations imposed the 200-mile exclusive economic zones (EEZs) in the 1970s and 1980s that increased incentives for international trade. Peru, Ecuador, and Chile implemented EEZs as early as 1952. By the time the United States declared its 200-mile EEZ in 1976, 37 nations had already extended their jurisdiction, and by the mid-1980s, nearly all coastal nations had imposed EEZs.[6] Countries with considerable distant-water fishing fleets, such as Spain and Japan, were negatively affected, as other coastal nations expanded their domestic fleets to exploit the fisheries within their 200-mile EEZs. As a result, countries that relied on harvesting within the 200-mile EEZ of foreign nations had to increase their imports to maintain domestic consumption at the same levels.

When seafood export quantity increases fourfold and export value only threefold, the unit value decreases. Lower unit values suggest seafood's competitiveness as a food source has increased and is a key factor explaining increased trade. Successful aquaculture species such as salmon and shrimp illustrate this phenomenon. Real prices for each are less than one-third of what they were 25 years ago, and internationally traded quantities have grown at an annual rate of 16% and 13%, respectively, during the period from 1985 to 2017.[5] The profitable expansion in the production of these species, despite decreasing prices, is due to a combination of lower production costs, improved production technologies, and lower distribution and logistics costs.

The trade patterns are widely different between exports and imports. The source of seafood exports was split almost equally between developing and developed countries in 2016, as shown in Figure 25.2.

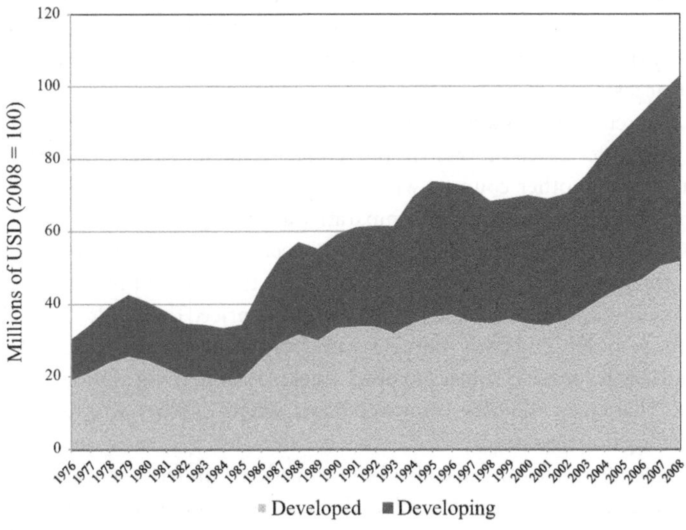

FIGURE 25.2 Real value of seafood exports from developing and developed countries: 1976–2016 (FAO FishStat).[5]

The value share of exports going from developing countries has increased from 37% in 1976, to 53% in 2016.[5] Alternatively, developed countries comprised 73% of all imports in 2016. Even though that share has declined from 86% in 1976, most of the increased trade pattern in seafood is to developed countries, and a considerable share is exported from developing countries. Japan and the United States are the two largest importers. However, if the European Union (EU) countries are aggregated, the EU is clearly the largest market.

It is certainly not arbitrary that developed countries account for most of the imports and that the EU, Japan, and the United States are the largest seafood importers. These regions are the world's wealthiest customers with the best ability to pay. Economic growth in China and Southeast Asia has led to substantial growth in seafood imports, with prospects for even further growth in demand.[7] However, the trade deficit of developing countries is significantly lower when measured by quantity, indicating a significant revenue surplus.[8] Importantly, even though some aquaculture species are highly export-oriented, the export share for aquaculture species on average is significantly lower than the share of wild fish.[9]

Economics of Seafood Markets and Trade

This section reviews the state of the art in the economic literature about markets, trade, and seafood. It covers the relationship between fisheries management and trade; the relationship between the value of seafood attributes and production practices value of seafood attributes; and the development of the Fish Price Index (FPI) by the UN Food and Agriculture Organization (FAO) to address food security concerns.

The Relationship between Fisheries Management and Trade

The impacts of trade liberalization will differ depending on several factors, including production method (capture or aquaculture) and domestic fisheries management policies. Modern theories of open access (and of fisheries management in general) are dynamic and account for the inseparable link between harvest now and the stock of fish in future periods. Economists began to grapple with the dynamic nature of fishery resources in the 1950s[10] and developed a fully dynamic theory of open access in the late 1960s.[10–12] These results have important implications for conventional models of international trade. Trade policies that decrease (or increase) prices or costs may have the opposite effect of what one might see in other resource or agricultural sectors depending on the level of exploitation of the fishery resource.[13] The basic intuition is that any short-run increase in harvest of the renewable resource sparked by trade liberalization will lead to long-run decreases in harvest when the resource is governed by open access.

A well-known theoretical analysis considers trade between two countries with equal natural resource endowments but that differ based on their resource management institutions: one country with open-access management and another country with a conservationist approach.[14] By allowing for excess exploitation, open access can be a source of comparative advantage and lead to exporting the resource to the conservationist country. Society benefits for the conservationist country but declines for the open-access country in the long run. However, it is also possible that the well-managed resource in the conservationist country can lead to this country having a lower price. The result is that the conservationist country exports the resource, and both countries gain from trade. Brander and Taylor aptly summarize that "when a renewable resource is subject to open access, or something approaching it, then free trade may not be the tide that raises all boats. Improved management of renewable resources may be a necessary precondition for gains from trade."[15]

The effects of trade liberalization have also been considered under other fisheries management scenarios. Trade liberalization may negatively affect the resource under open access or even under limited access fisheries management regimes.[16] For the society at large, trade will always be beneficial if the resources are well managed, although there will be redistribution effects within each society.

International Markets: Seafood Safety

There are many food safety issues involved in the seafood markets and trade. Some stem from the food system and others arise from the natural environment from which the products are raised or captured. Health risks from seafood consumption are varied and include the potential for a wide range of episodic events—including bacterial and parasitic illnesses and histamine poisoning—as well as concerns about repeated exposure to high levels of mercury and polychlorinated biphenyls (PCBs). The economic issues which result are equally varied, complex, and important for public and private decision-making (see Cato[17] for an overview). Some health issues stem from water quality problems, e.g., shellfish contamination from algal blooms or local waste water treatment difficulties. Other seafood-borne illnesses result from processing, handling, inadequate refrigeration, and spoilage. The public is concerned about the health benefits and risks involved in seafood consumption.[18] However, consumers have difficulty estimating health risks in the food they consumed even when given seafood advisories.[19] Consumers are flooded with information and misinformation related to the health risks of farmed seafood consumption, affecting both purchase decisions and consumer welfare.[18–20] Evidence is almost universal that information widely available to consumers is inadequate to promote informed consumption of farmed seafood.[21] Government agencies and nonprofit organizations often provide conflicting guidance, and research repeatedly demonstrates suboptimal consideration of long-term health risks and benefits related to farmed seafood.[21,22] Indeed, a recent National Academies study concludes that "research is needed to develop and evaluate more effective communication tools for use when conveying the health benefits and risks of seafood consumption."[22] Also, there is some experimental evidence that public communication programs could reduce the uncertainty by providing the accurate health benefit and risk information.[23] Yet, public communication programs do not appear to succeed in assisting consumers to make seafood consumption choices that balance health risks and benefits, regardless of how the message is framed or whether the sources of information are public or private groups.[24]

International Seafood Markets: Ecolabeling

Seafood ecolabeling programs that certify fisheries based on sustainability allow consumers to signal preferences for healthier global oceans each time they purchase labeled seafood. Ecolabels can, thus, create an economic incentive for environmental improvements.[3,25,26] Ecolabeling is an increasingly important tool used in the promotion of sustainable fishery products around the world. Whether the consumer is actually paying a price premium for ecolabeled products is of fundamental importance as it indicates a return on the investment of sustainable practices, providing an incentive for producers to undertake such practices.[27] A portion of consumers in the United States and select European countries have a statistically significant preference for seafood with ecolabels relative to unlabeled seafood.[28–33] In a study of consumers' noneconomic motivations to purchase ecolabeled seafood in Europe, there is a relationship between consumers' preference for ecolabeled fish and stated beliefs about the level of fisheries regulation and fish stocks.[34] Moreover, French consumers' taste for ecolabeled seafood is a function of perceptions of the fishing industry.[35] And most notably, there is a statistically significant premium around 14% for ecolabeled frozen processed whitefish products in the UK market.[36–38] Altogether, these studies jointly imply the potential for successful market differentiation for ecolabeling programs and possible incentives for sustainable fisheries practices.

Whether these modest price premiums translate into actual changes in fishing practices is partly dependent upon whether the price premiums at the consumer level are reflected at the wholesale level and for fishermen or fish farmers. Studies show either no or small premiums are present at the ex-vessel level across a variety of certified fisheries globally.[39–41] Certification appears to have some effects on international seafood markets, as prices of certified products may become less sensitive to competition from noncertified seafood products.[42]

International Seafood Markets: Attributes

Consumers value a wide array of seafood attributes, ranging from species, taste, texture, color, freshness, nutritional content, and product form to less tangible attributes such as country of origin or the presence of an ecolabel.[29,31,43] Moreover, attributes of fish also have value for the producer.[44–52] Some individual attributes may be more highly valued than others, and particular combinations of attributes lead to higher or lower valued fish at the producer level. For example, quality (resulting from handling during catch) or size of fish may segment fish within a particular fishery into different fresh or processed product forms.[52] Results of several of these studies have shown that quality matters; in particular, that there is a higher return for the raw fish product that is of sufficiently high quality to be put into a higher valued market segment. Quality issues can, thus, be an additional source of economic inefficiency in fisheries if poor fish quality results from management systems that provide incentives for long trips and poor treatment of fish. Under regulated open access, a larger share of the raw fish product can flow toward the inferior low-value market rather than the high-value fresh market because vessels race to fish and collapse the fishing season.[53] When fisheries are managed with rights-based systems, the race to fish slows down, creating more opportunities to land high-value fresh products.[54] An alternative to management reform is for processors to add value by breading into fish sticks to cover poor fish quality and create some direct employment in fish processing.[52] However, this strategy may reduce the overall value of the fishery for coastal communities and the fishermen when the fish is relegated to lower-valued product forms.

Overall, globalization and the expansion of the international seafood trade have often led to market integration. Competition from aquaculture, in particular, has tended to decrease prices, an undeniably good outcome for consumers but a challenge to capture fisheries.[55] This competition creates an even greater sense of urgency for wild fisheries to adopt effective management institutions that organize production efficiently and produce sustainable profits. Market integration also reduces opportunities for price compensation from ecological and technological disasters. Hypoxia (low dissolved oxygen), e.g., has decreased profits for the shrimp fishery in North Carolina, which makes up a small portion of the US total domestic shrimp production.[56] In the much larger Gulf of Mexico shrimp fishery, which also experiences hypoxia and was affected by the recent Deepwater Horizon oil spill, price data indicate that the market is integrated with imports; producer losses from recent supply disruptions likely have not been offset even partially by price increases.[57,58] Pushing back against market integration are attempts to segment the seafood market through country-of-origin labeling, ecolabeling, differentiating wild from farmed fish, and promotion of local seafood.

FAO Fish Price Index

Seafood is a critical contributor to livelihoods and food security.[3] As world food prices hit an all-time high in February 2011 and are still almost two and a half times those in 2000, there is an increased awareness that prices are an important indicator of food scarcity. The FAO, which has for a long time compiled prices for other major food categories, recently began tracking seafood prices, working together with a team of researchers to develop the FPI.[59] The FPI can facilitate understanding of seafood crises and may assist in averting them. The index uses a similar format as price indices for other foodstuffs, all of which are published in the FAO Food Outlook (www.fao.org/giews/english/fo/index.htm).

The FPI relies on trade statistics because seafood is a highly traded commodity, these data are easily updated, and trade data can provide a timely proxy for domestic seafood prices that are difficult to observe in many regions and costly to update with global coverage. Calculations of the extent of price competition in different countries support the plausibility of reliance on trade data. The FPI can also be separated into subindices according to production technology, fish species, or region, providing a valuable tool for understanding and tracking development in different segments of the seafood market. In Figure 25.3, the FPI indices for capture fisheries and aquaculture. Figure 25.3 suggests increased scarcity of capture fishery resources in recent years but growth in aquaculture that is keeping pace with demand.

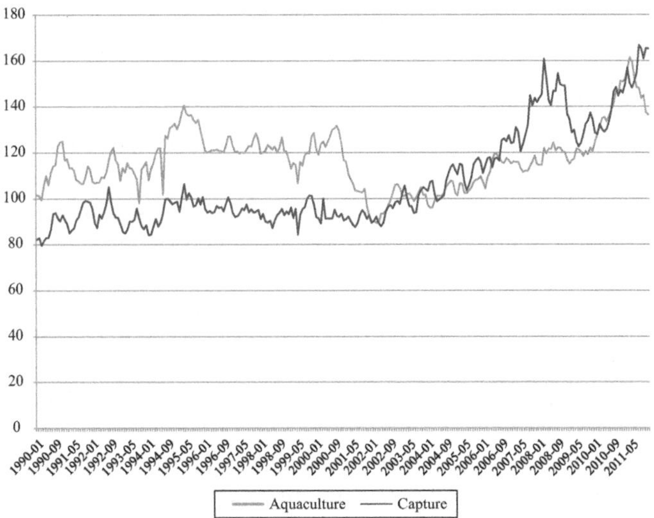

FIGURE 25.3 FAO fish price index, monthly 1990–2011. Aquaculture v. capture (Tverterås et al.)[59]

Conclusions

The international seafood trade will likely continue to grow. As aquaculture production expands in developing countries, it will continue to be exported to developed countries and will likely experience growth in exports to consumers in developing countries. Production from capture fisheries is unlikely to grow overall, although improvements in fisheries management in some countries (and continued degradation in fisheries stocks in other countries) may change the relative mix of seafood from capture fisheries available on the international market. Increasing affluence of consumers in emerging markets and increasing populations will lead to continued demand growth. Thus, to promote a future in which the international markets maintain access to seafood and developing countries have seafood as a source of food security, a challenge will be to create economic incentives for sustainability. In other words, institutions are required that allow the market to operate efficiently such that natural resource prices reflect the costs of sustainability.

References

1. Food and Agricultural Organization (FAO). *The State of the World Fisheries and Aquaculture.* Rome, 2018.
2. Anderson, J.L., Asche, F., and Garlock, T.M. Globalization and Commoditization: The Transformation of the Seafood Market. *J. Commod. Mark.*, 2018, *12*, 2–8.
3. Smith, M.D.; Roheim, C.A.; Crowder, L.B.; Halpern, B.S.; Turnipseed, M.; Anderson, J.L.; Asche, F.; Bourillón, L.; Gut-tormsen, A.G.; Khan, A.; Liguori, L.A.; McNevin, A.; O'Connor, M.; Squires, D.; Tyedmers, P.; Brownstein, C.; Carden, K.; Klinger, D.H.; Sagarin, R.; Selkoe. K.A. Sustainability and Global Seafood. *Science*, 2010, *327*, 784–786.
4. Asche, F. Farming the Sea. *Mar. Resour. Econ.*, 2008, *23* (4), 527–547.
5. Food and Agricultural Organization (FAO). *Rome. FISHSTAT,* 2019. www.fao.org/fishery/topic/16140/en.
6. Anderson, J.L.; Asche, F.; Tveterås, S. World Fish Markets. In *Handbook of Marine Fisheries Conservation and Management.* Grafton, R.Q.; Hilborn, R.; Squires, D.; Tait, M.; Williams, M. Eds.; Oxford: Oxford University Press, 2010.

7. Kobayashi, M.; Msangi, S.; Batka, M.; Vannuccini, S.; Dey, M.M; Anderson, J.L. Fish to 2030: The Role and Opportunity for Aquaculture. *Aquac. Econ. Manag.*, 2015, *19* (3), 282–300.
8. Asche, F.; Bellemare, M.; Roheim, C.A.; Smith, M.D.; Tveteras, S. Fair Enough? Food Security and the International Trade of Seafood. *World Dev.*, 2015, *67*, 151–160.
9. Belton, B.; Bush, S.R.; Little, D. Not Just for the Wealthy: Rethinking Farmed Fish Consumption in the Global South. *Glob. Food Sec.*, 2018, *16*, 85–92.
10. Scott, A. The Fishery: The Objectives of Sole Ownership. *J. Political Econ.*, 1955, *63*, 116–124.
11. Smith, V.L. Economics of Production from Natural Resources. *Am. Econ. Rev.*, 1968, *58*, 409–431.
12. Smith, V.L. On Models of Commercial Fishing. *J. Political Econ.*, 1969, *77*, 181–198.
13. Anderson, J.L. *The International Seafood Trade*; Cambridge: Woodhead Publishing, 2003.
14. Brander, J.A.; Taylor, M.S. International Trade and Open-Access Renewable Resources: The Small Open Economy Case. *Can. J. Econ.*, 1997a, *30*, 526–552.
15. Brander, J.A.; Taylor, M.S. International Trade between Consumer and Conservationist Countries. *Resour. Energy Econ.*, 1997b, *19*, 267–298.
16. OECD. *Liberalising Fisheries Markets: Scope and Effects*; Paris: OECD, 2003.
17. Cato, J.C. *Seafood Safety – Economics of Hazard Analysis and Critical Control Point (HACCP) Programmes.* FAO Fisheries Technical Paper – 381. Rome: Food and Agriculture Organization of the United Nations, 1998, www.fao.org/docrep/003/x0465e/x0465e00.htm.
18. Roosen, J.; Marette, S.; Blancemanche, S. Does Health Information Matter for Modifying Consumption? A Field Experiment Measuring the Impact of Risk Information on Fish Consumption. *Rev. Agric. Econ.*, 2009, *31*, 2–20.
19. Shimshack J.P.; Ward, M.B.; Beatty, T.K.M. Mercury Advisories: Information, Educations, and Fish Consumption. *J. Environ. Econ. Manag.*, 2007, *53*, 158–179.
20. Roheim, C.A. An Evaluation of Sustainable Seafood Guides: Implications for Environmental Groups. *Mar. Resour. Econ.*, 2009, *24*, 301–310.
21. Nesheim, M.C.; Yaktine, A.L., Eds., *Seafood Choices: Balancing Benefits and Risks; National Academy of Sciences, Committee on Nutrient Relationships in Seafood: Selections to Balance Benefits and Risks, Food and Nutrition Board, M.C.*; Washington, DC: National Academies Press, 2007.
22. Leiss, W.; Nicol, A.M. A Tale of Two Food Risks: BSE and Farmed Salmon in Canada. *J. Risk Res.*, 2006, *9*, 891–910.
23. Marette, S.; Roosen, J.; Blanchemanche, S. The Choice of Fish Species: An Experiment Measuring the Impact of Risk and Benefit Information. *J. Agric. Resour. Econ.*, 2008, *33*, 1–18.
24. Uchida, H.; Roheim, C.A.; Johnston, R. Balancing the Health Risks and Benefits of Seafood: How Does Available Guidance Affect Consumer Choices? *Am. J. Agric. Econ.*, 2017, *99* (4), 1056–1077.
25. Marine Stewardship Council (MSC). *Marine Stewardship Council Annual Report 2010/2011*, 2011. www.msc.org/documents/msc-brochures/annual-report-archive/annual-report-2010-11-english/view (accessed April 2012).
26. Roheim, C. The Economics of Ecolabelling, In *Seafood Ecolabelling: Principles and Practice*, Ward, T.; Phillips, B. Eds.; Oxford, UK: Blackwell Publishing, 2008, 38–57.
27. Roheim, C.R.; Bush, S.R.; Asche, F.; Sanchirico, J.; Uchida, H. Evolution and Future of the Sustainable Seafood Market. *Nat. Sustain.*, 2018, *1*, 392–398.
28. Wessells, C.R., Johnston, R.; Donath, H. Assessing Consumer Preferences for Ecolabeled Seafood: The Influence of Species, Certifier, and Household Attributes. *Am. J. Agric. Econ.*, 1999, *81*, 1084–1089.
29. Johnston, R.J.; Wessells, C.R.; Donath, H.; Asche, F. Measuring Consumer Preferences for Ecolabeled Seafood: An International Comparison. *J. Agric. Resour. Econ.*, 2001, *26*, 20–39.
30. Johnston, R.J.; Roheim, C.A. A Battle of Taste and Environmental Convictions for Ecolabeled Seafood: A Contingent Ranking Experiment. *J. Agric. Resour. Econ.*, 2006, *31*, 283–300.

31. Jaffry, S.; Pickering, H.; Ghulam, Y.; Whitmarsh, D.; Wattage, P. Consumer Choices for Quality and Sustainability Labelled Seafood Products in the UK. *Food Policy*, 2004, *29*, 215–228.
32. Bronnmann, J.; Asche, F. Sustainable Seafood from Aquaculture and Wild Fisheries: Insights from a Discrete Choice Experiment in Germany. *Ecol. Econ.*, 2017, *142*, 113–119.
33. Wakamatsu, H.; Anderson, C.M.; Uchida, H.; Roheim, C.A. Pricing Ecolabeled Seafood Products with Heterogenous Preferences: An Auction Experiment in Japan. *Mar. Resour. Econ.*, 2017, *32* (3), 277–294.
34. Brécard, D.; Hlaimi, B.; Lucas, S.; Perraudeau, Y.; Salladarré, F. Determinants of Demand for Green Products: An Application to Ecolabel Demand for Fish in Europe. *Ecol. Econ.*, 2009, *69*, 115–125.
35. Salladarré, F.; Guillotreau, P.; Perreudeau, Y.; Monfort, M.C. The Demand for Seafood Ecolabels in France. *J. Agricu. Food Ind. Organ.*, 2010, *8* (1), Article 10. doi:10.2202/1542-0485.1308.
36. Roheim, C.; Asche, F.; Santos, J. The Elusive Price Premium for Ecolabeled Products: Evidence from Seafood in the U.K. Retail Sector. *J. Agric. Econ.*, 2011, *62*, 655–668.
37. Sogn-Grundvåg, G.; Larsen, T.A.; Young, J.A. The Value of Line-Caught and other Attributes: An Exploration of Price Premiums for Chilled Fish in UK Supermarkets. *Marine Policy*, 2013, *38*, 41–44.
38. Asche, F.; Larsen, T.A.; Smith, M.D.; Sogn-Grundvåg, G.; Young, J.A. Pricing of Eco-labels with Retailer Heterogeneity. *Food Policy*, 2015, *67*, 82–93.
39. Wakamatsu, H. The Impact of MSC Certification on a Japanese Certified Fishery. *Mar. Resour. Econ.*, 2014, *29*, 55–67.
40. Blomquist, J.; Bartolino, V.; Waldo, S. Price Premiums for Providing Eco-labelled Seafood: Evidence from MSC-certified Cod in Sweden. *J Agri. Econ.*, 2015, *66*, 690–704.
41. Stemle, A.; Uchida, H.; Roheim, C.A. Have Dockside Prices Improved after MSC Certification? Analysis of Multiple Fisheries. *Fish. Res.*, 2017, *182*, 116–123.
42. Roheim, C.A.; Zhang, D. Sustainability Certification and Product Substitutability: Evidence from the Seafood Market. *Food Policy*, 2018, *79*, 92–100.
43. Holland, D.; Wessells, C.R. Predicting Consumer Preferences for Fresh Salmon: The Influence of Safety Inspection and Production Method Attributes. *Agricu. Resour. Econ. Rev.*, 1998, *27* (1), 1–14.
44. Gates, J. Demand Price, Fish Size and the Price of Fish. *Can. J. Agric. Econ.*, 1974, *22* (1), 1–22.
45. Anderson, L. Optimal Intra- and Interseasonal Harvesting Strategies When Price Varies with Individual Size. *Mar. Resour. Econ.*, 1989, *6*, 145–162.
46. Larkin, S.; Sylvia, G. Intrinsic Fish Characteristics and Production Efficiency. *Am. J. Agric. Econ.*, 1999, *81* (1), 29–43.
47. McConnell, K.; Strand, I. Hedonic Prices for Fish: Tuna Prices in Hawaii. *Am. J. Agric. Econ.*, 2000, *82*, 133–144.
48. Carroll, M.; Anderson, J.; Martinez-Garmendia, J. Pricing U.S. North Atlantic Bluefin Tuna and Implications for Management. *Agribusiness*, 2001, *17*, 243–254.
49. Asche, F.; Hannesson, R. Allocation of Fish between Markets and Product Forms. *Mar. Resour. Econ.*, 2002, *17*, 225–238.
50. Fong, Q.S.W.; Anderson, J.L. International Shark Fin Markets and Shark Management: An Integrated Market Preference-Cohort Analysis of the Blacktip Shark (*Carcharhinus limbatus*). *Ecol. Econ.*, 2002, *40*, 117–130.
51. Kristofersson, D.; Rickertsen, K. Efficient Estimation of Hedonic Inverse Input Demand Systems. *Am. J. Agric. Econ.*, 2004, *86* (4), 1127–1137.
52. Roheim, C.; Gardiner, L.; Asche, F. Value of Brands and Other Attributes: Hedonic Analysis of Retail Frozen Fish in the UK. *Mar. Resour. Econ.*, 2007, *22* (3), 239–253.
53. Homans, F.R.; Wilen, J.E. Markets and Rent Dissipation in Regulated Open Access Fisheries. *J. Environ. Econ. Manag.*, 2005, *49*, 381–404.

54. Birkenbach, A.M.; Kaczan, D.J.; Smith, M.D. Catch Shares Slow the Race to Fish. *Nature*, 2017, *544.7649* (2017), 223–226.
55. Anderson, J.L. Aquaculture and the Future: Why Fisheries Economists Should Care. *Mar. Resour. Econ.*, 2002, *17*, 133–151.
56. Huang, L.; Nichols, L.A.B.; Craig, J.K.; Smith, M.D. Measuring Welfare Losses from Hypoxia: The Case of North Carolina Brown Shrimp. *Mar. Resour. Econ.*, 2012, *27*, 3–23.
57. Asche, F.; Bennear, L.S.; Oglend, A.; Smith, M.D. U.S. Shrimp Market Integration. *Mar. Resour. Econ.*, 2012, *27*, 181–192.
58. Smith, M.D.; Oglend, A.; Kirkpatrick, J.; Asche, F.; Bennear, L.S.; Craig, J.; Nance, J.M. Seafood Prices Reveal Impacts of a Major Ecological Disturbance. *Proc. Natl. Acad. Sci.*, 2017, *114* (7), 1512–1517.
59. Tveterås, S.; Asche, F.; Bellemare, M.F.; Smith, M.D.; Guttormsen, A.G.; Lem, A.; Lien, K.; Vannuccini, S. Fish Is Food – The FAO's Fish Price Index. *PLoS ONE*, 2012, *7* (5), e36731. doi:10.1371/journal.pone.0036731.

26

Water Cycle: Ocean's Role

Don P. Chambers
University of South Florida

Introduction

The water, or hydrological, cycle, describes the processes that move water from one area to another on the Earth's surface. A significant part of the water cycle involves the transformation of water from its liquid state on the Earth's surface to the gaseous state in the atmosphere (*evaporation*) and then back to its liquid state (*precipitation*). Although the amount of water that can be stored at any time in the atmosphere is only a fraction of the amount that is stored on land or in the ocean (Table 26.1),[1] it can transport water great distances, and therefore is a vital component of the water cycle. An additional component of the water cycle is the liquid water transported via rivers, or from melting of frozen water in glaciers and ice sheets (*runoff*).

Because of the importance of water to human life, much of the study of the water cycle has focused on understanding measurements of precipitation, evaporation, and runoff over the land. However, when placed in perspective of the amounts of water available for evaporation and the relative amounts of water being exchanged with the atmosphere, it is clear that the ocean largely controls the water cycle.[2] The oceans contain ~1.3 billion cubic km (km^3) of water, which is 51 times more than the amount of water stored as ice (mainly in ice sheets over Greenland and Antarctica), 85 times more than the amount of water stored in all the underground aquifers on land, 4,000 times more than the water stored in the soil and rivers/lakes, and more than 133,000 times more than the water stored in the atmosphere (Table 26.1).

The amount of water evaporated from the ocean every year is also much larger than the amount evaporated from land, by a factor of more than 5. Although about 90% of this returns relatively quickly as precipitation over the oceans, the other 10% is transported via the atmosphere to precipitate over land. Thus, precipitation over land exceeds evaporation, and this is only possible because of the water transported from the oceans. This excess land precipitation cannot be stored permanently on land, and most of it flows back into the ocean via rivers resulting in a near balance in the water mass storage and transport on global scales.[1,3]

However, the balance is not exact on time-scales from months to several years, and changes in the climate can move the balance father from equilibrium, especially regionally. Some climate models predict that in a warming climate, there will be an intensification of both precipitation and evaporation,[4] with a result that that wet areas become wetter and dry areas become dryer. Although several studies have attempted to quantify trends in specific components of the water cycle (e.g., precipitation, evaporation, or runoff) over large regions or even globally, assessing the change in the balance over time is difficult, primarily because estimates of the uncertainty of yearly means of precipitation, evaporation, and runoff

TABLE 26.1 Estimated Average Volumes of Water Stored in Various Reservoirs on the Earth

Reservoir	Water Volume Storage (in millions of km^3)
Oceans	1,335
Ice	26.4
Groundwater (aquifers)	15.3
Soil	0.12
Rivers/Lakes	0.18
Atmosphere	0.01

Source: Adapted from Trenberth, Smith, et al.[1]

are all of order 3000 km^3, which means a balance cannot be determined to better than this.[1] Imbalances at this level in terms of the ocean volume (sea level) would imply more than 8 mm per year of sea-level rise, which is far larger than observed changes.[5] Because of this, most studies of the global water cycle budget assume a perfect closure, and fix one estimate (often runoff) to perfectly balance the other two.[1]

Because the ocean is a large component of the water cycle and is the primary source of cycling water, several important quantities of the ocean state can provide important clues to the size of water cycle variability and how it is changing. These are salinity, sea level, and ocean mass variations. The following sections will briefly describe how these are related to the water cycle and how measurements of changes in each quantity are providing important clues to how the water cycle is changing.

Ocean Salinity

The ocean is saline due to dissolved salts that have accumulated in the oceans over time. Oceanographers can measure the ratio of mass of dissolved salts to the mass of water and express this in terms of salinity. The global average salinity of the ocean is 35 parts per thousand, but can locally be more or less than this. It is clear from comparing the pattern of mean ocean surface salinity with that from evaporation (E) minus precipitation (P) over the ocean (Figure 26.1), that high salinity regions occur where there is excess evaporation (E-P positive) and low salinity regions occur where there is excess precipitation (E-P negative). Although the differences in E–P over the ocean basins are generally small for particular latitude bands, the difference in salinity between the difference basins are much larger. This is due mainly to how water is transported through the atmosphere and the humidity of air blowing across the oceans.[9] For example, the air over the narrow Atlantic tends to be dryer because it blows from the continents; therefore, the internal precipitation tends to be less than that of the Pacific.

While salts are being added to the ocean every year from river outflow, it is a fraction of the amount of total salts in the ocean, and would change the ocean salinity by <1 part in 17 million over the course of a year,[10] which is far too low to measure. Changes in precipitation and evaporation, along with river runoff and transport of different water masses within the ocean, however, can change the surface salinity locally by amounts that are several times larger than the measurement precision of current in situ instruments. Thus, changes in ocean surface salinity can provide insight into the changing water cycle.

Several studies have attempted to do this.[11,12] They find that since the 1950s, the SSS has increased in areas of high salinity and decreased in areas of low salinity. Although there is still uncertainty due to incomplete knowledge of oceanic transports of salt and mapping of sparse data, the results are consistent with the hypothesis of an increasing water cycle. Recently, two different ocean observing satellites have been launched to measure SSS from space for the first time: the soil moisture ocean salinity (SMOS) mission from the European Space Agency and the Aquarius/SAC-D mission from NASA and Argentina.[13] Although neither will match the precision of in situ salinity measurements, it is hoped that their near global coverage will provide better insight to into the changing global water cycle from an ocean perspective.

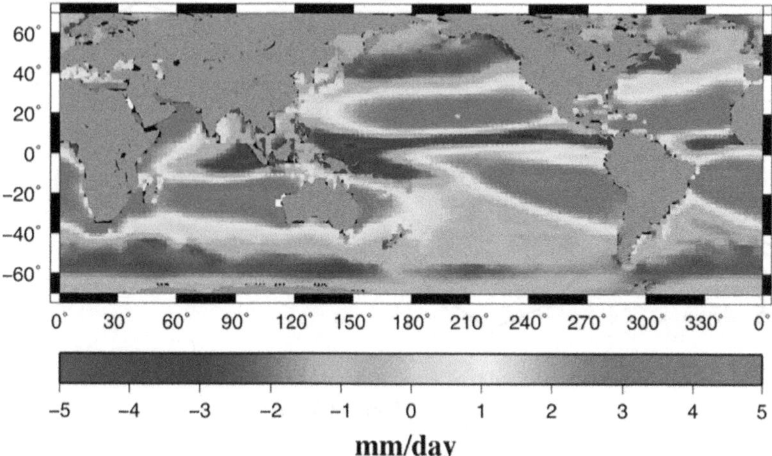

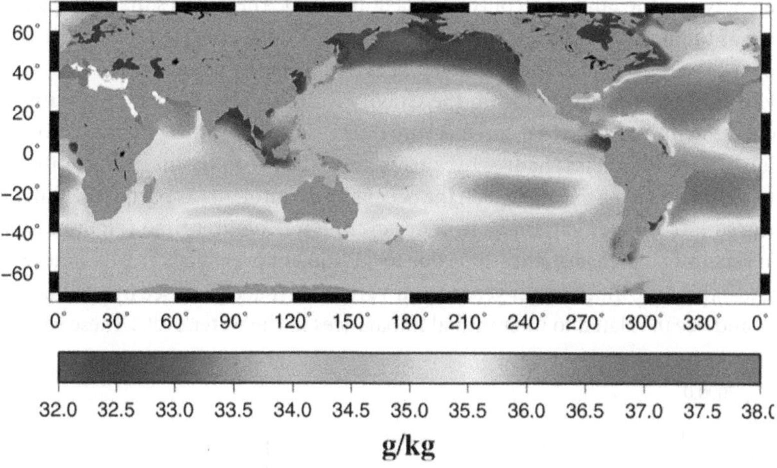

FIGURE 26.1 (See color insert.) (**a**) Mean Evaporation - Precipitation in mm/day and (**b**) SSS. Evaporation data are from OAflux (http://oaflux.whoi.edu/),[6] averaged 1979–2009. Precipitation data from Global Precipitation Climatology Project (GPCP) (http://precip.gsfc.nasa.gov/),[7] averaged 1979–2009. Sea surface salinity data from World Ocean Atlas 2009 (http://www.nodc.noaa.gov/OC5/WOA09/pr_woa09.html).[8]

Global Mean Sea Level and Ocean Mass

The water transported between the ocean and continents in the water cycle will be reflected in both changes in global mean sea level (GMSL) and the mass of the ocean. If there is more water leaving the ocean than being added, GMSL and the global mass of the ocean will drop. If more water is entering the ocean either from increased precipitation or runoff from land, or less is leaving via decreased evaporation, then GMSL will increase. GMSL will also be affected by ocean warming/cooling,[14,15] and so will only provide limited information on changes in the global water cycle without other information. With measurements of in situ temperatures, it has been possible to use GMSL measurements to extract some information about changes in the water cycle, especially at seasonal periods.[14,16] Beginning in the late 1990s, it became evident that there are large changes in the global ocean mass on seasonal time-scales as

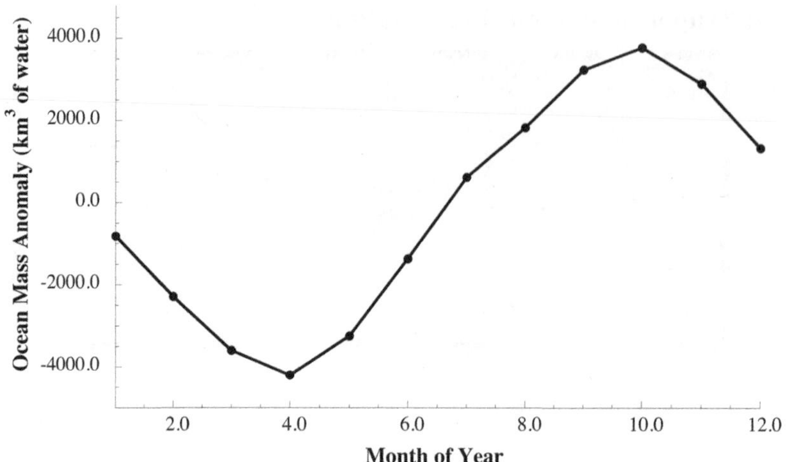

FIGURE 26.2 Mean seasonal ocean mass variations in km^3 of water, computed from 8 years of GRACE data.

water is cycled between the land and continents (Figure 26.2). The ocean's mass drops during the early part of the year as the water is cycled from the ocean to the continents (around April). Six months later (around October), the ocean gains mass as water is cycled from the continents.

This has been confirmed with data from the Gravity Recovery and Climate Experiment (GRACE).[17] GRACE is a novel satellite mission that can measure changes in the Earth's gravity at monthly time-scales accurately enough to resolve mass changes over areas as small as river basins.[18] In addition to confirming the seasonal exchange of water mass between the oceans and continents as part of the water cycle, GRACE has also measured interannual variations in water cycling for the first time (Figure 26.3). While ocean mass has been generally increasing since 2003 due to addition of previously frozen water from Greenland and Antarctica,[20] there are much larger year to year variations that are caused by cycling of water between the continents and ocean related to interannual imbalances in the water cycle. These variations can often

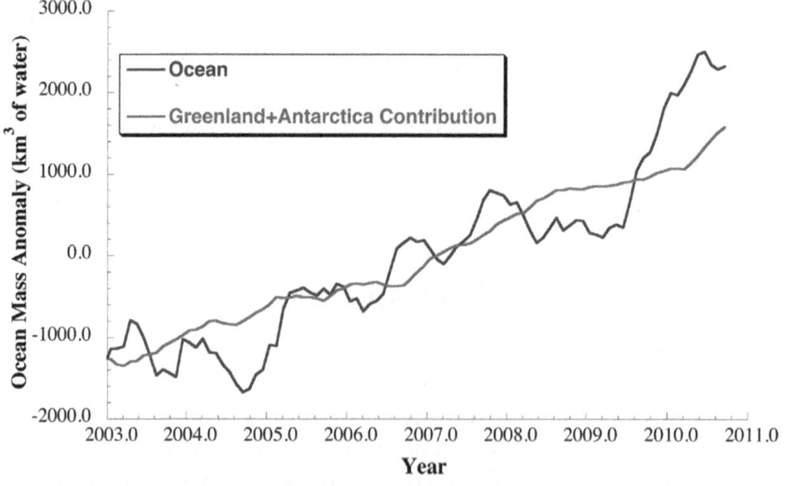

FIGURE 26.3 Non-seasonal ocean mass variations from GRACE (dark line), along with contribution to ocean mass change from Greenland and Antarctica (gray line), both in km^3 of water. Both time-series have been smoothed with a 6-month running mean filter.
Source: Chambers & Schröter,[19] Velicogna.[20]

account for much of the interannual GMSL variations.[21,22] For instance, there are significant deviations in GMSL associated with El Nino and La Nina events,[23,24] and evidence suggests that much of this is due to changes in the cycling of water between the ocean and continents.[25]

Conclusion

Most of the water that eventually returns to the Earth as precipitation is first evaporated from the oceans. Thus, the ocean has a vital role in the global water cycle. Understanding how the water cycle is changing is important for predicting future water resources for humanity.[2] However, while much has been learned about the relative sizes and variability of precipitation, evaporation, and water storage over large basins, much is still unknown. Because of its important role in the water cycle, ocean properties such as SSS, GMSL, and mean ocean mass are proving to be useful probes into the water cycle, and can aid us in understanding how the cycling of water is changing over time. Although results are still preliminary, new observations, especially from space-borne sensors, will likely improve our understanding in the future.

References

1. Trenberth, K.E.; L. Smith; T. Qian; A. Dai; J. Fasullo. Estimates of the global water budget and its annual cycle using observational and model data. J. Hydrometeor. **2007**, *8*, 758–769.
2. Oki, T.; Entekhabi, D.; Harrold; T.I. The global water cycle. *The State of the Planet: Frontiers and Challenges in Geophysics,* Sparks, R.S.J., Hawkesworth, C.J. Eds.; Geophysical Monograph 150, IUGG: 2004; Vol. 19, 225–237, doi:10.1029/150GM1829.
3. Oki, T. Global water cycle. *Global Energy and Water Cycles;* Browning, K., Gurney, R., Eds.; Cambridge University Press: 10–27, 1999.
4. Held, I.M.; Soden, B.J. Robust responses of the hydrological cycle to global warming. J. Clim. **2006**, *19*, 5686–5699.
5. Cazenave, A.; Llovel, W. Contemporary sea level rise. Ann. Rev. Marine Sci. **2010**, *2*, 145–174 pp.
6. Yu, L.; Jin, X.; Weller, R.A. *Multidecade Global Flux Datasets from the Objectively Analyzed Air-Sea Fluxes (OAFlux) Project;* Latent and sensible heat fluxes, ocean evaporation, and related surface meteorological variables. Woods Hole Oceanographic Institution, OAFlux Project Technical Report. 0A-2008-01, Woods Hole, Massachusetts, 2008; 64pp.
7. Adler, R.F.; Huffman, G.J.; Chang, A.; Ferraro, R.; Xie, P.; Janowiak, J.; Rudolf, B.; Schneider, U.; Curtis, S.; Bolvin, D.; Gruber, A.; Susskind, J.; Arkin. P.; Nelkin, E. The version 2 global precipitation climatology project (GPCP) monthly precipitation analysis (1979-Present). J. Hydrometeor. **2003**, *4*, 1147–1167.
8. Antonov, J.I.; Seidov, D.; Boyer, T.P.; Locarnini, R.A.; Mis-honov, A.V.; Garcia, H.E.; Baranova, O.K.; Zweng, M.M.; Johnson, D.R. *World Ocean Atlas 2009, Volume 2: Salinity.* Levitus, S. Ed. NOAA Atlas NESDIS 69, U.S. Government Printing Office: Washington, D.C., 2010; 184 pp.
9. Schmitt, R.W. Salinity and the global water cycle. Oceanography, **2008**, *21* (1), 12–19.
10. Pickard, G. L.; Emery, W.J. *Descriptive Physical Oceanography*; Pergamon Press: Oxford, England, 5th Edition, 1990.
11. Curry, R.; Dickson, B.; Yashayaev, I. A change in the freshwater balance of the Atlantic Ocean over the past four decades. Nature **2003**, *426*, 826–829.
12. Durack, P.J.; Wijffels, S.E. Fifty-year trends in global ocean salinities and their relationship to broad-scale warming. J. Clim. **2010**, *23*, 4342–4362, doi: 10.1175/2010JCLI3377.1
13. Lagerloef, G.; Font, J. SMOS and Aquarius/SAC-D missions: The era of spaceborne salinity measurements is about to begin. Chapter 3 in *Oceanography from Space;* Barale, V., Gower, J.R.F., Alberotanza, L. Eds.; Springer: 2010.
14. Chen, J.L.; Wilson, C.R.; Chambers, D.P; Nerem, R.S. Seasonal global water mass balance and mean sea level variations. Geophys. Res. Lett. **1998**, *25*, 3555–3558.

15. Levitus, S.; Antonov, J.; Boyer, T. Warming of the world ocean, 1955–2003, Geophys. Res. Lett. **2005**, 32, L02604, doi:10.1029/2004GL021592.
16. Minster, J.F.; Cazenave, A.; Rogel, P. Annual cycle in mean sea level from Topex-Poseidon and ERS-1: Inference on the global hydrological cycle. Global Planet. Change **1999**, *20,* 57–66.
17. Chambers, D.P.; Wahr, J.; Nerem, R.S. Preliminary observations of global ocean mass variations with GRACE, Geophys. Res. Lett. **2004**, *31,* L13310, doi:10.1029/2004GL020461.
18. Tapley, B.D.; Bettadpur, S.; Watkins, M.; Reigber, C. The gravity recovery and climate experiment: Mission overview and early results. Geophys. Res. Lett. **2004**, *31,* L09607, doi:10.1029/2004GL019920.
19. Chambers, D.P.; Schröter, J. Measuring ocean mass variability from satellite gravimetry. J. Geodyn. **2011**, *52* (5), 333–343.
20. Velicogna, I. Increasing rates of ice mass loss from the Greenland and Antarctic ice sheets revealed by GRACE. Geophys. Res. Lett. **2009**, *36,* L19503, doi: 10.1029/2009GL040222.
21. Willis, J.K.; Chambers, D.P.; Nerem, R.S. Assessing the globally averaged sea level budget on seasonal to interannual time scales. J. Geophys. Res. **2008**, *113,* C06015, doi:10.1029/2007JC004517.
22. Leuliette, E.W; Willis, J.K. Balancing the sea level budget. Oceanography **2011**, *24* (2),122–129, doi:10.5670/oceanog.2011.32.
23. Nerem, R.S.; Chambers, D.P.; Leuliette, E.; Mitchum, G.T.; Giese, B.S. Variations in global mean sea level during the 1997–98 ENSO event, Geophys. Res. Lett. **1999**, *26,* 3005–3008.
24. Chambers, D.P.; Mehlhaff, C.A.; Urban, T.J.; Fujii, D; Nerem, R.S. Low-frequency variations in global mean sea level: 1950–2000, J. Geophys. Res., **2002**, *107* (C4), 3026, doi:10.1029/2001JC001089.
25. Ngo-Duc, T.; Laval, K.; Polcher, J.; Cazenave, A. Contribution of continental water to sea level variations during the 1997–1998 El Niño–Southern Oscillation event: Comparison between Atmospheric Model Intercomparison Project simulations and TOPEX/Poseidon satellite data. J. Geophys. Res. **2005**, *110,* D09103, doi:10.1029/2004JD004940.

27

Climate Change Impacts and Adaptation for Coastal Transportation Infrastructure: A Sustainable Development Challenge for SIDS in the Caribbean and Beyond

Regina Asariotis[1]
United Nations Conference on Trade and Development (UNCTAD)

Introduction

Transport is a critical enabler of global trade and an engine for global trade-led development. With 80% of the volume of global trade (70% by value) carried by sea, shipping and seaports provide crucial linkages in the network of closely interconnected global supply chains. Seaports in developing countries account for more than 60% of goods loaded and unloaded at the global level (UNCTAD, 2019), illustrating the interconnectedness and interdependence of economies and of key transport nodes and networks. Transport is particularly important for the development prospects of vulnerable nations such as the Small Island Developing States (*SIDS*).[2]

SIDS are a diverse group of island countries in terms of location, land area, population size and markets, gross domestic product (*GDP*), and development level. Nevertheless, *SIDS* share common

[1] The views presented are those of the author and do not necessarily represent the views of the United Nations.
[2] For further information on Small Island Developing States, see http://unohrlls.org/about-sids/.

features that distinguish them as a special sustainable development case. These include insularity and remoteness, "smallness" of populations and economies, heavy import reliance, large exposure to external shocks, and vulnerabilities to environmental hazards and degradation. As is the case for all islands, access to the global community and markets is exclusively facilitated by seaports and airports. While maritime transport is the predominant mode used to carry freight, as well as accommodate cruise ship tourism, air transport is relied upon for passenger and tourist transport and for regional and domestic inter-island mobility.

Many islands are also major international tourism destinations according to the "Sun, Sea and Sand (*3S*)" tourism model (Phillips and Jones, 2006). Generally, tourism is considered as a viable and, in many cases, the main means of island economic growth; e.g., coastal tourism accounts for 11%–79% of the GDP of the Caribbean *SIDS* (UNECLAC, 2011). Given the isolated geographical location of many *SIDS* in relation to the major tourist markets, and the dependency of external trade/tourism on transport, resilient and well-maintained transport infrastructure is critical for their sustainable development (UNCTAD, 2014a; Pratt, 2015).

SIDS already face a number of transport-related challenges, including: small cargo volumes and trade imbalances; remoteness and limited access to global shipping networks; limited inter-island regional and domestic shipping connectivity; high dependency on energy imports; and disproportionately higher transport costs (UNCTAD, 2014a). For example, the cost of freight transport and insurance in the Caribbean basin is 30% higher than the world average (Pinnock and Ajagunna, 2012), and for some *SIDS*, including the Seychelles, Solomon Islands, Grenada, and Comoros, the 10-year average expenditure (2004–2013) on international transport as a percentage of the import value was more than double of the world average (UNCTAD, 2014b).

At the same time, most *SIDS* are located in disaster-prone regions and are also challenged by the impacts of climate variability and change (*CV & C*); this has prompted the inclusion of a temperature rise cap of 1.5°C above pre-industrial levels (the 1.5°C specific warming level—*SWL*) (IPCC, 2018) as an aspirational target in the 2015 Paris Agreement (Art. 2a). *SIDS* face increasing coastal hazards from *CV & C*, particularly from those associated with mean sea level rise (*SLR*), potential increases in the intensity of cyclones, increasing air/sea surface temperatures, more frequent heat waves, and coastal water acidification (Nurse et al., 2014; Stephenson and Jones, 2017; Mora et al., 2017). *SIDS* are particularly exposed due to the high concentration of populations, infrastructure, and services at the coast (Rhiney, 2015) and, given their limited capacity to respond due to terrain constraints, unfavorable economies of scale, and lack of financial and human resources, their vulnerability to *CV & C* is deemed high.

Transportation infrastructure/operations located at the coast are likely to be particularly impacted by *CV & C*. There will be direct impacts from increasing coastal hazards for seaports (freight and cruise) and coastal airports, in terms of damage, delay, and disruptions, as well as indirect impacts, arising from climate-driven changes in the demand for freight and passenger transportation (UNECE, 2013). In this context, it should be noted that while the tourism sector and its supporting services and imports in SIDS create significant demand for transportation, tourism itself is dependent on the aesthetics and environmental "health" of the "sandy" shores, i.e. the beaches, the primary natural island resource supporting *3S* tourism (Ghermandi and Nunes, 2013). Beaches and backshore infrastructure/assets fronted by beaches are particularly exposed to coastal erosion/flooding under *SLR* and extreme storm events (Scott et al., 2012; Wong et al., 2014; Monioudi et al., 2017). Cruise ship tourism, an increasingly important economic sector for many *SIDS*, may be affected too, as high winds and swell waves, which may also change under *CV & C*, can compromise the ability of vessels to berth safely; this may lead to lost ship calls/reduced income from cruise-tourism passenger spending and, in the longer term, could affect the commercial viability of cruise ship ports of call.

In order to estimate transportation risks under a changing climate and take effective adaptation response measures, targeted risk assessments are required, which depend on the spatiotemporal scales and resolution of the application and the available information (Christodoulou et al., 2018; Koks et al., 2019).

The present contribution presents an approach to assess risks to transportation in *SIDS* under *CV & C*, highlighting the nexus between climate, transport infrastructure/operations, and sustainable trade and tourism, to illustrate and help advance the case for urgent and concerted action to build climate resilience. To this end, an assessment of the potential marine inundation and other operational disruptions from *CV & C* of the coastal international airports and seaports of the Caribbean island of Saint Lucia is presented, together with an estimation of the beach erosion risk as a "proxy" of indirect climate impacts on transportation.

CV & C Risks for Seaports, Coastal Airports, and Connecting Road Networks in *SIDS*

Coastal transport infrastructure and operations face a range of climate-related risks (Table 27.1). Seaports will be incrementally impacted by *SLR* and storm events; their quays, jetties, and breakwaters may require redesigning and/or strengthening (Becker et al. 2013, 2016; Asariotis et al., 2017). Increasing sea levels could also alter nearshore flows, inducing port scouring and/or silting, whereas changes in the wind and wave regimes may require specific adaptation measures (UNECE, 2013). Coastal flooding may overwhelm low-lying coastal airports (Monioudi et al., 2018). Heavy rainfall (downpours) may also affect services, by inducing flash floods and landslides that can flood/damage connecting transport networks (road, tunnels, and bridges). Flooding can render transportation systems unusable for the duration of the event and damage terminals, intermodal facilities, freight villages, storage areas, and cargo and, thus, disrupt supply chains for longer periods (Fay et al., 2017; Koks et al., 2019).

Changes in other climatic factors can also have significant impacts on transportation infrastructure, services, and operations. For airports, higher mean temperatures and more frequent heat waves can affect runways (heat buckling) and aircraft lift, resulting in payload restrictions and disruptions (Coffel et al., 2017); thus, runways may require relocation and/or extension, which may not be always feasible in island settings due to topographic constraints. Moreover, the combination of higher temperatures and humidity will increase the heat index (http://www.nws.noaa.gov/om/heat/heat_index.shtml), resulting in increasing health risks for outdoor activities; this may affect not only staff and schedules at transport facilities but could also impact the attractiveness of the tourist destinations (De Freitas et al., 2008). Potential changes in the intensity/frequency of extreme precipitation events (downpours) and windstorms can also create problems for transport operations (Table 27.1).

At the same time, *SLR* and extreme storms and waves will have detrimental effects on "sandy coasts" (beaches) by inducing coastal erosion and inundation (Wong et al., 2014). Island beaches are particularly threatened, due to their limited dimensions and decreasing sediment supply (Peduzzi et al., 2013; Monioudi et al., 2017). As beaches form the pillars of the island *3S* tourism industry, assessments of climate-driven beach erosion are necessary to inform the design and timely implementation of the requisite technical adaptation measures (Ranasinghe, 2016; Narayan et al., 2016) to mitigate impacts on the tourism industry and, thus, the demand for transportation.

Coastal Risk Assessment of *SIDS* International Coastal Transportation Assets: Saint Lucia

Saint Lucia (Figure 27.1) is located at the southern section of the 850 km long, Lesser Antilles volcanic "double" arc (Macdonald et al., 2000). The island has an area of 616 km^2, a population of about 180,000, a GDP of 1.71 billion USD (2017), and a road network with an overall length of about 1,380 km (Koks et al., 2019). It faces moderate to high climatic risks being ranked as 49th out of 180 countries for the period 1996–2015 (Kreft et al., 2016), having been impacted by tropical storms (UNCTAD, 2017). Mean annual temperature has been increasing by about 0.16°C/decade and *SLR* has accelerated in 2005–2016, exceeding the Caribbean basin average (http://www.psmsl.org/data/obtaining/stations/1942.php). Winds are

TABLE 27.1 Major Climate Change Impacts on Coastal Transportation Infrastructure and Operations in SIDS

Stressor	Impacts on Seaports, Airports, and Connecting Roads
Sea Level (Mean and Extreme) and Waves	
Mean *SLR* Increases in *ESLs* Increases/changes in storm wave energy and direction	Damages to port infrastructure/cargo facilities from incremental and/or episodic inundation and wave regime changes; higher port construction/maintenance costs; sedimentation/dredging issues for port/navigation channels; increased risks of flood damages/wash-outs for coastal roads; increased flood risks for airport runways/taxiways, airport facilities and equipment; increased costs of technical adaptation measures; relocation of people and businesses; insurance issues
Temperature	
Higher mean temperatures Heat waves Increased variability in temperature extremes	Damages and asset lifetime reduction for infrastructure, equipment, and cargo; higher energy consumption for cooling (passengers and freight); occupational health and safety issues during extreme temperatures; decreased airport runway traction and payload restrictions due to reduced lift during aircraft takeoff; thermal pavement loading/degradation and asphalt rutting in connecting roads; increased landslides; increased construction and maintenance costs; potential changes in demand for passenger and freight transport
Precipitation	
Changes in the intensity/frequency of extreme events	Seaport facility flooding/damages; flooding of airport runways/taxiways and damage to airport structures and equipment; decreased traction on airport runways; connecting transportation network impacts (e.g., road inundation/washouts) and vital node damage (e.g., bridges); increased landslides/mudslides; poor visibility and reduced vehicle traction that can induce accidents; delays/disruptions; and changes in demand
Windstorms	
Changes in frequency and intensity of events	Issues with vessel navigation and berthing; cancellations/delays at airports; damages to airport terminals and equipment; increased risks for connecting road structures and accidents

Note: List is not exhaustive.

typically from E-ENE and E-S, with the average annual rainfall being about 1,300 mm at the coast but about three times as much in the island interior (ESL, 2015).

As is the case for all SIDS, Saint Lucia's connectivity with the global community and markets relies on its seaports and airports: the George Charles International Airport (*GCIA*) and Castries seaport (*CSP*) located in the capital Castries, and the Hewanorra International Airport (*HIA*) and Vieux Fort Seaport (*VFSP*) situated at the southern tip of the island (Figure 27.1). Both airports are built at low-lying coasts, close to the shoreline. *HIA* facilitates about 77% of air traffic (840,000 passengers in 2016), serving as the gateway for international long-haul flights, whereas *GCIA* handles regional flights. *CSP* and *VFSP* seaports handle a significant fraction of the total of the Eastern Caribbean States' (*OECS*) container traffic, with Port Castries being also a major cruise ship destination (677,400 arrivals in 2016). *VFSP* is

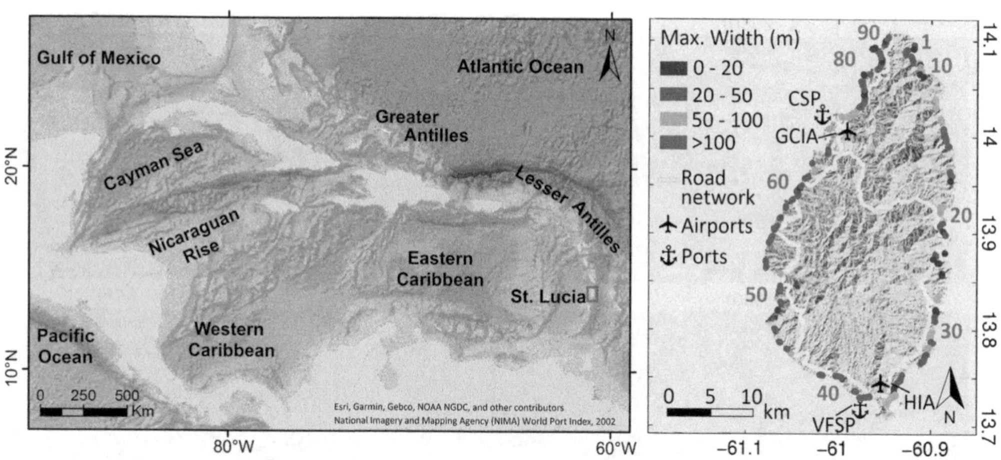

FIGURE 27.1 (See color insert.) Saint Lucia: Main transport network and nodes and recorded sandy shore (beach) locations and maximum widths (*BMWs*). DEM data from Shuttle Radar Topography Mission (SRTM) DTM and bathymetric data from GEBCO_2014 Grid. Key: HIA, Hewanorra International Airport; VFSP, Vieux Fort Seaport; GCIA, George Charles International Airport; and CSP, Castries Seaport.

backed by low-lying land and is approached by a narrow coastal road, not ideal for containerized traffic (UNCTAD, 2017).

CV & C impacts on transportation infrastructure, operations, and demand were estimated by: (i) modeling the coastal inundation of the above critical transportation infrastructure due to extreme sea levels (*ESLs*) under the IPCC *RCP4.5* (moderate) and *RCP8.5* ("business-as-usual") climatic scenarios; (ii) assessing direct impacts of changing climatic factors on operations, using an *operational thresholds* approach; and (iii) modeling the erosion of beaches, the primary natural resource supporting *3S* tourism, as a "proxy" for (indirect) impacts on demand.

Direct Impacts on Infrastructure and Operations

Impacts were assessed under the 1.5°C *SWL*, projected to be reached by early 2030s (Monioudi et al., 2018), as well as for later periods in the century. Extreme coastal flooding is driven by *ESLs* (i.e., the compound effect of mean sea levels, tides, storm surges, and coastal wave setups (Wong et al., 2014)) and extreme waves. The driving *ESLs* and resulting inundation for different return periods (e.g., for the 1–50 and 1–100-year events) under *RCP4.5* and *RCP8.5* were projected for 2020, 2030, 2040, 2050, 2060, 2080, and 2100, using dynamic flood simulations according to the approaches described in Vousdoukas et al. (2018). Simulations were based on a high-resolution Digital Elevation Model (*DEM*) (http://www.charim-geonode.net/layers/geonode:dem), with land hydraulic roughness derived from land use maps (https://www.esa-landcover-cci.org/).

The simulations showed that the *ESL* return periods, which form fundamental design parameters for any proposed technical adaptation measures, will significantly decrease over time. Under the 1.5°C *SWL* (2030), the baseline (1995) 100-year event will occur about every 10 years, occurring every year after the early 2040s. By comparison, the 100-year event projected for 2030 will occur every about 20 years by 2050 under *RCP4.5* (Monioudi et al., 2018).

Asset inundation projections (Figure 27.2) show that the low-lying areas, where the critical assets are located, are under increasing flood risk. In 2030, *GCIA* appears vulnerable to the 100-year extreme sea level (ESL_{100}), mostly at its northern side which backs a low-lying "sandy" shoreline (*Vigie* beach); as the century progresses, its exposure will increase as the beach itself is projected to face substantial beach erosion (Section 3.2). Hewanorra airport (*HIA*) also appears vulnerable at its (eastern) seaward edge.

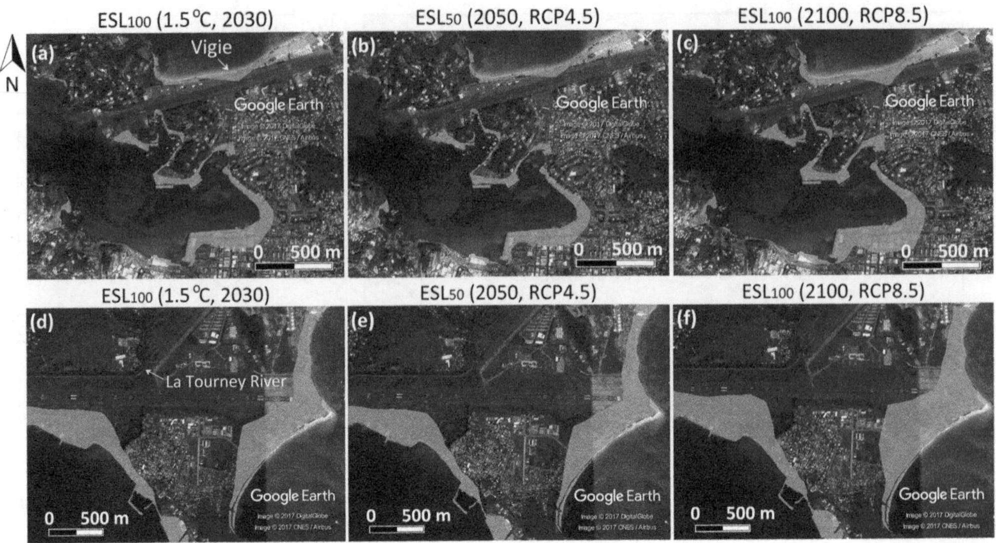

FIGURE 27.2 Inundation of GCIA and CSP (a–c) and HIA and VFSP (d–f). Projections are for the 100-year event (ESL100) in 2030 at 1.5°C SWL (a and d), the 50-year event (ESL50) in 2050 under RCP4.5 (b and e), and the 100-year event (ESL100) in 2100 under RCP8.5 (c and f). (After Monioudi et al., 2018.)

There, the runway is projected to be inundated at lengths of about 150, 130, and 380 m from the 1–100-year event (ESL_{100}) in 2030, the 1–50-year event (ESL_{50}) in 2050 (*RCP4.5*) and the ESL_{100} in 2100 (*RCP8.5*), respectively. Until now, *HIA* has been impacted by flash floods/overflowing of the redirected *La Tourney* river (Figure 27.2).

CSP, which has been damaged during previous extreme events, has been projected to be severely affected. In 2030 (under 1.5°C *SWL*), coastal flooding will impact docks, inundate berths, cargo sheds, and cruise ship facilities; there may also be flooding of the low-lying parts of the nation's capital, Castries. Later in the century, under both *RCP* scenarios tested, coastal flooding is projected to increase, in the absence of effective technical adaptation measures. Finally, the *VFSP* appears also vulnerable to coastal flooding, a marked deterioration from the present situation reported by facility managers (UNCTAD, 2017).

The Saint Lucian road network will also be impacted by *CV & C*. In addition to the coastal roads which are threatened by coastal flooding, inland roads will also be impacted (Koks et al., 2019). Empirical estimation of the potential landslides under a severe tropical storm event (Monioudi et al., 2019) has indicated that, in the absence of technical works to armor cliffs/embankments, landslide risks will remain high along the roads connecting the major transport gateways (ports and airports) with the major tourist destinations. The main road from HIA, the main international airport, to the capital Castries, where approximately 30% of the population reside, and to main tourist destinations in the north, has been assessed to be at particular risk; along this main road, landslide densities were found to be up to 1.75 landslides per road km (average density 0.75).

Generic operational thresholds were used to determine the (extreme) climatic conditions (e.g., extreme temperatures, rainfall, and winds) under which facility operations could be severely impaired (UNCTAD, 2017). These were then compared with projections of climatic factors under *CV & C*. The Caribbean Community Climate Change Centre's (*CCCCC*) downscaled climate projections from the *RCM PRECIS* (available for the *SRES A1B* scenario) were used (http://clearinghouse.caribbeanclimate.bz); in terms of both emissions and potential impacts, *A1B* approximates *RCP6.0*, i.e., a medium high emission scenario (van Vuuren and Carter, 2014).

The safety of airport/seaport staff when working outdoors depends on the *Heat Index*, i.e., a combination of temperature and relative humidity; according to the National Oceanic and Atmospheric Administration (NOAA), temperatures over 30.6°C and 32.5°C, combined with 80% relative humidity, present "high" and "very high" risks to human health, respectively (http://www.nws.noaa.gov/om/heat/heat_index.shtml). In 2030, staff working outdoors at the transportation facilities are projected to be at "high" risk for 2 days/year, under the *SRES A1B* scenario; such days could increase to 55 days/year by 2081–2100. Higher temperatures will also reduce aircraft lift, which gives rise to a need for reduced payloads or may require runway extension. It was found that in line with projections, by 2050, aircraft servicing these airports (e.g. Boeing 737-500) would have to decrease their takeoff load for about 11 days per year (assuming no technological advances). As these projections refer to medium size/range aircraft that require shorter runways than the long-haul aircraft servicing the intercontinental Caribbean routes, these projections are likely to be conservative (Monioudi et al., 2018).

Extreme heat can also raise energy demand/costs for facility heating, ventilation, and cooling (HVAC) systems. It was found that under the 1.5°C *SWL* (in 2030), the baseline (1986–2005 average) energy requirements will increase by 4% for 168 days/year, whereas a 3.7°C temperature rise (2081–2100) will increase energy requirements by 15% for 183 days/year. By contrast, intense (> 20 mm/day) and very heavy (> 50 mm/day) rainfall (downpours) that can inhibit operations (e.g. crane operations) are projected not to change significantly over the course of the century. Strong winds, which can affect aircraft takeoff/landing as well as the maneuverability/berthing of vessels at seaports, have also been found not to appreciably impact airport and seaport operations, on the basis of generic thresholds (Monioudi et al., 2018, but see also Nurse (2013)).

It appears that apart from the impacts of coastal inundation, most operational problems for the critical coastal transportation assets will be due to rising temperatures, with rainfall and wind effects projected to have no appreciable impacts. However, these projections do not include effects of tropical storms/hurricanes and, thus, might be considered as conservative. Also, facility-specific operational sensitivities which cannot be captured by generic thresholds (e.g., effects of wind/wave directional changes on ship berthing) may further increase operational disruptions.

Potential (Indirect) Impacts from Beach Erosion

Beaches are the primary natural resources supporting tourism in Saint Lucia, with coastal resorts designed so as to offer ocean views and immediate access to the beaches. Beach quality plays an important role in the selection of tourist destinations and the price guests are willing to pay (Ghermandi and Nunes, 2013).

The geo-spatial characteristics of all Saint Lucian beaches (e.g., length, beach maximum width—*BMW*, and area) were recorded from images and other related optical information available in the *Google Earth Pro* application (Monioudi et al., 2019). "Dry" beaches were defined as the low-lying coastal sediment bodies bounded on their landward side by either natural boundaries (vegetated dunes and/or cliffs) or permanent artificial structures (e.g., coastal embankments, seawalls, roads, and buildings) and on their seaward side by the shoreline, i.e., the median line of the foaming swash zone shown on the imagery. Other relevant information was also recorded, including the presence of natural and/or artificial beach features (e.g., coastal works) and the density of the backshore infrastructure/assets as a percentage of the beach length. Following this procedure, the geospatial characteristics of all 91 Saint Lucian beaches were recorded (Figure 27.1).

In order to estimate the erosion hazard, the *ESL* and extreme wave projections used to assess the coastal asset inundation were utilized to drive two cross-shore (1-D) morphodynamic model ensembles: a "long-term" ensemble consisting of the analytical models *Bruun, Dean,* and *Edelman* and a "short-term" ensemble comprising the numerical *SBEACH, Leont'yev, XBEACH,* and *Boussinesq* models (for model details and validation see Monioudi et al., 2017). The former was used to assess beach erosion under *SLR* and the latter to assess beach erosion due to extreme storms. To project future storm-induced erosion, the "long"- and "short-term" ensembles were used consecutively.

Given the spatiotemporal scales of the application, the input seabed slope and sediment data could not be based on observations. Models were set up using linear beach profiles, with various bed slopes (1/10, 1/15, 1/20, 1/25, and 1/30) and median beach sediment sizes (d_{50} of 0.2, 0.33, 0.50, 0.80, 1, 2, and 5 mm), and erosion was assessed for the 1–10-, 1–50-, and 1–100-year *ESLs* (ESL_{10}, ESL_{50}, and ESL_{100}) projected for 2030 (the 1.5°C *SWL*), 2050, and 2100 under the *RCP4.5* and *RCP8.5* scenarios and various extreme wave conditions. Due to the different conditions used in the model setups, the models produced ranges of beach erosion projections.

The results show that the *RCP 4.5* ESL_{100} in 2050 and 2100 will result in maximum storm erosion of up to about 62 and 73 m, respectively; under *RCP8.5*, about 64 and 84 m of maximum beach erosion is projected, respectively. Comparison of these projections with the current beach maximum widths (*BMWs*) suggests that, according to the most conservative projections (the 10th percentile, i.e., lowest projections) about 47% of all beaches will lose at least 50% of their current *BMWs* and 25% will be completely eroded under the *RCP4.5* ESL_{100} in 2050 (Figure 27.3). In terms of backshore asset exposure, at least 16% of beaches presently fronting infrastructure assets are projected to be overwhelmed

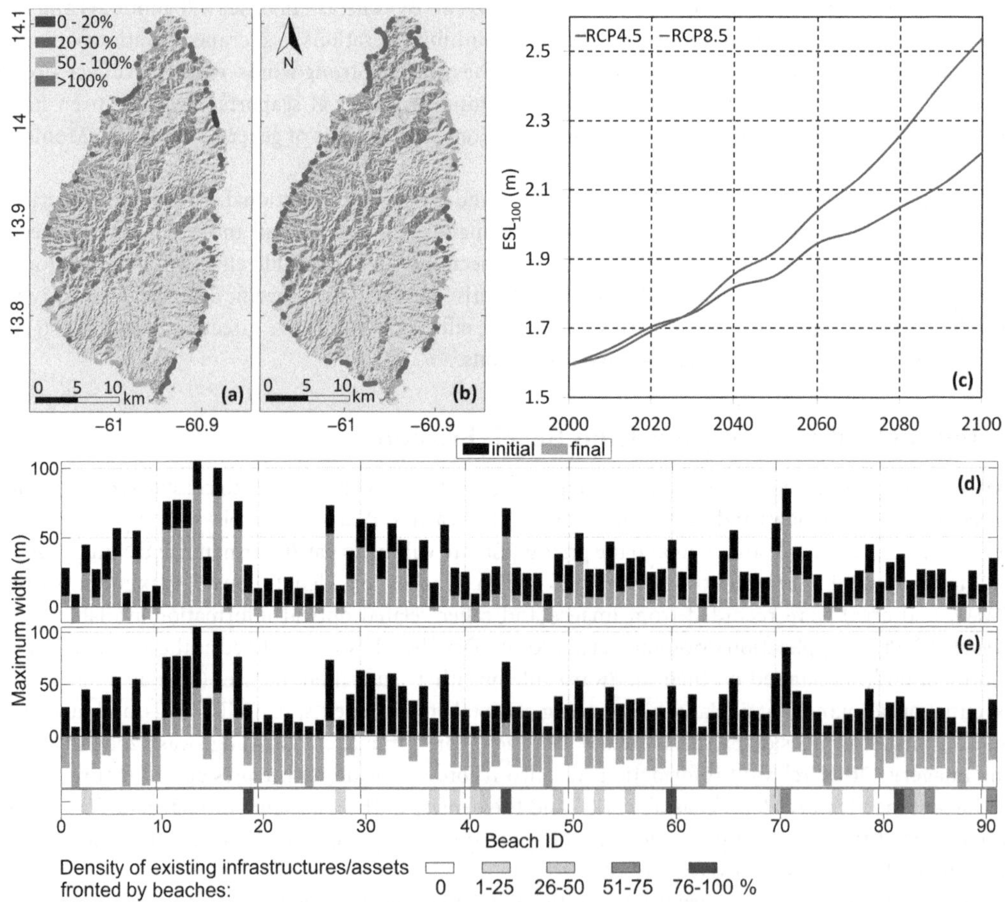

FIGURE 27.3 **(See color insert.)** (**a**) The 10th and (**b**) 90th percentiles of beach retreat projections under the ESL_{100} (for the year 2050 and RCP4.5), showing beaches projected to retreat (erode) by distances equal to different percentages of their initial maximum widths (*BMWs*). (**c**) Projected time evolution of ESL_{100} in the 21st century under RCP4.5 and RCP8.5. The current (initial) *BMWs* (black bars) are compared with those resulting from0 (**d**) the 10th and (**e**) 90th percentiles of beach erosion projections (blue bars); negative values indicate total beach erosion. The recorded density of the frontline backshore assets (as a percentage of the beach length) is also shown.

during the extreme event. By 2100, impacts could be devastating. Under the *RCP4.5* ESL_{100}, 24%–80% of all Saint Lucian beaches (and 23%–95% of the beaches fronting assets) are projected to be completely (at least temporarily) eroded under the low (10th percentile) and high (90th percentile) model projections, respectively. Erosion under the *RCP8.5* ESL_{100} represents an even worse scenario for the beaches and resorts; most resorts could sustain large damages, even in the case of a partial (or total) post-storm beach recovery as they are located within the beach erosion-recovery envelope (Monioudi et al., 2019).

The results show that beaches in Saint Lucia are vulnerable to *ESLs* and storm waves. Under increasing beach erosion/retreat, the long-term beach recreational value and carrying capacity, as well as the value of associated assets may fall considerably (Gopalakrishnan et al., 2011). Against this background, it appears that plans to respond effectively to the projected beach erosion risk should be urgently drawn up, with different adaptation options analyzed. Options based on the ecosystem approach should be considered first to protect both beaches and backshore infrastructure/assets (Peduzzi et al., 2013), although "hard" works might, in some cases, be still deemed necessary. However, the significance of beaches as economically critical natural resources and the limited effectiveness of hard coastal works (e.g. breakwaters) to protect beaches from *SLR* indicate that beach nourishment schemes will also be required, at least for the economically most important beaches.

Discussion and Concluding Remarks

Much of the international debate and policy action in relation to climate change and international transport is with a focus on the issue of mitigation, i.e., on efforts to reduce carbon emissions. By comparison, much less emphasis has so far been placed on the other side of the "climate change coin," that is to say the assessment of potential climate change impacts on transport infrastructure and operations and the development of adaptation measures.

The above results relate to the situation in one *SIDS*, Saint Lucia. However, they illustrate the urgent and important need for the development of adaptation measures and resilience building, informed by targeted risk assessments at local scale, for all *SIDS* and other vulnerable coastal/island economies that face similar challenges. In view of the long service life of transport infrastructure, effective adaptation requires rethinking established approaches and practices early. Moreover, a good understanding of risks and vulnerabilities is required for the development of well-designed adaptation measures that minimize the adverse effects of climatic factors. This, however, constitutes a major challenge. The potential adverse impacts of *CV & C* may be wide ranging, but they vary considerably by physical setting, climate forcing, and mode of transport, as well as other factors.

For the purposes of risk assessment and with a view to developing effective adaptation measures, dissemination of more tailored data and information is important, as are targeted case studies and effective multidisciplinary and multi-stakeholder collaboration. Infrastructure inventories, higher resolution data, including better DEMs, as well as a better understanding of coastal processes under *CV & C* are required for effective risk assessment and adaptation planning; and detailed technical studies at facility level are required to avoid maladaptation. Technical adaptation measures are widely needed, but these should involve innovative and efficient designs to avoid overengineering; ecosystem enhancement can play a significant role in reducing risks. Increased investment in local/regional human resources and skills (in particular skilled coastal scientists/engineers) will be critical for successful adaptation and resilience building in the future, as will be the mainstreaming of climate change considerations into ordinary transport planning, operations, and management.

Guidance, best practices, checklists, methodologies, and other tools in support of adaptation are urgently required, and targeted capacity building is going to be critical, especially for the most vulnerable countries, such as SIDS, which depend on their ports and airports for food and energy needs, external trade, and—crucially—tourism, which typically accounts for a major share of GDP. In this context, it is important to explore ways to generate the necessary financial resources and to consider how best to highlight and integrate the above considerations as part of Nationally Determined Contributions (NDCs) under the Paris Climate Agreement and in National Adaptation Plans.

Bearing in mind the potentially important trade and development implications of climate-related damage, delay, and disruption, enhancing the climate resilience of transport infrastructure is also going to be crucial for the implementation of several of the sustainable development goals and targets which collectively make up the international community's 2030 Agenda for Sustainable Development (UNGA Res 70/1, 25 September 2015). Appropriate policies, standards, and regulatory approaches have an important role to play, particularly in the context of infrastructure planning and coastal zone management. Examples include the Climate Change Policy Framework for Jamaica (2015), which provides for cross-sectoral mainstreaming of climate change considerations, the amended EU Directive on Environmental Impact Assessment (Directive 2014/52/EU), in force since May 2017, which requires climate change impacts to be taken into account as part of environmental impact assessments for large infrastructure projects, and the recently adopted ISO Standard 14090 (Adaptation to climate change—Principles, requirements, and guidelines, 2019), which provides a framework to enable organizations to prioritize and develop effective, efficient, and deliverable adaptation tailored to the specific climate change challenges they face and using a consistent, structured, and pragmatic approach.

As this brief contribution illustrates, multifaceted approaches to adaptation and resilience building for coastal infrastructure assets and important natural resources (beaches) will be required to ensure the sustainable transport, trade, tourism, and development prospects of *SIDS* under a changing climate. The case for action is clear and urgent.

Acknowledgments

This work was carried out as part of an UNCTAD technical assistance project, implemented in collaboration with a number of partners (see https://SIDSport-ClimateAdapt.unctad.org). The author gratefully acknowledges the assistance and contribution of all experts, collaborators, and stakeholders involved and, particularly of Dr. Isavela Monioudi, University of the Aegean. Finally, the financial assistance by the UN Development Account (UNDA 1415O) is gratefully acknowledged.

References

Asariotis R., Benamara H., Mohos-Naray V., 2017. Port industry survey on climate change impacts and adaptation. UNCTAD/SER.RP/2017/18. https://unctad.org/en/PublicationsLibrary/ser-rp-2017d18_en.pdf.

Becker A., Acciaro M., Asariotis R. et al., 2013. A note on climate change adaptation for seaports: a challenge for global ports, a challenge for global society. *Climatic Change* 120, 683–695. doi: 10.1007/s10584-013-0843-z.

Becker A., Chase N.T.L., Fischer M. et al., 2016. A method to estimate climate-critical construction materials applied to seaport protection. *Global Environmental Change* 40, 125–136. doi: 10.1016/j.gloenvcha.2016.07.008.

Christodoulou, A., Christidis, P., Demirel, H., 2018. Sea-level rise in ports: a wider focus on impacts. *Maritime Economics and Logistics* 21, 482–496.doi: 10.1057/s41278-018-0114-z.

Coffel E.D., Thompson T.R., Horton, R.M., 2017. The impacts of rising temperatures on aircraft takeoff performance. *Climatic Change* 144, 381–388. doi: 10.1007/s10584-017-2018-9.

De Freitas C.R., Scott D., McBoyle G., 2008. A second generation climate index for tourism (CIT): specification and verification. *International Journal of Biometeorology* 52, 399–407.

ESL, 2015. Impact assessment and national adaptation strategy and action plan to address climate change in the tourism sector, Saint Lucia. Caribbean Community Climate Change Centre (CCCCC) and the Government of Saint Lucia (GoSL), EU-GCCA Caribbean Support Project. https://www.climatechange.govt.lc/wp-content/uploads/2017/10/Impact-Assessment-National-Adaptation-Strategy-and-Action-Plan-in-Tourism-Sector.pdf

Fay M., Andres L.A., Fox C.J.E. et al., 2017. *Rethinking infrastructure in Latin America and the Caribbean: spending better to achieve more.* Washington, DC: World Bank Group. http://documents.worldbank.org/curated/en/676711491563967405/Rethinking-infrastructure-in-Latin-merica-and-the-Caribbean-spending-better-to-achieve-more.

Ghermandi A., Nunes P.A.L.D., 2013. A global map of coastal recreation values: results from a spatially explicit meta-analysis. *Ecological Economics* 86, 1–15.

Gopalakrishnan S., Smith M.D., Slott J.M. et al., 2011. The value of disappearing beaches: a hedonic pricing model with endogenous beach width. *Journal of Environmental Economics and Management* 61, 297–310.

IPCC, 2018. Summary for policy makers. In: V. Masson-Delmotte et al. (eds.), *Global warming of 1.5°C. An IPCC Special Report on the impacts of global warming of 1.5°C above pre-industrial levels.* Geneva, Switzerland: World Meteorological Organization. https://www.ipcc.ch/site/assets/uploads/sites/2/2019/05/SR15_SPM_version_report_HR.pdf.

Koks E.E., Rozenberg I., Zorn C. et al., 2019. A global multi-hazard risk analysis of road and railway infrastructure assets. *Nature Communications* 10, 2677. doi: 10.1038/s41467-019-10442-3.

Kreft S., Eckstein D., Melchior I., 2016. Global climate risk index 2017. Who suffers most from extreme weather events? Weather-related loss events in 2015 and 1996 to 2015. https://germanwatch.org/en/download/16411.pdf.

Macdonald R., Hawkesworth C.J., Heath E., 2000. The Lesser Antilles volcanic chain: a study in arc magmatism. *Earth Science Reviews* 49, 1–76. doi: 10.1016/S0012-8252(99)00069-0.

Monioudi I., Velegrakis A.F., Chatzipavlis A.E. et al., 2017. Assessment of island beach erosion due to sea level rise: the case of the Aegean Archipelago (Eastern Mediterranean). *Natural Hazards and Earth System Science* 17, 449–466. doi: 10.5194/nhess-17-449-2017.

Monioudi I., Asariotis R., Becker A. et al., 2018. Climate change impacts on critical international transportation assets of Caribbean Small Island Developing States (SIDS): the case of Jamaica and Saint Lucia. *Regional Environmental Change* 18, 2211–2225. doi: 10.1007/s10113-018-1360-4.

Monioudi I., Nikolaou A., Asariotis R. et al., 2019. Beach erosion hazard under Climate Variability and Change – the case of Saint Lucia (Eastern Caribbean). International Conference DMPCo, Athens.

Mora C., Dousset B., Caldwell I.R. et al., 2017. Global risk of deadly heat. *Nature Climate Change* 7, 501–507. doi: 10.1038/NCLIMATE3322.

Narayan S., Beck M.W., Reguero B.G., Losada I.J. et al., 2016. The effectiveness, costs and coastal protection benefits of natural and nature-based defences. *PLoS ONE* 11(5), e0154735. doi: 10.1371/journal.pone.0154735.

Nurse L., 2013. Climate change impacts and risks: the challenge for Caribbean ports, STC-13, April 15–18 Georgetown, Guyana.

Nurse L.A., McLean R.F., Agard J. et al., 2014. Climate change 2014: impacts, adaptation, and vulnerability. Part B: Regional aspects. Contribution of working group II to the fifth assessment report of the IPCC. Cambridge, UK and New York, NY: Cambridge University Press, pp. 1613–1654.

Peduzzi P., Velegrakis A.F., Estrella M. et al., 2013. Integrating the role of ecosystems in disaster risk and vulnerability assessments: lessons from the risk and vulnerability assessment methodology development project (RiVAMP) in Negril Jamaica. In F.G. Renaud et al., (eds.), *The role of ecosystems in disaster risk reduction.* Tokyo, Japan: United Nations University Press (ISBN 978-9280812213), pp. 109–139.

Phillips M.R., Jones A.L., 2006. Erosion and tourism infrastructure in the coastal zone: problems, consequences and management. *Tourism Management* 27(3), 517–524.

Pinnock F., Ajagunna I., 2012. The Caribbean maritime sector: achieving sustainability through efficiency. Caribbean Paper No. 13. https://www.cigionline.org/publications/caribbean-maritime-transportation-sector-achieving-sustainability-through-efficiency.

Pratt S., 2015. The economic impact of tourism in SIDS. *Annals of Tourism Research* 52, 148–160.

Ranasinghe, R., 2016. Assessing climate change impacts on open sandy coasts: a review. *Earth-Science Reviews.* 160, 320–332. doi: 10.1016/j.earscirev.2016.07.011

Rhiney K., 2015. Geographies of Caribbean vulnerability in a changing climate: issues and trends. *Geography Compass* 9(3), 97–114.

Scott D., Simpson M.C., Sim R., 2012. The vulnerability of Caribbean coastal tourism to scenarios of climate change related sea level rise. *Journal of Sustainable Tourism* 20(6), 883–898.

Stephenson T.S., Jones J.J., 2017. Impacts of climate change on extreme events in the coastal and marine environments of Caribbean small island developing states (SIDS). *Caribbean Marine Climate Change Report Card: Science Review* 2017, 10–22.

UNCTAD, 2014a. Small island developing states: challenges in transport and trade logistics. Note by the UNCTAD secretariat. Multi-year expert meeting on transport, trade logistics and trade facilitation, third session. 24–26/11/2014. Trade and Development Commission. http://unctad.org/meetings/en/SessionalDocuments/cimem7d8_en.pdf.

UNCTAD, 2014b. Review of Maritime Transport 2014. https://unctad.org/rmt.

UNCTAD, 2017. Climate change impacts on coastal transportation infrastructure in the Caribbean: enhancing the adaptive capacity of small island developing states (SIDS), SAINT LUCIA: a case study. UNDA Project 1415O. https://unctad.org/en/PublicationsLibrary/dtltlb2018d3_en.pdf. https://SIDSport-ClimateAdapt.unctad.org.

UNCTAD, 2019. Review of Maritime Transport 2019. https://unctad.org/rmt.

UNECE, 2013. Climate change impacts and adaptation for international transport networks. Expert Group Report, ITC, UN Economic Commission for Europe ECE/TRANS/238. 223 pages. http://www.unece.org/fileadmin/DAM/trans/main/wp5/publications/climate_change_2014.pdf.

UNECLAC, 2011. An assessment of the economic impact of climate change on the transportation sector in Barbados. UN Economic Commission for Latin America and the Caribbean ECLAC Technical Report LC/CAR/L309. 44 pages.

van Vuuren D.P., Carter T.R., 2014. Climate and socio-economic scenarios for climate change research and assessment: reconciling the new with the old. *Climatic Change* 122, 415–429.

Vousdoukas M.I., Mentaschi L., Voukouvalas E. et al., 2018. Global probabilistic projections of extreme sea levels show intensification of coastal flood hazard. *Nature Communications* 9, 2360. doi: 10.1038/s41467-018-04692-w.

Wong P.P., Losada I.J., Gattuso J.-P. et al., 2014. Coastal systems and low-lying areas. In C.B. Field et al., (eds.), *Climate change 2014, impacts, adaptation, and vulnerability. Part A: global and sectoral aspects. Contribution of working Group II to the Fifth IPCC Assessment Report.* Cambridge, UK: Cambridge University Press, pp. 361–409.

III

Coastal Change and Monitoring

28

Coastal Environments: Remote Sensing

Yeqiao Wang
University of Rhode Island

Introduction

Coastal zone includes areas of continental shelves, islands or partially enclosed seas, estuaries, bays, lagoons, beaches, and terrestrial and aquatic ecosystems within watersheds that drain into coastal waters. Coastal zone is the most dynamic interface between land and sea and represents the most challenging frontier between human civilization and environmental conservation. Worldwide, over 38% of the human population lives in the coastal zones.[1] In the United States, about 53% of the human population lives in the coastal counties.[2] An increasing proportion of the global population lives within the coastal zones of all major continents that require increasing attention to agricultural, industrial, and other human-related effects on coastal habitats and water quality and their impacts on ecological dynamics, ecosystem health, and biological diversity. Climate change and sea level rise (SLR) impose significant impacts and uncertainty on ecosystems and infrastructures along coastal zones.

Coastal environments contain a wide range of natural habitats, such as sand dunes, barrier islands, tidal wetlands and marshes, mangrove forests, coral reefs, and submerged aquatic vegetation (SAV) that provide foods, shelters, and breeding grounds for terrestrial and marine species. Coastal habitats also provide irreplaceable services such as filtering pollutants and retaining nutrients, maintaining water quality, protecting shoreline, and absorbing flood waters. Coastal habitats are facing intensified natural and anthropogenic disturbances by direct impacts such as hurricane, tsunami, harmful algae bloom, and storm surge and cumulative and secondary impacts, such as climate change, SLR, oil spill, and urban development (Figure 28.1). Inventory and monitoring of coastal environments become one of the most challenging tasks of the society in resource management and humanity administration.[3] Remote sensing science and technologies have profoundly changed the practice in the monitoring and understanding of the dynamics of coastal environments.

Remote sensing refers to art, science, and technology for Earth system data acquisition through nonphysical contact sensors or sensor systems mounted on space-borne, airborne, and other types of platforms; data processing and interpretation from automated and visual analysis; information generation under computerized and conventional mapping facilities; and applications of generated data and information for societal benefits and needs. Remote sensing data can be collected passively or actively

FIGURE 28.1 (See color insert.) The field photos illustrate the characteristics of coastal landscape. The photos show shorebirds in Cape May National Wildlife Refuge, New Jersey (**a**); a simple wooden bridge that crosses over an inlet in a coastal village of Ghana (*photographed in September 2011*) (**b**); a mangrove site along the Pangani River mouth in Tanzania coast (*photographed in January 2002*) (**c**); a tidewater cypress swamp close to Jean Lafitte National Historical Park and Preserve out of New Orleans (**d**); a *S. patens* salt marsh site in Jamaica Bay of the New York City (**e**); a barrier island sand dune on Fire Island National Seashore, New York (**f**); a damaged site by the East Japan earthquake and tsunami that devastated the Sendai and Sanriku Coast on March 11, 2011 (*photographed in August 2011*) (**g**); an overwash site on Fire Island National Seashore caused by Hurricane Sandy on October 24, 2012 (*photographed in May 2016*) (**h**); wild horses graze on salt marsh along the Chincoteague National Wildlife Refuge in Virginia (**i**); a salt marsh restoration site in Jamaica Bay (*photographed in October 2014*), New York (**j**). (Photos: Yeqiao Wang.)

through selected bandwidths of spectral ranges across the spectrum of the electromagnetic radiation with different spatial resolutions. The multispectral capabilities of remote sensing allow observation and measurement of biophysical characteristics of the Earth system components, while the multi-temporal and multi-sensor capabilities allow tracking of changes in the characteristics over time. Remote sensing has been broadly applied in physical world (e.g., the atmosphere, water, soil, rock), its living inhabitants (e.g., flora, fauna), and the processes at work (e.g., erosion, deforestation, urban sprawl, effects of climate and environmental change).

Coarse spatial resolution remote sensing data have been applied for coastal studies. For example, high concentrations of suspended particulate matter in coastal waters directly affect water column and benthic processes such as phytoplankton productivity,[4,5] coral growth,[6–9] productivity of SAV,[10] and nutrient dynamics.[11]

The Sea-Viewing Wide Field-of-View Sensor (SeaWiFS) system acquires radiance data from the Earth in eight spectral bands with a maximum spatial resolution of 1 km at nadir. The global 9 km^2 spatial resolution Level 3 data provide daily, 8-day, monthly, and annual standard data products. Acker et al.[12] employed monthly chlorophyll-a data from the 8-year SeaWiFS mission and data from Moderate Resolution Imaging Spectroradiometer (MODIS) to analyze the spatial pattern of chlorophyll concentrations and seasonal cycle. The data indicate that large coral reef complexes may be sources of either nutrients or chlorophyll-rich detritus and sediment, enhancing chlorophyll-a concentration in waters adjacent to the reefs.

Landsat data have been used in coastal applications for decades.[13–15] Landsat and similar types of imagery data have been applied in inventory mapping, change detection, and management of mangrove forests through visual interpretation,[16,17] through vegetation index,[18] and to map seagrass coverage,[19,20] among others[21–24] As archived Landsat images have been made available at no cost to user communities since early 2009,[25] coastal applications can take advantages from such type of data.

Active Remote Sensing of Coastal Environments

Active remote sensors are the data acquisition systems that transmit electromagnetic pulses with a specific wavelength (λ) and measure the time of return (translated to distance) of the signal reflected from a target. For example, light detection and ranging (LiDAR) systems generally transmit pulses at wavelength around the visible domain (λ in nanometers), whereas radar pulses are in the microwave domain (λ in centimeters). LiDAR and interferometric synthetic aperture radar (InSAR) are among new developments in active remote sensing that are the particular interests in coastal studies.

Active radar sensors are well known for their all-weather and day-and-night imaging capabilities, which are effective for mapping coastal habitats over cloud-prone tropical and subtropical regions. The synthetic aperture radar (SAR) backscattering signal is composed of intensity and phase components. The intensity component of the signal is sensitive to terrain slope, surface roughness, and dielectric constant. Studies have demonstrated that SAR intensity images can map and monitor forested and nonforested wetlands occupying a range of coastal and inland settings.[26,27] SAR intensity data have been used to monitor floods and dry conditions, temporal variations in the hydrological conditions of wetlands, including classification of wetland vegetation at various geographic settings.[28–32] When the phase components of two SAR images of the same area acquired from similar vantage points at different times are combined through InSAR processing, an interferogram can be constructed to depict range changes between the radar and the ground and can be further processed with a digital elevation model (DEM) to produce an image with centimeter to subcentimeter vertical precision. InSAR has been extensively utilized to study ground surface deformation associated with volcanic, earthquake, landslide, and land subsidence processes. Alsdorf et al.[33] found that an interferometric analysis of L-band (wavelength of 24 cm) Shuttle Imaging Radar-C (SIR-C) and Japanese Earth Resources Satellite (JERS-1) SAR imagery can yield centimeter-scale measurements of water level changes throughout inundated floodplain vegetation.

LiDAR technology includes small- and large-footprint laser scanners.[34] LiDAR has shown promising results for assessing coastal habitats.[35–37] Due to the rapid laser firing, LiDAR pulses can penetrate vegetation cover, which makes LiDAR technology well suited to measure topography in coastal salt marsh areas.[38] Airborne LiDAR data have been tested for identifying coral colonies on patch reefs.[39] Shuttle Radar Topography Mission (SRTM) data have been applied to explore 3D modeling of mangrove forests. Although the SRTM was designed to produce a global DEM of the Earth's surface, the SRTM InSAR measurement of height z is the sum of the ground elevation and the canopy height contribution. Therefore, SRTM DEM can be used to measure vegetation height as well as ground topography.[40]

Reported studies integrate LiDAR data, digital historical maps, and orthophotos to measure long-term coastal change rates, to map erosion hazard areas, and to understand coastal processes on a variety of timescales from storms to seasons to decades and longer. The studies in barrier islands along the US Gulf Coast and in rocky coastal environments of the US West Coast show how similar types of data can be used to map coastal hazards in a variety of geographic settings and at a variety of spatial scales. Datasets that have commonly been used for coastal zone assessments can be integrated with newer data sources to modernize and update analyses. Such assessments will continue to be critical to coastal management and planning, especially with the currently predicted rates of SLR through the 21st century.[41]

Hyperspectral Remote Sensing of Coastal Environments

Hyperspectral technology brings new insights about remote sensing of coastal environments. Hyperion sensor onboard of EO-1 satellite is a representative space-borne hyperspectral system. The Hyperion Imaging Spectrometer collects data in 30-m ground sample distance over a 7.5-km swath and provides 10-nm (sampling interval) 220 contiguous spectral bands of the solar reflected spectrum from 400 nm to 2,500 nm. Airborne Visible Infrared Imaging Spectrometer (AVIRIS) is a representative airborne hyperspectral sensor system. The AVIRIS whiskbroom scanner collects data in the same spectral interval and range as the Hyperion system but with 20-m spatial resolution. Hyperspectral remote sensing has advantages in coastal wetlands characterization due to its large number of narrow, contiguous spectral bands as well as high horizontal resolution from airborne platforms. It has been used to map habitat heterogeneity[42] to determine plant cover distribution in salt marshes.[43] Hyperspectral images have been used to separate vigor types by detecting slight differences in coloration due to stress factors, infestation, or displacement by invading species.[44] A study mapped the onset and progression of coastal *Spartina alterniflora* marsh dieback using hyperspectral image data at the plant leaf, canopy, and satellite levels without *a priori* information on where, when, or how long the dieback had proceeded.[45] Hyperspectral data offer an enhanced ability to determine dieback onset and track progression.

Increased population and urban development have contributed significantly to environmental pressures along many areas of the US coastal zone. The pressures have resulted in substantial physical changes to beaches, loss of coastal wetlands, and decline in ambient water and sediment quality, and the addition of higher volumes of nutrients (primarily nitrogen and phosphorus) from an urban, nonpoint source runoff. Algal growth is stimulated when nutrient concentrations, from sources such as stream and river discharges, wastewater sewage facilities, and agricultural runoff, are increased beyond the natural background levels of estuaries and other coastal receiving waters. These excess nutrients and the associated increased algal growth can also lead to a series of events that can decrease water clarity, cause benthic degradation, and result in low concentrations of dissolved oxygen. Recent research interests include quantification of the effects of sampling design and measurement accuracy, frequency, and resolution on the ability to improve our quantitative knowledge of coastal water quality. A study determined the ecological condition of numerous individual embayments and estuaries along the southern New England coast, as well as the adjoining coastal ocean, over an annual cycle using airborne hyperspectral remote sensing data and the criteria for assessing chlorophyll *a* concentrations.[46] The assessments would have been sample intensive and expensive if conducted over an annual period using traditional field-based monitoring.

High Spatial Resolution Remote Sensing of Coastal Environments

High spatial resolution remote sensing data provide much needed spatial details and variations at the submeter level for mapping dynamic coastal habitats. For example, besides changing in areas, degradations of mangrove forests due to changing environment, and selective harvesting have significant effects on ecosystem integrity and functions. Accurate and effective mapping of mangrove forests is essential for monitoring changes in spatial distribution and species composition. Advancement of high spatial resolution, multispectral remote sensing data makes such an inventory and monitoring possible.[47] High spatial resolution multispectral satellite and airborne digital remote sensing data have been employed to evaluate benthic habitats.[20,48–51]

Assessment of the quantity of impervious surface areas (ISAs) in landscapes has become increasingly important with growing concern of its impact on the environment.[52] This is particularly true for coastal areas due to the impacts of ISA on aquatic systems and its role in transportation and concentration of pollutants. Urban runoff, mostly over impervious surface, is the leading source of pollution in US estuaries, lakes, and rivers.[53] Precise data of ISA in spatial coverage and distribution patterns in association with landscape characterizations are critical for providing the key baseline information for effective coastal management and science-based decision-making. An example of studies extracted information on urban ISA from true-color digital orthophotography data to reveal the intensity of urban development surrounding the Narragansett Bay, Rhode Island.[54]

Remote sensing applications in disaster management have become critically important to support preparation through response to natural and human-induced hazards and events affecting human populations in coastal zones. Increased exposure and density of human settlements in coastal regions increase the potential loss of life, property, and commodities that are at risk from intense coastal hazards. Remote sensing has a long history of being used to capture towns, harbors, and coastal lines affected by disasters and played a key role in recovery efforts post disasters. The history can be traced back from the earthquake and tsunami that struck Alaska on March 1964, to the five hurricanes of Dennis, Katrina, Ophelia, Rita, and Wilma, which directly devastated the US coastal regions in 2005.[55] High spatial resolution images with timely coverage were proved to be efficient and useful to make rescue and recovery plans.

Integration of Remote Sensing and In Situ Measurement Data

Integration of multispectral, multi-temporal, multi-sensor airborne and space-borne remote sensing data with Global Positioning System (GPS)-guided in situ observations becomes necessary for effective and timely inventory and monitoring of the coastal environments. For example, a significant amount of the coastal wetlands along the Long Island Sound in the northeastern United States has been lost over the past century due to urban development, filling, and dredging or damaged due to human disturbance and modification. Beyond the physical loss of marshes, the species composition of marsh communities is changing. With the mounting pressures on coastal wetland areas, it is becoming increasingly important to identify and inventory the current extent and condition of coastal marshes located on the Long Island Sound estuary, implement a cost-effective way to track changes in wetlands over time, and monitor the effects of habitat restoration and management. Identification of distribution and health of individual marsh plant species like *Phragmites australis* using remote sensing is challenging because vegetation spectra are generally similar to one another throughout the visible to near-infrared (VNIR) spectrum. The reflectance of a single species may vary throughout the growing season due to variations in the amount and ratios of plant pigments, leaf moisture content, plant height, canopy effects, leaf angle distribution, and other structural characteristics.

A study addressed the use of multi-temporal field spectral data, satellite imagery, and LiDAR top-of-canopy data to classify and map common salt marsh plant communities.[56] VNIR reflectance spectra were measured in the field to assess the phenological variability of the dominant species of *Spartina patens*, *P. australis*, and *Typha* spp. The field spectra and single-date LiDAR canopy height data were used to define an object-oriented classification methodology for the plant communities in multi-temporal QuickBird imagery. The classification was validated using an extensive field inventory of marsh species.

Mapping of SAV is inherently difficult, as most species are subtidal. Traditionally, mapping of SAV has been accomplished using a combination of aerial photographs and in situ sampling by divers using scuba. A reported study created a video reference database as field reference of benthic habitats.[57] During the study, an underwater video camera was attached to a sled and the sled was towed to a vessel. The position, heading, and speed of the vessel were outputs from a GPS receiver located on the vessel and overlaid on the live video image. To increase the spatial accuracy of the coordinates, the study placed a GPS receiver in a kayak and positioned directly over the sled. A VCR/TV combo installed on the towing vessel recorded SAV conditions and geographic coordinates on video files. The recorded video files were related to the coordinates captured on the towed GPS unit through the time stamp. Based on speed, the amount of video equaling the required 30 m was recorded. The corresponding video files to each line segment were linked to the geographic information system database. Each video file was reviewed for the percent of SAV within each transect. The percent was estimated by dividing the number of seconds in which SAV was observed and the dominant SAV species in each video transect was noted. The line coverage was then overlaid on top of the satellite images to aid our classification or conduct accuracy assessment.

High-resolution imagery data from Quickbird-2 and Worldview-2/3 satellites were used in salt marsh mapping and change analysis. The study revealed and analyzed changes caused by natural and anthropogenic forces, e.g., Hurricane Sandy and salt marsh restoration. The study mapped salt marsh change in Jamaica Bay, New York, from 2003 to 2013.[58] In addition, it has been recognized that salt marsh vegetation extent and zonation is often controlled by bottom-up factors determined in part by frequency and duration of tidal inundation. Tidal inundation during time of remote sensing data acquisition can affect the resulting image classification and change analysis. A study utilizes topobathymetric LiDAR data and bathtub models of tidal stage at 5-cm intervals from Mean Low Water (MLW) to Mean High Water (MHW) and determines the impact of tidal variation in salt marsh mapping within Jamaica Bay, New York. Tidal inundation models were compared to Worldview-2 and Quickbird-2 imagery acquired at a range of tidal stages. The study found that at 0.6 m above MLW, only 3.5% of *S. alterniflora* is inundated. The study demonstrated that incremental modeling of tidal stage is important for understanding areas most at risk from SLR and inform management decisions.[59]

Conclusions

Coastal environments service the boundaries of the most dynamic areas of the Earth. Tidal wetlands, e.g., are positioned at the interface between land and sea where two of the most powerful forces acting on the planet's waters collide. At the mouths of the rivers, edges of embayments, and lagoons, these two great forces interact in complex ways to form the hydrodynamic framework for the development of tidal wetlands. The complexity of the interactions makes characterizing tidal wetlands a difficult task.[60] Increased recognition of the global importance of salt marshes as "blue carbon" sinks has led to concern that salt marshes could release large amounts of stored C into the atmosphere (as CO_2) if they continue undergoing disturbance.[61] The dynamics of coastal constituents imposes significant challenges in application of remote sensing technologies due to the nature of spatial and temporal variation of coastal habitats, as well as the management decisions address issues ranging from salt marsh migration and resilience to emergency responses and post-disaster actions.[62–66]

Remote sensing of coastal environments straddles the technical boundary of land and ocean satellite remote sensing. Satellite radiometers optimized for remote sensing of terrestrial environments, as well as those for the open ocean, are not well suited for quantitative remote sensing of phytoplankton biomass, sediment, and other constituents of coastal waters. Remote sensing of coastal waters requires high spatial resolution to resolve characteristically small features as well as daily or high-frequency coverage to understand many of the important coastal processes of these dynamic regions. The ideal satellite radiometer, for measurements of coastal waters would, e.g., have a finer spatial resolution and higher frequency of revisit time. The future promises many exciting advances in new and improved sensors for use as well as improved geospatial processing tools and computing technology to study coastal ecosystems.

References

1. Crossett, K.M.; Culliton, T.J.; Wiley, P.C.; Goodspeed, T.R. *Population Trends along the Coastal United States: 1980–2008*; Coastal Trends Report Series; National Oceanic and Atmospheric Administration, National Ocean Service Management and Budget Office: Silver Spring, MD, 2004; 54 p.
2. Small, C.; Cohen, J.E. Continental physiography, climate, and the global distribution of human population. *Curr. Anthropol.* 2004, *45* (2), 269–277.
3. Wang, Y. Coastal environments: Remote sensing. In *Encyclopedia of Natural Resources: Land*; Wang, Y., Ed.; Taylor and Francis: New York, 2014; 100–105 pp.
4. Cole, B.E.; Cloern, J.E. An empirical model for estimating phytoplankton productivity in estuaries. *Marine Ecology. Prog. Ser.* 1987, 36, 299–305.
5. Cloern, J.E. Turbidity as a control on phytoplankton biomass and productivity in estuaries. *Continental Shelf Res.* 1987, *7* (11), 1367–1381.
6. Dodge, R.E.; Aller, R.; Thompson, J. Coral growth related to resuspension of bottom sediments. *Nature* 1974, *247*, 574–577.
7. Miller, R.L.; Cruise, J.F. Effects of suspended sediments on coral growth: Evidence from remote sensing and hydrologic modeling. *Remote Sensing Environ.* 1995, *53*, 177–187.
8. Torres, J.L.; Morelock, J. Effect of terrigenous sediment influx on coral cover and linear extension rates of three Caribbean massive coral species. *Caribb. J. Sci.* 2002, *38* (3–4), 222–229.
9. McLaughlin, C.J.; Smith, C.A.; Buddemeier, R.W.; Bartley, J.D.; Maxwell, B.A. Rivers, runoff and reefs. *Glob. Planetary Change* 2003, *39* (1–2), 191–199.
10. Dennison, W.C.; Orth, R.J.; Moore, K.A.; Stevenson, J.C.; Carter, V.; Kollar, S. Assessing water quality with submersed aquatic vegetation. *Bioscience* 1993, *43*, 86–94.
11. Mayer, L.M.; Keil, R.G.; Macko, S.A.; Joye, S.B.; Ruttenberg, K.C.; Aller, R.C. The importance of suspended particulates in riverine delivery of bioavailable nitrogen to coastal zones. *Global Biogeochem. Cycles* 1998, *12*, 573–579.
12. Acker, J.; Leptoukh, J.; Shen, S.; Zhu T.; Kempler, S. Remotely-sensed chlorophyll a observations of the northern Red Sea indicate seasonal variability and influence of coastal reefs. *J. Mar. Syst.* 2008, *69*, 191–204.
13. Munday, J.C.; Jr.; Alfoldi, T.T. Landsat test of diffuse reflectance models for aquatic suspended solids measurements. *Remote Sensing Environ.* 1979, *8*, 169–183.
14. Bukata, R.P.; Jerome, J.H.; Bruton, J.E. Particulate concentrations in Lake St. Clair as recorded by a shipborne multispectral optical monitoring system. *Remote Sensing Environ.* 1988, *25*, 201–229.
15. Ritchie, J.C.; Cooper, C.M.; Shiebe, F.R. The relationship of MSS and TM digital data with suspended sediments, chlorophyll and temperature in Moon Lake, Mississipi. *Remote Sensing Environ.* 1990, *33*, 137–148.

16. Gang, P.O.; Agatsiva, J.L. The current status of mangroves along the Kenyan coast, a case study of Mida Creek mangroves based on remote sensing. *Hydrobiologia* 1992, *247*, 29–36.
17. Wang, Y.; Bonynge, G.; Nugranad, J.; Traber, M.; Ngusaru, A.; Tobey, J.; Hale, L.; Bowen, R.; Makota, V. Remote sensing of mangrove change along the Tanzania coast. *Mar. Geodesy* 2003, *26* (1–2), 35–48.
18. Jensen, J.R.; Ramset, E.; Davis, B.A.; Thoemke, C.W. The measurement of mangrove characteristics in south-west Florida using SPOT multispectral data. *Geocarto Int.* 1991, *2*, 13–21.
19. Armstrong, R.A. Remote sensing of submerged vegetation canopies for biomass estimation. *Int. J. Remote Sensing* 1993, *14* (3), 621–627.
20. Lathrop, R.G.; Montesano, P.; Haag, S. A multi-scale segmentation approach to mapping seagrass habitats using airborne digital camera imagery. *Photogramm. Eng. Remote Sensing* 2006, *72* (6), 665–675.
21. Gao, J. A hybrid method toward accurate mapping of mangroves in a marginal habitat from SPOT multispectral data. *Int. J. Remote Sensing* 1998, *19*, 1887–1899.
22. Green, E.P.; Mumby, P.J.; Edwards, A.J.; Clark, C.D.; Ellis, A.C. The assessment of mangrove areas using high resolution multispectral airborne imagery. *J. Coastal Res.* 1998, *14*, 433–443.
23. Rasolofoharinoro, M.; Blasco, F.; Bellan, M.F.; Aizpuru, M.; Gauquelin, T.; Denis, J. A remote sensing based methodology for mangrove studies in Madagascar. *Int. J. Remote Sensing* 1998, *19*, 1873–1886.
24. Pasqualini, V.; Iltis, J.; Dessay, N.; Lointier, M.; Gurlorget, O.; Polidori, C. Mangrove mapping in North-Western Madagascar using SPOT-XS and SIR-C radar data. *Hydrobiologia* 1999, *413*, 127–133.
25. Woodcock, C.E.; Allen, R.; Anderson, M.; Belward, A.; Bindschadler, R.; Cohen, W.; Gao, F.; Goward, S.N.; Helder, D.; Helmer, E.; Nemani, R.; Oreopoulos, L.; Schott, J.; Thenkabail, P.S.; Vermote, E.F.; Vogelmann, J.; Wulder, M.A.; Wynne, R. Free access to Landsat imagery. *Science* 2008, *320*, 1011.
26. Ramsey, E.W., III. Monitoring flooding in coastal wetlands by using radar imagery and ground-based measurements. *Int. J. Remote Sensing* 1995, *16* (13), 2495–2502.
27. Ramsey, E., III.; Rangoonwala, A. Canopy reflectance related to marsh dieback onset and progression in coastal Louisiana. *Photogramm. Eng. Remote Sensing* 2006, *72* (6), 641–652.
28. Hess, L.L.; Melack, J.M. Mapping wetland hydrology and vegetation with synthetic aperture radar. *Int. J. Ecol. Environ. Sci.* 1994, *20*, 197–205.
29. Hess, L.L.; Melack, J.M.; Filoso, S.; Wang, Y. Delineation of inundated area and vegetation along the Amazon floodplain with the SIR-C synthetic aperture radar. *IEEE Trans. Geosci. Remote Sensing* 1995, *33* (4), 896–904.
30. Kasischke, E.S.; Bourgeau-Chavez, L.L. Monitoring south Florida wetlands using ERS-1 SAR imagery. *Photogramm. Eng. Remote Sensing* 1997, *63*, 281–291.
31. Simard, M.; De Grandi, G.; Saatchi, S.; Mayaux, P. Mapping tropical coastal vegetation using JERS-1 and ERS-1 radar data with a decision tree classifier. *Int. J. Remote Sensing* 2002, *23* (7), 1461–1474.
32. Kiage, L.M.; Walker, N.D.; Balasubramanian, S.; Babin, A.; Barras, J. Applications of RADARSAT-1 synthetic aperture radar imagery to assess hurricane-related flooding of coastal Louisiana. *Int. J. Remote Sensing* 2005, *26* (24), 5359–5380.
33. Alsdorf, D.; Melack, J.; Dunne, T.; Mertes, L.; Hess, L.; Smith, L. Interferometric radar measurements of water level changes on the Amazon floodplain. *Nature* 2000, *404*, 174–177.
34. Lefsky, M.A.; Cohen, W.B.; Parker, G.G.; Harding, D.J. LiDAR remote sensing for ecosystem studies. *Bioscience* 2002, *52*, 19–30.
35. Brock, J.C.; Sallenger, A.H.; Krabill, W.B.; Swift, R.N.; Wright, C.W. Recognition of fiducial surfaces in LiDAR surveys of coastal topography. *Photogramm. Eng. Remote Sensing* 2001, *67*, 1245–1258.

36. Brock, J.C.; Wright, C.W.; Sallenger, A.H.; Krabill, W.B.; Swift, R.N. Basis and methods of NASA Airborne Topographic Mapper LiDAR surveys for coastal studies. *J. Coastal Res.* 2002, *18*, 1–13.
37. Nayegandhl, A.; Brock, J.C.; Wright, C.W.; O'Connell, J. Evaluating a small footprint, waveform-resolving LiDAR over coastal vegetation communities. *Photogramm. Eng. Remote Sensing* 2006, *72* (12), 1407–1417.
38. Montane, J.M.; Torres, R. Accuracy assessment of LiDAR saltmarsh topographic data using RTK GPS. *Photogramm. Eng. Remote Sensing* 2006, *72* (8), 961–967.
39. Brock, J.; Wright, C.W.; Hernandez, R.; Thompson, P. Airborne LiDAR sensing of massive stony coral colonies on patch reefs in the northern Florida reef tract. *Remote Sensing Environ.* 2006, *104* (2006), 31–42.
40. Simard, M.; Zhang, K.; Rivera-Monroy, V.H.; Ross, M.S.; Ruiz, P.L.; Castañeda-Moya, E.; Twilley, R.R.; Rodriguez, E. Mapping height and biomass of mangrove forests in Everglades National Park with SRTM elevation data. *Photogramm. Eng. Remote Sensing* 2006, *72* (3), 299–311.
41. Hapke, C.J. Integration of LiDAR and historical maps to measure coastal change on a variety of time and spatial scales (Ch. 4). In *Remote Sensing of Coastal Environments*; Wang, Y., Ed.; CRC Press: Boca Raton, FL, 2009; 61–78 pp.
42. Artigas, F.J.; Yang, J. Hyperspectral remote sensing of habitat heterogeneity between tide-restricted and tide-open areas in New Jersey Meadowlands. *Urban Habitat* 2004, *2*, 1.
43. Belluco, E.; Camuffo, M.; Ferrari, S.; Modenese, L.; Silvestri, S.; Marani, A.; Marani, M. Mapping salt marsh vegetation by multispectral and hyperspectral remote sensing. *Remote Sensing Environ.* 2006, *105*, 54.
44. Artigas, F.J.; Yang, J. Hyperspectral remote sensing of marsh species and plant vigor gradient in the New Jersey Meadowland. *Int. J. Remote Sensing* 2005, *26*, 5209.
45. Ramsey, E. III.; Rangoonwala, A. Mapping the onset and progression of marsh dieback (Ch. 5). In *Remote Sensing of Coastal Environments*; Wang, Y., Ed.; CRC Press: Boca Raton, FL, 2009; 123–149 pp.
46. Keith, D.J. Estimating chlorophyll conditions in Southern New England Coastal Waters from hyperspectral aircraft remote sensing (Ch. 7). In *Remote Sensing of Coastal Environments*; Wang, Y., Ed.; CRC Press: Boca Raton, FL, 2009; 151–172 pp.
47. Wulder, M.A.; Hall, R.J.; Coops, N.C.; Franklin, S.E. High spatial resolution remotely sensed data for ecosystem characterization. *Bioscience* 2004, *6*, 511–521.
48. Su, H.; Kama, D.; Fraim, E.; Fitzgerald, M.; Dominguez, R.; Myers, J.S.; Coffland, B.; Handley, L.R.; Mace, T. Evaluation of eelgrass beds mapping using a high-resolution airborne multispectral scanner. *Photogramm. Eng. Remote Sensing* 2006, *72* (7), 789–797.
49. Deepak, M.; Narumalani, S.; Rundquist, D.; Lawson, M. Benthic habitat mapping in tropical marine environments using QuickBird multispectral data. *Photogramm. Eng. Remote Sensing* 2006, *72* (9), 1037–1048.
50. Wang, L.; Sousa, W.P.; Gong, P.; Biging, G.S. Comparison of IKONOS and QuickBird images for mapping mangrove species on the Caribbean coast of Panama. *Remote Sensing Environ.* 2004, 91, 432–440.
51. Wang, L.; Silvan-Cardenas, J.L.; Sousa, W.P. Neural network classification of mangrove species from multi-seasonal IKONOS imagery. *Photogramm. Eng. Remote Sensing* 2008, *74* (7), 921–927.
52. Civco, D.L.; Hurd, J.D.; Wilson, E.H.; Arnold, C.L.; Prisloe, S. Quantifying and describing urbanizing landscapes in the Northeast United States. *Photogramm. Eng. Remote Sensing* 2002, *68*, 1083–1090.
53. Booth, D.B.; Jackson, C.R. Urbanization of aquatic systems: Degradation thresholds, stormwater detection, and the limits of mitigation. *J. Am. Water Resources Assoc.* 1997, *35*, 1077–1090.
54. Zhou, Y.; Wang, Y.; Gold, A.J.; August, P.V. Modeling watershed rainfall-runoff using impervious surface-area data with high spatial resolution. *Hydrogeol. J.* 2010, *18* (6), 1413–1423.

55. White, S.; Aslaken, M. NOAA's use of direct georeferencing to support emergency response. *PERS* 2006, *72* (6), 623–627.
56. Gilmore, M.S.; Wilson, E.H.; Barrett, N.; Civco, D.L.; Prisloe, S.; Hurd, J.D.; Chadwick, C. Integrating multitemporal spectral and structural information to map wetland vegetation in a lower Connecticut River tidal marsh. *Remote Sensing Environ.* 2008, *112*, 4048–4060.
57. Wang, Y.; Traber, M.; Milstead, B.; Stevens, S. Terrestrial and submerged aquatic vegetation mapping in fire Island National seashore using high spatial resolution remote sensing data. *Mar. Geodesy* 2006, *30* (1), 77–95.
58. Campbell, A.; Wang, Y.; Christiano, M.; Stevens, S. Salt marsh monitoring in Jamaica Bay, New York from 2003 to 2013: A decade of change from restoration to Hurricane Sandy. *Remote Sensing* 2017, *9*, 131. doi:10.3390/rs9020131.
59. Campbell, A.; Wang, Y. Examining the influence of tidal stage on salt marsh mapping using high spatial resolution satellite remote sensing and topobathymetric LiDAR. *IEEE Trans. Geosci. Remote Sensing* 2018, *56* (9), 5169–5176.
60. Conner, W.H.; Krauss, K.W.; Baldwin, A.H.; Hutchinson, S. Wetlands: Tidal. In *Encyclopedia of Natural Resources: Land*; Wang, Y., Ed.; Taylor and Francis: New York, 2014; 575–588 pp.
61. Macreadie, P.I.; Hughes, A.R.; Kimbro, D.L. Loss of 'Blue Carbon' from coastal salt marshes following habitat disturbance. *PLOS One* 2013, *8* (7), e69244. doi:10.1371/journal.pone.0069244.
62. Yin, J.; Schlesinger, M.E.; Stouffer, R.J. Model projections of rapid sea-level rise on the northeast coast of the United States. *Nat. Geosci.* 2009, *2*, 262–266.
63. Anderson, M.G.; Clark, M.; Olivero Sheldon, A. Resilient sites for terrestrial conservation in the Northeast and Mid-Atlantic Region. The Nature Conservancy, Eastern Conservation Science, 2012, 168 pp.
64. Anderson, M.G.; Barnett, A. Resilient coastal sites for conservation in the Northeast and Mid-Atlantic US. The Nature Conservancy, Eastern Conservation Science, 2017 (www.nature.org/resilientcoasts).
65. Watson, E.B.; Raposa, K.B.; Carey, J.C.; Wigand, C.; Warren, R.S. Anthropocene survival of Southern New England's salt marshes. *Estuaries Coasts* 2017, *40*, 617–625.
66. Takano, T. Coastal natural disasters: Tsunamis on the Sanriku Coast. In *Encyclopedia of Natural Resources: Water*; Wang, Y., Ed.; Taylor and Francis: New York, 2014; 668–674 pp.

29

Submerged Aquatic Vegetation: Seagrasses

Alyssa B. Novak
Boston University

Frederick T. Short
University of New Hampshire

Introduction

Coastal waters are among the most productive yet highly threatened ecosystems in the world. They are vulnerable and at risk from development, overexploitation, physical alteration, and destruction of habitat, as well as climate change-related stress. Submerged aquatic vegetation, or SAV, is an important component of coastal ecosystems. SAV are comprised of nonflowering and flowering macrophytes that grow completely underwater in freshwater, estuarine, and marine habitats. The following chapter focuses on seagrasses, an ecological group of flowering plants (angiosperms) that thrive in estuarine and shallow marine environments to 70 m. The chapter provides an overview of seagrasses and their ecosystem services, evolutionary history, adaptations, classification, distribution, status and potential threats, as well as strategies to ensure their future existence in a globally changing environment.

An Essential Natural Resource

Seagrasses are angiosperms that have adapted to exist fully submerged in estuarine and marine environments. They grow in shallow coastal waters and form extensive meadows that provide ecosystem services more valuable than salt marshes, mangroves, and coral reefs [1]. Seagrass meadows exhibit high primary and secondary production and are considered a significant global carbon sink, a key to combatting global climate change [2]. While seagrasses comprise only 2% of the ocean's area, they are responsible for more than 15% of the carbon annually buried there, with some of the largest documented organic carbon stores found in sediment mattes produced by the Mediterranean seagrass *Posidonia oceanica* [2]. Figure 29.1 shows an erosional escarpment of *P. oceanica*. Seagrasses also serve an important role in trophic transfer and export 24% of their net production ($0.6 \times 1{,}015$ g C/year) to adjacent ecosystems [3], including the nutrient-poor deep sea [4]. In addition to their roles in primary production, carbon storage, and export,

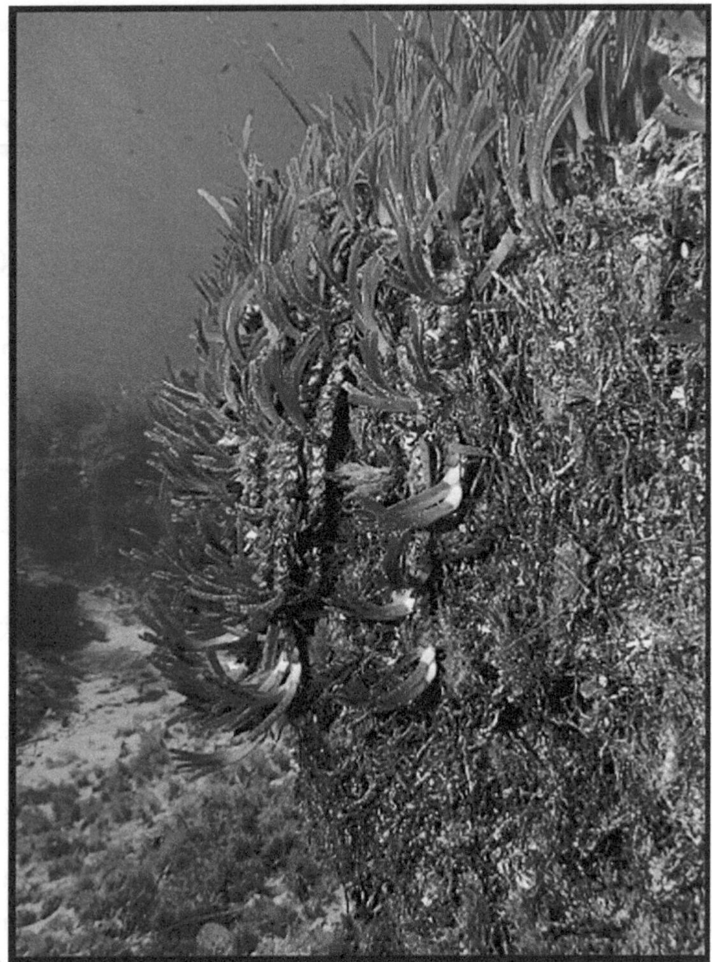

FIGURE 29.1 (See color insert.) An erosional escarpment in a *Posidonia oceanica* meadow with organic-rich soils in Calvi Bay, Corsica Island, France. The exposed face of the matte has a thickness of 2 m. (Arnaud Abadie.)

seagrass meadows serve as filters and improve water quality and clarity of coastal ecosystems through the direct trapping of suspended particles and the retention of organic matter [5–7]. Seagrass meadows are also hotspots of biodiversity, providing food and habitat to a variety of organisms including microbes, invertebrates, and vertebrates that are often endangered, such as dugongs, or commercially important, such as fish and shrimp [8–11]. Finally, seagrasses are considered biological indictors of coastal conditions because they are vulnerable and rapidly respond to anthropogenic stressors including eutrophication, sedimentation, oil spills, and commercial fishing [12,13].

Origin, Adaptations, and Classification

Seagrasses evolved 40–100 million years ago in the late Cretaceous Period. They arose from terrestrial monocotyledons that reinvaded the sea and developed into three separate lineages (Cymodoceaceae complex, Hydrocharitaceae, and Zosteraceae [14,15]. The colonization of the sea required seagrasses to grow and reproduce while enduring the osmotic effects of salt water, changes in the availability of dissolved CO_2, changes in the intensity and quality of light, and the density and mechanical drag of water [16–18]. Seagrasses have a number of unique adaptations that allow for survival in these conditions, including a well-developed horizontal rhizome that anchors plants into the substrate and roots extending from the rhizome that assist in anchoring and nutrient uptake. Leaves are flexible and offer little resistance to wave action. They also function in nutrient uptake and as receptors of light, with the epidermis of blades serving as the main site for photosynthesis. Tissues (aerenchyma) extending through the roots, rhizomes, and leaves facilitate internal gas and solute transport while regularly arranged air spaces (lacunae) give plants buoyancy. Finally, all species of seagrasses exhibit hydrophilous pollination while completely submerged except for *Enhalus acoroides* and *Ruppia* spp., which pollinates at the water surface [16–20].

Seagrasses are classified as a functional rather than a taxonomic group of angiosperms. Historically, species designations were based on ecological, reproductive, and vegetative characteristics including: leaf and flowering characteristics, vein numbers, fiber distributions, epidermal cells, and roots and rhizomes to create a complete taxonomic description [19,20]. Advances in technology now allow seagrass species designation to be based on genetic difference [21]. As of 2019, there are 72 seagrass species belonging to 6 families and 13 genera [19,21,22]. Five genera are placed in the family Cymodoceaceae (*Amphibolis, Cymodocea, Halodule, Syringodium,* and *Thalassodendron*), three in Hydrocharitaceae (*Enhalus, Halophila,* and *Thalassia*), one in Posidoniaceae (*Posidonia*), one in Ruppiaceae (*Ruppia*), one in Zannichelliaceae (*Lepilaena*), and two in Zosteraceae (*Phyllospadix* and *Zostera*; Table 29.1; [19,21,22]. Although many species superficially resemble terrestrial grasses, seagrasses exhibit a diversity of size and morphologies and may be as small as 1 cm (*Halophila minor*) or as long as 7 m (*Zostera caulescens*). Likewise, blades may be strap-like, cylindrical, ovate, or ovate-linear, with some species resembling ferns and others clover [16,17,19].

TABLE 29.1 List of the 72 Seagrass Species of the World [19,21]

Family	Genus: *Species*
Cymodocaceae	**Amphibolis C. Agardh:**
	Amphibolis antarctica (Labillardière) Sonder et Ascherson
	Amphibolis griffithii (J.M. Black) den Hartog
	Cymodoceaceae König in König et Sims:
	Cymodocea angustata Ostenfeld
	Cymodocea nodosa (Ucria) Ascherson
	Cymodocea rotundata Ehrenber et Hemprich ex Ascherson
	Cymodocea serrulata (R. Brown) Ascherson et Magnus
	Halodule Endlicher:
	Halodule beaudettei (den Hartog)
	Halodule bermudensis den Hartog
	Halodule emarginata den Hartog
	Halodule pinifolia (Miki) den Hartog
	Halodule uninervis (Forsskål) Ascherson
	Halodule wrightii Ascherson
	Syringodium Kützing in Hohenacker:
	Syringodium filiforme Kützing in Hohenacker
	Syringodium isoetifolium (Ascherson) Dandy
	Thaslassodendron den Hartog:
	Thalassodendron ciliatum (Forsskål) den Hartog
	Thalassodendron pachyrhizum den Hartog

(*Continued*)

TABLE 29.1 (*Continued*) List of the 72 Seagrass Species of the World [19,21]

Family	Genus: *Species*
Hydrocharitaceae	**Enhalus L.C. Richard:**
	Enhalus acoroides (Linnaeus *f*.) Royle
	Halophila Du Petit Thours:
	Halophila australis Doty et Stone
	Halophila baillonii Ascherson ex Dixie in J.D. Hooker
	Halophila beccarii Ascherson
	Halophila capricorni Larkum
	Halophila decipiens Ostenfeld
	Halophila engelmannii Ascherson
	Halophila euphlebia Makino
	Halophila hawaiiana Doty et Stone
	Halophila johnsonii Eiseman in Eiseman et McMillan
	Halophila minor (Zollinger) den Hartog
	Halophila nipponica Kuo
	Halophila ovalis (R. Brown) J.D. Hooker
	Halophila ovata Gaudichaud in Freycinet
	Halophila spinulosa (R. Brown) Ascherson
	Halophila stipulacea (Forsskål) den Hartog
	Halophila sulawesii Kuo
	Halophila tricostata Greenway
	Thalassia Banks ex König in König et Sims:
	Thalassia hemprichii (Ehrenberg) Ascherson in Petermann
	Thalassia testudinum Banks ex König in König et Sims
Posidoniaceae	**Posidonia König in König et Sims:**
	Posidonia angustifolia Cambridge et Kuo
	Posidonia australis J.D. Hooker
	Posidonia coriacea Cambridge et Kuo
	Posidonia denhartogii Kuo et Cambridge
	Posidonia kirkmanii Kuo et Cambridge
	Posidonia oceanica (Linnaeus) Delile
	Posidonia ostenfeldii den Hartog
	Posidonia sinuosa Cambridge et Kuo
Ruppiaceae	**Ruppia Linnaeus:**
	Ruppia cirrhosa (Petagna) Grande
	Ruppia filifolia (Phil.) Skottsb.
	Ruppia maritima L.
	Ruppia megacarpa R. Mason
	Ruppia polycarpa R. Mason
	Ruppia tuberosa J.S. Davis & Toml.
Zannichelliaceae	**Lepilaena Frummond ex Harvey:**
	Lepilaena australis Harv.
	Lepilaena marina E.L Robertson
Zosteraceae	**Phyllospadix W.J. Hooker:**
	Phyllospadix iwatensis Makino
	Phyllospadix japonicus Makino
	Phyllospadix scouleri W.J. Hooker
	Phyllospadix serrulatus Ruprecht ex Ascherson
	Phyllospadix torreyi S. Watson
	Zosetera Linnaeus:
	Zostera asiatica Miki
	Zostera caespitosa Miki
	Zostera capensis Setchell
	Zostera capricorni Ascherson

(*Continued*)

TABLE 29.1 (*Continued*) List of the 72 Seagrass Species of the World [19,21]

Family	Genus: *Species*
	Zostera caulescens Miki
	Zostera chilensis Kuo
	Zostera geojeensis Shin.
	Zostera japonica Ascherson et Graebner
	Zostera marina Linnaeus
	Zostera muelleri Irmisch ex Ascherson
	Zostera nigricaulis Kuo
	Zostera noltti Hornemann
	Zostera pacifica L.
	Zostera polychlamis Kuo
	Zostera tasmanica (Marten ex Ascherson) den Hartog
	Zostera nigricaulis
	Zostera noltti Hornemann
	Zostera pacifica S. Watson

Distribution

Seagrasses meadows are found in coastal waters along every continent except Antarctica, growing from the intertidal zone to depths between 60 and 70 m in the clearest ocean waters [23]. The dominant factor controlling distribution is light, with most seagrasses requiring a minimum of 10%–25% of incident surface radiation for survival [24]. Other parameters that influence seagrass geographic and depth distribution include: water clarity, temperature, salinity, current and wave patterns, nutrients, and substrate [25].

The distribution of seagrass species across the globe is divided into six geographic bioregions (four temperate and two tropical), based on assemblages of taxonomic groups in temperate and tropical areas and the physical separation of oceans [26]. Figure 29.2 shows a map of the six geographic bioregions. Within each bioregion, seagrass species are further distributed according to the physical habitat (e.g., lagoon, estuary, surf zone, back reef, deep water) and/or different successional roles.

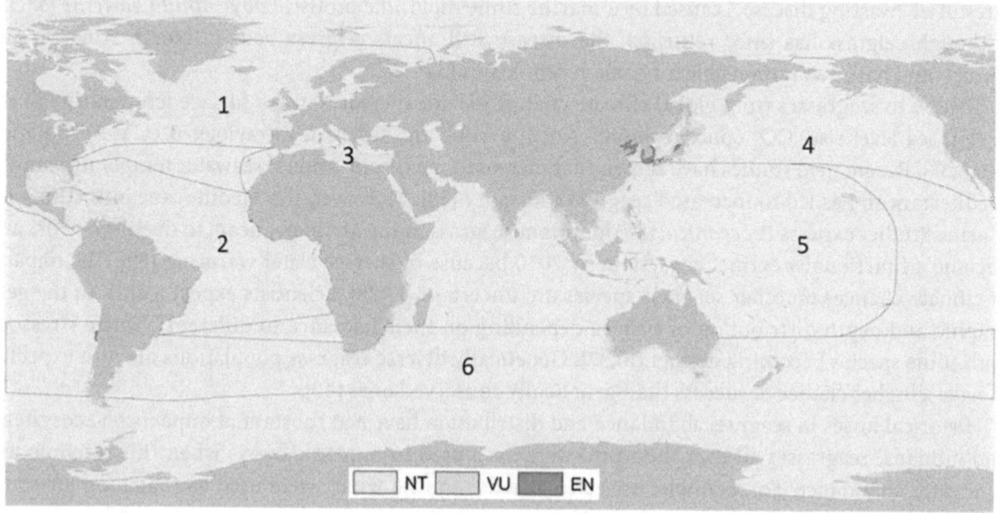

FIGURE 29.2 Number of seagrass species in stable and declining population trends across the globe. Numbers 1–6 indicate bioregions. (© ELSEVIER BV 2011 as included in "Extinction risk assessment of the world's seagrass species", *Biol. Conserv.* 144(7). Reprinted with permission.)

According to the bioregional model, there are approximately the same number of genera and species in temperate and tropical areas although the Tropical Indo-Pacific bioregion has the greatest seagrass species diversity. The most widely distributed species is *Ruppia maritima*, a species that can tolerate a wide range of salinities, and is found in both temperate and tropical bioregions in a variety of habitats [26].

Status

Seagrass ecosystems are facing a global crisis [13]. There has been a tenfold increase in reports of seagrass declines in recent decades, and Waycott et al. [27] estimated that a minimum of 29% of the known global extent of seagrass area has been lost since 1879. Moreover, the International Union for Conservation of Nature (IUCN) Red List of Threatened Species recently classified ten species of seagrass at elevated risk for extinction and three species as endangered [21]. Figure 29.2 shows a map of stable and declining population trends across the globe.

Many stressors, including global climate change, have been responsible for the decline in the distribution of seagrasses. Threats may be localized or regional, and some seagrass populations are being exposed to multiple stressors [21,28]. The greatest anthropogenic threats have been eutrophication and sedimentation from urban and agricultural runoff, as well as aquaculture practices [6,10,27,28]. Both eutrophication and sedimentation decrease the amount of light available to seagrasses for photosynthesis. Moreover, in systems with high nutrient loadings, epiphytes and fast-growing macroalgae outcompete seagrasses since they uptake nutrients more effectively and have relatively lower light requirements to sustain growth [29,30]. The relationship between human watershed activity and seagrass declines is well documented: SeagrassNet, a global monitoring program, recently captured the dramatic decrease in percent cover and shoot density of seagrass meadows in Placencia, Belize as coastal housing and tourism development surged in the region [31]. Other anthropogenic activities that have had direct impacts on the distribution of seagrasses by reducing water clarity and/or uprooting plants include dredge and fill, land reclamation, and dock and jetty construction [6,10,27].

The direct loss of seagrasses by organisms other than humans has also occurred through overgrazing (e.g., by dugongs, urchins, sea turtles), bioturbation, and disease [10,13,31]. For example, in the early 1930s, 90% of all *Zostera marina* (eelgrass) populations from Canada to North Carolina disappeared as a result of "wasting disease," caused by a marine slime mold-like protist, *Labyrinthula zosterae* [32,33]. Although eelgrass has since returned, the disease still affects eelgrass beds in North America and Europe and has been responsible for some recent losses [34–36].

Threats to seagrasses from global climate change include increases in sea surface temperature of the ocean, sea level rise, CO_2 concentrations, storm events, and levels of ultraviolet-B (UV-B) radiation [10,13,37]. Recent field studies have shown that increased maximum annual seawater temperature in the Mediterranean has led to increased seagrass mortality [38]. Moreover, the Mediterranean Institute for Marine Studies expects *P. oceanica*, the dominant seagrass in the Mediterranean, to decline by 90% and become a functionally extinct ecosystem by 2050 because of surface water warming [39]. The impacts of climate changes on other seagrass species are uncertain [13,28]. Scientists expect a shift in the geographic and depth distribution of species depending on their tolerance to different climate stressors, with some species becoming extinct [10,37]. Genetically diverse seagrass populations are also expected to have a higher chance of success than genetically conserved ones [40].

Historical losses in seagrass abundance and distribution have had substantial impacts on ecosystems and humans. Seagrasses all over the world were once used in numerous ways when this resource was in greater abundance. For example, leaves collected from the wrack were used as thatching for roofs: *Z. marina* in the Netherlands through the 1800s, and *Phyllospadix japonicus* in northern China for generations. In China, whole villages are still mostly thatched with seagrass, but it is old seagrass collected from meadows of *P. japonicus* in the 1960s. Similarly, the vast meadows of eelgrass, *Z. marina*, that once existed in the Canadian Maritimes supported an industry involving the collection of eelgrass wrack from the beach which was dried and used in a commercial home insulation product (the "Cabot's Quilt").

Ensuring the Future of Seagrasses

The long-term survival of seagrasses in a globally changing environment depends upon reducing human impacts to the coastal oceans. Researchers and managers can support seagrass ecosystem resilience by continuing efforts to better understand the status and health of seagrasses and using the information gained to develop effective management strategies. Actions that support the maintenance of healthy seagrass ecosystems include: establishing mapping and monitoring programs; improving habitat conditions; protecting populations in addition to restoring; and raising awareness about the value, status, and threats to seagrasses [40–42].

Establishing Mapping and Monitoring Programs

Mapping and monitoring programs provide crucial information on the health, status, and trends of populations, as well as environmental conditions. Aerial photography, remote sensing, and GPS can be used to delineate and map seagrass beds within a given area and to assess changes in seagrass abundance and distribution. Monitoring programs collect more detailed information on selected seagrass and environmental parameters. SeagrassNet, a global monitoring network composed of scientists, managers, and stakeholders, uses a standardized protocol to detect changes in seagrass habitat, monitoring both seagrass parameters (e.g., percent cover, shoot density, canopy height) and environmental variables (e.g., changes in water clarity; [43,44]). The information is used to evaluate trends in the distribution, structure, and function of seagrasses habitats, identify potential threats, and adapt or implement management strategies. Figure 29.3 shows a diver collecting information for SeagrassNet.

Improving Habitat Conditions

Seagrasses growing along coastlines that are stressed by human activities are the most vulnerable to generalized disturbances such as climate change. Improving habitat conditions in these areas can enhance seagrass health for better resistance or adaptation to future conditions. Poor water quality and clarity can be improved by encouraging land use better practices such as developing coastal buffer zones to decrease nutrient and sediment runoff, reducing the use of fertilizers and pesticides, and increasing filtration of effluent. Moreover, the effects of disturbances such as dredging, fishing, and boating activities on seagrass habitats can be minimized by establishing codes of conduct [40–42]. In systems where seagrass habitats are exposed to multiple anthropogenic threats acting at broad scales, spatially explicit assessments can be conducted to identify "hot spots" for prioritizing seagrass management actions [45].

Protecting Resilient Populations

Seagrass populations exhibit different responses to stressors, with some populations exhibiting higher tolerances to poor water quality [46], increased temperatures [47], and/or elevated levels of UV-B [48–50]. Identifying and protecting seagrass communities that are potentially resilient to stressors is an important strategy because it supplies a source of seeds to repopulate following disturbances [40]. Priority should be given to resilient populations growing in different types of conditions and from a wide geographical range. Patterns of connectivity between seagrass beds and adjacent habitats such as salt marshes, mangroves, and coral reefs should also be identified to promote ecological linkages and shifts in species distribution [40]. Once resilient populations are identified, their status and health should be monitored and habitat conditions conducive for their growth maintained or improved if needed.

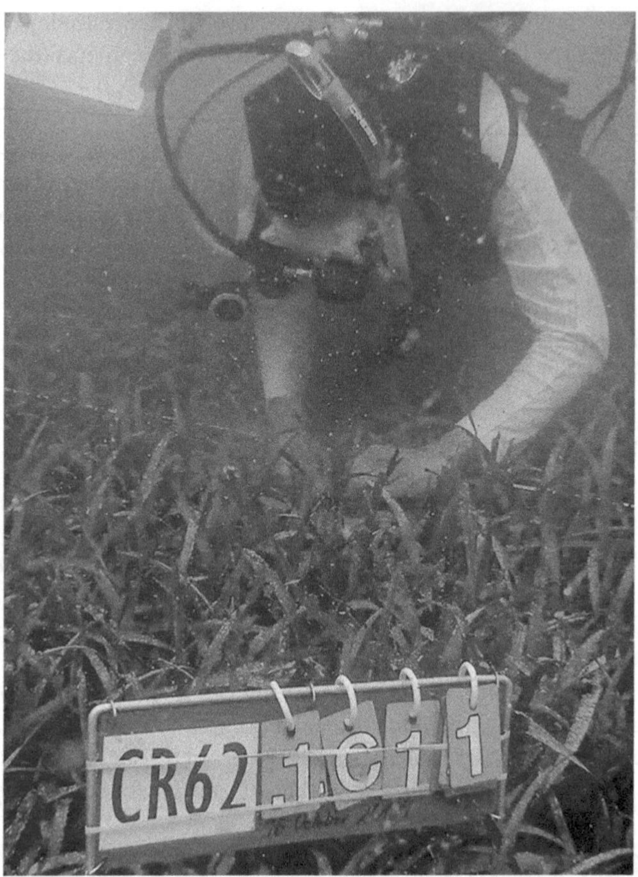

FIGURE 29.3 **(See color insert.)** SeagrassNet sampling in *Thalassia testudinum* meadows at Caulker Marine Reserve, Belize. (Fred Short.)

Restoring

Restoration of seagrass beds is considered to be a viable strategy for increasing seagrass habitats although transplanting techniques are often costly and not always successful [51–53]. Two factors critical to transplanting success are the selection of suitable transplant sites and robust donor populations [52,54]. Transplant sites that are conducive to establishment and growth have good water quality and clarity, appropriate substrate (type and size), low exposure (wind, waves, dessication), and are free of conflicting uses [52]. Donor populations that are from sites with similar conditions as transplant sites and have high genetic diversity increase transplanting success by contributing to seagrass productivity and recovery potential [55–61]. All transplanting efforts should be assessed through long-term mapping and monitoring programs. Figure 29.4 shows a diver in Nantucket Harbor, MA, USA transplanting eelgrass.

Raising Awareness

Seagrasses have received less attention from the media than other marine ecosystems despite their higher economic value [13]. Engaging local communities and stakeholders in seagrass programs is an essential conservation strategy that raises awareness about the importance of these valuable ecosystems and the threats to their survival. Building awareness of seagrasses ultimately improves coastal

FIGURE 29.4 **(See color insert.)** A diver transplanting eelgrass in Nantucket Harbor, MA, USA. (Eric Savetsky.)

habitats and encourages habitat protection. Community monitoring programs (e.g., Seagrass-Watch) or community-based restoration efforts (e.g., TERFS) can reinforce the value of seagrass habitats while collecting information about the condition of seagrasses [40,62–64].

Conclusions

Seagrasses are angiosperms adapted to exist fully submerged in estuarine and near-shore marine environments. They form extensive meadows that provide important ecological and economic services in coastal systems throughout the world by stabilizing and enriching sediments, trapping and cycling nutrients, maintaining water quality and clarity, and providing habitat for organisms in a vast food web. Moreover, seagrass meadows have the ability to mitigate climate change through carbon sequestration and storage. Despite their many values, seagrass meadows are rapidly declining, and there is no evidence that seagrass recoveries are compensating for large-scale losses. The future of seagrass meadows is dependent upon immediate actions to implement strong science-based strategies that will maintain and increase seagrass habitats in a globally changing environment while considering the interaction between land and sea processes in policy, planning, and management decisions. Their survival also relies on increasing our understanding and awareness of these species, the relationships between species, their genetic variability, unique adaptive traits, and ability to naturally recolonize. Beyond research and monitoring, management is needed to secure the future of seagrass ecosystems so they continue to deliver important ecological and ecosystem services to the world's oceans and people.

References

1. Costanza, R.; d'Arge, R.; de Groot, R; Farber, S.; Grasso, M; Hannon, B; Naeem, S; Limburg, K; Paruelo, J; O'Neill, R.V.; Raskin, R.; Sutton, P.; van den Belt, M. The value of the world's ecosystem services and natural capital. *Nature*. 1997, 387, 253–260.
2. Fourqurean, J.W.; Duarte, C.M.; Kennedy, H.; Marba, N.; Holmer, M.; Mateo, M.A.; Apostolaki, E.T; Kendrick, G.A.; Krause-Jensen, D.; McGlathery, K.J.; Serrano, O. Seagrass ecosystems as a globally significant carbon stock. *Nature Geosci*. 2012, 5, 505–509.

3. Duarte, C.M.; Cebrian, J. The fate of autotrophic production in the sea. *Assoc. Sci. Limnol. Oceanogr.* 1996, 41, 1758–1766.
4. Suchanek, T.H.; Williams, S.W.; Ogden, J.C.; Hubbard, D.K.; Gill, I.P. Utilization of shallow-water seagrass detritus by Caribbean deep-sea macrofauna:$\delta^{13}C$ evidence. *Deep Sea Res.* 1985, 32, 2201–2214.
5. Heck, K.L.; Able, K.W.; Roman, C.T.; Fahay, M.P. Composition abundance, biomass, and production of marcrofauna in a New England estuary: Comparisons among eelgrass meadows and other nursery habitats. *Estuar.* 1995, 18(2), 379–389.
6. Short, F.T.; Wyllie-Echeverria, S. Natural and human-induced disturbance of seagrasses. *Environ. Conserv.* 1996, 23(1), 17–27.
7. Terrados, J.; Duarte, C.M. Experimental evidence of reduced particle resuspension within a seagrass (*Posidonia oceanica* L.) meadow. *Exper. Mar. Biol. Ecol.* 2000, 243, 45–53.
8. Fry, B.; Parker, P.L. Animal diet in Texas seagrass meadows. C evidence for the importance of benthic plants. *Estuar. Coast. Mar. Sci.* 1979, 8, 499–509.
9. Hemminga, M.A.; Duarte, C.M. *Seagrass Ecology*. Cambridge University Press: Cambridge, UK; New York, NY, 1991.
10. Duarte, C.M. The future of seagrass meadows. *Environ. Conserv.* 2002, 29(2), 192–206.
11. Heck, K.L.; Hays, C.; Orth, R.J. A critical evaluation of the nursery role hypothesis for seagrass meadows. *Mar. Ecol. Prog. Ser.* 2003, 253, 123–136.
12. Bricker, S.B.; Ferreira, J.G.; Simas, T. An integrated methodology for assessment of estuarine trophic status. *Ecol. Model.* 2003, 169, 39–60.
13. Orth, R.J.; Carruthers, T.J.B.; Dennison, W.C.; Duarte, C.M.; Fourqurean, J.W.; Heck, K.L. Jr.; Hughes, A.R.; Kendrick, G.A.; Kenworthy, W.J.; Olyarnik S.; Short, F.T.; Waycott, M.; Williams, S.L. A global crisis for seagrass ecosystems. *BioSci.* 2006, 56(12), 987–996.
14. Les, D.H.; Cleland, M.A.; Waycott, M. Phylogenetic studies in the Alismatidae, II: Evolution of the marine angiosperms (seagrasses) and hydrophily. *Syst. Bot.* 1997, 22, 443–463.
15. Waycott, M.; Procaccini, G.; Les, D.H.; Reusch, T.B.H. Seagrass evolution, ecology and conservation: A genetic perspective. In *Seagrasses: Biology, Ecology and Conservation*; Larkum, A.W.D.; Orth, R.J.; Duarte, C.M.; Eds; Springer: Dordrecht, 2006; 25–50.
16. den Hartog, C. Seagrasses of the World. Verh. Kon. Ned. Akad. Wetens. Afd. Naturk. Ser. 2 59, 1–275+31 plates, 1970.
17. Phillips, R.C.; Meñez, E.G. Seagrasses. Smiths. Contr. Mar. Sci. Number 34, Smithsonian Institution Press, Washington, DC, 1998.
18. Dawes, C.J. *Marine Botany*. 2nd edition. John Wiley & Sons: New York, 1998.
19. Kuo, J.; den Hartog, C. Seagrass taxonomy and identification key. In *Global Seagrass Research Methods*; Short, F.T.; Coles, R.G., Eds; Elsevier Science B.V: Amsterdam, 2001; 31–58.
20. Kuo, J.; den Hartog, C. Morphology, anatomy, and ultrastructure. In *Seagrasses: Biology, Ecology and Conservation*; Larkum, A.W.D.; Orth, R.J.; Duarte, C.M., Eds; Springer Verlag: Berlin, 2006; 51–88.
21. Short, F.T.; Polidoro, B.; Livingstone, S.R.; Carpenter, K.E.; Bandeira, S.; Bujang, J.S.; Calumpong, H.P.; Carruthers, T.J.; Coles, R.G.; Dennison, W.C.; Erftemeijer, P.L.A.; Fortes, M.D.; Freeman, A.S.; Jagtap, T.G.; Kamal, A. H. M.; Kendrick, G.A.; Kenworthy, W.J.; La Nafie, Y.A.; Nasution, I.M.; Orth, R.J.; Prathep, A.; Sanciangco, J.C.; van Tussenbroek, B.; Vergara, S.G.; Waycott, M.; Zieman, J.C. Extinction risk assessment of the world's seagrass species. *Biol Conserv.* 2011, 144, 1961–1971.
22. Moore, K.A.; Short, F.T. Zostera: Biology, ecology and management. In *Seagrasses: Biology, Ecology and Conservation*; Larkum, A.W.D.; Orth, R.J.; Duarte, C.M., Eds; Springer Verlag: Berlin, 2006; 361–386.
23. Coles, R.G.; McKenzie, L.J.; De'ath, G.; Roelofs, A.J.; Lee Long, W. Spatial distribution of deepwater seagrass in the inter-reef lagoon of the Great Barrier Reef World Heritage Area. *Mar. Ecol. Prog. Ser.* 2009, 392, 57–68.

24. Dennison, W.C.; Orth, R.J.; Moore, K.A.; Stevenson, J.C.; Carter, V.; Kollar, S.; Bergstrom, P.W.; Batuik R.A. Assessing water quality with submersed aquatic vegetation. *BioScience*. 1993, 43, 86–94.
25. Short, F.T.; Coles, R.G.; Pergent-Martini, C. Global seagrass distribution. In *Global Seagrass Research Methods*; Short, F.T.; Coles, R.G., Eds; Elsevier Science B.V: Amsterdam, 2001; 5–30.
26. Short, F.T.; Carruthers, T.; Dennison, W.; Waycott, M. Global seagrass distribution and diversity: A bioregional model. *J. Exper. Mar. Biol. Ecol.* 2007, 350, 3–20.
27. Waycott, M.; Duarte, C.M.; Carruthers, T.J.B.; Orth, R.J.; Dennison, W.C.; Olyarnik, S.; Calladine, A.; Fourqurean, J.W.; Heck, K. Jr.; Hughes, A.R.; Kendrick, G.A.; Kenworthy, W.J.; Short, F.T.; Williams, S.L. Accelerating loss of seagrasses across the globe threatens coastal ecosystems. *Proc. Nat. Acad. Sci.* 2009, 106, 12377–12381.
28. Grech, A.; Chartrand-Miller, K.; Erftemeijer, P.; Fonseca, M.; McKenzie, L.; Rasheed, R.; Taylor, H.; Coles, R.G. A comparison of threats, vulnerabilities and management opportunities in global seagrass bioregions. *Environ. Res. Lett.* 2012, 7, 024006.
29. Harlin, M.M.; Thorne-Miller, B. Nutrient enrichment of seagrass beds in a Rhode Island coastal lagoon. *Mar. Biol.* 1981, 65, 221–229.
30. Twilley, R.R.; Kemp, W. M.; Staver, K.W.; Stevenson, J. C.; Boynton, W.R. Nutrient enrichment of estuarine submersed vascular plant communities. 1 Algal growth and effects on production of plants and associated communities. *Mar. Ecol. Prog. Ser.* 1985, 23, 179–219.
31. Short, F.T.; Koch, E.; Creed, J.C.; Magalhaes, K.M. SeagrassNet monitoring of habitat change across the Americas. *Biol. Mar. Medit.* 2006, 13, 272–276.
32. Rasmussen, E. The wasting disease of eelgrass (*Zostera marina*) and its effects on environmental factors and fauna. In Seagrass Ecosystems. McRoy, C.P.; Helfferich, C., Eds; Marcel Dekker: New York, 1977; 1–51.
33. Ralph, P.J.; Short, F.T. Impact of the wasting disease pathogen, Labyrinthula zosterae, on the photobiology of eelgrass, *Zostera marina*. *Mar. Ecol. Prog. Ser.* 2002, 228, 265–271.
34. Short, F.T.; Mathieson, A.C.; Nelson, J.I. Recurrence of the eelgrass wasting disease at the border of New Hampshire and *Maine, USA*. *Mar. Ecol. Prog. Ser.* 1986, 29, 89–92.
35. Short, F.T.; Ibelings, B.W.; den Hartog, C. Comparison of a current eelgrass disease to the wasting disease of the 1930's. *Aquat. Bot.* 1988, 30, 295–304.
36. den Hartog, C. Wasting disease and other dynamic phenomena in *Zostera* beds. *Aquat. Bot.* 1987, 27, 3–14.
37. Short, F.T.; Neckles, H.A. The effects of global climate change on seagrasses. *Aquat. Bot.* 1999, 63, 169–196.
38. Marbà, N.; Duarte, C.M. Mediterranean warming triggers seagrass (Posidonia oceanica) shoot mortality. *Global Change Biol.* 2010, 16, 2366–2375.
39. Jordà, G.; Marbà, N.; Duarte, C.M. Mediterranean seagrass vulnerable to regional climate warming. *Nat. Clim. Change*. 2012, 2, 821–824. doi:10.1038/nclimate1533.
40. Björk, M.; Short F.T.; Mcleod E.; Beer, S. *Managing Seagrasses for Resilience to Climate Change.* IUCN: Gland, Switzerland, 2008.
41. Long, W.L.; Thom, R.M. Seagrass Habitat Conservation-Improving seagrass habitat quality. In *Global Seagrass Research Methods*; Short, F.T.; Coles, R.G., Eds; Elsevier Science B.V: Amsterdam, 2001; 407–445.
42. Coles, R.C.; Fortes, M. Protecting seagrass—Approaches and methods. In *Global Seagrass Research Methods*; Short, F.T.; Coles, R.G., Eds; Elsevier Science B.V: Amsterdam, 2001; 445–463.
43. Short, F.T.; McKenzie, L.G.; Coles, R.G.; Vidler, K.P; Gaeckle, J.L. *SeagrassNet Manual for Scientific Monitoring of Seagrass Habitat - Worldwide Edition.* University of New Hampshire Publication, Durham, NH, 2006.
44. SeagrasNet Official Site. www.SeagrassNet.org.

45. Grech, A.; Coles, R.G.; Marsh, H. A broad-scale assessment of the risk to coastal seagrasses from cumulative threats. *Mar. Policy.* 2011, 35(5) 560–567.
46. Short, F.T.; Burdick, D.M.; Moore, G.E. *The Eelgrass Resource of Southern New England and New York: Science in Support of Management and Restoration Success.* The Nature Conservancy, Arlington, VA, 2012.
47. Franssen, S.U.; Gu, J.; Bergmann, N.; Winters, G.; Klostermeier, U.C.; Rosenstiel, P.; Bornberg-Bauer, E.; Reusch, T.B.H. Transcriptomic resilience to global warming in the seagrass *Zostera marina*, a marine foundation species. *Proc. Nat. Acad. Sci.* 2011, 19276–19281.
48. Trocine, R.P.; Rice, J.D.; Wells, G.N. Inhibition of seagrass photosynthesis by Ultraviolet-B radiation. *Plant Physiol.* 1981, 68, 74–81.
49. Dawson, S.P.; Dennison, W.C. Effects of ultraviolet and photosynthetically active radiation on five seagrass species. *Mar. Biol.* 1996, 124, 629–638.
50. Novak, A.B; Short, F.T. UV-B induces leaf reddening and contributes to the maintenance of photosynthesis in the seagrass *Thalassia testudinum*. *J. Exp. Mar. Biol. Ecol.* 2011, 409, 136–142.
51. Fonseca, M.S.; Kenworthy,W.J.; Thayer, G.W. Guidelines for conservation and restoration of seagrass in the United States and adjacent waters. NOAA/NMFS Coastal Ocean Program and Decision Analysis Series, No. 12. NOAA Coastal Ocean Office: Silver Spring, MD, 1998.
52. Short, F.T.; Davis, R.C.; Kopp, B.S.; Short, C.A.; Burdick, D.M. Site-selection model for optimal transplantation of eelgrass *Zostera marina* in Northeastern US. *Mar. Ecol. Prog. Ser.* 2002, 227, 253–267.
53. Cunha, A. H.; Marbá, N. N.; van Katwijk, M. M.; Pickerell, C.; Henriques, M.; Bernard, G.; Ferreira, M. A.; Garcia, S.; Garmendia, J. M.; Manent, P. Changing paradigms in seagrass restoration. *Rest. Ecol.* 2012, 20, 427–430.
54. van Katwijk, M. M.; Bos, A.R.; de Jonge, V.N.; Hanssen, L.S.A.M.; Hermus, D.C.R.; de Jong, D.J. Guidelines for seagrass restoration: Importance of habitat selection and donor population, spreading of risks, and ecosystem engineering effects. *Mar. Pollut. Bull.* 2009, 58(2), 179–188.
55. Procaccini, G.; Piazzi, L.; Genetic polymorphism and transplantation success in the Mediterranean seagrass *Posidonia oceanica*. *Rest. Ecol.* 2001, 9, 332–338.
56. Williams, S.L. Reduced genetic diversity in eelgrass transplantations affects both population growth and individual fitness. *Ecol. Appl.* 2001, 11, 1472–1488.
57. Hughes, A.R.; Stachowicz, J.J. Genetic diversity enhances the resistance of a seagrass ecosystem to disturbance. *Proc. Nat. Acad. Sci.* 2004, 101, 8998–9002.
58. Reusch, T.B.H.; Ehlers, A.; Hammerli, A.; Worm, B. Ecosystem recovery after climatic extremes enhanced by genotypic diversity. *Proc. Nat. Acad. Sci.* 2005, 102, 2826–2831.
59. Reusch, T.B.H.; Hughes, A.R. The emerging role of genetic diversity for ecosystem functioning: Estuarine macrophytes as models. *Estuar. Coasts.* 2006, 29, 159–164.
60. Novak, A.B.; Hays, C.; Plaisted, H.K.; Hughes, A.R. Limited effects of source population identity and number on seagrass transplant performance. *PeerJ.* 2017, 5, e2972.
61. Plaisted, H.K.; Novak, A.B.; Weigel, S.; Klein, A.S.; Short, F.T. Eelgrass genetic diversity influences resilience to stress associated with eutrophication. *Estuar. Coasts.* 2019, doi:10.1007/s12237-019-00669-0.
62. McKenzie, L.J.; Lee Long, W.J.; Coles, R.G.; Roder, C.A. Seagrass-watch: Community based monitoring of seagrass resources. *Biol Mar. Mediterr.* 2000, 7(2), 393–396.
63. SeagrassWatch Official Site. www.Seagrasswatch.org.
64. Short, F.T.; Short, C.A.; Burdick-Whitney, C.A. *Manual for Community-Based Eelgrass Restoration.* Sponsored by NOAA Restoration Center: University of New Hampshire, Durham, NH, 2002.

30

Wetlands: Coastal, InSAR Mapping

Zhong Lu
U.S. Geological Survey (USGS)

Jinwoo Kim and C. K. Shum
The Ohio State University

Introduction

Coastal wetlands include hydrologic and other processes that are fundamental to understanding ecological and climatic changes.[1,2] Characterizing wetland types and changes is critical for monitoring human, climatic, and other effects on wetland systems including their restoration. Measuring changes in water level over wetlands, and consequently changes in water storage capacity, provides a governing parameter in hydrologic models and is required for comprehensive assessment of flood hazards.[3] With frequent coverage over wide areas, satellite sensors can provide accurate measurements of coastal wetland characteristics and underneath-water storage.

A synthetic aperture radar (SAR) is an advanced radar system that utilizes image-processing techniques to synthesize a large virtual antenna, which provides much higher spatial resolution than is practical using a real-aperture radar.[4] A SAR system transmits electromagnetic waves at a wavelength that can range from a few millimeters to tens of centimeters. Because a SAR actively transmits and receives signals backscattered from the target area and the radar signals with long wavelengths are mostly unaffected by weather clouds, a SAR can operate effectively during day and night under most weather conditions. Through SAR processing, both the intensity and phase of the radar signal backscattered from each ground resolution element (typically meters to submeters in size) can be calculated and combined to form a complex-valued SAR image that represents the radar reflectivity of the ground surface.[4] The intensity (or strength) of the SAR image is determined primarily by the terrain slope, surface roughness, and dielectric constant. The phase of the complex-valued SAR image is related to the apparent distance from the satellite to a ground resolution element as well as the interaction between radar waves and scatterers within a resolution element of the imaged area.

Interferometric SAR (InSAR) utilizes the phase components of two coregistered SAR images of the same area acquired from similar vantage points to extract the landscape topography and patterns of surface change, including ground displacement.[5,6] The spatial separation between two SAR antennas or two vantage points of the same SAR antenna used to acquire the two SAR images is called the InSAR baseline. Two antennas can be displaced in the along-track direction to produce an along-track

InSAR image if the two SAR images are acquired with a short time lag.[5,7] Along-track InSAR can be used to measure the velocity of targets moving toward or away from the radar to determine sea ice drift, ocean currents, river flows, and ocean wave parameters. Along-track InSAR applications in the past have been limited to technology demonstration in experimental studies from airborne SAR observations. Two antennas can be displaced in the cross-track direction to produce cross-track InSAR. For cross-track InSAR, the two antennas can be mounted on a single platform for simultaneous, single-pass InSAR observation, which is the usual implementation for airborne systems and spaceborne systems (e.g., Shuttle Radar Topography Mission) for generating high-resolution, precise digital elevation models (DEMs) over large regions.[8] Alternatively, InSAR images can be formed by using a single antenna on an airborne or spaceborne platform in nearly identical repeating flight lines or orbits for repeat-pass InSAR.[9,6] In this case, even though successive observations of the target area are separated in time, the observations will be highly correlative if the backscattering properties of the surface have not changed in the interim. In this way, InSAR is capable of measuring surface displacement with subcentimeter vertical precision for X-band and C-band sensors ($\lambda = 2–8$ cm) or few-centimeter precision for L-band sensors ($\lambda = 15–30$ cm), in both cases at a spatial horizontal resolution of tens of meters over an image swath (width) of a few tens of kilometers up to ~100 km. This is the typical implementation for past, present, and future spaceborne SAR sensors (Table 30.1). The required wavelength to monitor the earth surface varies among applications. Among SAR systems with similar repeat intervals, spatial resolutions, and imaging geometries, long-wavelength (such as L-band) SAR is generally a better choice for studying coastal wetlands due to the penetration of the long-wavelength SAR signal into vegetation to capture canopy characteristics and reveal the water surface beneath.

Interactions of radar waves with wetlands can be complicated.[10] Over flooded vegetation, the radar backscattering consists of contributions from the interactions of radar waves with the canopy surface, canopy volume, and water surface. On the basis of a canopy backscattering model for continuous tree canopies,[11] the total radar backscattering over wetland can be approximated as the incoherent

TABLE 30.1 Satellite SAR Sensors Capable of InSAR Mapping

Mission	Agency	Period of Operation[1]	Orbit Repeat Cycle (days)	Band/Frequency (GHz)	Wave-Length (cm)	Incidence Angle (°) at Swath Center	Resolution (m)
Seasat	NASA	06/1978 to 10/1978	17	L-band/1.275	23.5	23	25
ERS-1	ESA	07/1991 to 03/2000	3, 168, and 35[1]	C-band/5.3	5.66	23	30
JERS-1	JAXA	02/1992 to 10/1998	44	L-band/1.275	23.5	39	20
ERS-2	ESA	04/1995 to 07/2011	35	C-band/5.3	5.66	23	30
Radarsat-1	CSA	11/1995 to 2013	24	C-band/5.3	5.66	10 to 60	10 to100
Envisat	ESA	03/2002 to 04/2012	35, 30[2]	C-band/5.331	5.63	15 to 45	20 to 100
ALOS	JAXA	01/2006 to 05/2011	46	L-band/1.270	23.6	8 to 60	10 to 100
TerraSAR-X	DLR	6/2007 to present	11	X-band/9.65	3.1	20 to 55	1 to 16
Radarsat-2	CSA	12/2007 to present	24	C-band/5.405	5.55	10 to 60	3 to 100
COSMO-SkyMed	ASI	6/2007 to present	1, 4, 5, 7, 8, 9, 12 and 16[3]	X-band/9.6	3.1	20 to 60	1 to 100

Abbreviations: ALOS, Advanced Land Observing Satellite; ASI, Agenzia Spaziale Italiana (Italian Space Agency); COSMO-SkyMed, COnstellation of small Satellites for the Mediterranean basin Observation; CSA, Canadian Space Agency; DLR, Deutsches Zentrum für Luft-und Raumfahrt e.V. (German Space Agency); Envisat, Environmental Satellite; ERS, European Remote Sensing Satellite; ESA, European Space Agency; JAXA, Japanese Aerospace Exploration Agency; JERS, Japanese Earth Resources Satellite; NASA, National Aeronautics and Space Administration.

[1] To accomplish various mission objectives, the ERS-1 repeat cycle was 3 days from July 25, 1991, to April 1, 1992, and from December 23, 1993, to April 9, 1994; 168 days from April 10, 1994, to March 20, 1995; and 35 days at other times.

[2] ENVISAT repeat cycle was 35 days from March, 2002 to October, 2010, and 30 days from November, 2010 to April 8, 2012.

[3] A constellation of 4 satellites, each of which has a repeat cycle of 16 days, can collectively produce repeat-pass InSAR images at intervals of 1 day, 4 days, 5 days, 7 days, 8 days, 9 days, and 12 days, respectively.

summation of contributions from a) canopy surface backscattering, b) canopy volume backscattering that includes backscattering from multiple-path interactions of canopy–water, and c) double-bounce trunk-water backscattering (i.e., the radar signal is initially reflected away from the sensor by the water's surface, toward a tree bole or other vertical structure and is then directly reflected toward the sensor). The relative contributions from surface backscattering, volume backscattering, and double bounce backscattering are controlled primarily by radar parameters including wavelength, polarization, incidence angle, and spatial resolution as well as wetland characteristics including vegetation type (and structure), vegetation leaf on/off condition, canopy closure, and other environmental factors.[10,12]

InSAR Studies of Coastal Wetlands

Typical InSAR processing includes precise registration of an interferometric SAR image pair, a common Doppler/bandwidth filter to improve InSAR coherence, interferogram generation, removal of the curved earth phase trend, adaptive filtering, phase unwrapping, precision estimation of interferometric baseline, generation of a surface displacement (deformation) image (or a DEM map), estimation of interferometric correlation, and rectification of interferometric products.[6,13–15] Using a single pair of SAR images as input, a typical InSAR processing chain outputs two SAR intensity images, a displacement or DEM map, and an interferometric correlation map.

When more than two scenes of SAR images are available over a study area, multiple-temporal InSAR images can be produced. Multi-interferogram InSAR processing should then be employed to improve the accuracy of displacement measurement (or other InSAR products).[16–20] The goal of multi-interferogram InSAR processing is to characterize the spatial and temporal behaviors of the displacement signal and various artifacts and noise sources (atmospheric delay anomalies including radar frequency-dependent ionosphere refraction and non-dispersive troposphere delay of the radar signals, orbit errors, DEM-induced artifacts) in individual interferograms and then to remove the artifacts and anomalies to retrieve time-series displacement measurements at the SAR pixel level.

SAR Intensity Image

In its simplest form, a SAR intensity image can be regarded as a thematic layer containing information about specific surface characteristics and is particularly sensitive to terrain slope, surface roughness, and the target's dielectric constant. Surface roughness refers to the SAR wavelength-scale variation in the surface relief. Dielectric constant is an electric property of material that influences radar return strength and is controlled primarily by moisture content of the imaged surface. SAR backscattering intensity images have been used in characterizing types, conditions, and flooding of wetlands[21–25] (Figure 30.1).

InSAR Coherence Image

An InSAR coherence image is a cross correlation product derived from two coregistered complex-valued (both intensity and phase components) SAR images.[26,27] It depicts changes in backscattering characteristics on the scale of radar wavelength. Loss of InSAR coherence is often referred to as decorrelation. Decorrelation over wetlands can be caused by the combined effects of a) thermal decorrelation caused by uncorrelated noise sources in radar instruments, b) geometric decorrelation resulting from imaging a target from different look angles, c) volume decorrelation caused by volume backscattering effects, and d) temporal decorrelation due to environmental changes over time.[28–30,12] On the one hand, decorrelation renders an InSAR image useless for measuring surface displacement or constructing a topographic map. On the other hand, the pattern of decorrelation within an image can indicate surface modifications caused by flooding, wildfire, or other processes.

Geometric decorrelation increases as the perpendicular baseline length increases until a critical length is reached at which InSAR coherence is lost.[26] For surface backscattering, most of the effect of baseline

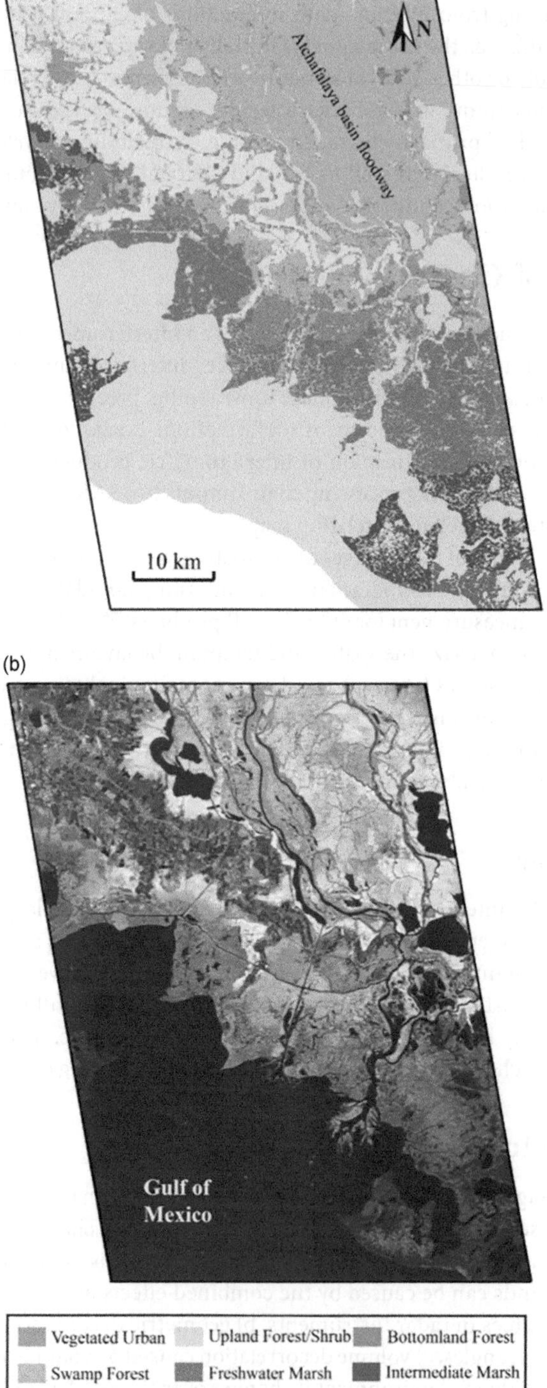

FIGURE 30.1 (**a**) Thematic map showing major land cover classes of coastal forests in southeastern Louisiana, U.S.A. (**b**) L-band (wavelength of ~24 m) SAR intensity image produced from Advanced Land Observing Satellite (ALOS) Phased Array type L-band Synthetic Aperture Radar (PALSAR) image acquired on January 12, 2007.

geometry on the measurement of interferometric coherence can be removed by the common spectral band filtering.[31] Volume backscattering describes multiple scattering of the radar wave within a distributed volume over wetlands and can be significantly affected by the InSAR baseline geometry configuration. As a result, volume decorrelation is most often coupled with geometric decorrelation and is a complex function of vegetation canopy structure. Generally, the contribution of volume backscattering is controlled by the proportion of transmitted signal that penetrates the surface and the relative two-way attenuation from the surface to the volume element and back to the sensor.[30] Because both surface backscattering and volume backscattering consume and attenuate the transmitted radar signal, they determine the proportion of radar signal that is available to produce double-bounce backscattering. Hence, surface backscattering and volume backscattering over wetlands combine to lower InSAR coherence and reduce double-bounce backscattering that is utilized to sense water surface beneath wetlands.

Temporal decorrelation describes any event that changes the physical orientation, composition, or scattering characteristics and spatial distribution of scatterers within an imaged volume. Over wetlands, these decorrelations are primarily caused by wind changing the leaf orientations, moisture condensation and rain changing the dielectric constant, and flooding changing the dielectric and roughness of canopy background, and seasonal phenology, as well as anthropogenic activities such as cultivation and timber harvesting.[30] Temporal decorrelation is the net effect of changes in radar backscattering and therefore depends on the stability of the scatterers, the canopy penetration depth of the transmitted pulse, and the response to the changing conditions with respect to the SAR wavelength.

DEM

A precise DEM can produce a very important data set for characterizing and monitoring coastal wetlands. The ideal SAR configuration for DEM production is a single-pass (simultaneous) two-antenna system.[8] However, repeat-pass single-antenna InSAR also can be used to produce useful DEMs.[32] Either technique is advantageous in areas where the photogrammetric approach to DEM generation is hindered by persistent clouds or other factors.[33] There are many sources of error in DEM construction from repeat-pass SAR images, e.g., inaccurate determination of the InSAR baseline, atmospheric delay anomalies, possible surface displacement due to groundwater movement, or other geological processes during the time interval spanned by the images, etc. To generate a high-quality DEM, these errors must be identified and corrected using a multi-interferogram approach.[33] A data fusion technique can be used to combine DEMs from several interferograms with different spatial resolution, coherence, and vertical accuracy to generate the final DEM product.

For vegetated terrains, InSAR analysis based on fully polarized SAR images (i.e., radar signals are transmitted and received with both vertical and horizontal polarizations) is often employed to estimate vegetation characteristics and underlying topography.[34–36] Polarization signatures of the vegetation canopy, the bulk volume of vegetation, and the ground are different and can be separated using polarimetric analysis. An optimization procedure can be employed to maximize interferometric coherence between two polarimetric radar images, thereby reducing the effect of baseline and temporal decorrelation in the interferogram. Then, using a coherent target decomposition approach that separates distinctive backscattering returns from the canopy top, the bulk volume of vegetation, and the ground surface, one can derive height differences between physical scatterers with differing scattering characteristics.[34] Physical radar backscattering models for different vegetation types including wetlands can be developed to calculate the canopy height, the bare earth topography, and other parameters based on measurements from polarimetric InSAR images.

InSAR Displacement Image

An InSAR displacement image is derived from phase components of two SAR images. SAR is a side-looking sensor; so an InSAR displacement image depicts surface displacements in the SAR line-of-sight

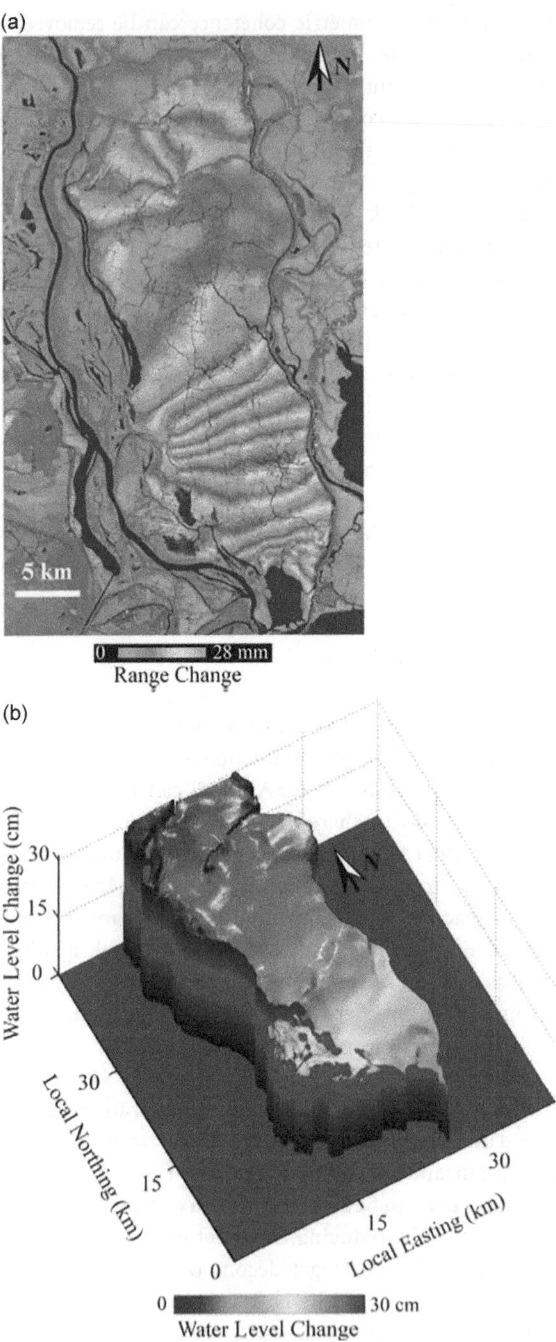

FIGURE 30.2 **(See color insert.)** **(a)** C-band (wavelength of ~5.7 cm) interferometric synthetic aperture radar (InSAR) image produced from RADAR-SAT-1 images, showing heterogeneous water level changes in swamp forests in southeastern Louisiana, U.S.A between May 22 and June 15, 2003. The interferometric phase image is draped over the radar intensity image. Each fringe (full-color cycle) represents a 2.8 cm LOS range change or 3.1 cm vertical change in water level. **(b)** Three-dimensional view of water volume changes derived from the InSAR image combined with radar altimeter and water level gauge measurements. Presumably, the uneven distribution of water level changes results from dynamic hydrologic effects in a shallow wetland with variable vegetation cover.

(LOS) direction, which include both vertical and horizontal motion. Typical look angles for satellite-borne SARs are less than 45° from vertical; so LOS displacements in InSAR displacement images are more sensitive to vertical motion (uplift or subsidence) than horizontal motion.

When double-bounce backscattering dominates the returning radar signal, a repeat-pass InSAR image can be sufficiently coherent to allow the measurement of water level changes from the interferometric phase values—a significant research front on coastal wetland studies with SAR. Alsdorf et al.[37,38] discovered that interferometric analysis of L-band (wavelength of ~24 cm) SAR imagery can yield centimeter-scale measurements of water level changes throughout inundated floodplain vegetation over the Amazon. Their work confirmed that scattering elements for L-band radar consist primarily of the water surface and vegetation trunks, which allows double-bounce backscattering returns. Later, Wdowinski et al.[39,40] applied L-band images to study water level changes over the Everglades in Florida. Using C-band SAR images, Lu et al.[41] found that the generated InSAR images maintained adequate coherence to measure phase change over swamp forests in southeastern Louisiana (Figure 30.2a and b). This finding was unexpected because the radar signal with wavelengths shorter than L-band, such as C-band (wavelength of ~5.7 cm), was thought to backscatter from the upper canopy of swamp forests rather than the underlying water surface, and a double-bounce travel path could only occur over inundated macrophytes and small shrubs.[42,43] Later, detailed studies of C-band and X-band SAR images confirmed that the dominant radar backscattering mechanism for selected coastal wetlands is double-bounce backscattering, implying that even C-band and X-band InSAR images can be used to estimate water level changes over wetlands.[12,44–46]

InSAR studies of coastal wetlands suggest that water level changes can be dynamic and spatially heterogeneous and cannot be represented by readings from sparsely distributed gauge stations. However, InSAR phase measurements are relative[15] and are often disconnected by structures and other barriers in wetlands.[12] This makes it difficult to obtain absolute water level measurements over coastal wetlands from InSAR phase measurements alone. To address this issue, Kim et al.[47] and Lu et al.[48] successfully combined radar altimetry and InSAR to estimate absolute water level changes in the swamp forests of the Atchafalaya basin, Louisiana, and showed agreement between InSAR-estimated and altimeter-observed wetland water level changes over the smaller Helmand River wetland, Afghanistan, respectively. Therefore, high-resolution InSAR images, calibrated by satellite radar altimeter or other gauge measurements, can provide precise estimates of absolute water level changes over coastal wetlands.

Conclusion

InSAR analysis can provide timely observations of vegetation characteristics and water level changes over coastal wetlands at high spatial resolution. Future InSAR analysis of fully polarimetric radar images will offer the capability of imaging 3-D structure of coastal wetlands on a large scale for improved characterization and management purposes. With more operational satellite radar platforms available for timely data acquisitions, InSAR will continue to provide solutions to many scientific questions related to the monitoring of coastal wetlands and is poised to become a critical tool in wetland restoration.

Acknowledgments

Radarsat-1 SAR images and Advanced Land Observing Satellite (ALOS) Phased Array type L-band Synthetic Aperture Radar (PALSAR) images are copyright © 2003 Canadian Space Agency and 2007 Japan Aerospace Exploration Agency/Ministry of Economy, Trade and Industry (METI), respectively. All SAR images were provided by the Alaska Satellite Facility. The authors thank O. Kwoun, E. Ramsey, and R. Rykhus for help and contribution to this project. Funding from USGS and NASA is acknowledged. The Ohio State University component of this work is partially supported by grants from USGS (No. G10AC00628) and by the Chinese Academy of Sciences/SAFEA International Partnership Program for Creative Research Teams (KZZD-EW-TZ-05).

References

1. Prigent, C.; Matthews, E.; Aires, E.; Rossow, W. Remote sensing of global wetland dynamics with multiple satellite datasets. Geophys. Res. Lett. **2001**, *28*, 4631-4634.
2. Alsdorf, D.; Rodriguez, E.; Lettenmaier, D. Measuring surface water from space. Rev. Geophys. **2007**, *45*, RG2002, doi: 10.1029/2006RG000197.
3. Coe, M. A linked global model of terrestrial hydrologic processes: Simulation of the modern rivers, lakes, and wetlands. J. Geophys. Res. **1998**, *103*, 8885–8899.
4. Curlander, J.C.; McDonough, R.N. *Synthetic Aperture Radar: Systems and Signal Processing*, 1st Ed.; Wiley series in remote sensing and image processing; Publisher: Wiley-Interscience, 1991.
5. Goldstein, R.; Zebker, H. Interferometric radar measurements of ocean surface currents. Nature **1987**, *328*, 707–709.
6. Massonnet, D.; Feigl, K. Radar interferometry and its application to changes in the Earth's surface. Rev. Geophys. **1998**, *36*, 441–500.
7. Romeiser, R.; Thompson, D. Numerical study on the along-track interferometric radar imaging mechanism of oceanic surface currents. IEEE Trans. Geosci. Remote Sensing *38*, 446–458.
8. Farr, T.G.; Rosen, P.A.; Caro, E.; Crippen, R.; Duren, R.; Hensley, S.; Kobrick, M.; Paller, M.; Rodriguez, E.; Roth, L.; Seal, D.; Shaffer, S.; Shimada, J.; Umland, J.; Werner, M.; Oskin, M.; Burbank, D.; Alsdorf, D. The shuttle radar topography mission. Rev. Geophys. **2007**, *45*, RG2004, doi: 10.1029/2005RG000183.
9. Gray, L.; Farris-Manning, P. Repeat-pass interferometry with airborne synthetic aperture radar. IEEE Trans. Geosci. Remote Sensing **1993**, *31*, 180–191.
10. Ramsey III, E.W. Radar remote sensing of wetland. In *Remote Sensing Change Detection;* Lunetta, R.S., Elvidge, C.D. Eds.; Ann Arbor Press: Chelsea, Michigan, 1999; 211–243.
11. Sun, G. *Radar Backscattering Modeling of Coniferous Forest Canopies* Ph.D. Dissertation; The University of California at Santa Barbara; California, 1990; 121 p.
12. Lu, Z.; Kwoun, O. Radarsat-1 and ERS interferometric analysis over southeastern coastal Louisiana: implication for mapping water-level changes beneath swamp forests. IEEE Trans. Geosci. Remote Sensing **2008**, *46*, 2167–2184.
13. Bamler, R.; Hart, P. Synthetic aperture radar interferometry. Inverse Probl. **1998**, *14*, R1–R54.
14. Rosen, P.; Hensley, S.; Joughin, I.R.; Li, F.K; Madsen, S.N.; Rodriguez, E.; Goldstein, R.M. Synthetic aperture radar interferometry. Proc. IEEE **2000**, *88*, 333–380.
15. Lu, Z. InSAR imaging of volcanic deformation over cloud-prone areas - Aleutian Islands. Photogram. Eng. Remote Sensing **2007**, *73*, 245–257.
16. Ferretti, A.; Prati, C.; Rocca, F. Permanent scatterers in SAR interferometry. IEEE Trans. Geosci. Remote Sensing *39*, 8–20.
17. Ferretti, A.; Savio, G.; Barzaghi, R.; Borghi, A.; Musazzi, S.; Novali, F.; Prati, C.; Rocca, F. Submillimeter accuracy of InSAR time series: Experimental validation. IEEE Trans. Geosci. Remote Sensing **2007**, *45*, 1142–1153.
18. Berardino, P.; Fornaro, G.; Lanari, R.; Sansosti, E. A new algorithm for surface deformation monitoring based on small baseline differential SAR interferograms. IEEE Trans. Geosci. Remote Sens. **2002**, *40*, 2375–2383.
19. Hooper, A.; Segall, P.; Zebker, H. Persistent scatterer inter-ferometric synthetic aperture radar for crustal deformation analysis, with application to Volcán Alcedo, Galápagos. J. Geophys. Res. **2007**, *112*, B07407, doi: 10.1029/2006 JB004763.
20. Rocca, F. Modeling interferogram stacks. IEEE Trans. Geosci. Remote Sensing **2007**, *45*, 3289–3299.
21. Bourgeau-Chavez, L.L.; Smith, K.B.; Brunzell, S.M.; Kasischke, E.S.; Romanowicz, E.A.; Richardson, C.J. Remote monitoring of regional inundation patterns and hydroperiod in the greater Everglades using synthetic aperture radar. Wetlands **2005**, *25*, 176–191.

22. Kasischke, E.S.; Smith, K.B.; Bourgeau-Chavez, L.L.; Romanowicz, E.A.; Brunzell, S.M.; Richardson, C.J. Effects of seasonal hydrologic patterns in south Florida wetlands on radar backscatter measured from ERS-2 SAR imagery. Remote Sensing Environ. **2003**, *88*, 423–441.
23. Kiage, L.M.; Walker N.D.; Balasubramanian, S.; Babin, A.; Barras, J. Applications of Radarsat-1 synthetic aperture radar imagery to assess hurricane-related flooding of coastal Louisiana. Int. J. Remote Sensing **2005**, *26*, 5359–5380.
24. Rykhus, R.; Lu, Z. *Hurricane Katrina Flooding and Possible Oil Slicks Mapped with Satellite Imagery. USGS Circular 1306;* Science and the Storms - The USGS Response to the Hurricanes of 2005: 2007; 50–53.
25. Kwoun, O.; Lu, Z. Multi-temporal RADARSAT-1 and ERS backscattering signatures of coastal wetlands at southeastern Louisiana. Photogram. Eng. Remote Sensing **2009**, *75*, 607–617.
26. Zebker, H.; Villasenor, J. Decorrelation in interferometric radar echoes. IEEE Trans. Geosci. Remote Sensing **1992**, *30*, 950–959.
27. Lu, Z.; Freymueller, J. Synthetic aperture radar interferometry coherence analysis over Katmai volcano group, Alaska. J. Geophys. Res. **1998**, *103*, 29887–29894.
28. Hagberg, J.; Ulander, L.; Askne, J. Repeat-pass SAR interferometry over forested terrain. IEEE Trans. Geosci. Remote Sensing. **1995**, *33*, 331–340.
29. Wegmüller, U.; Werner, C. Retrieval of vegetation parameters with SAR interferometry. IEEE Trans. Geosci. Remote Sensing **1997**, *35* (1), 18–24.
30. Ramsey III, E.W.; Lu, Z.; Rangoonwala, A.; Rykhus, R. Multiple baseline radar interferometry applied to coastal land cover classification and change analyses. GISci. Remote Sensing **2006**, *43*, 283–309.
31. Gatelli, F.; Guarnieri, A.M.; Parizzi, F.; Pasquali, P.; Prati, C.; Rocca, F. The wavenumber shift in SAR interferometry. IEEE Trans. Geosci. Remote Sensing **1994**, *32*, 855–865.
32. Lu, Z.; Jung, H.S.; Zhang, L.; Lee, W.J.; Lee, C.W.; Dzurisin, D. DEM generation from satellite InSAR. In *Advances in Mapping from Aerospace Imagery: Techniques and Applications,* Yang, X., Li, J., Eds.; CRC Press: 2013; 119–144.
33. Lu, Z.; Fielding, E.; Patrick, M.; Trautwein, C. Estimating lava volume by precision combination of multiple baseline spaceborne and airborne interferometric synthetic aperture radar: The 1997 eruption of Okmok Volcano, Alaska. IEEE Trans. Geosci. Remote Sensing **2003**, *41*, 1428–1436.
34. Cloude S.; Papathanassiou, K. Polarimetrie SAR interferometry. IEEE Trans. Geosci. Remote Sensing **1998**, *36*, 1551–1565.
35. Treuhaft, R.; Siqueira, P. Vertical structure of vegetated land surfaces from interferometric and polarimetric radar. Radio Sci. **2000**, *35*, 141–177.
36. Touzi, R.; Boerner, W.; Lee, J.; Lueneburg, E. A review of polarimetry in the context of synthetic aperture radar: concepts and information extraction. Can. J. Remote Sensing **2004**, *30*, 380–407.
37. Alsdorf, D.; Melack, J.; Dunne, T; Mertes, L.; Hess, L.; Smith, L. Interferometric radar measurements of water level changes on the Amazon floodplain. Nature **2000**, *404*, 174–177.
38. Alsdorf, D.; Birkett, C.; Dunne, T.; Melack, J.; Hess, L. Water level changes in a large Amazon lake measured with spaceborne radar interferometry and altimetry. Geophys. Res. Lett. **2001**, *28*, 2671–2674.
39. Wdowinski, S.; Amelung, F.; Miralles-Wilhelm, F.; Dixon, T.; Carande, R. Space-based measurements of sheet-flow characteristics in the Everglades wetland, Florida. Geophys. Res. Lett. **2004**, *31*, L15503, doi: 10.1029/2004GL020383.
40. Wdowinski, S.; Kim, S.; Amelung, F.; Dixon, T.; Miralles-Wilhelm, F.; Sonenshein, R. Space-based detection of wetlands surface water level changes from L-band SAR interferometry. Remote Sensing Environ. **2008**, *112* (3), 681–696.
41. Lu, Z.; Crane, M.; Kwoun, O.; Wells, C.; Swarzenski, C.; Rykhus, R. C-band radar observes water level change in swamp forests. EOS **2005**, *86* (14), 141–144.

42. Hess, L.; Melack, J.; Filoso, S.; Wang, Y. Delineation of inundated area and vegetation along the Amazon floodplain with SIR-C synthetic aperture radar. IEEE Trans. Geosci. Remote Sensing **1995**, *33*, 896–904.
43. Wang, Y.; Hess, L.; Filoso, S.; Melack, S. Understanding the radar backscattering from flooded and nonflooded Amazon forests: Results from canopy backscatter modeling. Remote Sensing Environ. **1995**, *54*, 324–332.
44. Lu, Z.; Kwoun, O. Interferometric synthetic aperture radar (InSAR) study of coastal wetlands over southeastern Louisiana. In *Remote Sensing of Wetlands,* Wang, Y.Q. Ed.; CRC Press; 2009: 25–60.
45. Hong, S.-H.; Wdowinski, S.; Kim, S.-W. Evaluation of TerraSAR-X observations for wetland InSAR application. IEEE Geosci. Remote Sensing **2010a**, *48*, 864–873.
46. Hong, S.-H.; Wdowinski, S.; Kim, S.-W. Space-based multi-temporal monitoring of wetland water levels: Case study of WCA1 in the Everglades. Remote Sensing Environ. **2010b**, doi: 10.1016/j.rse.2010.05.019.
47. Kim, J.W.; Lu, Z.; Lee, H.K.; Shum, C.K.; Swarzenski, C.M.; Doyle, T.W. Integrated analysis of PALSAR/Radar-sat-1 InSAR and ENVISAT altimeter for mapping of absolute water level changes in louisiana wetland. Remote Sensing Environ. **2009**, *113*, 2356–2365.
48. Lu, Z.; Kim, J.W.; Lee, H.K.; Shum, C.K.; Duan, J.; Ibaraki, M.; Akyilmaz, O.; Read, C. Helmand river hydrologic studies using ALOS PALSAR InSAR and ENVISAT altimetry, Mar. Geodesy **2009**, *32* (3), 320–333, doi: 10.1080/ 01490410903094833.

31

A Hybrid Approach for Mapping Salt Marsh Vegetation

Yeqiao Wang
University of Rhode Island

Lin Chen
University of Rhode Island
Northeast Institute of Geography and Agroecology, Chinese Academy of Sciences
University of Chinese Academy of Sciences

Introduction

Salt marshes and the environment are constantly changing with impacts from anthropogenic and nature forces such as sand deposition, hurricane and storm-driven over wash, salt spray, surface water variation, land use and urbanization, climate change and sea level rise (SLR) (Leonardi, et al., 2016; Watson et al, 2017; Thorne et al., 2018). Mapping of salt marsh and tracking the change are among important tasks that always challenge the resource managers. Remote sensing has been applied as one of the primary data sources for mapping and analyzing salt marsh change (Campbell and Wang, 2019). Challenges in salt marsh monitoring and mapping include understanding of the relationship between driving factors and the consequences of the change; determining the current status of salt marshes and the trends of change; determining the effects of tidal change on salt marsh vegetation mapping; and understanding the history of salt marsh change and the migration and resilience analysis toward the decision-making in response to SLR. With the rapid change of remote sensing science and technologies, salt marsh mapping and change analysis are advancing toward finer spatial and spectral resolutions and with improved field data support. The reported studies regularly introduce newly developed and improved algorithms and methods in image classification and analysis (Mui et al., 2015; Sun et al., 2016; Hird et al., 2017; Campbell and Wang, 2018; Watson et al., 2018). Increasingly, the availability of ancillary and geospatial data can provide a larger suite of environmental variables used to describe landscapes (Gross et al., 2009; Kennedy et al., 2009). From the practical consideration, despite the increased exposure and appreciation of improved sensors and data quality, it's always a challenge to incorporate data from existing or previous research efforts and integrate geospatial data from multiple sources into a new mapping and change analysis process. In order to accomplish the change analysis effectively, it is necessary to apply a mapping approach that can produce results quicker, accurate, and consistent with previous mapping efforts and existing data. The routine mapping exercises should be conducted as a way to discover changed areas, e.g., where critical salt marsh vegetation or communities are being lost due to a variety of human-induced or natural forces. Making such discoveries in a timely fashion means the difference between mitigation and recovery or just documenting the loss (Wang et al., 2007). This chapter introduces a hybrid approach, which engages multisource data including Landsat-8 Operational Land Imagers (OLI), land cover data with estuarine vegetation classes from NOAA's Coastal Change Analysis Program (C-CAP) data, and the fine spatial resolution geospatial data as references. The image processes include object-based image analysis (OBIA), stratified image classification, visual interpretation, and manual selection from OBIA-generated segments of salt marsh patches. The purpose of this chapter is for introducing the hybrid approach for salt marsh vegetation mapping only. It has no intension for producing a reference map for the identified area.

Method

This chapter takes the salt marsh vegetation in the lagoon and estuarine environment of the Long Island, New York, as an example site (Figure 31.1). Within the study area includes the Jamaica Bay, the Great South Bay, and the Fire Island National Seashore, all adjacent to the New York City. Mapping of salt marsh and change using very high resolution (VHR) Worldview-2 satellite data have been reported for

FIGURE 31.1 (See color insert.) (**a**) Landsat OLI pseudo color image display (Bands 5, 6, 4 in RGB) acquired in July 2018 shows the locations of Jamaica Bay and Great South Bay along the Long Island shoreline, New York; (**b**) a field site of saltmeadow cordgrass (*S. patens*) in Jamaica Bay (photographed October 2016); (**c**) a field site of smooth cordgrass (*S. alterniflora*) in Jamaica Bay (photographed August 2015); (**d**) a breached site by Hurricane Sandy in Fire Island National Seashore in October 2012 (photographed May 2016).

the Jamaica Bay and the Fire Island National Seashore (Campbell et al., 2017; Campbell and Wang, 2019). The resulted information provided insights and references for the presentation of this hybrid approach in salt marsh vegetation mapping of this chapter.

This approach took the estuarine classes from the 2010 NOAA's C-CAP land cover data, i.e., *estuarine emergent wetland* and *estuarine scrub/shrub wetland*, as the Phase I data. The cloud-free Landsat OLI images, acquired in July 10 and July 19, 2018, were used as the Phase II data. For keeping the classification consistent, we adapted the classification scheme used by the Phase I map. For the interest of this study, we grouped and renamed categories associated with salt marshes into a single category. As the approach was focused on salt marsh vegetation, it excluded other land cover categories defined in the Phase I classification scheme.

The workflow of this hybrid approach involved segmentation process by OBIA using the software eCognition Developer 8.64 (Definiens, 2011). The 2018 OLI image data were processed to produce segmentations based on the spectral similarity of salt marsh patches. Substantial studies have been empirically demonstrated that the OBIA approach can overcome the problem of salt-and-pepper effects by traditional per pixel approaches (Blaschke, 2010; Mao et al., 2018). There are many segmentation procedures, but multi-resolution segmentation used in this study is the most popular and important segmentation algorithm (Jia et al., 2018; Mao et al., 2018). It is a region-growing image segmentation algorithm that merges individual pixels with their most similar neighbors, until the threshold of the within-object heterogeneity is reached (Benz et al., 2004; Elmqvist et al., 2008). Three criteria are defined in the eCognition software to constrain the pixel-growing algorithm, namely scale, shape/color, and compactness/smoothness (Duro et al., 2012). In this study, the paired values of shape/color and compactness/smoothness represent weightings between shape and salt marsh information and compactness and smoothness of the object borders. Generally, color information has higher weight than shape; compactness and smoothness are weighted equally. When the size of a growing region exceeds the threshold defined by the scale parameter, the merging process stops (Li and Shao, 2014). After a "trial and error" process for testing the segmentation parameters, a satisfactory match between image objects and salt marsh features can be achieved.

The stratified digital image classification technique was used to extract approximate areas of salt marsh vegetation from the OLI imagery data using the Phase I data as the reference. This approach requires the support from an existing map, which defined approximate boundaries of salt marshes. The stratified classification helped reduce errors caused by the similar spectral features among some common terrestrial vegetation types. The Watershed Boundary Dataset for 12-Digit Hydrologic Units (HUC-12) polygon boundary was used to define a general study area and then to extract the salt marsh categories as defined from the estuarine wetland categories by NOAA C-CAP 2010 map. We then used visual identification and manual selection through the segmented objects in the processed OLI image, i.e., the Phase II data. For identification and validation purpose, fine resolution images from Google Earth were referenced. The Phase I map was referenced as a thematic layer, and corresponding segmented objects of salt marsh vegetation obtained from OLI imagery were examined. The changes of agreement and disagreement of segmented objects of salt marsh vegetation with the Phase I map were manually selected to modify the delineation and determination of the update.

From visual inspections, a satisfactory match between image objects and salt marsh features was achieved when the scale, shape/color, and compactness/smoothness parameters were set to 150, 0.1/0.9, and 0.5/0.5, respectively. The classification results of salt marsh vegetation were shown in Figures 31.2 and 31.3. In particular, Figure 31.2 illustrates the result derived from manual identification and delineation of OBIA segments on the 2018 OLI images, with the 2010 NOAA C-CAP land cover data defined two estuarine classes, i.e., *estuarine emergent wetland* and *estuarine scrub/shrub wetland*, as the guideline. The manual delineation fine-tuned the C-CAP salt marsh relevant categories with modifications based on the 2018 OLI data. Some of the updates and commission and omission errors of the 2010 C-CAP data were corrected in the process. This was considered as the 2018

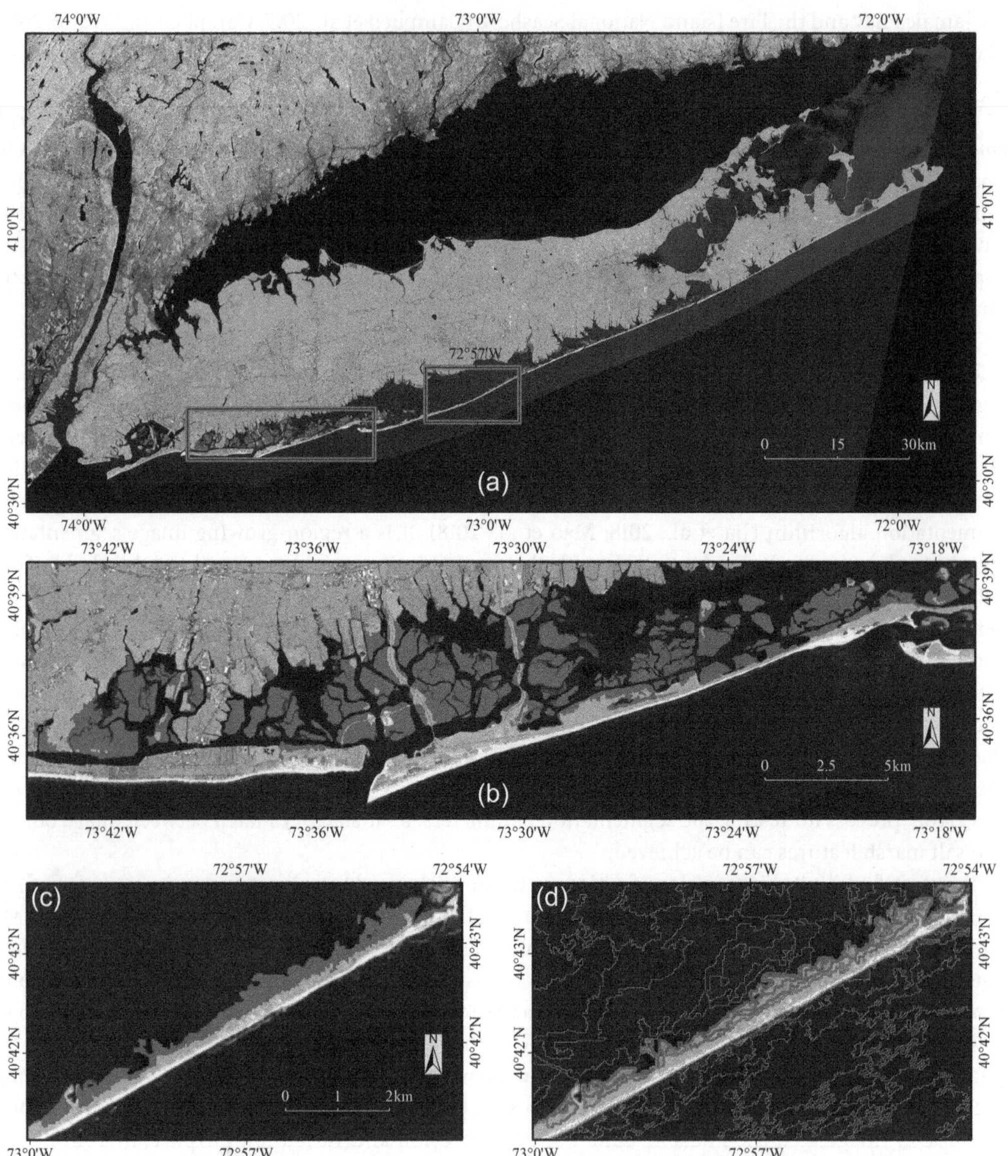

FIGURE 31.2 (**a**) The distribution of salt marsh vegetation of 2018 along the Long Island shoreline, New York with the Landsat OLI image of July 2018 displayed as RGB = Band 4, 5, and 3; (**b**) the distribution of salt marsh vegetation of 2018 along the western section of Great South Bay; (**c**) the distribution of salt marsh vegetation of 2018 along the section of Fire Island National Seashore; (**d**) the segmentation result of the Landsat OLI image of July 2018 with 150, 0.1, and 0.5 for values of scale, shape, and compactness, respectively.

update of the salt marsh map of the study site. Figure 31.3 illustrates a mapping update for a section in the Fire Island National Seashore, where the barrier island was breached by Hurricane Sandy in October 29, 2012. This super storm damaged section affected the salt marsh in the Great South Bay as originally protected by the barrier island. The hybrid approach mapped the salt marsh vegetation with 2018 OLI to update the 2010 C-CAP salt marsh area and provided the update for reference in post-hurricane change analysis and for management practice.

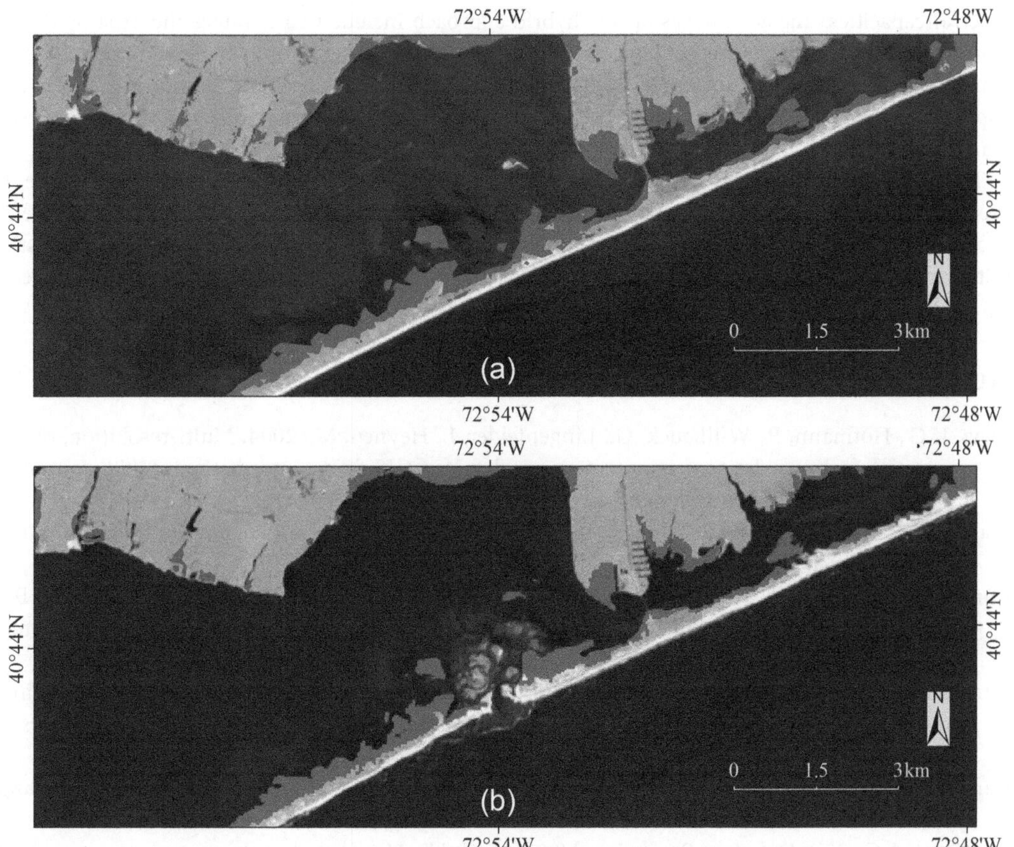

FIGURE 31.3 (**a**) The distribution of salt marsh vegetation of 2010 along the section of Fire Island National Seashore with the Landsat OLI image of July 2010 displayed as RGB = Band 3, 4, and 2; (**b**) the distribution of salt marsh vegetation of 2018 along the section of Hurricane Sandy breached area in Fire Island National Seashore with the Landsat OLI image of July 2018 displayed as RGB = Band 4, 5, and 3.

Conclusion

Monitoring the dynamics of salt marsh vegetation is a critical component for protecting habitats on coastal and barrier islands. In order to accomplish this task effectively, it is necessary to take a hybrid mapping protocol that can produce results quicker, accurate, and consistent with previous and existing mapping efforts. The timely salt marsh mapping should be conducted as a way to discover problem areas where critical communities are being lost due to a variety of human-induced disturbances and natural forces, such as storm surge, hurricane impacts, and SLR.

The object-based segmentation identified salt marsh habitats within the study area at the time of OLI imagery acquisition. The 2010 C-CAP estuarine wetland categories provided reference data. The process fine tune the salt marsh mapping with updated remote sensing imagery. The changed pixels and segments can be labeled as new categories, while unchanged pixels and segments can be kept no change. The multispectral OLI data possess the same spatial resolution as the C-CAP reference data and are valid for updating the changes of salt marshes between the years of the newly acquired remote sensing imagery and the C-CAP data.

With the quick change of remote sensing technology, a practical salt marsh vegetation mapping protocol should be flexible to deal with data from changing sensors, such as finer spatial resolution and spectral capacities. This advantages of this hybrid approach include that it values the existing coastal land cover mapping efforts, keeps the classification scheme consistent, meets the goal of updating salt marsh maps for a particular coastal section, and is valid for change analysis and monitoring. This hybrid approach is practical to work on future mapping replications. This is particularly helpful for coastal management practice.

This approach requests existence of baseline data in GIS format as the Phase I reference to guide the process. Creditable geospatial data from existing mapping exercises could be used as the Phase I reference. On the other hand, this approach may not be appropriate to map areas where there is no existing salt marsh vegetation or land cover map with estuarine categories available in the required level of details for both thematic and spatial resolutions.

References

Benz, U.C., Hofmann, P., Willhauck, G., Lingenfelder, I., Heynen, M., 2004. Multi-resolution, object-oriented fuzzy analysis of remote sensing data for GIS-ready information. *ISPRS Journal of Photogrammetry and Remote Sensing*, 58, 239–258.

Blaschke, T., 2010. Object-based image analysis for remote sensing. *ISPRS International Journal of Photogrammetry and Remote Sensing*, 65, 2–16.

Campbell, A., Wang, Y., Christiano, M., Stevens, S., 2017. Salt marsh monitoring in Jamaica Bay, New York from 2003 to 2013: A decade of change from restoration to hurricane sandy. Remote Sensing, 9, 131. doi:10.3390/rs9020131.

Campbell, A., Wang, Y., 2018. Examining the influence of tidal stage on salt marsh mapping using high spatial resolution satellite remote sensing and topobathymetric LiDAR. *IEEE Transactions on Geoscience and Remote Sensing*, 56, 5169–5176.

Campbell, A., Wang, Y., 2019. High spatial resolution remote sensing for salt marsh mapping and change analysis at Fire Island National Seashore. *Remote Sensing*, 11, 1107.

Definiens, A.G., 2011. Definiens Professional 8.6 User Guide. Munchen, Germany.

Duro, D.C., Franklin, S.E., Dube, M.G., 2012. Multi-scale object-based image analysis and feature selection of multi-sensor earth observation imagery using random forests. *International Journal of Remote Sensing*, 33, 4502–4526.

Elmqvist, B.; Ardo, J., Olsson, L., 2008. Land use studies in drylands: An evaluation of object-oriented classification of very high resolution panchromatic imagery. *International Journal of Remote Sensing*, 29, 7129–7140.

Gross, J. E., Goetz, S. J., Cihlar, J., 2009. Application of remote sensing to parks and protected area monitoring: Introduction to the special issue. *Remote Sensing of Environment*, 113, 1343–1345.

Hird, J.N., DeLancey, E.R., McDermid, G.J., Kariyeva, J., 2017. Google earth engine, open-access satellite data, and machine learning in support of large-area probabilistic wetland mapping. *Remote Sensing*, 9, 1315.

Jia, M., Wang, Z., Zhang, Y., Mao, D., Wang, C., 2018. Monitoring loss and recovery of mangrove forests during 42 years: The achievements of mangrove conservation in China. *International Journal of Applied Earth Observation and Geoinformation*, 73, 535–545.

Kennedy, R.E., Townsend, P.A., Gross, J.E., Cohen, W.B., Bolstad, P., Wang, Y., Adams, P., 2009. Remote sensing change detection tools for natural resource managers: Understanding concepts and tradeoffs in the design of landscape monitoring projects. *Remote Sensing of Environment*, 113, 1382–1396.

Leonardi, N., Ganju, N.K., Fagherazzi, S., 2016. A linear relationship between wave power and erosion determines salt-marsh resilience to violent storms and hurricanes. *Proceedings of the National Academy of Sciences*, 113, 64–68.

Li, X., Shao, G., 2014. Object-based land-cover mapping with high resolution aerial photography at a county scale in midwestern USA. *Remote Sensing*, 6, 11372–11390.

Mao, D., Wang, Z., Wu, J., Wu, B., Zeng, Y., Song, K., Yi, K., Luo, L., 2018. China's wetlands loss to urban expansion. *Land Degradation and Development*, 29, 2644–2657.

Mui, A., He, Y., Weng, Q., 2015. An object-based approach to delineate wetlands across landscapes of varied disturbance with high spatial resolution satellite imagery. *ISPRS Journal of Photogrammetry and Remote Sensing*, 109, 30–46.

Sun, C., Liu, Y., Zhao, S., Zhou, M., Yang, Y., Li, F., 2016. Classification mapping and species identification of salt marshes based on a short-time interval NDVI time-series from HJ-1 optical imagery. *International Journal of Applied Earth Observation and Geoinformation*, 45, 27–41.

Thorne, K., MacDonald, G., Guntenspergen, G., Ambrose, R., Buffington, K., Dugger, B., Freeman, C., Janousek, C., Brown, L., Rosencranz, J., Holmquist, J., Smol, J., Hargan, K., Takekawa, J., 2018. U.S. Pacific coastal wetland resilience and vulnerability to sea-level rise. *Science Advances*, 4, eaao3270.

Wang, Y., Traber, M., Milestead, B., Stevens, S., 2007. Terrestrial and submerged aquatic vegetation mapping in Fire Island National Seashore using high spatial resolution remote sensing data. *Marine Geodesy*, 30, 77–95.

Watson, E.B., Raposa, K.B., Carey, J.C., Wigand, C., Warren, R.S., 2017. Anthropocene survival of southern New England's salt marshes. *Estuaries and Coasts*, 40, 617–625.

Watson, E.B., Hinojosa Corona, A., 2018. Assessment of blue carbon storage by Baja California (Mexico) tidal wetlands and evidence for wetland stability in the face of anthropogenic and climatic impacts. *Sensors*, 18, 32.

32

Remote Sensing of Coastal Wetlands of the Yellow River Delta Region

Lin Chen
Northeast Institute of Geography and Agroecology, Chinese Academy of Sciences
University of Chinese Academy of Sciences
University of Rhode Island

Chunying Ren
Northeast Institute of Geography and Agroecology, Chinese Academy of Sciences

Introduction

Coastal wetlands provide essential ecosystem functions and services, such as carbon sequestration, erosion and water quality control, wildlife habitat, and sustaining biodiversity [1–6]. Due to the poor accessibility of coastal wetlands, the efficiency of traditional field surveys is limited. Active and passive remote sensing-based approaches demonstrate advantages to monitoring coastal wetlands and developing time series maps.

Optical remote sensing data from a number of platforms, e.g., IKONOS, Quickbird, Worldview, ZY-3, SPOT, Sentinel-2, Landsat, and MODIS, are first applied and found favor in imaging horizontal structure of offshore regions [7]. The circulation was observed using Landsat Multispectral Scanner (MSS) images to plot suspended sediment contours in 1974 [8]. Active sensor data, e.g., Radar and LiDAR can capture vertical characteristics, e.g., RADARSAT, Terra-SAR, ALOS PALSAR, and ICESat/GLAS (Ice, Cloud, and Land Elevation Satellite/Geoscience Laser Altimeter System), are unique for the all-weather and day-and-night imaging capability [9–11]. ENVISAT Advanced SAR imagery were applied for the flooded area estimation of Sudd wetland in the Nile River Basin from 2007 to 2011, and the analyses showed reasonable spatiotemporal consistency with available lines of evidence [12]. The total area, canopy height distributions, and aboveground biomass of mangrove forests were estimated in Africa, with a combination of mangrove maps derived from Landsat Enhanced Thematic Mapper Plus (ETM+), LiDAR canopy height estimates from ICESat/GLAS, and elevation data from SRTM (Shuttle Radar Topography Mission) [13]. Although various coastal wetlands characteristics can be monitored and retrieved by multisource remote sensing data, the classification of wetlands types as well as their changes are the basis for further analyses.

Coastal wetlands of the Yellow River Delta in China have been experiencing serious degradations by human activities, especially in recent 40 years [14,15]. It results in serious environmental problems, e.g., wetlands pollution and habitat loss [16,17]. Using multi-temporal Landsat images, it was found that anthropogenic activities led to the reduction of water and sediment discharge, shrinking of tidal flats, and sinking of land surface in the Yellow River Delta [18]. Based on a topographic map, Landsat Thematic Mapper (TM) images, and filed experiments, it was revealed that the area of marsh wetland reduce by 65.09 km^2 during the period from 1986 to 2005. The change had resulted in loss of soil nutrient storage [19]. To address urgent needs of data for implementation of management strategies for wetland conservation and restoration, this article provides an overview of spatiotemporal patterns of coastal wetlands in the Yellow River Delta during from 1976 to 2015 using Landsat images and visual interpretation.

Coastal Yellow River Delta

Yellow River Delta (36°55′–38°16′N, 117°31′–120°32′E) is located in north east of Shandong province, China, and on the southern bank of the Bohai Sea. The total area of the region is about 12,050 km^2 (Figure 32.1), including Dongying City and counties of Kenli, Guangrao, Zhanhua, Lijin, Shouguang, and Bayi. This area has a temperate monsoon climate, with the annual mean temperature and precipitation of 11.9°C and 640 mm, respectively. The interactions between land and ocean in this alluvial plain create a vast area of coastal wetlands. The delta region sustains abundant wetland vegetation (e.g., *Phragmites australis* and *Suaeda heteroptera*) and aquatic organisms, as well as crucial habitat for breeding and wintering endangered migratory bird species (e.g., *Grus japonensis*, *Larus saundersi*). As the critical national agricultural and aquaculture area, the Yellow River Delta has been experiencing continuous large-scale reclamation of coastal marshes and tidal flats. Meanwhile, oil exploitation and expansion of impervious surface settlement also dramatically impacted natural wetlands.

Methods and Data

In this study, mapping of land cover types was carried out through visual interpretation of Landsat imageries acquired by MSS (1976 and 1985), TM (1990, 2000, and 2010), and Operational Land Imager (OLI; 2015). Total 14 land cover types were interpreted, including five types associated with natural

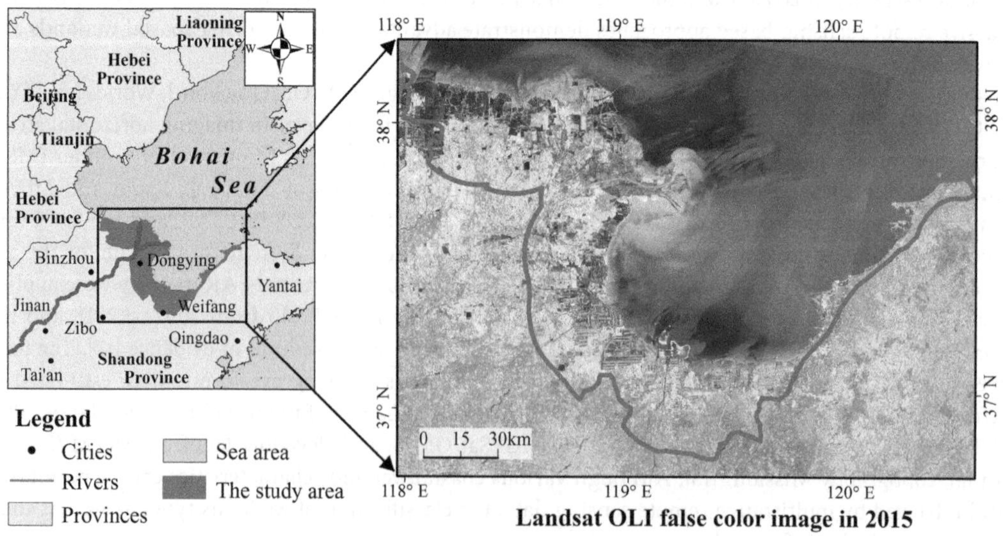

FIGURE 32.1 Location of Yellow River Delta Region.

wetlands, i.e., *river, freshwater marsh, salt marsh, mudflat,* and *beach*, four types associated with man-made wetlands, i.e., *graff, reservoirs and ponds, salt pan,* and *aquaculture*, as well as farmland, forest, grassland, settlement, and unused land.

Fragstats software was used to quantify the landscape structure of coastal wetlands in the Yellow River Delta region. Four metrics were calculated, including patch number, largest patch index, mean patch area, and mean patch fractal dimension. Those indices are important to reflecting various ecological processes which convey information on the spatial heterogeneity, fragmentation, and human disturbance [20]. For example, the patch number of a particular wetland type shows the extent of subdivision or fragmentation of this wetland type. The largest patch index equals the percentage of the landscape comprised by the largest patch and is a measure of dominance. The mean patch area equals the sum area of a particular wetland type divided by the patch number of this wetland type. The mean patch fractal dimension equals two times the logarithm of patch perimeter divided by the logarithm of patch area, which reflects shape complexity.

Areal Changes of Wetlands

The farmland, mudflat, and salt marsh categories were identified as major land cover types in 1976 with their larger areas (Table 32.1). During the past 40 years, farmland and salt marsh have continued to decrease, while settlement, reservoirs and ponds, aquaculture, and salt pan areas have been increasing. In 2015, farmland occupied as the largest area among land cover types across the inland of the region (Figure 32.2). Aquaculture, as the second largest in land cover type, was mostly located in the northwestern Yellow River Delta, within the Dongying City and counties of Lijin and Kenli. Salt pan was situated in the southern part, such as Shouguang and Bayi.

Coastal natural wetlands have declined sharply in each phase from 1976 to 2015, whereas man-made wetlands showed an opposite trend (Figure 32.3). The fastest decrease of natural wetlands occurred from 1990 to 2000, in which all types of natural wetlands shrunk except for beach category (Table 32.2). Man-made wetlands showed the largest increase from 1985 to 1990 (Table 32.2).

TABLE 32.1 The Areas of Various Land Cover Types of the Yellow River Delta Region from 1976 to 2015 (km^2)

Year	1976	1985	1990	2000	2010	2015
River	128	132	134	99	112	108
Freshwater marsh	284	221	211	136	117	66
Salt marsh	991	599	461	324	175	135
Mudflat	1,436	2,007	1,736	1,345	975	843
Beach	20	24	36	64	41	79
Graff	42	102	168	72	62	114
Reservoirs and ponds	10	115	228	301	332	516
Salt pan	111	213	436	640	1,048	1,049
Aquaculture	0.12	40	438	681	977	1,375
Farmland	4,769	4,374	4,247	4,154	3,783	3,517
Forests	288	99	130	76	92	85
Grassland	202	378	209	290	324	181
Settlement	180	486	522	580	780	934
Unused land	283	19	53	65	75	59

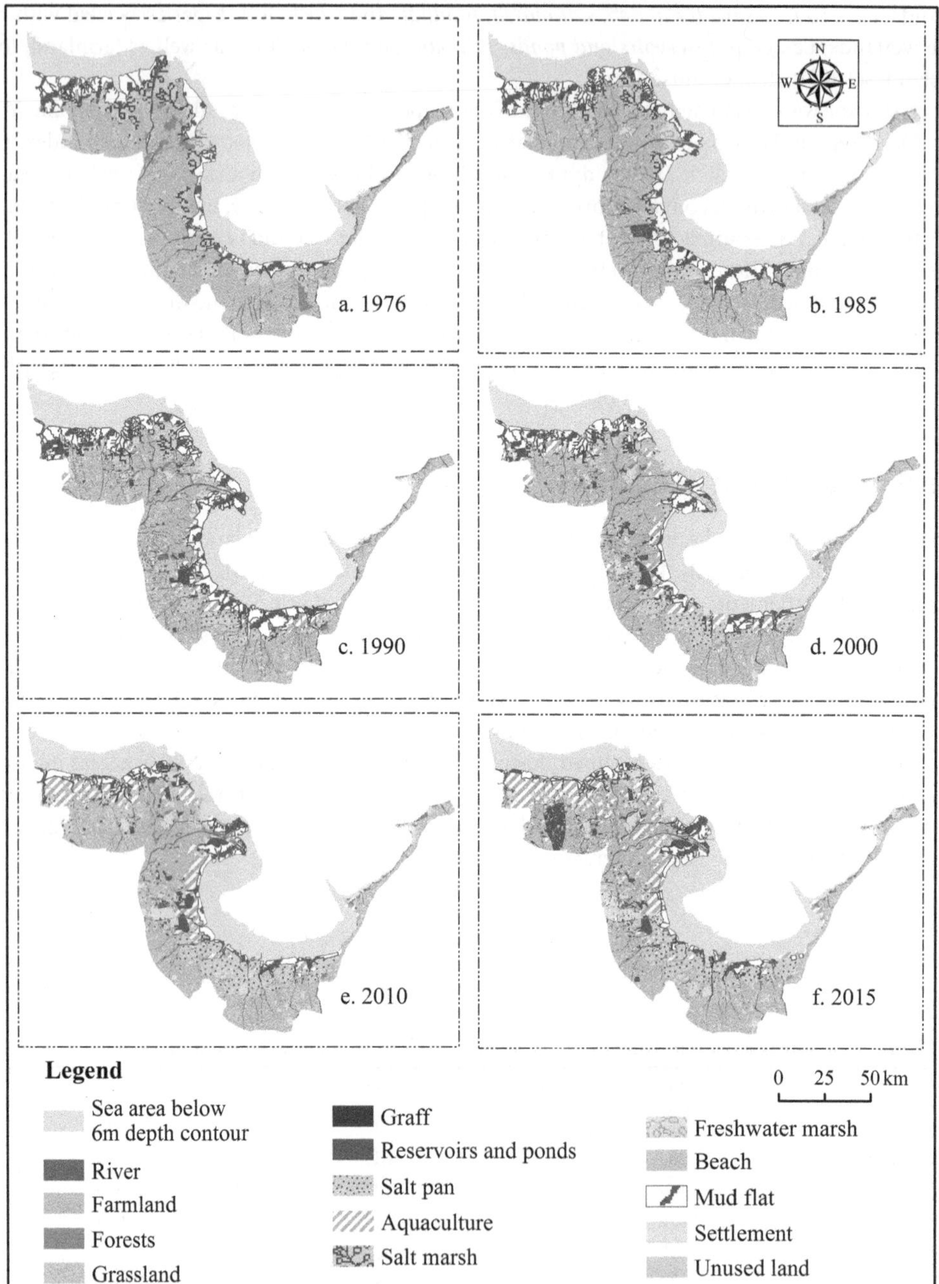

FIGURE 32.2 Land cover maps of coastal Yellow River Delta region from 1976 to 2015.

Conversions between Wetlands and Other Land Cover Types

From 1976 to 2015, farmland, aquaculture, and salt pan types have significantly encroached into natural wetlands areas (Table 32.3). River, freshwater marsh, and salt marsh types have mainly been converted into farmland and aquaculture areas. Meanwhile, beach and mudflat areas have been transferred into salt pan and aquaculture types.

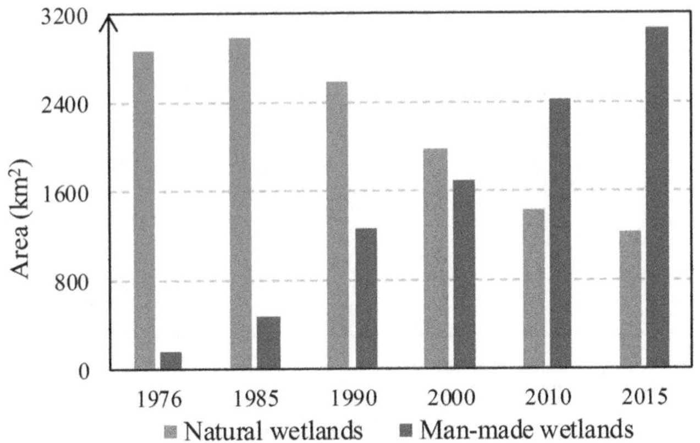

FIGURE 32.3 Areas of the natural and man-made wetlands.

TABLE 32.2 The Dynamic Degrees (%) of Coastal Wetlands in the Yellow River Delta Region from 1976 to 2015

Types	1976–1985	1985–1990	1990–2000	2000–2010	2010–2015
River	0.4	0.3	−2.6	1.4	−0.8
Salt marsh	−4.4	−4.6	−3.0	−4.6	−4.6
Freshwater marsh	−2.5	−0.9	−3.6	−1.4	−8.6
Beach	2.4	9.7	8.0	−3.6	18.8
Mudflat	4.4	−2.7	−2.3	−2.8	−2.7
Natural wetlands	0.5	−0.1	−3.4	−2.8	−2.7
Graff	16.0	13.1	−5.7	−1.4	17.2
Reservoirs and ponds	118.7	19.8	3.2	1.0	11.1
Salt pan	10.1	21.0	4.7	6.4	0.02
Aquaculture	3,709.0	197.1	5.6	4.3	8.1
Man-made wetlands	20.9	34.1	3.3	4.3	5.3

TABLE 32.3 The Transfer Matrix (%) between Wetlands and Other Land Cover Types from 1976 to 2015 in the Yellow River Delta region

2015 \ 1976	River	Freshwater Marsh	Salt Marsh	Beach	Mudflat	Graff	Reservoirs and Ponds	Salt Pan	Aquaculture
Farmland	31.1	27.2	20.2	48.0	3.3	57.3	37.5	3.4	0
Grassland	0	14.6	1.7	0	1.0	1.8	0	0	0
Settlement	0.4	1.8	4.6	2.0	6.7	1.8	8.3	1.9	0
Unused land	0	0	1.4	0	0.5	0	0	0.4	0
River	20.3	0.8	0.9	0	1.8	2.7	0	0	0
Freshwater marsh	0.4	4.8	0.6	0	0.3	0.9	0	0	0
Salt marsh	0.8	0.6	5.8	0	1.9	0	0	0	0
Beach	5.6	0.2	1.1	4.0	1.9	0	0	0.8	0
Mudflat	18.7	8.7	8.7	2.0	28.6	0.9	0	0	0
Graff	2.0	0.5	1.0	20.0	0.9	21.8	4.2	2.6	0
Reservoirs and ponds	0.8	8.1	5.1	0	2.1	3.6	16.7	0	0
Salt pan	5.2	9.5	14.2	12.0	11.5	3.6	16.7	84.6	0
Aquaculture	13.2	20.5	34.1	12.0	39.2	2.7	12.5	6.4	1.0

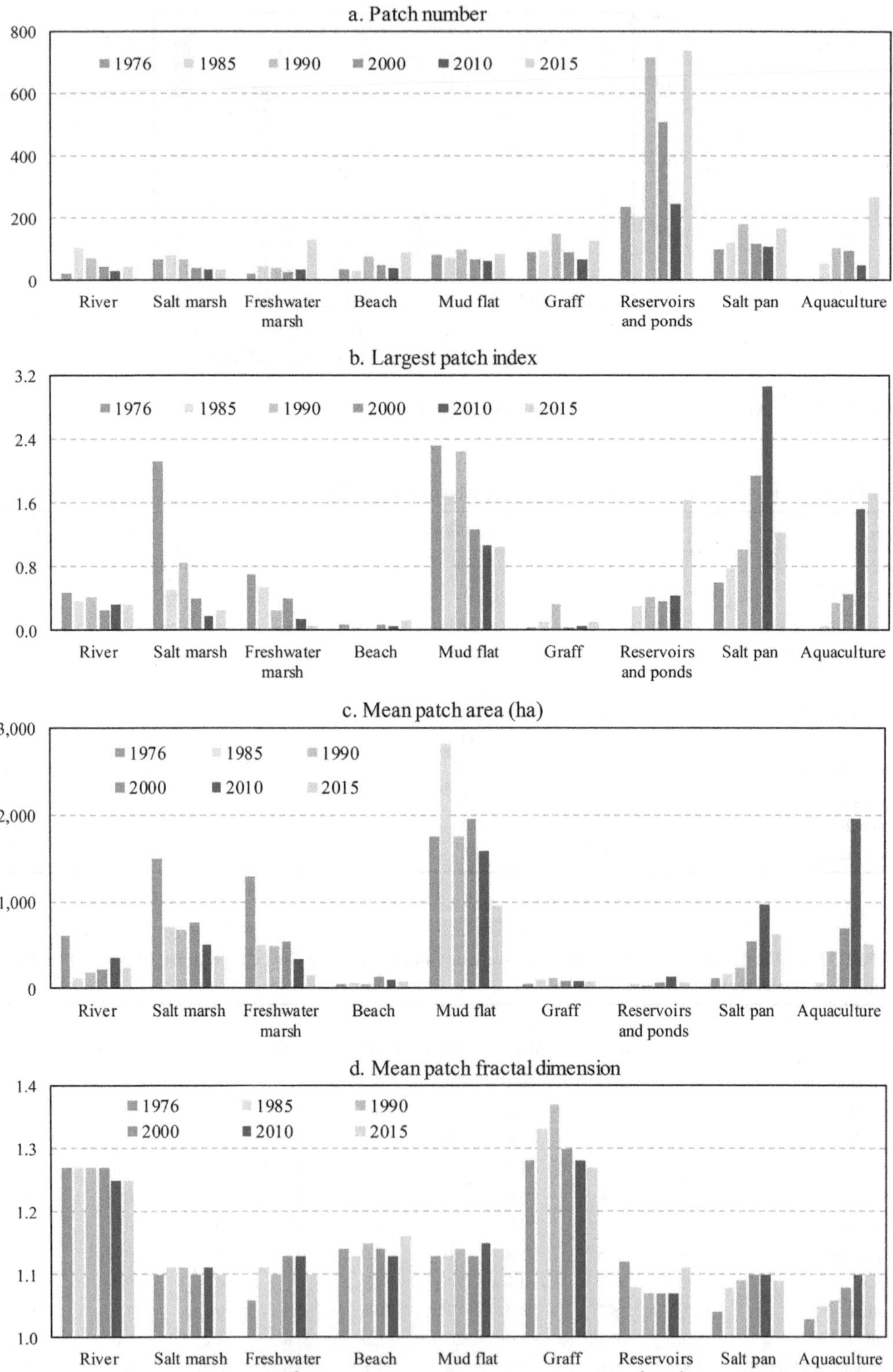

FIGURE 32.4 The four landscape indices of various wetlands types.

Landscape Pattern and Change of Wetlands

For the past 40 years, the patch number of coastal wetlands has been growing (Figure 32.4a). Natural wetland types have been sharply reduced from 1976 to 2015 (Table 32.1), showing the change tendency of the landscape fragmentation for natural wetlands. The largest patch index of man-made wetlands increased, especially for the aquaculture (Figure 32.4b). It showed that there have been more large-scale human activities in fish farming, salt extraction, and water conservancy facilities. Intensive human activities have also been indicated by the enlarged measure of mean patch area metrics (Figure 32.4c). As a regular patch shape is measured and represented by mean patch fractal dimension equals 1, more regular shapes of salt pans and aquaculture sites were developed than other types of coastal wetlands (Figure 32.4d). The result showed that human disturbances were more active in fish farming and salt extraction practices.

Driving Factors of Wetlands Changes

Natural factors and human disturbances together contributed to degradation of natural wetlands in the coastal Yellow River Delta region. The Yellow River has become a seasonal river with frequent occurrence of drying up, which is a typical threat to wetlands shrinkage. The artificial water consumption upstream shows a significant increasing trend and is the main cause of the drying up of the Yellow River [21,22]. Although the Yellow River Delta is renowned for the most sediment-filled plain on Earth, sediment discharge in this region constantly reduces in recent years [23,24]. It causes a slow rate of silting for the Yellow River Delta and a regional erosion recession of mudflat. Besides, the storm surge, frequent disasters at the coastal Yellow River Delta, happened with the seawater intrusion and inundation (Table 32.4). The natural disturbances damaged vegetation, water, and soil quality of coastal wetlands. Furthermore, the repeated break flow of Yellow River makes the lack of the freshwater, and natural wetlands drying.

Human activities had primary impacts on coastal wetland dynamics in the Yellow River Delta region. Mudflat, as a common land resource for reclamation, is exploited to build aquaculture ponds and salt pans in this area (Table 32.5). Natural wetlands also reclaimed to farmland in the Yellow River Delta, with long-time draining and construction. Shengli oil field, e.g., is one of the main petroleum producing bases of China. Its exploitation promoted economic and urbanization developments, as well as the construction of sea wall, road, ports, and housing projects (Table 32.6). However, new oil fields exploited

TABLE 32.4 The Distance and Flooded Area because of Seawater Intrusion in Dongying City

Year	1980	1982	1992	1997	2003
Seawater intrusion distance (km)	50	10	20	—	—
Seawater inundation area (km^2)	23	90	960	1,417	410

TABLE 32.5 The Aquaculture Areas (km^2) of Mudflat in Donging City

Year	1984	1990	1995	2002
Dongying district	0	7	7	52
Hekou district	0	7	17	19
Kenli county	0	18	13	67
Lijin county	0	15	12	38
Guangrao county	0	6	1	19
Shengli oil field	0	21	18	19
Total	0	73	68	381

TABLE 32.6 Social Economic Indicators of Shengli Oil Field

Year	1996	1997	1998	1999	2000	2001	2002	2003
Per capita living area (m^2)	404.7	402.2	206.1	409.2	411.7	476.6	417.2	495.2
Per capita road area (m^2)	142.9	143.9	144.1	143.5	145.7	146.7	151.1	152.8
Average annual growth rate of GDP (%)	34.7	9.0	13.7	−12.1	13.8	14.7	16.9	23.9
Urbanization rate (%)	18.4	18.3	19.5	20.4	20.9	31.9	32.3	34.9
Population density (people/km^2)	94.3	94.6	95.3	95.7	95.9	96.2	96.2	96.4

after 1984 are all located in the coastal Yellow River Delta. Reclamation, oil exploitation, and settlement construction directly result in natural wetlands loss and then water pollution and habitats loss.

Conclusions

In the Yellow River Delta, natural wetlands decline by 1,627 km^2 with landscape fragmentation, which turn into farmlands, aquaculture, and salt pan during 1976 to 2015. In contrast, man-made wetlands convert from salt marsh and mudflat increased by 2,891 km^2 with large-scale precision management. Natural factors tide erosion including ever-decreasing sedimentation and the constant drying up of Yellow River lead to natural wetland shrinkage. Primary driving forces for natural wetlands degradation are human activities such as reclamation, oil exploitation, and settlement construction.

References

1. Duarte, C.M.; Losada, I.J.; Hendriks, I.E.; Mazarrasa, I.; Marbà, N. The role of coastal plant communities for climate change mitigation and adaptation. *Nat. Clim. Change.* 2013, 3, 961–968.
2. Jankowski, K.L.; Törnqvist, T.E.; Fernandes, A.M. Vulnerability of Louisiana's coastal wetlands to present-day rates of relative sea-level rise. *Nat. Commun.* 2017, 8, 14792.
3. Barbier, E.B.; Hacker, S.D.; Kennedy, C.; Koch, E. W.; Stier A.C.; Silliman, B.R. The value of estuarine and coastal ecosystem services. *Ecol. Monogr.* 2011, 81, 169–193.
4. Costanza, R.; Pérez-Maqueo, O.; Martinez, M.L.; Sutton, P.; Anderson, S.J.; Mulder, K. The value of coastal wetlands for hurricane protection. *Ambio: A J. Human Environ.* 2008, 37, 241–248.
5. Mcleod, E.; Chmura, G.L.; Bouillon, S.; Salm, R.; Björk, M.; Duarte, C.M; Lovelock, C.E; Schlesinger, W.H; Silliman, B.R. A blueprint for blue carbon: Toward an improved understanding of the role of vegetated coastal habitats in sequestering CO_2. *Front. Ecol. Environ.* 2011, 9, 552–560.
6. Mitsch, W.J.; Gosselink, J.G. The value of wetlands: Importance of scale and landscape setting. *Ecol. Econ.* 2000, 35, 25–33.
7. Mahdavi, S.; Salehi, B.; Granger, J.; Amani, M.; Brisco, B.; Huang, W. Remote sensing for wetland classification: A comprehensive review. *GISci. Remote Sens.* 2018, 55, 623–658.
8. Rouse, L.J.; Coleman, J.M. Circulation observations in the Louisiana Bight using Landsat imagery. *Remote Sens. Environ.* 1976, 5, 55–66.
9. Kwoun, O.; Lu, Z. Multi-temporal RADARSAT-1 and ERS backscattering signatures of coastal wetlands in southeastern Louisiana. *Photogramm. Eng. Rem. Sens.* 2009, 75, 607–617.
10. Kandus, P.; Minotti, P.G.; Morandeira, N.C.; Grimson, R.; Trilla, G.G; González, E.B.; Martín, L.C. Gayol, M.P. Remote sensing of wetlands in South America: Status and challenges. *Int. J. Remote Sens.* 2018, 39, 993–1016.
11. Mohammadimanesh, F.; Salehi, B.; Mahdianpari, M.; Brisco, B.; Motagh, M. Multi-temporal, multi-frequency, and multi-polarization coherence and SAR backscatter analysis of wetlands. *ISPRS J. Photogramm. Remote Sens.* 2018, 142, 78–93.

12. Wilusz, D.C.; Zaitchik, B.F.; Anderson, M.C.; Hain, C.R.; Yilmaz, M.T.; Mladenova, I.E. Monthly flooded area classification using low resolution SAR imagery in the Sudd wetland from 2007 to 2011. *Remote Sens. Environ.* 2017, 194, 205–218.
13. Fatoyinbo, T.E.; Simard, M. Height and biomass of mangroves in Africa from ICESat/GLAS and SRTM. *Int. J. Remote Sens.* 2013, 34, 668–681.
14. Liu, G.; Zhang, L.; Zhang, Q.; Musyimi, Z.; Jiang, Q. Spatio-Temporal dynamics of wetland landscape patterns based on remote sensing in Yellow River delta, China. *Wetlands.* 2014, 34, 787–801.
15. Kuenzer, C.; Ottinger, M.; Liu, G.; Sun, B.; Baumhauer, R.; Dech, S. Earth observation-based coastal zone monitoring of the Yellow River Delta: dynamics in China's second largest oil producing region over four decades. *Appl. Geogr.* 2014, 55, 92–107.
16. Wang, M.; Qi, S.; Zhang, X. Wetland loss and degradation in the Yellow River Delta, Shandong Province of China. *Environ. Earth Sci.* 2012, 67, 185–188.
17. Xu, X.; Lin, H.; Fu, Z. Probe into the method of regional ecological risk assessment—A case study of wetland in the Yellow River Delta in China. *J. Environ. Manage.* 2004, 70, 253–262.
18. Fan, H.; Huang, H.; Zeng, T. Impacts of anthropogenic activity on the recent evolution of the Huanghe (Yellow) River Delta. *J. Coastal Res.* 2006, 22, 919–929.
19. Huang, L.; Bai, J.; Chen, B.; Zhang, K.; Huang, C.; Liu, P. Two-decade wetland cultivation and its effects on soil properties in salt marshes in the Yellow River Delta, China. *Ecol. Inform.* 2012, 10, 49–55.
20. McGarigal, K.; Marks, B.J. FRAGSTATS: Spatial pattern analysis program for quantifying landscape structure. Portland (OR): USDA Forest Service, Pacific Northwest Research Station; General Technical Report PNW-GTR-351, 1995.
21. Cong, Z.; Yang, D.; Gao, B.; Yang, H.; Hu, H. Hydrological trend analysis in the Yellow River Basin using a distributed hydrological model. *Water Resour. Res.* 2009, 45, W00A13.
22. Zhang, K.; Pan, S.; Zhang, W.; Xu, Y.; Cao, L.; Hao, Y.; Wang, Y. Influence of climate change on reference evapotranspiration and aridity index and their temporal-spatial variations in the Yellow River Basin, China, from 1961 to 2012. *Quatern. Int.* 2015, 4, 380–381.
23. Wang, S.; Fu, B.; Piao, S.; Lü, Y.; Ciais, P.; Feng, X.; Wang, Y. Reduced sediment transport in the Yellow River due to anthropogenic changes. *Nat. Geosci.* 2016, 9, 38–41.
24. Wei, Y.; Jiao, J.; Zhao, G.; Zhao, H.; He, Z.; Mu, X. Spatial-temporal variation and periodic change in streamflow and suspended sediment discharge along the mainstream of the Yellow River during 1950–2013. *Catena.* 2016, 140, 105–115.

33

Coastal Change: Remote Sensing of Wetlands in Yangtze River Estuary

Lin Chen
Northeast Institute of Geography and Agroecology, Chinese Academy of Sciences
University of Chinese Academy of Sciences
University of Rhode Island

Chunying Ren
Northeast Institute of Geography and Agroecology, Chinese Academy of Sciences

Introduction

The coastal area, a small fraction of the total land area of the Earth, disproportionately supports more than one-third of the population in the world [1,2]. Wetlands in those areas are vulnerable and sensitive to climatic changes, sea level dynamics, and human threats [3,4]. Management of coastal wetlands presents challenges from the growing global population and changing climate [5]. It requires for the quick evaluation on the status and trends of changing coastal wetlands over larger areas. Thus, satellite remote sensing techniques have been widely used and successfully improved management capabilities in coastal wetlands.

Abundant vital basins and delta regions worldwide have been monitored. The spectral response of Landsat-5 Thematic Mapper (TM) images was evaluated for the interpretation of different wetlands and associated environments at the mouth of the Amazon River by a spectral angle mapper classifier [6]. Dual-season and full-polarimetric Radarsat-2 images to map vegetation of the Amazon várzea wetlands in 2011, with Producer's and User's accuracies between 80% and 90% [7]. In Mississippi River region, the landscape-level phenology of marshes was quantified using Landsat-derived Normalized Difference Vegetation Index of 1999, 2005, and 2007 [8]. Synthetic aperture radar (SAR) data acquired by the Uninhabited Aerial Vehicle SAR between 2009 and 2012 were adopted to quantify land loss along the marsh edge and found that petroleum exposure substantially increased shoreline in the Mississippi River Delta [9]. As for Africa, the changes in distribution and decrease area of mangrove forests along the mainland Tanzania coast were extracted by the interpretation of both 1988–1990 Landsat TM and 2000 Enhanced Thematic Mapper Plus (ETM^+) images [10]. Geomorphic changes along the Nile Delta coastline during 1945–2015 were assessed and observed erosion processes occurring after the construction of the Aswan High Dam, using topographic maps and Landsat satellite images [11]. In Asia, Landsat TM and ETM+ from 1996 to 2013 were applied to estimate suspended sediment dynamics in mega deltas of the Mekong floodplains [12].

Yangtze Estuary experiences intensive reclamation and natural wetlands loss due to booming economic development near the Shanghai Municipality. It has also caused several wetland environmental problems [13,14]. Upstream dam construction, estuarine engineering, land reclamation, and ecological engineering led to 35% wetland loss based on the Landsat data from 2000 to 2010 in the Yangtze Delta [15]. Based on GF-1 satellite, Landsat, and SPOT-7 data, it was found that natural wetlands decreased 163 km^2 from 1979 to 2016, due to the conversion of coastal wetland into reservoirs, aquaculture ponds, and paddy fields [16]. With nearly 50 years of continuous imagery acquisition, the Landsat sensors recorded invaluable long-time series data for monitoring of coastal changes. This article provides an overview of spatiotemporal patterns of coastal wetland and reclamation in the Yangtze Estuary from 1960s to 2015 using Landsat and other type of historical data.

Yangtze Estuary

The Yangtze Estuary is situated between 30°50′55″N–32°19′12″N and 120°24′02″E–122°05′20″E as the area where the Yangtze River injects into the East China Sea (Figure 33.1). It is the third largest alluvial estuary of the world. The estuary region provides fresh water and developable waterpower resources in China. As one of the most densely populated regions on Earth, it has maintained the highest economic growth speed and urbanization scale of China in the past decades. This region is characterized by subtropical monsoon climate and sustains main wetland vegetation types such as *Phragmites autralis, Spartina alterniflora*, and *Scripus mariqueter.* It is the only avenue for upstream and catadromous migration of various species. The area is also the vital stopover site and wintering habitat for migratory waterfowls in the Asia-Pacific region.

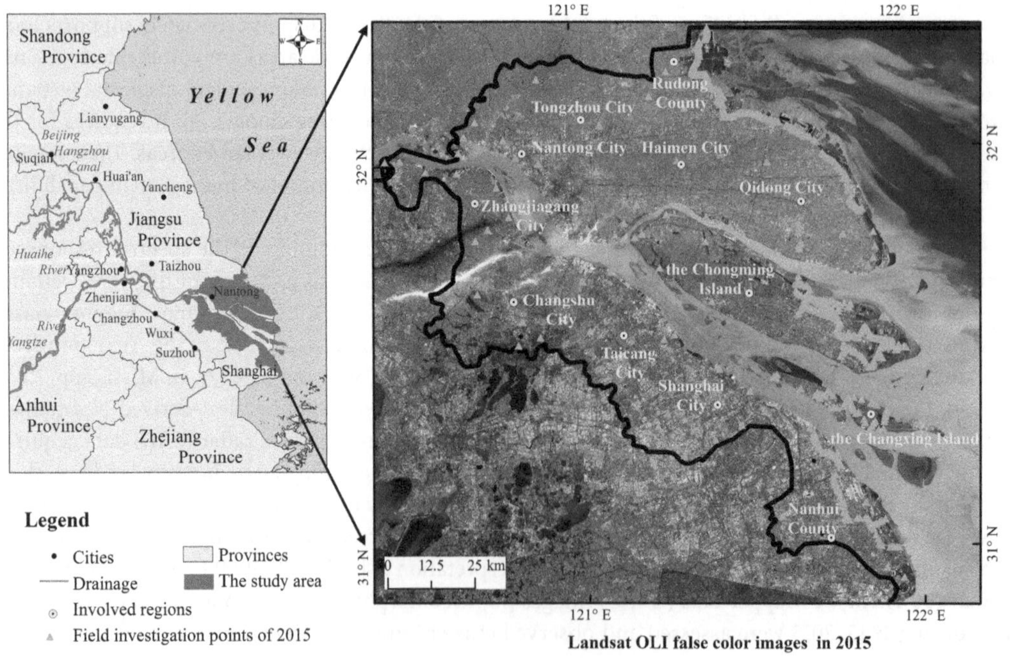

FIGURE 33.1 The location of the study area and field investigation points.

Methods and Data

Land cover types in this area include forest, grassland, wetlands, farmland, settlement, and other land. Wetlands are divided into natural and man-made wetlands. Natural wetlands contain lake, salt marsh, freshwater marsh, mudflat, and beach, and man-made wetlands cover canal, reservoir/pond, aquaculture pond, and salt field. In this study, the boundary of the 1960s' land use was used as the baseline, and the areal change of the Yangtze Estuary in each phase compared with the basic region was divided into reclamation areas and natural wetlands. The land use types referenced in this study were obtained through visual interpretation and manual delineation from Landsat images (Figure 33.2), except that land cover types of 1960s were derived from topographic maps (1: 100,000) and historical local land use maps.

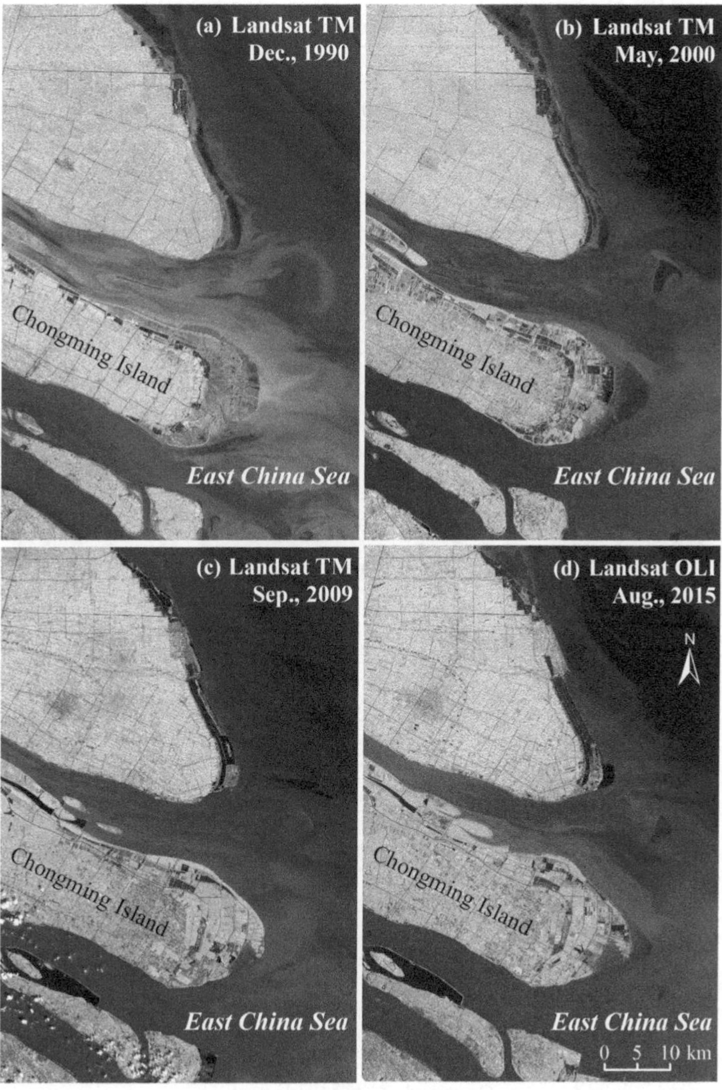

FIGURE 33.2 Landsat images of a section of the study area. The TM and OLI images are in Bands 5, 4, 3 and 6, 5, 4 in RGB displays, respectively.

Change Detection of Wetlands and Reclamation

In 2015, the area of coastal wetlands was 10.82% of the total study area. The natural wetlands accounted for about 37% of the coastal wetlands (Table 33.1). Natural wetlands were distributed mainly in the island area. Man-made wetlands were widely dispersed along the coastal line and in nearby cities (Figure 33.3). Most of the sea water and land reclamation were considered as man-made wetlands, settlements, and farmland, typically in the Chongming Island and the east coast of the Shanghai Municipality. From the 1960s to 2015, the net area of natural wetlands declined by 574.3 km^2, while man-made wetlands increased by 553.6 km^2. The area for land reclamation was 543.9 km^2. Wetlands experienced the largest change in the period of 1990–2000 when the land reclamation rate was about 15.7 km^2/yr. Man-made wetlands, settlement, and reclamation continued to increase, indicating that the Yangtze Estuary was subjected to continuous impacts by human activities.

TABLE 33.1 Areal Change of Coastal Wetlands and Reclamation in the Yangtze Estuary

Classes	Area in 1960s (km^2)	Area in 2015 (km^2)	Area Change (km^2)				
			1960s–1980	1980–1990	1990–2000	2000–2010	2010–2015
Natural wetlands	722.5	411.3	−249.3	−28.1	−132.8	−110.4	−53.8
Man-made wetlands	146.0	699.6	21.1	179.3	246.8	18.7	87.7
Reclamation areas	0	543.7	46.9	24.2	157.0	276.5	39.3

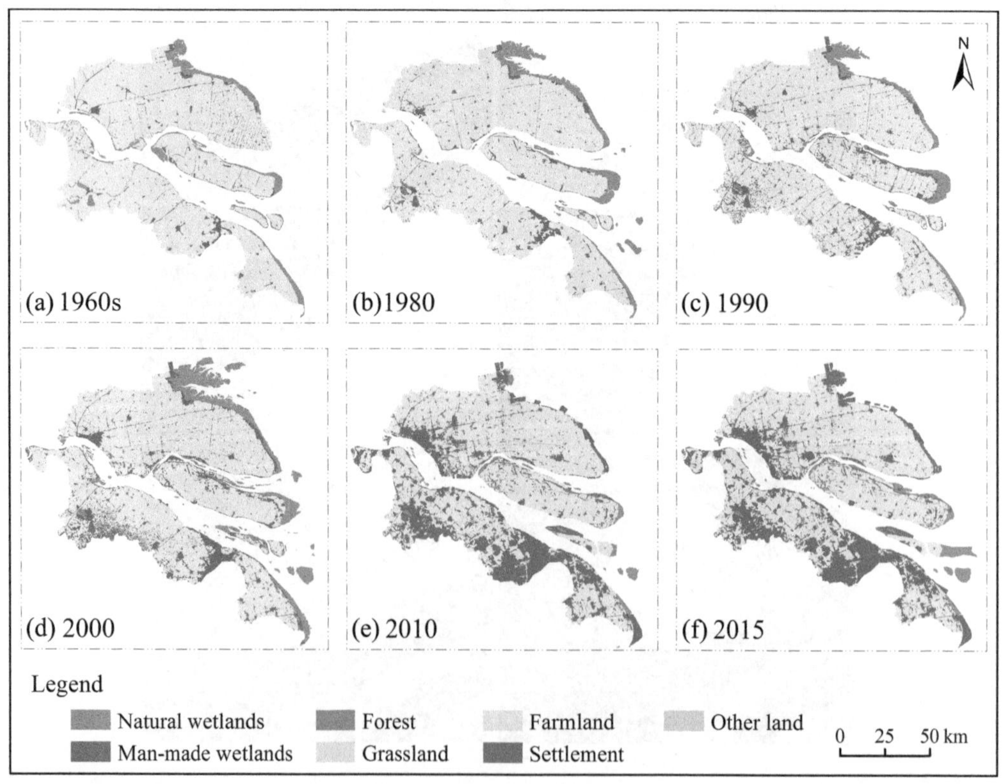

FIGURE 33.3 Coastal wetlands in the Yangtze Estuary.

Conversions of Wetlands and Reclamation

The area of different conversions between wetlands and other land cover types was calculated based on the land cover transfer matrix [17] (Figure 33.4). Conversions occurred primarily between farmland and wetlands and secondarily between settlement and wetlands. The anthropogenic disturbance to coastal wetlands was more intense in the Chongming Island and Changshu City, as well as the east coast of Shanghai Municipality.

The area changes of different land use types of reclamation during five study phrases are shown in Table 33.2. Farmland, settlement, and man-made wetlands showed more variation than others, and their temporal dynamics were especially remarkable in 2000–2010. The dominant reclamation change

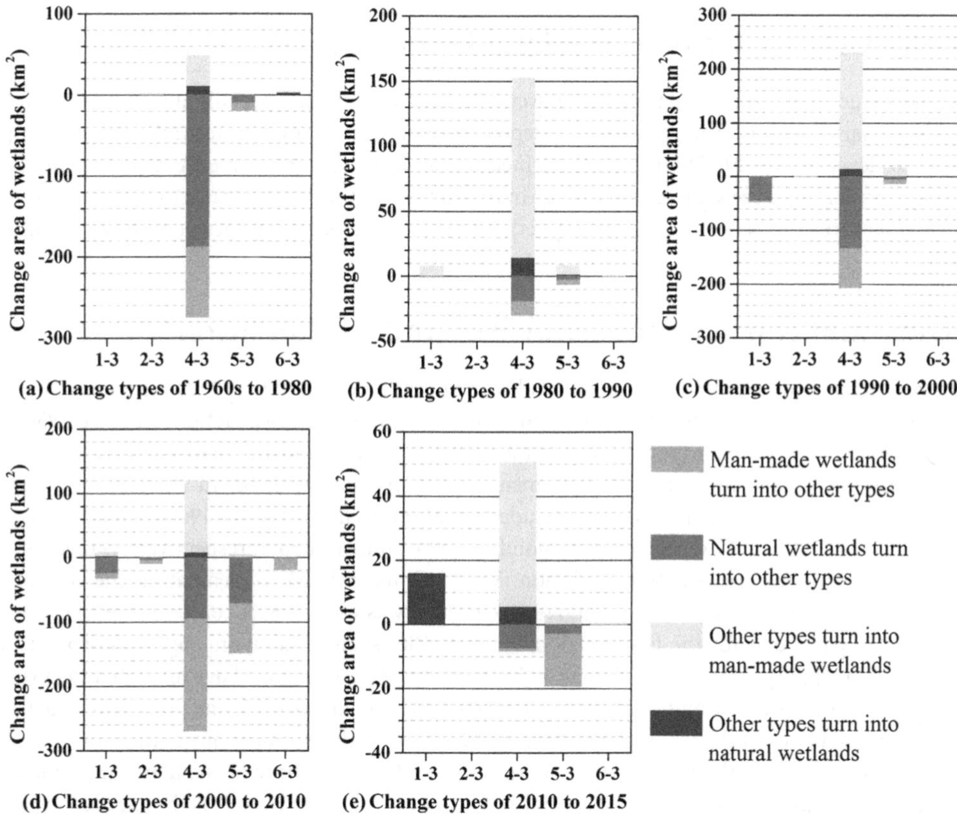

FIGURE 33.4 The change area between coastal wetlands and other land use types (1–3: between forest and wetlands, 2–3: between grassland and wetlands, 4–3: between farmland and wetlands, 5–3: between settlement and wetlands, 6–3: between other land and wetlands) in the Yangtze Estuary.

TABLE 33.2 The Change Area among Different Types of Reclamation in the Yangtze Estuary (km^2)

Phrase	Forest	Grassland	Farmland	Settlement	Other Land	Man-Made Wetlands
1960–1980	0.00	0.11	30.20	0.83	0.01	15.74
1980–1990	0.03	−0.11	−2.38	0.95	−0.01	25.68
1990–2000	13.31	1.09	80.57	3.73	0.00	58.28
2000–2010	16.88	9.90	106.27	85.05	0.96	57.39
2010–2015	0.47	0.00	−23.84	18.48	−0.61	44.75

occurred among aquaculture/salt extraction, farming, and construction during the study period. Reclamation of settlement and man-made wetlands escalated during the period from 1960s to 2015, but farmland reclamation varied (Table 33.2).

When a certain landscape type has the same growth or decline rate in every direction, its centroid remains almost invariant; otherwise, its centroid moves to the direction in which the landscape gains or reduces significantly [18]. Overall, the centroids of coastal wetlands and reclamation all shifted southward during the study period. The centroid of natural wetlands generally moved toward the East China Sea, indicating natural wetlands declined faster. However, the centroid movement of man-made wetlands was about equal in all directions and the inland. Meanwhile, the centroid path for reclamation showed that reclamation in the Yangtze Estuary was more intensive toward Shanghai Municipality.

Factors Driving Coastal Wetland Dynamics

Substantial studies have revealed that coastal wetlands are threatened, especially at lower elevation (e.g., Yangtze Estuary), because the rate mineral deposition and organic matter accumulation fail to pace up with sea level rise [19,20]. Hydrological engineering projects, e.g., the Three Gorges Dam and South Water to North, reduce the sediment discharge in the Yangtze Estuary. It has caused a slow rate of silting and a regional erosion recession. The coastal wetlands would be eroded, degenerated, and even vanished [21,22].

History and Impacts of Costal Wetland Reclamation

China has reclamation history in the past decades, including salt field in the 1950s, agriculture between 1960s and 1970s, aquaculture between 1980s and 1990s, and industrial constructions since the 1950s [23]. As for Yangtze Estuary, major types of reclamation have been farmland and man-made wetlands in the 1980s, 1990s, and 2000s and farmland, settlement, and man-made wetlands in 2010 and 2015, respectively. Main types of man-made wetlands included canals and salt field in the 1960s, aquaculture pond and salt field in 1980 and 1990, aquaculture pond in 2000, 2010, and 2015, respectively. It revealed that aquaculture dominated reclamation evolution. When China started implementing the reform and open-up policy in the late 1970s, the Yangtze River Delta was chosen as a coastal economic development zone, in particular for the first tier of open port cities such as Lianyungang, Nantong City, and Shanghai Municipality. Increased reclamation in the Yangtze Estuary may have mitigated the pressure of a high population density and expanded industrial development space. Many ports and wharfs have been constructed to meet the trading and shipping needs. In addition, this is one of the most developed tourism areas, especially the Chongming Island, so the settlements have been significantly increased (Table 33.2). Resource demands for rapid social and economic blossom as well as the speeded urbanization process stimulated multiple high-intensity reclamation activities. All contributed to the pattern of expansion of reclamation.

Conclusions

In the Yangtze Estuary, the net area of natural wetlands declined by 574.3 km^2, mainly turning into farmlands and settlement from 1960s to 2015. In contrast, man-made wetlands converted from farmlands and gained from reclamation, which increased by 553.6 km^2. The reclamation area expanded by 543.9 km^2 toward Shanghai Municipality. The Chongming Island and the east coast of Shanghai Municipality were the typical regions subject to the shrinkage of natural wetlands, growth of man-made wetlands, and reclamation activities. It demonstrated that anthropogenic driving forces dominated the change in declining of natural wetlands and reclamation boom in the Yangtze Estuary region.

References

1. Giosan, L; Syvitski, J; Constantinescu, S.; Day, J. Climate change: Protect the world's deltas. *Nature* 2014, 516, 31–33.
2. Tian, B.; Wu, W.; Yang, Z.; Zhou, Y. Drivers, trends, and potential impacts of long-term coastal reclamation in China from 1985 to 2010. *Estuarine, Coastal Shelf Sci.* 2016, 170, 83–90.
3. Cohen, M.C.L.; Behling, H.; Lara, R.J.; Smith, C.B.; Matos, H.R.S.; Vedel, V. Impact of sea-level and climatic changes on the Amazon coastal wetlands during the late Holocene. *Veg. Hist. Archaeobot.* 2009, 18, 425.
4. Twilley, R.R.; Bentley, S.J.; Chen, Q.; Edmonds, D.A.; Hagen, S.C.; Lam, N.S.N.; Willson, C.S.; Xu. K.; Braud, D.W.; Peele, R.H.; McCall, A. Co-evolution of wetland landscapes, flooding, and human settlement in the Mississippi River Delta Plain. *Sustainability Sci.* 2016, 11, 711–731.
5. McCarthy, M.J.; Colna, K.E.; El-Mezayen, M.M.; Laureano-Rosario, A.E.; Méndez-Lázaro, P.; Otis, D.B.; Toro-Farmer, G.; Vega-Rodriguez, M.; Muller-Karger, F.E. Satellite remote sensing for coastal management: A review of successful applications. *Environ. Manage.* 2017, 60, 323–339.
6. Cardoso, G.F.; Jr., C.S.; Souza-Filho, P.W.M. Using spectral analysis of Landsat-5 TM images to map coastal wetlands in the Amazon River mouth, Brazil. *Wetl. Ecol. Manag.* 2014, 22, 79–92.
7. Furtado, L.F.A.; Silva, T.S.F.; Novo, E.M.L.M. Dual-season and full-polarimetric C band SAR assessment for vegetation mapping in the Amazon várzea wetlands. *Remote Sens. Environ.* 2016, 174, 212–222.
8. Mo, Y.; Momen, B.; Kearney, M.S. Quantifying moderate resolution remote sensing phenology of Louisiana coastal marshes. *Ecol. Model.* 2015, 312, 191–199.
9. Rangoonwala, A.; Jones, C.E.; Ramsey III, E.R. Wetland shoreline recession in the Mississippi River Delta from petroleum oiling and cyclonic storms. *Geophys. Res. Lett.* 2016, 43, 11652–11660.
10. Wang, Y.; Bonynge, G.; Nugranad, J.; Traber, M.; Ngusaru, A.; Tobey, J.; Hale, L. Bowen, R.; Vedast, M. Remote sensing of mangrove change along the Tanzania coast. *Mar. Geod.* 2003, 26 (1–2), 35–48.
11. Darwish, K.; Smith, S.E.; Torab, M.; Monsef, H.; Hussein, O. Geomorphological changes along the Nile Delta coastline between 1945 and 2015 detected using satellite remote sensing and GIS. *J. Coast. Res.* 2017, 33, 786–794.
12. Dang, T.D.; Cochrane, T.A.; Arias, M.E. Quantifying suspended sediment dynamics in mega deltas using remote sensing data: A case study of the Mekong floodplains. *Int. J. Appl. Earth Obs.* 2018, 68, 105–115.
13. Cui, J.; Liu, C.; Li, Z.; Wang, L.; Chen, X.; Ye, Z.; Fang, C. Long-term changes in topsoil chemical properties under centuries of cultivation after reclamation of coastal wetlands in the Yangtze Estuary, China. *Soil Till. Res.* 2012, 123, 50–60.
14. Sun, Z.; Sun, W.; Tong, C.; Zeng, C.; Yu, X.; Mou, X. China's coastal wetlands: Conservation history, implementation efforts, existing issues and strategies for future improvement. *Environ. Int.* 2015, 79, 25–41.
15. Zhang, L.; Wu, B.; Yin, K.; Li, X.; Kia, K.; Zhu, L. Impacts of human activities on the evolution of estuarine wetland in the Yangtze Delta from 2000 to 2010. *Environ. Earth Sci.* 2015, 73, 435–447.
16. Sun, N.; Zhu, W.; Cheng, Q. GF-1 and Landsat observed a 40-year wetland spatiotemporal variation and its coupled environmental factors in Yangtze River estuary. *Estuarine, Coastal Shelf Sci.* 2018, 207, 30–39.
17. Wang, J.; Su, P.; Elena, A.G. Land cover change characteristics of north-south transect in Northeast Asia from 2001 to 2012. *J. Resour. Ecol.* 2016, 7, 36–43.
18. Jia, M.; Wang, Z.; Zhang, Y.; Ren, C.; Song, K. Landsat-based estimation of mangrove forest loss and restoration in Guangxi Province, China, influenced by human and natural factors. *IEEE J-STARS.* 2015, 8, 311–323.

19. Cui, L.; Ge, Z.; Yuan, L.; Zhang, L. Vulnerability assessment of the coastal wetlands in the Yangtze Estuary, China to sea-level rise. *Estuarine, Coastal Shelf Sci.* 2015, 156, 42–51.
20. Lovelock, C.; Cahoon, D.; Friess, D.A.; Guntenspergen, G.R.; Krauss, K.W.; Reef, R.; Rogers, K.; Saunders, M.L.; Sidik, F.; Swales, A.; Saintilan, N.; Thuyen, L.X.; Triet, T. The vulnerability of Indo-Pacific mangrove forests to sea-level rise. *Nature* 2015, 526, 559–563.
21. Song, C.; Wang, J. Erosion-accretion changes and controlled factors of the submerged delta in the Yangtze Estuary in 1982–2010. *Acta Geographica Sinica.* 2014, 69, 1683–1696.
22. Yang, S.; Zhang, J.; Zhu, J.; Smith, J.P.; Dai, S.; Gao, A.; Li, P. Impact of dams on Yangtze River sediment supply to the sea and delta intertidal wetland response. *J. Geophys. Res.* 2005, 110, F03006.
23. Wang, W.; Liu, H.; Li, Y.; Su, J. Development and management of land reclamation in China. *Ocean Coast. Manage.* 2014, 102, 415–425.

34

Remote Sensing of Mangrove Forests in an Environment of Global Change

Wilfrid Rodriguez
Smithsonian Environmental Research Center

Introduction

Mangrove forests are unique terrestrial ecosystems living in the intertidal zone of subtropical and tropical regions (30°N and 30°S latitude), with the largest concentration between 5°N and 5°S latitude [1]. Figure 34.1 shows the global geographical distribution of mangrove forests, and Figure 34.2 shows some examples of mangrove forests found around the world. The intertidal zone, also known as the coastline, is where land and sea meet between the high and low tides; it is the living matrix where we find mangrove habitats interrelating with mudflats, salt marshes, seagrasses, marine algae, and coral reefs. Mangroves are ancient species [2] that have evolved specialized processes and organs to survive the daily fluctuation of salinity, moisture, temperature, and turbulence that occur during daily tidal changes. Figure 34.3 shows three main species of mangroves and their location in the intertidal zone. The unique adaptations that mangrove forests have developed over time create a mangrove "biocomplexity" that differentiate them from other ecosystems within the intertidal zone [3]. Because of their position at the land–water interface, mangrove habitats form structural and biological linkages, which enhance several important processes like: nutrient cycling, sedimentation, carbon sequestration, reduction of impacts from hurricanes, and tsunamis [4,5]. Furthermore, mangrove habitats are home to a vast variety of organisms, including fish, shellfish, invertebrates, birds, reptiles, and mammals, thus providing the source of numerous essential services for the livelihood of many coastal communities around the world [6].

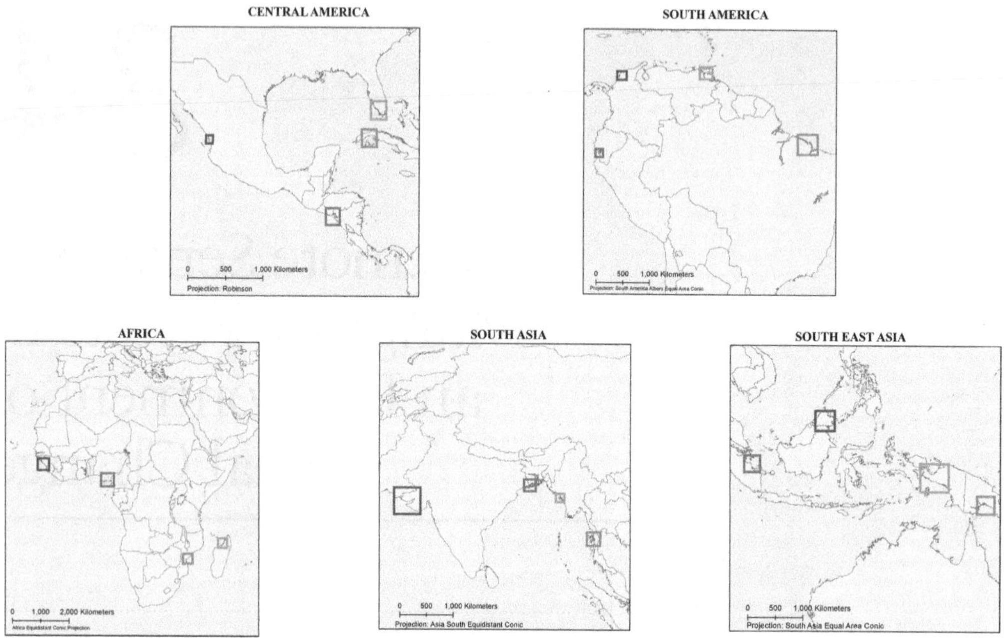

FIGURE 34.1 Landsat thematic mapper (TM) satellite 2000 data used to assess the geographical range of mangrove forests around the globe. (Present figure was created by the author to summarize maps presented by Giri et al., [37] at the Socioeconomic Data and Applications Center (SEDAC) website (http://sedac.ciesin.columbia.edu/data/set/lulc-global-mangrove-forests-distribution-2000/maps)).

FIGURE 34.2 **(See color insert.)** Some examples of mangrove forests around the world. (**a**) Mangroves of Ao Phang Nga National Park in Thailand; (**b**) mangroves of the Sundarbans National Park, a vast forest in the coastal region of the Bay of Bengal; (**c**) mangrove forest in the Esmeraldas-Pacific ecoregion located along the pacific coast of Colombia and Ecuador; (**d**) mangrove tunnels in Florida, United States. (Illustration created by author.)

FIGURE 34.3 (See color insert.) Mangrove species are numerous, but red, black, and white mangroves are the most common species found around the globe. Inset shows the salinity gradient they occupy in the intertidal zone. (**a**) Red mangrove (*Rhizophora mangle L.*) forests grow aerial roots that exude salt crystals and oxygenate the tree; (**b**) black mangrove (*Avicennia germinans*) in the salt marsh of northern Florida, United States. *Avicennia* leaves exude salt crystals and also have pencil size aerial roots (Pneumatophores) that help oxygenate the tree; (**c**) white mangrove (*Laguncularia racemosa*) grows to 12–18 m (39–59 ft.) tall. It typically grows inland of black and red mangroves, well above the high tide line. (Illustration created by author.)

Mangrove forests, the most productive ecosystems of the world [7,8], are being impacted by two main events: (i) high rates of deforestation that may lead to their extinction by the end of this century [5] and (ii) climate change drivers, i.e., extremes in precipitation and temperature patterns, ocean warming, and sea level rise [9–13]. For example, it has been found that mangrove forests have expanded and contracted in response to climate variability at their northern range limit along the Atlantic coast of North America [14]. Furthermore, while responses to shifts in temperature may have contributed to their poleward expansion along the northeast coast of Florida [15], low sea surface temperature may restrict mangrove propagation in their range limits in eastern South America [16].

Remote sensing methods and technologies are perfectly suited for the study, mapping and monitoring of mangrove forests at different spatial and temporal scales. Remote sensing methods include: interpretation of optical, infrared (IR), microwave, and Lidar imagery from Earth observation satellites; hyperspectral and thermal remote sensing from airborne platforms; aerial photography; and handheld and static detectors, balloons, and drones. In addition to the use of classical remote sensing from satellite and airborne platforms, the use of close-range remote sensing of individual leaves and canopies provides a wealth of data in the diagnosis and monitoring of nutritional and water plant stresses.

Basics in Remote Sensing of Vegetation

There are some fundamental concepts in the use of remote sensing of vegetation that need to be addressed before we move to its application in mangrove forests research. Here we present succinctly some of these concepts. First, what is remote sensing? Second, how plant material interacts with the sun's energy. Third, the specific characteristics of remote sensors which include: spatial, spectral, temporal, and radiometric resolution. Finally, limitations of sensors in tropical regions.

Remote Sensing Definition: An accepted definition is the following: "Remote sensing is the science and art of acquiring information about the Earth's surface without being in contact with it, by sensing and recording reflected or emitted energy and processing, analyzing, and applying that information." The energy used in remote sensing of vegetation is part of the electromagnetic (EM) spectrum, which includes visible light, IR, microwaves, radio waves, ultraviolet light, X-rays, and gamma rays.

Plant Interaction: A basic knowledge of how EM radiation interacts with plant material, i.e., how it is absorbed and reflected, provides us with the ability to use and interpret the ever-growing data we can obtain from remote sensing. The total range of the EM spectrum is not totally used in Earth observing systems for plant studies (https://imagine.gsfc.nasa.gov/science/toolbox/emspectrum1.html). Plant leaves reflect most of the radiation in the near-infrared (NIR) and IR regions and partially in some of the green (G) radiation in the visible region of the spectrum. Plant leaves absorb most of the blue (B), red (R), and G radiation in the visible region. Figure 34.4 illustrates interactions of radiation with plants. Remote sensing of vegetation relies on a thorough understanding of the biophysical and biochemical characteristics of plants and their canopies. For example, factors such as crown closure, leaf area index (LAI), canopy structure, chlorophyll content, foliar nutrients, and foliar water content are all important indicators of vegetation health.

Spatial Resolution: In remote sensing, resolution is the amount of details that can be observed in an image. A digital image consists of an array of picture elements or pixels. Each pixel contains information about a small area of the land surface, which is considered as a single object. Spatial resolution is expressed by the size of the pixel on the ground in meters. For example, Landsat data is considered a medium to high spatial resolution (15–30 m) multispectral system when compared with commercial Earth observation satellite sensors like IKONOS, GeoEye, QuickBird, and WorldView with very high spatial resolutions (4.0–0.35 m) (www.satimagingcorp.com/satellite-sensors/).

Spectral Resolution: It is the ability of a sensor to define fine wavelength intervals to characterize different constituents of Earth's surface. Wavelength in remote sensing is usually measured in micrometers (μm) or nanometers (nm). Table 34.1 presents three regions in the EM radiation spectrum most commonly used in remotely sensed data of vegetation; they are the visible, the IR, and radar in the microwave regions.

Furthermore, recent development of hyper-spectral sensors which detect hundreds of very narrow spectral bands advance the quantification of spectral libraries for vegetation in different states of growth and health.

Temporal Resolution: The ability to collect imagery of the same area of the Earth's surface at different periods of time. The changes in spectral characteristics of plants change daily and seasonally, and these changes can be detected by collecting and comparing multi-temporal imagery.

Radiometric Resolution: It is the ability of a remote sensor to discriminate very slight differences in reflected or emitted energy. Image data are recorded by sensors as positive digital numbers (DNs) which vary from 0 (displayed as black) to a selected power of 2. This range corresponds to the number of bits used for coding numbers in binary format. Each bit records an exponent of power 2 (i.e., 1 bit = 2). The maximum number of brightness levels depends on the number of bits used in representing the energy recorded. Image data is displayed in a range of gray tones, with black representing a DN of 0 and white representing the maximum value (i.e., 255 in 8-bit data; 65,535 in 16-bit data). There is a big difference in the level of detail between an 8-bit image with a 16-bit image. It is important to consider that preprocessing of multi-date sensor imagery requires a number of steps which include: image registration, geometric correction, mosaicking, and radiometric rectification [17,18]. Radiometric rectification is an essential, although time-consuming process; it is needed when: (i) comparing multi-date, multi-sensor images; (ii) estimating interactions between

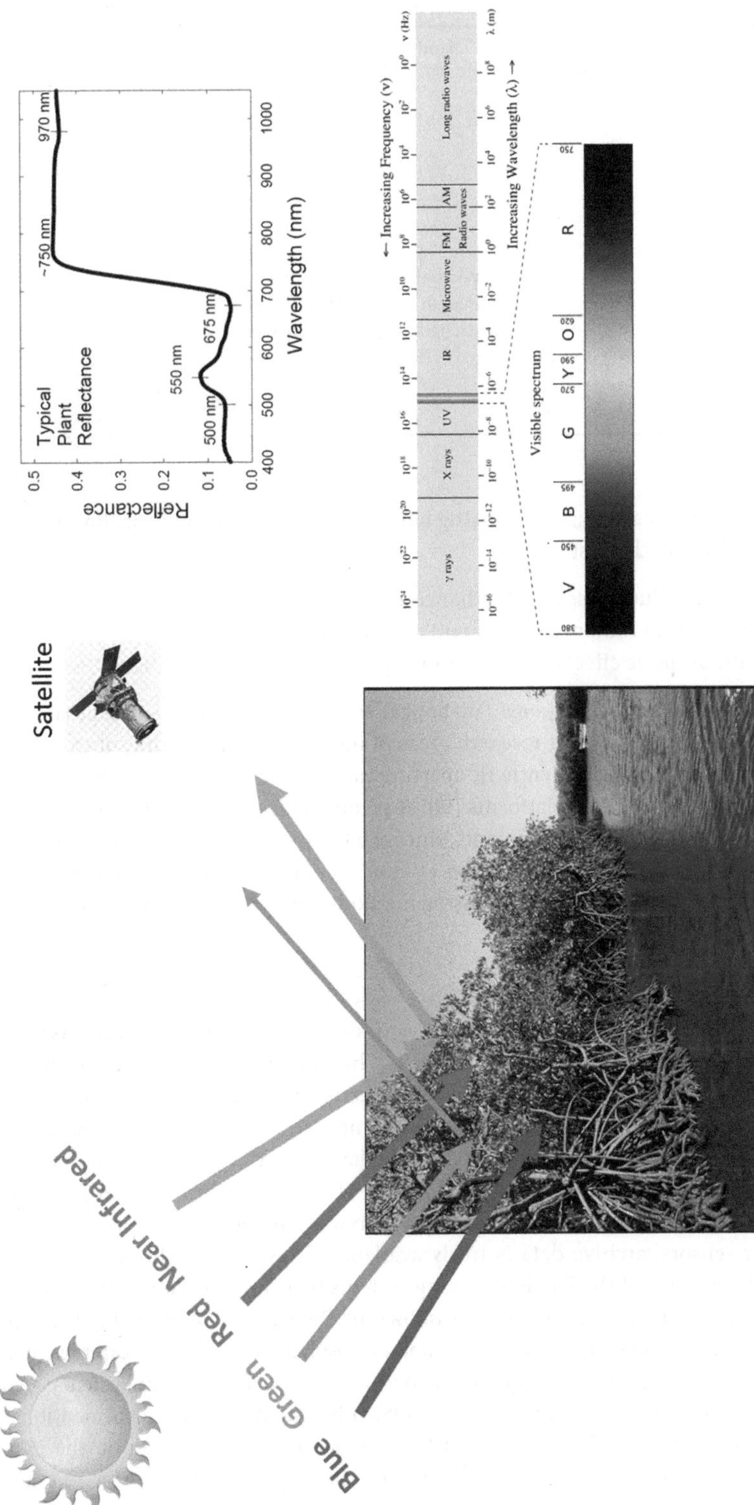

FIGURE 34.4 A general view of absorption and reflection of EM energy by plants in the visible and IR regions of the spectrum. The reflected radiation captured by an Earth observing satellite will be processed and converted to a digital image and transmitted back to ground receiving stations. (Illustration created by author.)

TABLE 34.1 Three of the Main Regions of the Electromagnetic Radiation Spectrum Applicable in Remote Sensing of Vegetation

Radiation	Band Name	Wavelength (nm)
Visible region	Violet	380–450
	Blue	450–495
	Green	495–570
	Yellow	570–590
	Orange	590–620
	Red	620–750
IR region	NIR	0.75–1.4 μm
	Short-wavelength infrared (SWIR)	1.4–3.0 μm
	Mid-wavelength infrared (MWIR)	3.0–8.0 μm
	Long-wavelength infrared (LWIR)	8.0–15 μm
	Far-infrared (FIR)	15–1,000 μm
Microwave region	C	3.75–7.5 cm
Radar	X	8–12 cm

EM radiation and surface features; or (iii) using band ratios for image analysis. Radiometric correction involves the following three steps:

1. Conversion of DN values to spectral radiance at the sensor
2. Conversion of spectral radiance to apparent reflectance
3. Removal of atmospheric effects due to the absorption and scattering of light.

Limitations of Sensors in Tropical Regions: For optical sensors, cloud cover is the greatest obstacle in the acquisition of clear image data for research areas of interest in tropical and subtropical coastlines. This problem can be overcome using synthetic aperture radar (SAR) sensors on ERS (European Remote Sensing Satellites) and RADARSAT platforms [19]. A second deficiency, found in most remote sensing sensors, is the lack of continuous measurements, since remotely sensed imagery are snapshots separated by daily or longer intervals. This last limitation is more important with radiation fluxes and surface temperature studies and may be remedied by using methods to interpolate between observations.

Earth Observing Platforms

A great deal of remote sensing has been done, and continue being done, using aircraft as the main platform. Aircraft platforms include: reconnaissance planes, helicopters, balloons, and recently, unmanned aerial vehicles (UAVs) and drones. With the advent of artificial satellites, aircraft platforms were put somewhat on the background. The spatial, spectral, temporal, and radiometric resolutions of satellite remote sensors have continued improving since the launch of the first Earth observing satellite in 1972 by a joint effort between the US Geological Survey (USGS) and the National Aeronautics Space Administration (NASA). This first Earth observing system was renamed Landsat after its launch. Today, Landsat and other sensors' archive data is freely available to researchers (https://earthexplorer.usgs.gov/). Table 34.2 shows some of the Earth observing systems that can be used in mangrove studies.

There are other Earth observing satellites not shown in Table 34.2 that are being used in the study of land use, climate change, ocean studies, water, and atmospheric chemistry. Examples include: Terra (https://terra.nasa.gov/), Aqua (https://aqua.nasa.gov/), and Aura (https://aura.gsfc.nasa.gov/); these satellites form NASA's Earth Observing System (EOS) mission. We should also mention one of the latest platforms, the International Space Station (ISS), which is collecting data on the Earth's atmosphere, oceans, and land surface through a number of current and future NASA missions (https://eol.jsc.nasa.gov/ESRS/ISS_Remote_Sensing_Systems/). Examples of ISS research applicable to mangrove forests include: using thermal infrared (TIR) measurements to study plant–water dynamics and

TABLE 34.2 Land Observation Satellite and Airborne Platforms and Sensors from the United States and Canada

Platform	Launch	Sensors	Bands	Spatial Resolution (m)
Landsat 1–3	1972, 75, 78	MSS	4	60 × 80
Landsat 4, 5	1982, 84	TM	7	30, 120
Landsat 7	1999	ETM+	8	15, 30, 60
Landsat 8	2013	OLI	9	15, 30
Digital Globe (USA)	2014	TIRS	2	100
		WorldView-3	29	0.34, 1.24, 4, 30
Airborne	1989	MEIS-II	8	1
RADARSAT (Canada)	1995	CASI	288	10, 30, 100
	2007	SAR	C	1, 100
		SAR	C, X	

WorldView-3 is a commercial very high spatial resolution sensor.
MSS, multispectral scanner; TM, thematic mapper; ETM+, enhanced thematic mapper plus; OLI, operational land imager; TIRS, thermal infrared sensor; SPOT, Systeme Pour l'Observation de la Terre; HRV, high resolution visible; MEIS-II, multispectral electro-optical imaging scanner; CASI, compact airborne spectrographic imager; SAR, synthetic aperture radar.

ecosystem changes (https://ecostress.jpl.nasa.gov/); and the GEDI Ecosystem Lidar mission launched in December 5, 2018 (https://gedi.umd.edu/) to study carbon dioxide sequestration, forest fragmentation, and biodiversity.

Remote Sensing Applications on Mangrove Forests

Areas of study where remote sensing technologies will continue to advance our knowledge of mangroves at the forest stand, ecosystem, regional, and global scales include: (i) effects of climate change; (ii) land use/land cover change and landscape pattern dynamics; (iii) vegetation stress detection; (iv) water content estimation; (v) spatial variability of energy fluxes; (vi) stand characteristics such as species composition, canopy closure and height, and standing biomass; (vii) net primary productivity (i.e., amount of carbon fixed annually); and (viii) ecosystem modeling. The use of multispectral, multi-temporal, multi-resolution, and multi-sensor remotely sensed data in the study of mangrove forests is a novel approach that integrates the capabilities of different sensors and methodologies. For example, mapping and classification of mangroves in Tanzania using Landsat TM and ETM+ data [20]; use of optical and radar data for global monitoring of mangroves at a global scale [21]; integration of microwave and optical sensors to quantify the resilience and resistance of mangrove forest to sea-level rise [22]; use of very high spatial resolution data able to discriminate different mangrove categories and improve visual interpretation of images for mapping, monitoring, and modeling of mangrove habitats. Latest experimentation with drone images (spatial resolution of 5.0 cm) have resulted in highest accuracy and image classification in comparison with satellite imagery [23].

Examples of other remote sensing applications that supplement our understanding of mangrove habitats and terrestrial ecosystems in general include: global monitoring and mapping of terrestrial and oceanic temperatures in climate change studies [10]. Figure 34.5 shows global maximum temperatures between 1961 and 1990; landscape dynamics of net primary production and phenology using data from the Advanced Very High Resolution Radiometer (AVHRR) sensor [24]; analysis of sea surface temperature from the Multi-scale Ultra-high Resolution (MUR)-Sea Surface Temperature (SST), which combines data from microwave and IR sensors [16]; use of optical and radar satellite imagery to quantify the rate of melting of Antarctica ice sheets, which help to predict up-to-date sea level rise data [13].

Data from Earth observing platforms has led to widespread use of its measurements to estimate a number of variables crucial in the understanding of land surface and atmospheric processes. These variables, better understood as biophysical variables, include: surface albedo, normalized difference vegetation index (NDVI), LAI, land surface temperature, photosynthetic active radiation, and above ground

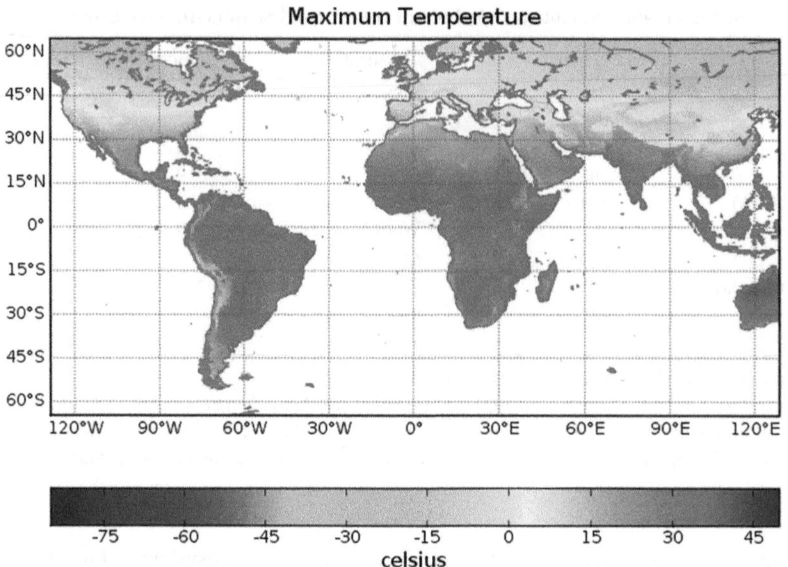

FIGURE 34.5 Climatology data obtained from the Intergovernmental Panel on Climate Change (IPCC) Data Distribution Center website (http://ipcc-data.org/index.html).

net primary productivity [25–32]. Here we explain two of the most common vegetation indices (VIs) used in mangrove research:

NDVI: One of the earliest VIs [33] widely used in vegetation studies today. NDVI responds to changes in amount of green biomass, LAI, percentage of vegetation cover, chlorophyll content, and canopy water stress. It is a nonlinear transformation of the spectral reflectance (ρ) in the visible red (R) and NIR bands. It is calculated as:

$$\text{NDVI} = \frac{\rho_{\text{NIR}} - \rho_R}{\rho_{\text{NIR}} + \rho_R}$$

The NDVI values range from −1 to +1. Healthy vegetation will have a high NDVI value. Bare soil and rock reflect similar levels of NIR and red and so will have NDVI values near zero. Clouds, water, and snow reflect more visible energy than IR energy, thus they will produce negative NDVI values. Figure 34.6 shows an example of mapping mangroves in Panama, South America, based on computation of NDVI values from Landsat and IKONOS satellite data.

LAI: LAI is the leaf area per unit ground area. It indicates how many leaf (or photosynthetically active) surfaces are in a column extended from, the ground area under the canopy diameter, up through the canopy. It can be estimated from NDVI as:

$$\text{LAI}_i = \text{LAI}_{\text{max}} \times \frac{(\text{NDVI}_i - \text{NDVI}_{\text{min}})}{(\text{NDVI}_{\text{max}} - \text{NDVI}_{\text{min}})}$$

Where max = maximum; min = minimum; i = values observed, respectively. Assuming a nonlinear relationship, LAI estimates from NDVI are highly dependent on leaf and soil optical properties, canopy geometry, sun position, and cloud coverage. Reviews of other significant VIs can be found in previous work [28,34,35].

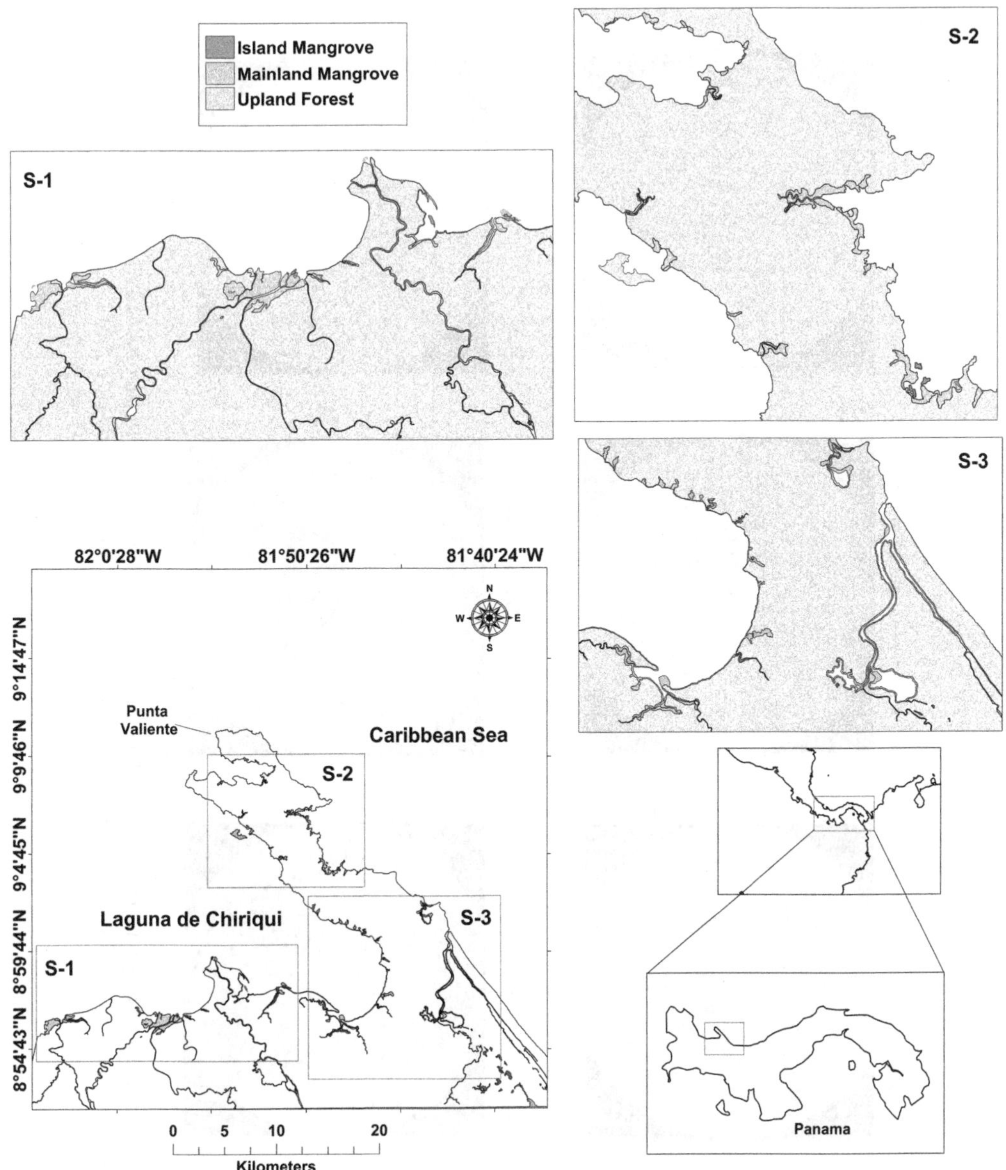

FIGURE 34.6 Mapping and classification of mangrove forests in Panama done by integrating Landsat ETM+ and QuickBird multispectral, multitemporal, remotely sensed imagery. (Unpublished data, Rodriguez and Feller, 2004.)

Case Study

Impact of Climate Change on a Mangrove–Salt Marsh Ecotone: Mangroves have sustained shifts in climate regimes, including sea level rise and temperature and precipitation changes, since their appearance in the late Cretaceous period, 99.6–93.6 million years ago, showing their expansion, contraction, extinction, and reappearance in response to those climatic forces [12,36]. Using historical black and white aerial photography and multispectral, multi-sensor, multi-temporal remotely sensed data, Rodriguez et al. [14] quantified the changes in space and time and the effect of climate variables on the expansion

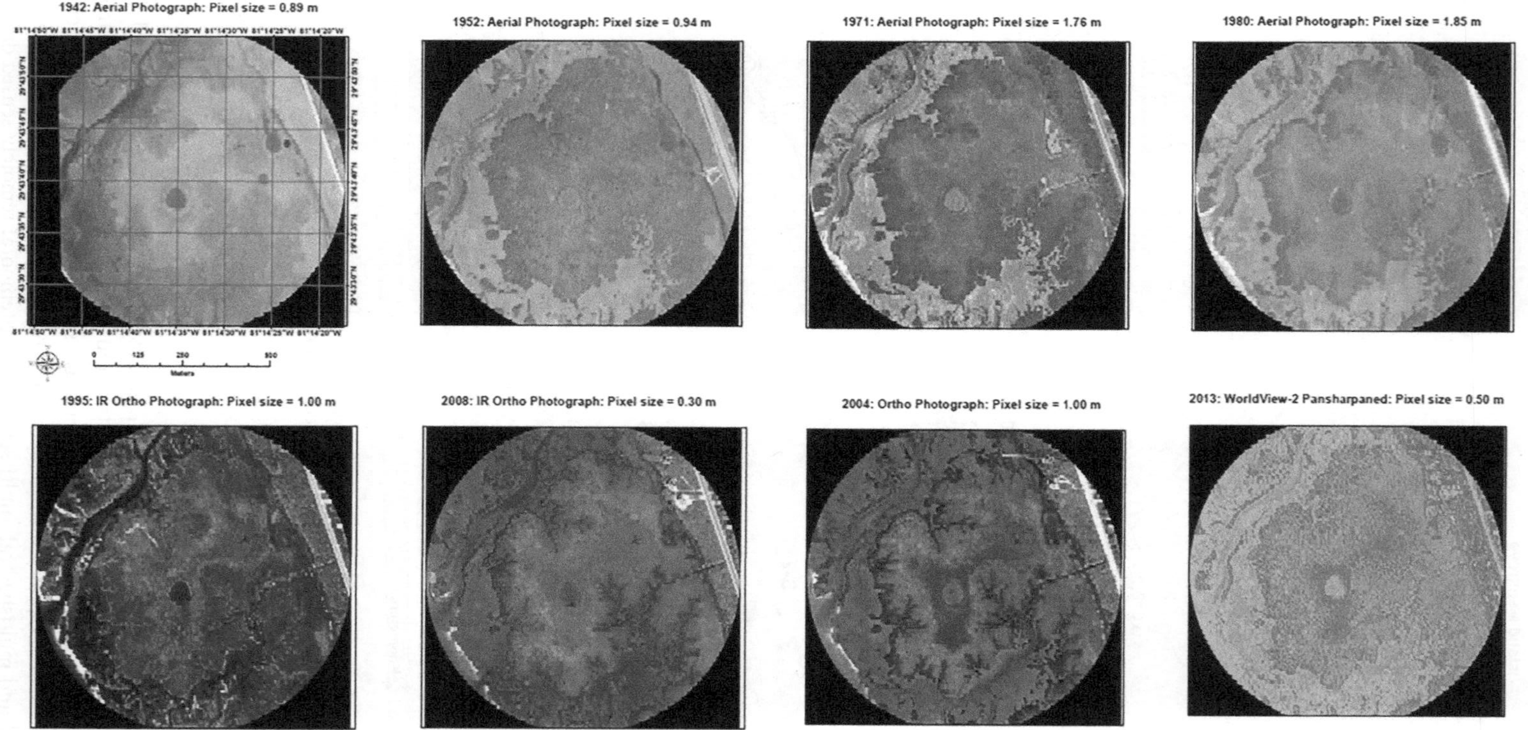

FIGURE 34.7 **(See color insert.)** Use of aerial black and white photography and satellite imagery to quantify changes of mangrove forests over time in northeastern Florida. Satellite data included orthophotos, IKONOS, QuickBird, and WorldView sensors. (Unpublished data, Rodriguez and Feller, 2004.)

and contraction of mangrove forests and salt marsh in the northeastern coast of Florida. The highlights of this research were:

- Significant mangrove expansion/contraction near their northern range limit.
- Significant effect of climate variables on habitat areal extent.
- Cyclical spatiotemporal dynamism over a 71-year period.
- Reversals in habitat dominance may point to complex biotic–abiotic interactions.
- Results may contribute to management/conservation strategies under climate change.

Vegetation mapping was done based on NDVI values computed from a very high resolution multispectral WorldView-2 image. Historical land cover analysis was based on photointerpretation and manual on-screen digitizing of black and white aerial photography dating from 1942 to 1980. Climate data consisted of temperature, precipitation, and mean sea level databases. Land cover change analysis was done using aerial black and white aerial photography and satellite data which included orthophotos, IKONOS, QuickBird, and WorldView-2 sensors. Figure 34.7 shows the imagery used for land cover change analysis. Results from this analysis indicate a shift in mangrove/salt marsh dominance over seven decades, from 1942 to 2013.

Figure 34.8 illustrates the percent of mangrove/salt marsh areal extent over time. Climate variables like precipitation and temperature explained more than 90% of the variation in areal extent in both mangrove and salt marsh habitats. In conclusion, these types of studies help the development of new management strategies for the conservation of these important coastal ecosystems at the regional and global scales, which are being impacted by climate change forces.

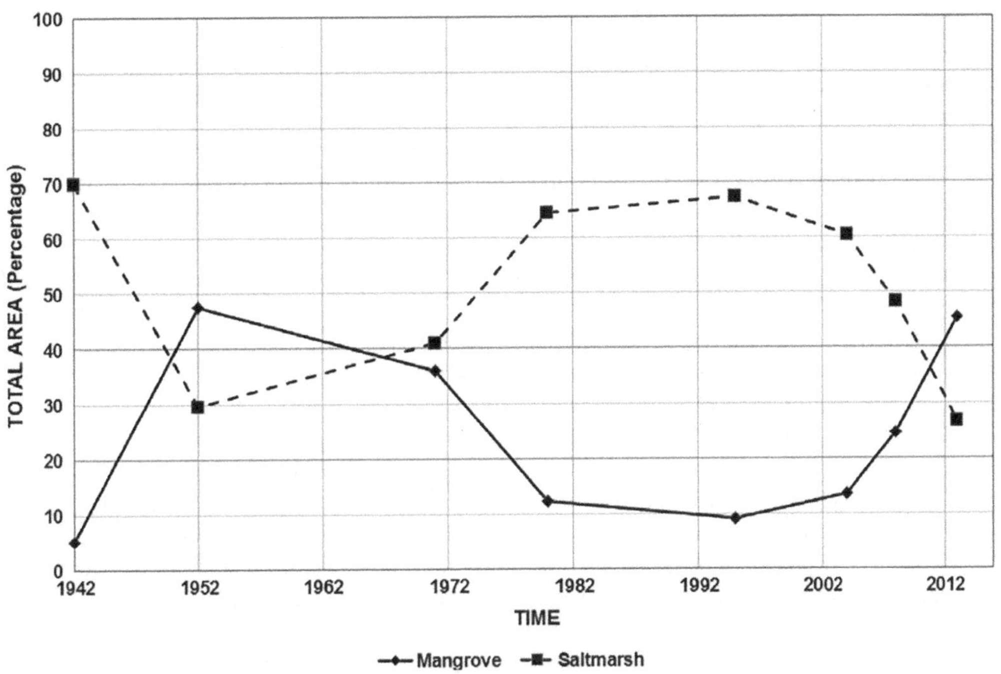

FIGURE 34.8 Expansion and contraction of land cover of mangroves and salt marsh from 1942 to 2013 in northeastern Florida [14].

Conclusion

The continuous advancement in remote sensing technology, computer software, and readily available imagery provide both students and researchers alike the necessary tools and methods to perform a range of studies of mangrove forests at local, regional, and global scales. Although change is an intrinsic phenomenon in nature, the present global rates of change in climate and land use and land cover are impacting terrestrial and marine ecosystems in unforeseen ways. Mangrove habitats in conjunction with other coastal ecosystems are essential providers and supporters of coastal communities; the need for their conservation and sustainable management is of paramount importance.

Further Reading

- Introduction and basics in remote sensing methods and technologies can be found in the following texts: *Remote Sensing and Image Interpretation* by Lillesand, Kiefer, and Chipman (2007); *Quantitative Remote Sensing of Land Surfaces* by Liang (2004); *Remote Sensing of Vegetation: Principles, Techniques, and Applications* by Jones and Vaughan (2010); *Introduction to Microwave Remote Sensing* by I. H. Woodhouse (Woodhouse, 2006); *Remote Sensing Handbook for Tropical Coastal Management* by Green, Mumby, Edwards, and Clark (2000).
- Recommended software for image analysis includes: ERDAS IMAGINE (www.hexagongeospatial.com/products/power-portfolio/erdas-imagine); ENVI (www.harrisgeospatial.com/Software-Technology/ENVI); and GRASS GIS (grass.osgeo.org/), which is a free and open source geospatial analysis program.

References

1. Giri C., Ochieng E., Tieszen L. L., Zhu Z., Singh A., Loveland T., Masek J., Duke N. 2010. Status and distribution of mangrove forests of the world using earth observation satellite data. *Global Ecology and Biogeography*. 1, 154–159.
2. Ellison A. M., Farnsworth E. J., Merkt R. E. 1999. Origins of mangrove ecosystems and the mangrove biodiversity anomaly. *Global Ecology and Biogeography*. 8, 95–115.
3. Feller I. C., Lovelock C. E., Berger U., McKee K. L., Joye S. B., Ball M. C. 2010. Biocomplexity in mangrove ecosystems. *Annual Review of Marine Science*. 2, 395–417.
4. Odum W. E., McIvor C. C., Smith III T. J. 1982. *The Ecology of the Mangroves of South Florida: A Community Profile*. U. S. Fish and Wildlife Service, Office of Biological Services, Washington, D. C. FWS/OBS-81/24. 144pp. Reprinted September 1985.
5. Alongi, D. M. 2002. Present state and future of the world's mangrove forests. *Environmental Conservation*. 29, 331–349.
6. UNEP. 2014. Chapter 1. Tropical Mangrove Ecosystems, pp. 33–38. In: *The Importance of Mangroves to People: A call to Action*. Van Bochove J., Sullivan E., Nakamura T. (Eds.). United Nations Environment Programme World Conservation Monitoring Centre, Cambridge. 128 pp.
7. Lee S. Y., Primavera J. H., Dahdouh-Guebas F., McKee K., Bosire J. O., Cannicci S., Diele K., Fromard F., Koedam N., Marchand C., Mendelssohn I. 2014. Ecological role and services of tropical mangrove ecosystems: a reassessment. *Global Ecology and Biogeography*. 23, 726–743.
8. Hyndes G. A., Nagelkerken I., McLeod R. J., Connolly R. M., Lavery P. S., Vanderklift M. A. 2014. Mechanisms and ecological role of carbon transfer within coastal seascapes. *Biological Reviews*. 89, 232–254.
9. IPCC. 2013. Chapter 1. Introduction. In: *Climate Change 2013: The Physical Science Basis. Contribution of Working Group I to the Fifth Assessment Report of the Intergovernmental Panel on Climate Change*. Stocker T. F., Qin D., Plattner G. -K., Tignor M., Allen S. K., Boschung J., Nauels A., Xia Y., Bex V., Midgley P. M. Eds.). Cambridge University Press, Cambridge, United Kingdom and New York, 1535 pp.

10. IPCC. 2014. Chapter 1. Introduction. In: *Climate Change 2014: Impacts, Adaptation, and Vulnerability. Part A: Global and Sectoral Aspects. Contribution of Working Group II to the Fifth Assessment Report of the Intergovernmental Panel on Climate Change.* Field C. B., Barros V. R., Dokken D. J., Mach K. J., Mastrandrea M. D., Bilir T. E., Chatterjee M., Ebi K. L., Estrada Y. O., Genova R. C., Girma B., Kissel E. S., Levy A. N., MacCracken S., Mastrandrea P. R., White L. L. (Eds.). Cambridge University Press, Cambridge, United Kingdom and New York, 1132 pp.
11. IPCC. 2018. Summary for Policymakers. In: *Global Warming of 1.5°C. An IPCC Special Report on the Impacts of Global Warming of 1.5°C Above Pre-Industrial Levels and Related Global Greenhouse Gas Emission Pathways, in the Context of Strengthening the Global Response to the Threat of Climate Change, Sustainable Development, and Efforts to Eradicate Poverty.* Masson-Delmonte V., Zhai P., Portner H. O., Roberts D., Skea J., Shukula P. R., Pirani A., Moufouma-Okia W., Pean C., Pidcock R., Connors S., Mathews J. B. R., Chen Y., Zhou X., Gomis M. I., Lonnoy E., Maycock T., Tignor M., Waterfield T. (Eds.). World Meteorological Organization, Geneva, Switzerland, 32 pp.
12. Alongi D. M. 2015. The impact of climate change on mangrove forests. *Current Climate Change Reports*. 1, 30–39.
13. Rignot E., Mouginot J., Scheuchl B., Van den Broeke M., Van Wessem M., Morlighem M. 2018. Four decades of Antarctic ice sheet mass balance from 1979 – 2017. *Proceedings of National Academy of Sciences*. doi:10.1073/pnas.1812883116.
14. Rodriguez W., Feller I. C., Cavanaugh K. C. 2016. Spatio-temporal changes of a mangrove-saltmarsh ecotone in the northeastern coast of Florida, USA. *Global Ecology and Conservation*. 7, 245–261.
15. Cavanaugh K. C., Kellner J. R., Forde A. J., Gruner D. S., Parker J. D., Rodriguez W., Feller I. C. 2014. Poleward expansion of mangroves is a threshold response to a decreased frequency of extreme cold events. *Proceedings of the National Academy of Sciences*. 111(2), 723–727.
16. Ximenes A. C., Ponsoni L., Lira C. F., Koedam N., Dahdouh-Guebas F. 2018. Does sea surface temperature contribute to determining range limits and expansion of mangroves in eastern South America (Brazil)? *Remote Sensing*. 10, 1787.
17. Coppin P., Jonckhere I., Nackaerts K., Mays B., Lambin E. 2004. Digital change detection methods in ecosystem monitoring: a review. *International Journal of Remote Sensing*. 25, 1565–1596.
18. Lunetta R. S., Elvidge C. D. 1998. *Remote Sensing Change Detection: Environmental Monitoring Methods and Applications*. Ann Arbor Press, Chelsea, MI.
19. Green E. P., Mumby P. J., Edwards A. J., Clark C. D. 2000. Part 1. Remote Sensing for Coastal Managers - An Introduction. In: *Remote Sensing Handbook for Tropical Coastal Management*. Edwards A. J. (Ed.). Coastal Management Sourcebooks 3. UNESCO, Paris. 316 pp.
20. Wang Y., Bonyge G., Nugranad J., Traber M., Ngusaru A., Tobey J., Hale L., Bowen R., Makota V. 2003. Remote sensing of mangrove change along the Tanzania Coast. *Marine Geodesy*. 26, 35–48.
21. Giri C. 2016. Observation and monitoring of mangrove forests using remote sensing: opportunities and challenges. *Remote Sensing*. 8, 783.
22. Duncan C., Owen H. J. F., Thompson J. R., Primavera J. H., Pettorelli N. 2018. Satellite remote sensing to monitor mangrove forest resilience and resistance to sea level rise. *Methods in Ecology and Evolution*. 9, 1837–1852.
23. Ruwaimana M., Satyanarayana B., Otero V., Muslim A. M., Syafiq M. A., Ibrahim S., Raymaekers D., Koedam N., Dahdouh-Guebas F. 2018. The advantages of using drones over space-borne imagery in the mapping of mangrove forests. *PLoS ONE*. 13(7), e0200288. doi:10.1371/journal.pone.0200288.
24. Wang Y., Zhao J., Zhou Y., Zhang H. 2012. Variation and trends of landscape dynamics, land surface phenology and net primary production of the Appalachian Mountains. *Journal of Applied Remote Sensing*. 6, 061708.
25. Qi J., Cabot F., Moran M. S., Dedieu G. 1995. Biophysical parameter estimations using multidirectional spectral measurements. *Remote Sensing of Environment*. 54, 71–83.

26. Treitz P. M., Howarth P. J. 1999. Hyperspectral remote sensing for estimating biophysical parameters of forest ecosystems. *Progress in Physical Geography.* 23, 359–390.
27. Morisette J. T., Privette J. L., Justice C. O. 2002. A framework for the validation of MODIS land products. *Remote Sensing of Environment.* 83, 77–96.
28. Liang S. 2004. *Quantitative Remote Sensing of Land Surfaces.* Wiley Inter-Science. John Wiley & Sons, Inc., Hobken, NJ. ISBN 0-471-28166-2.
29. Weng Q., Lu D., Liang B. 2006. Urban surface biophysical descriptors and land surface temperature variations. *Photogrammetric Engineering & Remote Sensing.* 72, 1275–1286.
30. Ghilain N., Arboleda A., Sepulcre-Canto G., Batelaan O., Ardo J., Gellens-Meulenberghs F. 2012. Improving evapotranspiration in a land surface model using biophysical variables derived from MSG/SEVIRI satellite. *Hydrology and Earth System Sciences.* 16, 2567–2583.
31. Borges C. K., de Raimundo R. M., Ribeiro R. E., dos Santos E. G., Carneiro, R. G., dos Santos, C. A. 2016. Study of biophysical parameters using remote sensing techniques to Quixere-CE region. *Journal of Hyperspectral Remote Sensing.* 6(6), 283–294.
32. Wocher M., Berger K., Danner M., Mauser W., Hank T. 2018. Physically-Based retrieval of canopy equivalent water thickness using hyperspectral data. *Remote Sensing.* 10(12), 1924.
33. Rouse Jr. J. W., Haas R., Schell J., Deering D. 1974. Monitoring vegetation systems in the great plains with ERTS. NASA Special Publication. 351, 309–317.
34. Jones H. G., Vaughan R. A. 2010. *Remote Sensing of Vegetation: Principles, Techniques, and Applications.* Oxford University Press. Inc., New York.
35. Xue, J., Su B. 2017. Significant remote sensing vegetation indices: A review of developments and applications. *Journal of Sensors.* 2017, Article ID 1353691, 17 pages. doi:10.1155/2017/1353691.
36. El-Saadawi W., Osman R., El-Faramawi M. W., Bkhat H., Kamal-El-Din M. 2016. On the Cretaceous mangroves of Bahariya Oasis, Egypt. *Taechlomia.* 36, 1–16.
37. Giri C., Ochieng E., Tieszen L. L., Zhu Z., Singh A., Loveland T., Masek J., Duke N. 2013. *Global Mangrove Forests Distribution, 2000.* NASA Socioeconomic Data and Applications Center (SEDAC), Palisades, NY. http://sedac.ciesin.columbia.edu/downloads/maps/lulc/lulc-global-mangrove-forests-distribution-2000.

35

Salt Marsh Mapping and Change Analysis: Remote Sensing

Anthony Daniel Campbell
University of Rhode Island
Yale University

Introduction

Land use and land cover mapping by remote sensing image classification has become more complex over the past 10 years with the ubiquity of high spatial and spectral resolution data. Scientific innovation has been necessary to address the glut of data including machine learning, object-based image analysis, and data fusion. Image classifications are further refined with ancillary data, e.g., LiDAR-derived elevations which elucidate the salt marsh elevation gradient an important factor in inundation period and in turn vegetation communities [1]. As a result, elevation is frequently an important component of salt marsh classification both in combination with spectral imagery [2] and on its own [3]. Remote sensing image classification has particular requirements from the high spatial and temporal resolution necessary to delineate the salt marsh landscape.

One major innovation is object-based classifications which utilize a pre-classification step of segmentation. Segmentation divides the imagery into meaningful image objects that are then classified based on spectral, textural, and geospatial attributes [4]. Object-based classifications address limitations of pixel-based methods, e.g., minimizing the salt and pepper effect and providing additional spatial context [5–7]. When visually compared to pixel-based approaches, object-based methods result in higher quality mapping and change analysis [8] and improved accuracy of salt marsh vegetation [9]. Quantitative comparisons of pixel- vs. object-based classifications have shown no significant increase in accuracy when using medium resolution sensors [8]; however, object-based methods with very high spatial resolution (VHR) imagery had higher accuracy in agricultural, wetland, and urban landscapes [7,10,11]. Past research has shown that object-based methods are suited for classifying complex landscapes with fine spatial resolution data. This chapter compares object-based and pixel-based methods with machine learning classifications for mapping heterogeneous salt marsh environments.

This study utilized two types of satellite data: (i) Worldview-2 to provide high spatial resolution data for object-based classification of the salt marsh, and (ii) Sentinel-2 for multi-temporal pixel-based classification and time series analysis. Worldview-2 is a commercial satellite launched in October of 2009 with eight spectral bands, including coastal blue, blue, yellow, green, red, red edge, NIR-1 and NIR-2, and a panchromatic spatial resolution of 0.46 m. Sentinel-2 was launched in June of 2015 by the European Space Agency (ESA). The Multispectral Instrument (MSI) sensor's four bands including blue, green, red, and NIR possess a 10 m spatial resolution.

Time series monitoring is common for many applications including forest change detection, farm abandonment, urban expansion, and wetland change [12–16]. The Landsat archive provides moderate temporal resolution and multiple decades of data, which are essential components for understanding long-term salt marsh change. Additionally, salt marsh mapping benefited from high spatial resolution (<1 m) to discern fine-scale changes in extent [17]. Salt marsh change analysis often utilizes high-resolution satellites or aerial imagery [18,19].

Salt marsh environments are difficult to conduct expansive field work due to tides, dense vegetation, and an unstable marsh platform. Understanding salt marsh composition and change is of particular importance for assessment and quantification of the variety of ecosystem services they provide such as carbon sequestration and storage, habitat for a diverse set of fish and bird species, and buffer against storm and flood impacts [20]. Remote sensing image classification fulfills a necessary gap in knowledge for salt marsh environments.

Advances in remote sensing classification have been applied to salt marsh environments including object-based methods [9], time series [21,22], and machine learning [2]. This chapter introduces three approaches to remote sensing mapping for a mid-Atlantic barrier island including object-based VHR classification, pixel-based time series classification, and a time series approach to understanding salt marsh change. The experiments elucidate the important methodology decisions that must be made when mapping to leverage remote sensing advances and the proliferation of satellite imagery to better understand salt marsh environments.

Methodology

Study Site

This chapter's three experiments are focused on Assateague Island National Seashore (ASIS), a 59.5 km barrier island on the border between Maryland and Virginia (38°4′N, 75°12′W) of the United States. The island is composed of three conservation areas including ASIS, Chincoteague National Wildlife Refuge, and Assateague State Park managed by the National Park Service, United States Fish and Wildlife, and the Maryland Department of Natural Resources, respectively. The United States Congress created the National Seashore in 1965 due in part to the destruction of the Ash Wednesday Storm of 1962 [23].

Image Classification

Assateague Island is a pristine barrier island system with three vegetation communities including salt marsh on the bayside of the island, forested upland, and dune and swale [24]. These communities were represented by land cover classes of *Spartina alterniflora*, high marsh, sand, dune vegetation, mudflat, water, developed, shrub, and forest. Field collected vegetation plot data gathered across the island were used to create training data. The data were collected from 1 m^2 vegetation plots. Percent cover was calculated with the Braun-Blanquet method [25]. Objects of the developed and water classes were visually interpreted from the imagery and knowledge from field observations.

Object-based methods with machine learning have produced higher image classification accuracies and more visually appealing representations of wetlands [26] and an improved differentiation between vegetation types in wetlands [27]. Random forest, one such machine learning algorithm, is a nonparametric ensemble learning algorithm that creates many classification trees and then uses majority votes to decide the class of an object [28]. Random forest has been shown to be an efficient and high performing algorithm for a variety of data [29]. Both pixel-based and object-based methods were classified with Random forest in this study.

Accuracy Assessment

The reference data were divided 70% and 30% between training and testing. Confusion matrices were computed for the 30% reserved for verification. The Kappa coefficient, Overall, User's, and Producer's Accuracy, and variable importance were calculated to understand classification accuracy for both Sentinel-2 and Worldiew-2 classifications. McNemar's test was also applied to compare classification accuracy. McNemar's test compares verification points that were correctly classified in one classification and not in the other utilizing the chi-squared distribution [30]. The producer's and user's accuracy represent error of omission and commission, respectively. To remain consistent across all classifications, the same training data were used to select training objects and as training points for the object and pixel-based classifications, respectively.

Time Series Change Analysis

The Sentinel-2 MSI image classification was used to select the low marsh extent and identify areas of salt marsh change across the Sentinel-2 archive utilizing Google Earth Engine time series approach. Breakpoint analysis of the NDVI time series was conducted with Breaks for Additive and Seasonal Trend (BFAST) and was applied to all pixels classified as low marsh. BFAST detects time periods that deviate from the expected harmonic model and flags those as changed. The monitoring period for the BFAST algorithm included 2015 and 2016, with possible disturbances occurring in 2017 and 2018. The time series analysis was conducted for low marsh on ASIS which accounted for an area of 1,887 ha. BFAST was chosen due to its ability to detect disturbances of a magnitude >0.1 NDVI within noisy time series [31].

Results

Sentinel-2 Image Classification

The multi-date classification included National Elevation Dataset (NED) 1/3 arc second DEM and National Wetland Inventory as ancillary data, which performed best when compared to other Sentinel-2 image classifications. The classification utilized nine scenes of cloud-free Sentinel-2 MSI images for 2016 acquired on 01/02/2016, 02/18/2016, 04/18/2016, 5/28/2016, 06/10/2016, 06/30/2016, 08/17/2016, 10/18/2016, and 11/17/2016. The classification utilizing all available images achieved a higher accuracy than any individual date (Figure 35.1). The 6/30/2016 image achieved the highest single date classification accuracy. The best single date classification was compared to the classification utilizing all available data with McNemar's test. The classification utilizing all dates and the 6/30/2016 classification had a significant difference in accuracy ($X^2(1, N = 270) = 17.01, p < 0.001$).

The classification included the region surrounding the study area contributing to a more comprehensive understanding of salt marsh extent in the region (Figure 35.2). The Sentinel-2 image classification achieved greater than 80% overall accuracy (Table 35.1). The classification performed well for the classes of most interest, i.e., high marsh and *S. alterniflora* categories. The Sentinel-2 MSI image classification performed poorly in mudflat delineation due to limitation in spatial resolution and the rapid change this class can experience during the tidal cycle. In general, the salt marsh classes of high marsh and *S. alterniflora* performed well with producer's and user's accuracy greater than 80%. The high accuracy

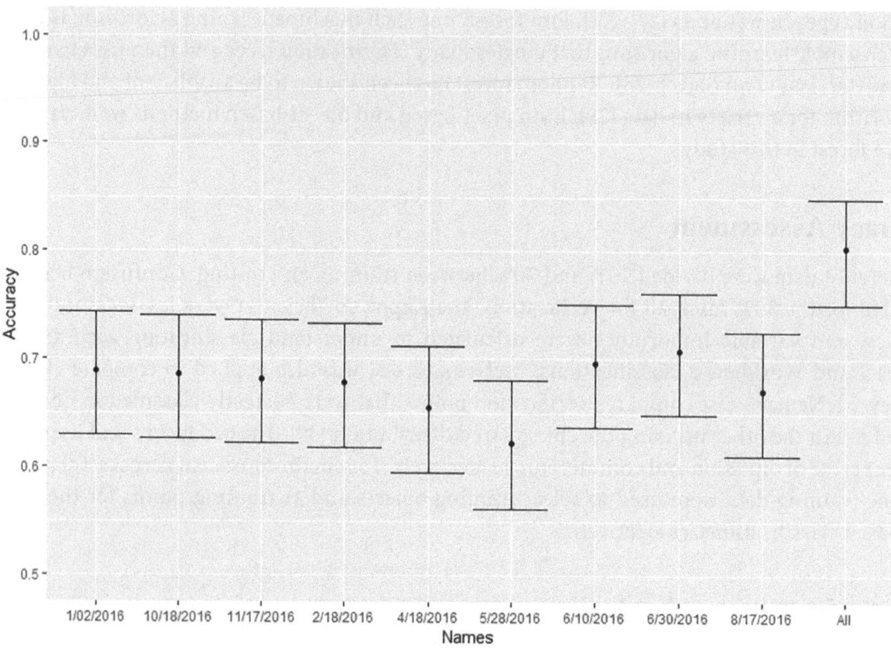

FIGURE 35.1 A comparison of the Sentinel-2 single date random forest model accuracy and multi-temporal model accuracy. Ancillary data was not included in these models.

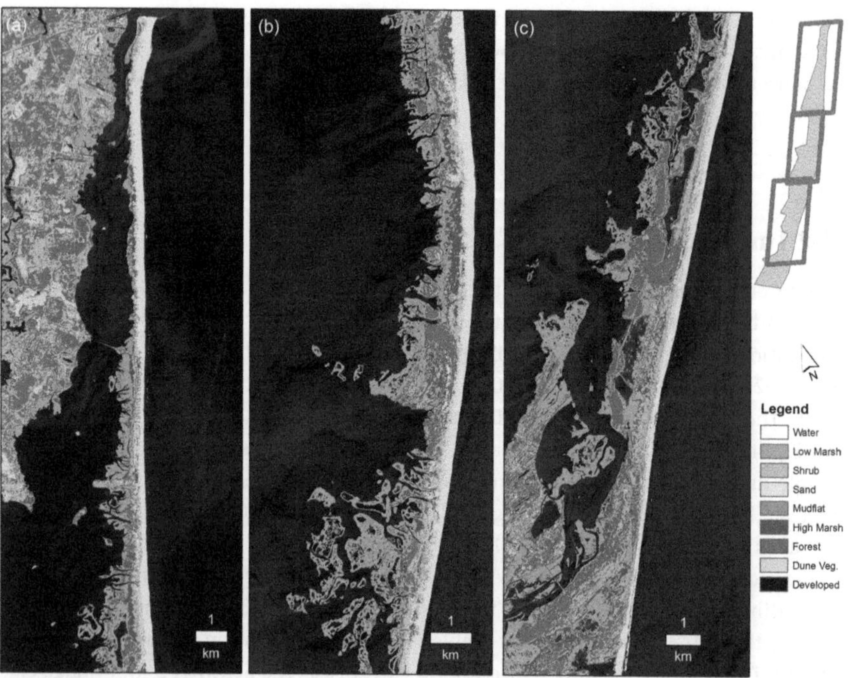

FIGURE 35.2 The classification using Sentinel-2, 1/3 arc second NED DEM, and National Wetland Inventory for ASIS. (**a**) The northern portion of Assateague Island dominated by overwash, (**b**) the center of the island with large amounts of *S. alterniflora* marshes, and (**c**) the southern portion of the island with more upland areas.

TABLE 35.1 Sentinel-2 Confusion Matrix of the Pixel-Based Classification

		Reference									
	Classes	Dev	DV	FOR	HM	Mud	Sand	Shrub	*SA*	Water	User's Accuracy
Predicted	Dev	23	1	0	0	0	0	0	0	1	92
	DV	1	20	0	1	0	1	3	1	0	74
	FOR	0	2	7	0	0	0	0	0	0	78
	HM	0	3	3	52	0	0	7	2	0	78
	Mud	0	0	0	0	9	0	0	1	2	75
	Sand	1	0	0	0	0	13	0	0	0	93
	Shrub	0	0	3	2	0	0	21	0	0	81
	SA	0	0	0	5	5	0	0	74	0	88
	Water	0	0	0	0	0	0	0	0	6	100
	Producer's Accuracy (%)	92	77	54	87	64	93	68	95	67	OA: 83.33 Kappa: 0.79

Column heading are abbreviated classes: Dev, developed; DV, dune vegetation; FOR, forest; HM, high marsh; Mud, mudflat; SA, *Spartina alterniflora.*

of the *S. alterniflora* class and importance for monitoring its changing health led to its use as the extent for the time series analysis.

Variable importance can be computed from the random forest model. Variable importance for the Sentinel-2 MSI image classification demonstrates agreement with the comparison between single date models (Table 35.2). The tidal stage was also verified for each image acquisition date. The variable importance suggests that tidal stage of the acquired images was an important component of variable utilization.

TABLE 35.2 The 20 Most Important Variables as Computed Internal by the Random Forest Algorithm for the Sentinel-2 Multi-Temporal Classification

Variable	Date	Importance	Tidal Stage (m)[a]
Digital elevation model	NA	100.00	NA
Near-infrared	02/18/2016	99.02	−0.03
NDWI	02/18/2016	64.43	−0.03
Near-infrared	01/02/2016	59.92	0.438
NDVI	2/18/2016	56.86	−0.03
NDVI	06/30/2016	51.16	0.158
NDWI	01/02/2016	49.94	0.438
Near-infrared	11/17/2016	49.31	0.857
National wetland inventory	NA	48.62	NA
Near-infrared	10/18/2016	47.78	0.88
Near-infrared	06/10/2016	47.12	0.599
NDWI	06/30/2016	46.80	0.158
Near-infrared	04/18/2016	44.19	0.368
Near-infrared	05/28/2016	36.67	0.512
NDWI	06/10/2016	33.63	0.599
Near-infrared	06/30/2016	33.15	0.158
Red	01/02/2016	31.42	0.438
Blue	06/30/2016	29.43	0.158
NDVI	06/10/2016	26.78	0.599
NDWI	10/18/2016	25.43	0.88

[a]Tidal stage above MLW at the Ocean City Inlet (Station ID 8570283).

Object-Based Image Analysis

The image classification achieved a visually appealing result with improved delineation between classes due to the higher spatial resolution of the Worldview-2 imagery (Figure 35.3). The object-based Worldview-2 image classification achieved >85% overall accuracy and high producer's and user's accuracy for all classes except the Forest class. Detailed performance metric for the classification is summarized in Table 35.3. The forest class had high variability due to changes following the training data

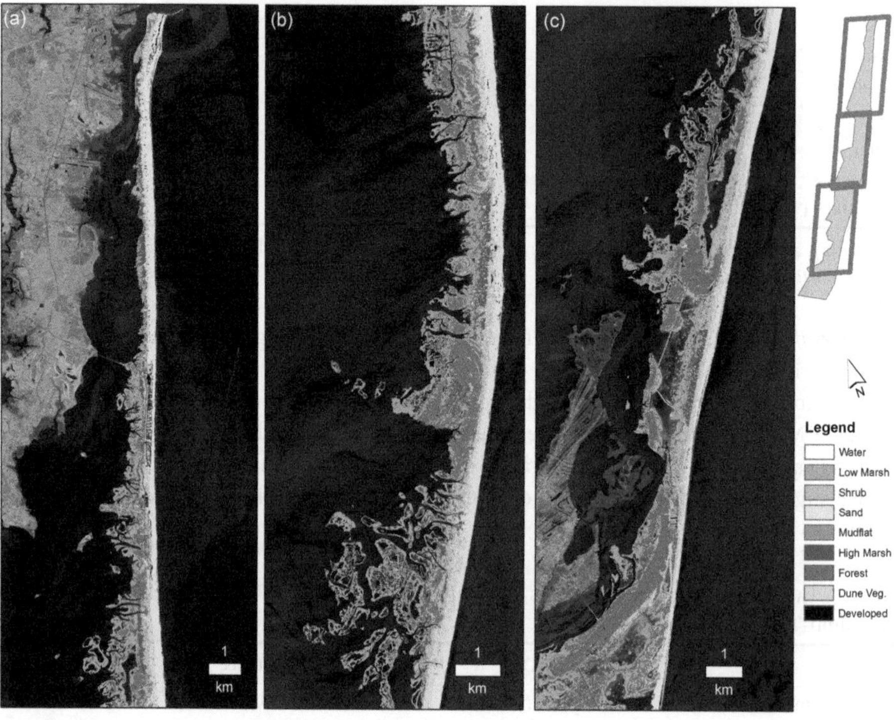

FIGURE 35.3 The Worldview-2 object-based classification of ASIS. Showing three sections of the island.

TABLE 35.3 Worldview-2 Confusion Matrix of the Object-Based Classification

		Reference									
	Classes	Dev	DV	For	HM	Mud	Sand	Shrub	SA	Water	User's Accuracy
Predicted	Dev	24	0	0	0	0	0	0	0	0	100
	DV	0	23	1	0	0	1	5	0	0	77
	For	0	0	6	0	0	0	0	0	0	100
	HM	0	0	1	53	0	0	0	2	0	95
	Mud	0	0	0	0	13	0	0	0	2	87
	Sand	1	0	0	0	0	13	0	0	0	93
	Shrub	0	3	5	3	0	0	26	0	0	70
	SA	0	0	0	4	1	0	0	76	0	94
	Water	0	0	0	0	0	0	0	0	7	100
	Producer's accuracy (%)	96	88	50	88	93	93	84	97	78	OA: 89.26 Kappa: 0.87

Abbreviated classes: Dev, developed; DV, dune vegetation; For, forest; HM, high marsh; Mud, mudflat; SA, *Spartina alterniflora*.

collection occurring from pine bark beetles outbreaks. The understory being exposed in these areas led to a spectral similarity with the shrub class. The high marsh and *S. alterniflora* performed well with all producer's and user's accuracy metrics >85%. Comparing the object and pixel-based classification with McNemar's test demonstrates that Worldview-2 image classification was significantly more accurate than the Sentinel-2 pixel-based approach ($X^2(1, N = 270) = 4.68, p < 0.05$).

Time Series Change Analysis

The NDVI time series analysis was conducted for 188,706 Sentinel-2 imagery pixels across Assateague Island. On average, 210 Sentinel-2 images were utilized for each time series spanning 2015–2018. The time series found an average trend of −0.03 NDVI for the study site. Of the area analyzed, 68% experienced a loss and 32% experienced an increase in NDVI. In total, 43.7% of *S. alterniflora* pixels experienced a disturbance of an absolute magnitude greater than 0.1 NDVI. Examples of both fragmented areas of low marsh with extensive NDVI increases (Figure 35.4a) and those with mostly NDVI declines (Figure 35.4b) are evident in the time series analysis.

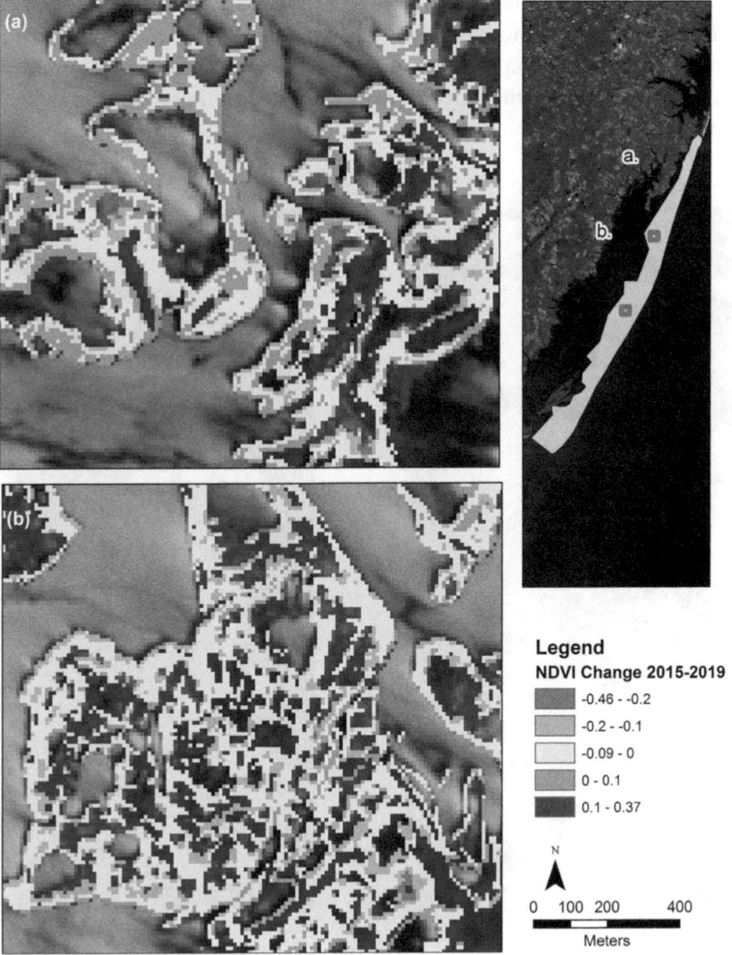

FIGURE 35.4 The figure shows two areas of the low marsh time series: (**a**) a northern section of the island with extensive increases in NDVI and (**b**) a heavily fragmented section of salt marsh with widespread declines in NDVI located in the center of Assateague Island.

Discussion and Conclusions

The three experiments reveal some important lessons evident from the literature, i.e., importance of high spatial resolution imagery when mapping the salt marsh extent, site-specific attributes (e.g., tidal range, landscape, and urbanization), extent of area being studied, and importance of ancillary datasets. The Worldview-2 image classification was more accurate and able to delineate class boundaries, e.g., between developed and vegetation classes, and the transition from intertidal mudflat to vegetated salt marsh. The salt marsh landscape is composed of near monocultures of *S. alterniflora*; however, the transition from *S. alterniflora* to mudflat or high marsh is often sudden making the higher spatial resolution necessary to delineate the salt marsh landscape. The temporal resolution of the medium resolution time series can provide a detailed understanding of seasonal dynamics and disturbance within the salt marsh environment.

The salt marsh focused classification defined meaningful vegetation classes within the barrier island for both spatial resolutions. In particular, the *S. alterniflora* and high marsh classes achieved high accuracy. The random forest classifier has been frequently applied in land cover change studies [27,32] and salt marsh environments [2,33]. The use of a common training dataset facilitates the comparison of the two classifications. However, when large-scale maps of the two classifications are compared, the differences become clear (Figure 35.5), including lack of spatial definition to interior salt marsh pannes, in transitions from land cover types, and in delineation of complex classes. The complexity and organic shapes of the landscape features within the salt marsh resulted in those differences due to limitation of spatial resolution by Sentinel-2 MSI.

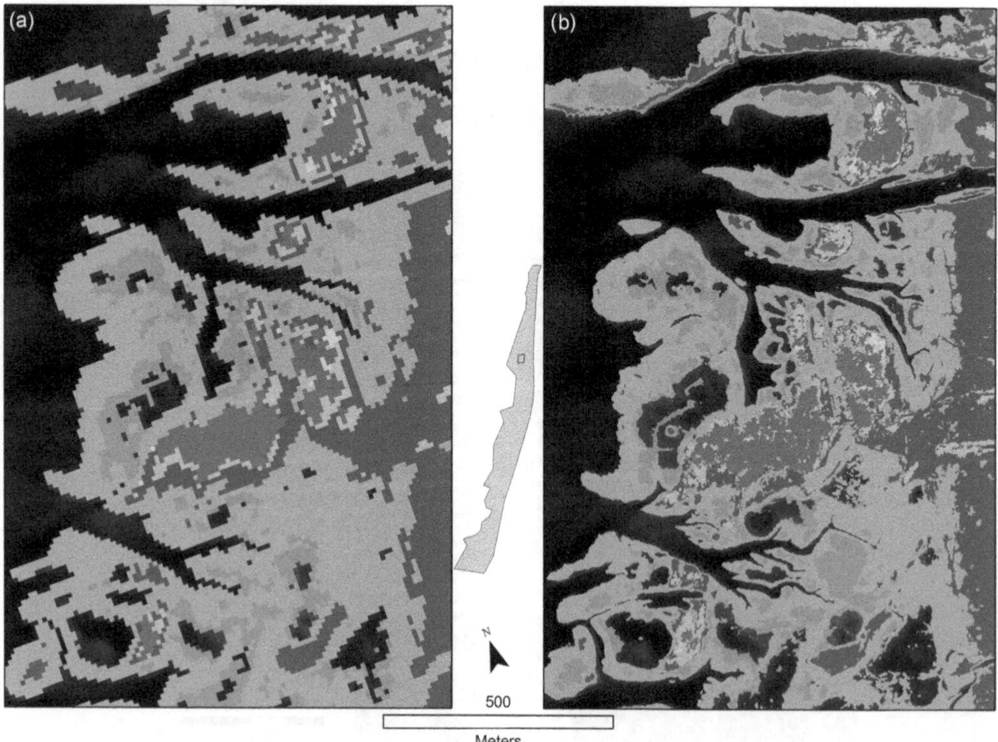

FIGURE 35.5 A comparison of (**a**) the Sentinel-2 multi-temporal classification and (**b**) the Worldview-2 object-based classification for an area of salt marsh to the north of the island.

The multi-temporal Sentinel-2 image classification demonstrated an appropriate use for pixel-based methods. The application of machine learning, ancillary data, and openly available satellite imagery is applicable to the entirety of the United States and many parts of the world. Single date images performed worse than the multi-temporal classification. Classifications derived from images outside the desired low tidal stage or growing season achieved low overall accuracy. However, when included within the multi-temporal model, these dates were often some of the most important as shown in Table 35.2. The multi-temporal classification minimizes the effect of tidal stage and acquires useful information from both low and high tidal stage images. In general, vegetation indices and the near-infrared band were the most important variables. The best performed single-date image represented a low tidal stage (0.158 m above Mean Low Water or MLW), within the growing season, and a cloud-free image. These are ideal characteristics in mapping salt marsh. A major limitation to this classifications accuracy was adapting it to be comparable to the object-based classification, which resulted in training points corresponding with a single object. This conversion resulted in the underrepresentation of certain homogenous classes such as water, i.e., represented by a few large objects. Additionally, the vegetation data from the field was collected in a grid often resulting in an object corresponding with more than one vegetation plot. Under normal circumstances, a pixel-based classification would have garnered multiple training locations in these situations. The other major limiting factor on accuracy was likely the lack of a cloud-free, low tidal stage, growing season images. The 6/30/2016 image was the closest to these requirements; however, tall-form *S. alterniflora* would be expected to increase dramatically in aboveground live biomass from June to August [34]. Due to these limitations, it is not surprising that the Sentinel-2 image classification achieved <85% overall accuracy. The major advantages of the Sentinel-2 classification were the applicability of the approach to the entire Sentinel-2 scenes which allowed for a classification of the Hydrological Unit Code 8 watershed.

The time series analysis with Sentinel-2 demonstrates promise. Sentinel-2 has only been operating since 2015 making the determination of trends and disturbance with the time series uncertain due to the short archive. However, given the revisit time and spatial resolution of Sentinel-2, the sensor is well suited to salt marsh time series analysis. Previous studies have utilized the harmonized Landsat-8 and Sentinel-2 time series to understand land-use dynamics [35]. The methods address many limitations of medium resolution sensors for time series analysis.

References

1. Levine, J.M., Brewer, J.S., Bertness, M.D., 1998. Nutrients, competition and plant zonation in a New England salt marsh. *Journal of Ecology* 86 (2), 285–292.
2. Campbell, A., Wang, Y., Christiano, M., Stevens, S., 2017. Salt marsh monitoring in Jamaica Bay, New York from 2003 to 2013: A decade of change from restoration to Hurricane sandy. *Remote Sensing* 9 (2), 131.
3. Collin, A., Long, B., Archambault, P., 2010. Salt-marsh characterization, zonation assessment and mapping through a dual-wavelength LiDAR. *Remote Sensing of Environment* 114 (3), 520–530.
4. Hay, G.J., Castilla, G., 2008. Geographic object-based image analysis (GEOBIA): A new name for a new discipline. In: *Object-based image analysis*, Blaschke, T., Lang, S. and Hay, G.J. (Eds.), Springer, Berlin, pp. 75–89.
5. Blaschke, T., 2010. Object based image analysis for remote sensing. *ISPRS Journal of Photogrammetry and Remote Sensing* 65 (1), 2–16.
6. Mui, A., He, Y., Weng, Q., 2015. An object-based approach to delineate wetlands across landscapes of varied disturbance with high spatial resolution satellite imagery. *ISPRS Journal of Photogrammetry and Remote Sensing* 109, 30–46.
7. Lantz, N.J., Wang, J., 2013. Object-based classification of Worldview-2 imagery for mapping invasive common reed, Phragmites australis. *Canadian Journal of Remote Sensing* 39 (4), 328–340.

8. Robertson, D.L., King, D.J., 2011. Comparison of pixel- and object-based classification in land cover change mapping. *International Journal of Remote Sensing* 32 (6), 1505–1529.
9. Ouyang, Z., Zhang, M., Xie, X., Shen, Q., Guo, H., Zhao, B., 2011. A comparison of pixel-based and object-oriented approaches to VHR imagery for mapping saltmarsh plants. *Ecological informatics* 6 (2), 136–146.
10. Castillejo-González, I.L., López-Granados, F., García-Ferrer, A., Peña-Barragán, J.M., Jurado-Expósito, M., de la Orden, M. S., González-Audicana, M. 2009. Object- and pixel-based analysis for mapping crops and their agro-environmental associated measures using QuickBird imagery. *Computers and Electronics in Agriculture* 68 (2), 207–215.
11. Myint, S.W., Gober, P., Brazel, A., Grossman-Clarke, S., Weng, Q., 2011. Per-pixel vs. object-based classification of urban land cover extraction using high spatial resolution imagery. *Remote Sensing of Environment* 115 (5), 1145–1161.
12. Kennedy, R.E., Townsend, P.A., Gross, J.E., Cohen, W.B., Bolstad, P., Wang, Y.Q., Adams, P. 2009. Remote sensing change detection tools for natural resource managers: Understanding concepts and tradeoffs in the design of landscape monitoring projects. *Remote Sensing of Environment* 113 (7), 1382–1396.
13. Estel, S., Kuemmerle, T., Alcántara, C., Levers, C., Prishchepov, A., Hostert, P., 2015. Mapping farmland abandonment and recultivation across Europe using MODIS NDVI time series. *Remote Sensing of Environment* 163, 312–325.
14. Song, X., Sexton, J.O., Huang, C., Channan, S., Townshend, J.R., 2016. Characterizing the magnitude, timing and duration of urban growth from time series of Landsat-based estimates of impervious cover. *Remote Sensing of Environment* 175, 1–13.
15. Kayastha, N., Thomas, V., Galbraith, J., Banskota, A., 2012. Monitoring wetland change using inter-annual landsat time-series data. *Wetlands* 32 (6), 1149–1162.
16. Hird, J.N., DeLancey, E.R., McDermid, G.J., Kariyeva, J., 2017. Google earth engine, open-access satellite data, and machine learning in support of large-area probabilistic wetland mapping. *Remote Sensing* 9 (12), 1315.
17. Klemas, V., 2011. Remote sensing of wetlands: Case studies comparing practical techniques. *Journal of Coastal Research* 27 (3), 418–427.
18. Watson, E.B., Wigand, C., Davey, E.W., Andrews, H.M., Bishop, J., Raposa, K.B., 2017. Wetland loss patterns and inundation-productivity relationships prognosticate widespread salt marsh loss for Southern New England. *Estuaries and Coasts* 40 (3), 662–681.
19. Watson, E.B., Hinojosa Corona, A., 2018. Assessment of blue carbon storage by Baja California (Mexico) tidal wetlands and evidence for wetland stability in the face of anthropogenic and climatic impacts. *Sensors* 18 (1), 32.
20. Barbier, E.B., Hacker, S.D., Kennedy, C., Koch, E.W., Stier, A.C., Silliman, B.R., 2011. The value of estuarine and coastal ecosystem services. *Ecological Monographs* 81 (2), 169–193.
21. Sun, C., Liu, Y., Zhao, S., Zhou, M., Yang, Y., Li, F., 2016. Classification mapping and species identification of salt marshes based on a short-time interval NDVI time-series from HJ-1 optical imagery. *International Journal of Applied Earth Observation and Geoinformation* 45, 27–41.
22. Sun, C., Fagherazzi, S., Liu, Y., 2018. Classification mapping of salt marsh vegetation by flexible monthly NDVI time-series using Landsat imagery. *Estuarine, Coastal and Shelf Science* 213, 61–80.
23. Mackintosh, B., 1982. *Assateague Island National Seashore: An Administrative History.* National Park Service. History Division, Washington, DC.
24. Stalter, R., Lamont, E.E., 1990. The vascular flora of Assateague Island, Virginia. *Bulletin of the Torrey Botanical Club* 117, 48–56.
25. Braun-Blanquet, J., 1964. *Plantzensoziologie.* Springer Verlag. Wien, NY.
26. Duro, D.C., Franklin, S.E., Dubé, M.G., 2012. A comparison of pixel-based and object-based image analysis with selected machine learning algorithms for the classification of agricultural landscapes using SPOT-5 HRG imagery. *Remote Sensing of Environment* 118 (0), 259–272.

27. Dronova, I., Gong, P., Clinton, N.E., Wang, L., Fu, W., Qi, S., Liu, Y. 2012. Landscape analysis of wetland plant functional types: The effects of image segmentation scale, vegetation classes and classification methods. *Remote Sensing of Environment* 127 (0), 357–369.
28. Breiman, L., 2001. Random forests. *Machine Learning* 45 (1), 5–32.
29. Fernández-Delgado, M., Cernadas, E., Barro, S., Amorim, D., 2014. Do we need hundreds of classifiers to solve real world classification problems. *Journal of Machine Learning Research* 15 (1), 3133–3181.
30. de Leeuw, J., Jia, H., Yang, L., Liu, X., Schmidt, K., Skidmore, A.K., 2006. Comparing accuracy assessments to infer superiority of image classification methods. *International Journal of Remote Sensing* 27 (1), 223–232.
31. Verbesselt, J., Hyndman, R., Newnham, G., Culvenor, D., 2010. Detecting trend and seasonal changes in satellite image time series. *Remote Sensing of Environment* 114 (1), 106–115.
32. Ghimire, B., Rogan, J., Galiano, V., Panday, P., Neeti, N., 2012. An evaluation of bagging, boosting, and random forests for land-cover classification in Cape Cod, Massachusetts, USA. *GIScience & Remote Sensing* 49 (5), 623–643.
33. Timm, B.C., McGarigal, K., 2012. Fine-scale remotely-sensed cover mapping of coastal dune and salt marsh ecosystems at Cape Cod national seashore using random forests. *Remote Sensing of Environment* 127, 106–117.
34. Gross, M.F., Hardisky, M.A., Wolf, P.L., Klemas, V., 1991. Relationship between aboveground and belowground biomass of *Spartina alterniflora* (smooth cordgrass). *Estuaries* 14 (2), 180–191.
35. Zhou, Q., Rover, J., Brown, J., Worstell, B., Howard, D., Wu, Z., Gallant, A. L., Rundquist, B., Burke, M. 2019. Monitoring landscape dynamics in Central US grasslands with harmonized Landsat-8 and Sentinel-2 time series data. *Remote Sensing* 11 (3), 328.

36

Tidal Effects in Salt Marsh Mapping: Remote Sensing

Anthony Daniel Campbell
University of Rhode Island
Yale University

Introduction

High accuracy mapping of salt marsh environment is necessitated by their uncertain response to climate change and extensive ecosystem services, e.g., sustaining biodiversity, carbon sequestration, water quality, and storm resilience [1]. Carbon sequestration is estimated to be an order of magnitude greater globally in salt marshes than terrestrial forests [2]. The loss of salt marsh land cover has the potential to exacerbate climate change. Salt marsh ecosystems are at risk due to natural and anthropogenic stressors including herbivory [3,4] invasive species [5], interior die-off from drought [6], eutrophication [7,8], and climate change and sea level rise (SLR) [9]. Uncertainty surrounds the projected response of salt marsh to SLR with studies suggesting a loss of between 20% and 45% of salt marshes by 2100 [10] or significant increases in salt marsh extent depending on salt marsh migration and management actions [11]. This disagreement in future projections necessitates accurate mapping of current salt marsh extents, which is essential to accurately monitor, forecast, and map change. Tidal inundation during remote sensing image collection complicates image classification and can result in lower accuracy [12]. The resulting classification can be impacted by tidal inundation leading to a reduction in the extent of the mapped salt marsh. This concern requires salt marsh mapping studies to understand and quantify the potential inundation of the imagery being utilized.

The incorporation of the tidal stage into salt marsh mapping has been evaluated and analyzed starting with the development of the Coastal Change Analysis Program (C-CAP) [12,13]. Finding that for the region analyzed a tidal range of between 0 and 0.9 m above Mean Low Water (MLW) was sufficient for mapping salt marsh [12]. However, given the expectation of improved accuracy for very high spatial resolution (VHR) mapping, the standards were reconsidered for VHR imagery and microtidal regions leading to the suggestion that individual sites should be evaluated when mapping salt marsh [14]. Given the effect of tidal range on the growth range of *Spartina alterniflora* [15], it is necessary to assess the effect of the tidal stage on salt marsh mapping within a particular site. A method for assessment using bathtub models, VDatum, and VHR imagery was reported [14]. Spectral indices have been used to filter coarse resolution imagery [16]. For medium resolution image, time series was used to determine and filter by a tidal threshold [17]. Both methods identified and removed inundated pixels. VHR salt marsh mapping has the highest expectation of accuracy; therefore, tidal inundation should be identified and mapped.

Topobathymetric LiDAR allows for the creation of a seamless digital elevation model (DEM) stretching across the tidal inundated landscape without concern for tidal stage at the time of LiDAR acquisition. Topobathymetric LiDAR can penetrate the water column retrieving bottom measurements from 2 to 3 times the Secchi depth [18]. Topobathymetric LiDAR can detect nearshore features of less than 1 m^2 [19]. The technology has been used to improve hydrological modeling resulting in accurate sub-grid models of inundation [20]. Topobathymetric LiDAR is an important development allowing for better understanding and management of the nearshore environments.

Methodology

Study Site

Cedar Island, located on the Delmarva Peninsula, was selected given its proximity to the Wachapreague tidal gauge (Station ID 8631044) [21] (Figure 36.1). Cedar Island is a barrier island off the coast of Virginia, with semidiurnal tides, and a tidal range from MLW to mean high water (MHW) of 1.23 m. This study site was selected given the availability of Topobathymetric LiDAR, salt marsh environment, and proximity to the National Oceanic and Atmospheric Administration (NOAA) tidal gauge as described above.

Data

This chapter utilizes a combination of Topobathymetric LiDAR collected in 2014 for the east coast of the United States. The LiDAR point data stored as LAS files were interpolated into a DEM using LAS datasets within ArcGIS, the average of all bathymetric bottom and ground points were utilized as the elevation with bins. The spatial resolution of the created DEM was 3 × 3 m to match the imagery data. The dataset was not converted into MLW with VDatum due to a lack of conversion extent for the study site. Binary inundation raters were computed for the site starting at MLW at Wachapreague which is −0.75 m North American Vertical Datum (NAVD) 1988. Inundation raster data were created at every 5 cm increment.

Planetscope imagery is comprised of four bands (blue, green, red, and near infrared or NIR) collected daily for many areas of the globe. The spatial resolution of Planetscope is effectively 3 m. Cloud-free images from 2017 during the growing season (July, August, September, and October) were downloaded as surface reflectance radiometrically corrected by Planet Labs. Normalized Difference Vegetation Index (NDVI) was computed for each of the Planetscope images. The tidal inundation at time of image collection was determined from the Wachapreague tidal gauge (Station ID 8631044) (Table 36.1).

National Wetland Inventory (NWI) data were used as the vegetation extent on the islands. The data suggested that nearly all vegetated areas on Cedar Island were salt marsh (Figure 36.1). This vegetation extent was utilized in the inundation model and compared to real inundation extents as derived from the Planetscope images. The accuracy of NWI is not reported, and as such, the uncertainty is unknown; however, wetlands that are 0.4 ha. are routinely delineated [22]. The study area of this chapter was mapped between 2008 and 2009. The aerial imagery utilized are higher resolution and should be one step closer to reality than extents derived from the Planetscope data.

Tidal Analysis

The DEM was georeferenced to the image acquired on September 23, 2017 to minimize geographic differences. All Planetscope images were orthorectified scenes with positional Root Mean Square Error (RMSE) of less than 10 m [23]. The scenes were examined for geolocational error. NDVI was calculated

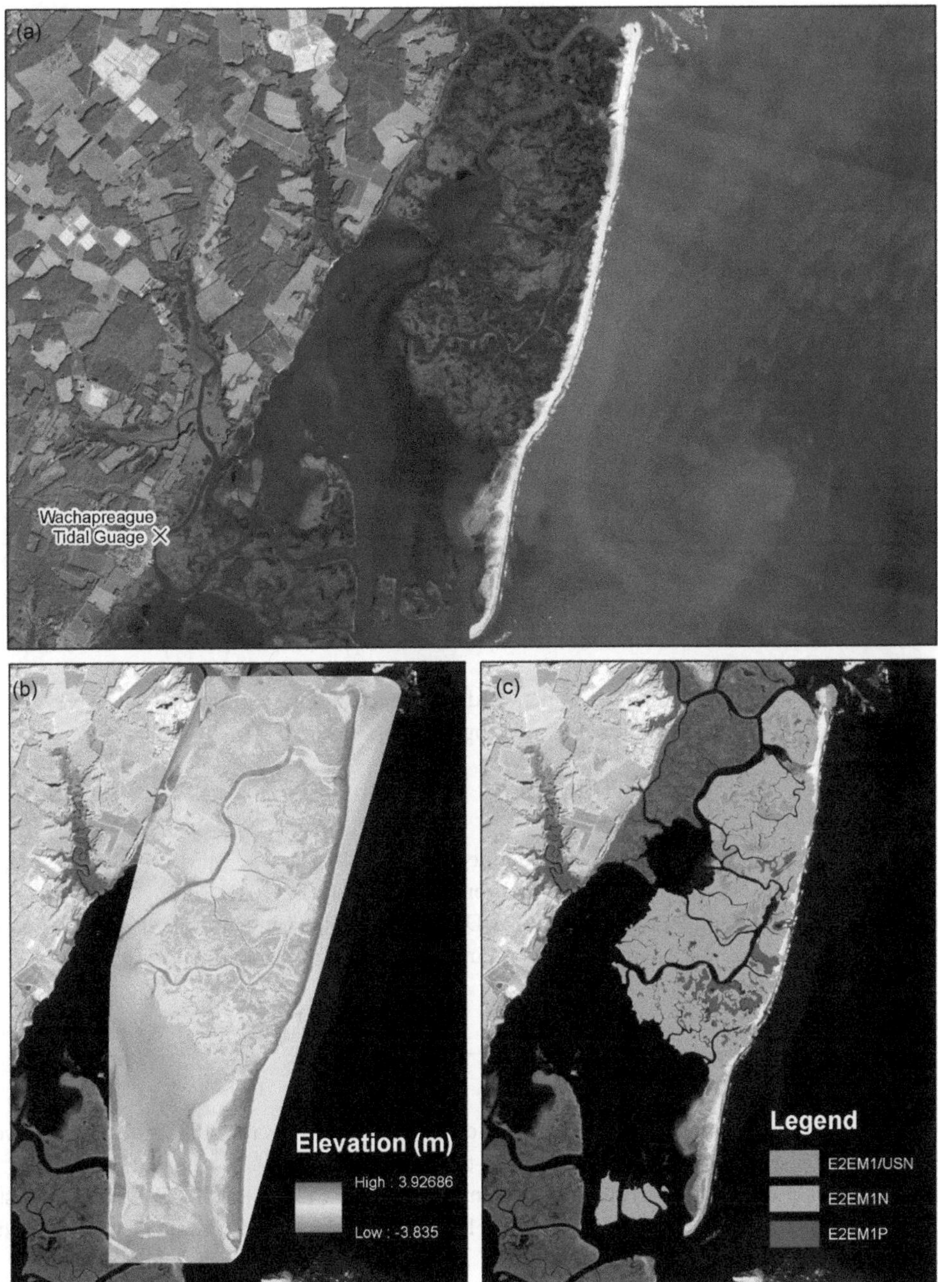

FIGURE 36.1 **(See color insert.)** (**a**) The Planetscope image of Cedar Island acquired on October 28, 2017 and surrounding area including the Wachapreague tidal gauge location. (**b**) The elevation range of the DEM. (**c**) NWI data for Cedar Island which were utilized for the analysis.

for each of the Planetscope images. Binary images were created with a threshold of >0.1 NDVI for July, August, and September and areas >0.0 NDVI for October. The threshold was determined by minimizing the difference in area between similar tidal stages regardless of the month.

Binary models of inundation were created for each tidal stage from MLW to 1.3 m above MLW or approximately MHW. These models represented inundated and non-inundated pixels for each of the

TABLE 36.1 Planetscope Imagery Acquisition Date, Time, and Tidal Stage

Date	Time (UTC)	Tidal Stage (m)[a]
07/08/2017	15:02:34	0.780
07/09/2017	15:02:11	0.983
07/30/2017	15:04:26	1.025
08/02/2017	16:05:43	0.422
08/06/2017	15:03:47	0.743
08/20/2017	15:03:57	0.587
08/24/2017	15:03:38	1.537
09/09/2017	15:07:15	1.479
09/20/2017	15:59:10	1.405
09/23/2017	15:59:15	1.682
09/29/2017	15:58:03	0.88
10/04/2017	15:57:53	0.572
10/20/2017	15:06:06	1.157
10/28/2017	15:08:21	0.613

[a] As determined at the Wachapreague tidal gauge relative to MLW.

5 cm increments. Percent inundation was calculated for each tidal stage within each of the NWI classes. The same was done for each of the NDVI Planetscope images.

Tidal Model

The salt marsh extent, for this chapter, was defined as the NWI classes E2EM1/USN, E2EM1N, and E2EM1P [24]. These classes represent a combination of salt marsh (E2EM1N, E2EM1P) and salt marsh combined with the unconsolidated shore (E2EM1/USN). The combination of these classes allows for an understanding of both salt marsh inundation and intertidal mudflat inundation.

Results

Inundation Model

The model's RMSE, when compared to the verification stages, was 17.6% across all three classes. However, when comparing just verification images with less than a meter above MLW, the RMSE was 9.5%. The RMSE of the maximum tidal model was slightly improved with a 12.9% RMSE. The RMSE of just the E2EM1N class was 11.1%, and the E2EM1/USN was 14.4% (Figure 36.2). The high tidal stage image dates had especially large error.

Satellite Image Verification

The inundation model with satellite imagery demonstrated a significant statistical relationship between inundation and areas within the E2EM1/USN class with <0.1 NDVI ($F_{1,12} = 166.1$, $p < 0.001$) and an R^2 of 0.93. The E2EM1N class also had a significant relationship between tidal stage and area with <0.1 NDVI ($F_{1,12} = 92.41$, $p < 0.001$) and a R^2 of 0.88. The E2EM1P class also had a significant relationship; however, the linear model did not explain as much of the class's variability ($F_{1,12} = 12.13$, $p < 0.05$) and a R^2 of 0.46. In general, area below the NDVI threshold increased as the tidal stage increased at the time of image acquisition (Figure 36.3).

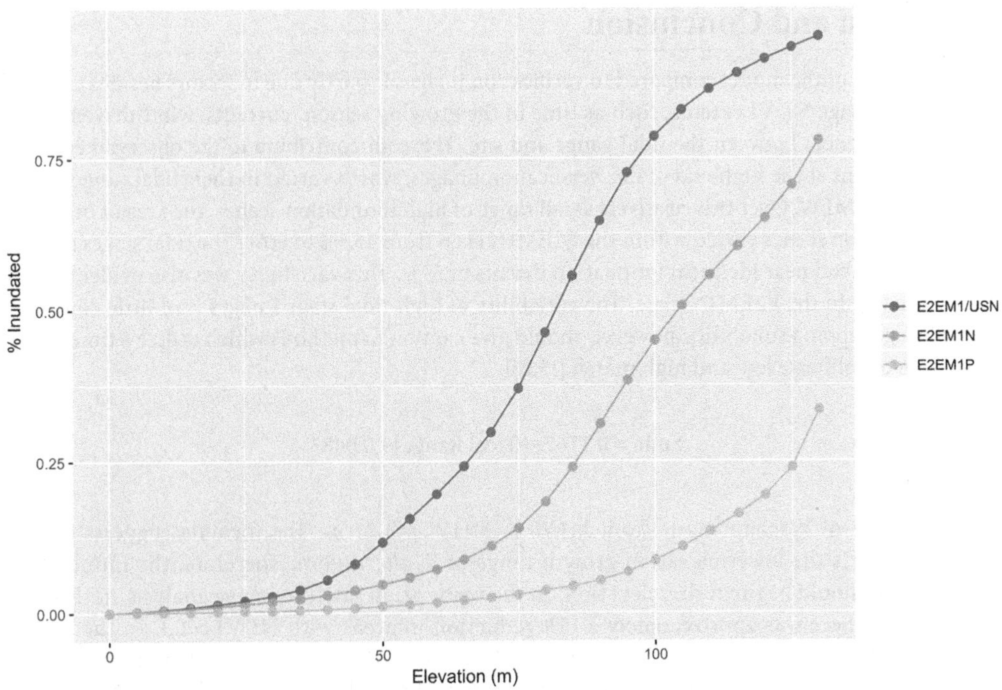

FIGURE 36.2 Modeled response of Cowardin et al. (1979) classes to tidal stage at time of acquisition for Cedar Island, Virginia.

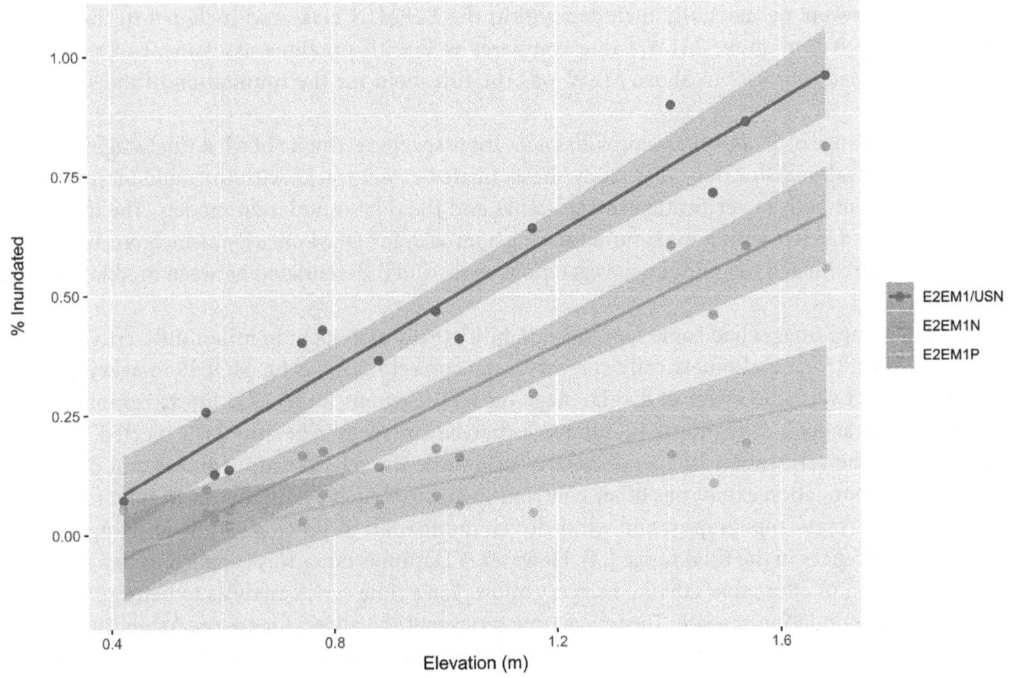

FIGURE 36.3 The tidal stage at the time of image acquisition and percent of the total area of each class inundated at that tidal stage.

Discussion and Conclusion

The high RMSE of the model compared to verification images is partly due to factors besides tidal stage determining image NDVI extents, such as time in the growing season, currents, wind-driven inundation, and differences between the tidal gauge and site. These all contribute to the observed error. It is especially evident at the high end of our verification images, which varied in their tidal range from 1.4 to 1.68 m above MLW. Over this relatively small range of high inundation scenes, the area of inundation in our verification scenes varied within the E2EM1N class from 45.5% to 80%. The tidal stages of 1.4 and 1.5 above MLW had near identical inundation extents of 59%. This variability was also evident at lower tidal stages within the E2EM1N class. The variability of high tidal stage images is of little concern for determining mapping suitability, however, should give caution to methods which utilize a time series of inundation to delineate low and high marsh [15,25].

$$z\min = 0.7167 * (\text{Tidal Range}) - 0.0483$$

The tidal range of Wachapreague from MLW to MHW is 1.23 m. The formula suggests that 0.83 m above MLW is the lower elevation growth range of *S. alterniflora*. Therefore, the inundation of *S. alterniflora* should begin at this elevation. In the verification satellite image analysis, at this elevation (0.83 m), there was approximately a 15% reduction in areas with NDVI > 0.1. In the modeled inundation, 8.3% of areas were inundated at 0.85 m above MLW. The model better approximates the expected outcome. However, it is likely that intertidal mudflats are a sizeable component of the units mapped as E2EM1N resulting in areas below the NDVI threshold that are not inundated at the corresponding tidal stage. This reasoning suggests that NDVI > 0.1 includes approximately 15% non-vegetated areas in the E2EM1N class. The discrepancy between model and verification images was partly due to differences between the metrics being utilized. The NDVI threshold excluded intertidal mudflats which were occasionally included within the E2EM1N class and included in the E2EM1/USN class. The 0.83 m above MLW value compares well with previous assessment of the region, which determined that 0.87 m above MLW was the threshold for the inundation of the salt marsh platform [17].

The growth range of *S. alterniflora* provides a location-specific estimate of what tidal stage will affect the mapping of salt marsh extent. The study site estimated extent tracks with the rapid increase in the inundation extent both in verification image scenes and the tidal inundation models. The inundation model performed better with a maximum binning approach for DEM creation, which created a DEM that better approximated the salt marsh vegetation height and differentiated between mudflat and vegetated pixels.

The Planetscope images had some spectral variability that was evident in minor differences between images. While all were radiometrically corrected, the correction relies on MODIS imagery [26]. The comparisons between the effect of a tidal stage on NDVI extent followed a linear response in the unconsolidated shore and regularly inundated estuarine emergent vegetation as expected. The variability within the verification images illustrates that while the relationship is linear and dominated by tidal stage inundation extent has other contributing factors such as differences in tidal range, vegetation type, currents, topography, and wind-driven inundation. A previous study utilizing VDatum addressed differences in the tidal range [14]. However, VDatum is not always applicable to a study site. The proliferation of affordable satellite imagery allows for a time series analysis to better understand potential tidal inundation at a site. The image time series method allows for an understanding of inundation and the variability at varying tidal stages. This method could be applied to other study sites to determine the suitability of imagery collected at a particular tidal stage and estimate inundated areas and their extent.

Acknowledgments

The author would like to thank the Planet Lab's ambassador program which allowed for free access to the remote sensing imagery utilized in this chapter.

References

1. Zedler, J.B.; Kercher, S. Wetland Resources: Status, Trends, Ecosystem Services, and Restorability. *Annu. Rev. Environ. Resour.* **2005**, *30*, 39–74.
2. Chmura, G.L. What Do We Need to Assess the Sustainability of the Tidal Salt Marsh Carbon Sink? *Ocean Coast. Manage.* **2013**, *83*, 25–31.
3. Altieri, A.H.; Bertness, M.D.; Coverdale, T.C.; Herrmann, N.C.; Angelini, C. A Trophic Cascade Triggers Collapse of a Salt-marsh Ecosystem with Intensive Recreational Fishing. *Ecology* **2012**, *93*, 1402–1410.
4. Holdredge, C.; Bertness, M.D.; Altieri, A.H. Role of Crab Herbivory in Die-Off of New England Salt Marshes. *Conserv. Biol.* **2009**, *23*, 672–679.
5. Silliman, B.R.; Bertness, M.D. Shoreline Development Drives Invasion of Phragmites Australis and the Loss of Plant Diversity on New England Salt Marshes. *Conserv. Biol.* **2004**, *18*, 1424–1434.
6. Alber, M.; Swenson, E.M.; Adamowicz, S.C.; Mendelssohn, I.A. Salt Marsh Dieback: An Overview of Recent Events in the US. *Estuar. Coast. Shelf Sci.* **2008**, *80*, 1–11.
7. Deegan, L.A.; Johnson, D.S.; Warren, R.S.; Peterson, B.J.; Fleeger, J.W.; Fagherazzi, S.; Wollheim, W.M. Coastal Eutrophication as a Driver of Salt Marsh Loss. *Nature* **2012**, *490*, 388–392.
8. Wigand, C.; Roman, C.T.; Davey, E.; Stolt, M.; Johnson, R.; Hanson, A.; Watson, E.B.; Moran, S.B.; Cahoon, D.R.; Lynch, J.C. Below the Disappearing Marshes of an Urban Estuary: Historic Nitrogen Trends and Soil Structure. *Ecol. Appl.* **2014**, *24*, 633–649.
9. Watson, E.B.; Wigand, C.; Davey, E.W.; Andrews, H.M.; Bishop, J.; Raposa, K.B. Wetland Loss Patterns and Inundation-Productivity Relationships Prognosticate Widespread Salt Marsh Loss for Southern New England. *Estuaries Coasts* **2017**, *40*, 662–681.
10. Craft, C.; Clough, J.; Ehman, J.; Joye, S.; Park, R.; Pennings, S.; Guo, H.; Machmuller, M. Forecasting the Effects of Accelerated Sea-Level Rise on Tidal Marsh Ecosystem Services. *Front. Ecol. Environ.* **2009**, *7*, 73–78.
11. Schuerch, M.; Spencer, T.; Temmerman, S.; Kirwan, M.L.; Wolff, C.; Lincke, D.; McOwen, C.J.; Pickering, M.D.; Reef, R.; Vafeidis, A.T. Future Response of Global Coastal Wetlands to Sea-Level Rise. *Nature* **2018**, *561*, 231.
12. Jensen, J.R., Cowen, D.J., Althausen, J.D., Narumalani, S.,Weatherbee, O. An evaluation of the CoastWatch change detection protocol in South Carolina. *Photogramm. Eng. Remote Sens.* **1993**, *59* (6), 1039–1044.
13. Jensen, J.R., Cowen, D.J., Althausen, J.D., Narumalani, S., Weatherbee, O. The Detection and Prediction of Sea Level Changes on Coastal Wetlands Using Satellite Imagery and a Geographic Information System. *Geocarto Int.* **1993**, *8* (4), 87–98.
14. Campbell, A.; Wang, Y. Examining the Influence of Tidal Stage on Salt Marsh Mapping using High-Spatial-Resolution Satellite Remote Sensing and Topobathymetric Lidar. *IEEE Trans. Geosci. Remote Sens.* **2018**, *56*, 5169–5176.
15. Mckee, K.L.; Patrick, W.H. The Relationship of Smooth Cordgrass (Spartina Alterniflora) to Tidal Datums: A Review. *Estuaries* **1988**, *11*, 143–151.
16. O'Connell, J.L.; Mishra, D.R.; Cotten, D.L.; Wang, L.; Alber, M. The Tidal Marsh Inundation Index (TMII): An Inundation Filter to Flag Flooded Pixels and Improve MODIS Tidal Marsh Vegetation Time-Series Analysis. *Remote Sens. Environ.* **2017**, *201*, 34–46.

17. Sun, C.; Fagherazzi, S.; Liu, Y. Classification Mapping of Salt Marsh Vegetation by Flexible Monthly NDVI Time-Series Using Landsat Imagery. *Estuarine Coastal Shelf Sci.* **2018**, *213*, 61–80.
18. Brock, J.C.; Purkis, S.J. The Emerging Role of Lidar Remote Sensing in Coastal Research and Resource Management. *J. Coastal Res.* **2009**, Special Issue 53, 1–5.
19. Andersen, M.S.; Gergely, Á.; Al-Hamdani, Z.; Steinbacher, F.; Larsen, L.R.; Ernstsen, V.B. Processing and Performance of Topobathymetric Lidar Data for Geomorphometric and Morphological Classification in a High-Energy Tidal Environment. *Hydrol. Earth Syst. Sci.* **2017**, *21*(1), 43–63.
20. Loftis, J.D.; Wang, H.V.; DeYoung, R.J.; Ball, W.B. Using Lidar Elevation Data to Develop a Topobathymetric Digital Elevation Model for Sub-Grid Inundation Modeling at Langley Research Center. *J. Coastal Res.* **2016**, *76*(spl), 134–148.
21. National Oceanic and Atmospheric Administration (NOAA), Wachapreague, VA - Station ID: 8631044. 2019. https://tidesndcurrents.noaa.gov/stationhome.html?id=8631044.
22. Dahl, T.E. 2011. *Status and Trends of Wetlands in the Conterminous United States 2004 to 2009.* U.S. Department of the Interior; Fish and Wildlife Service, Washington, DC, 108 pp.
23. Planet Labs. Planet Imagery Product Specifications. 2018. www.planet.com/products/satellite-imagery/files/Planet_Combined_Imagery_Product_Specs_December2017.pdf
24. Cowardin, L.M.; Carter, V.; Golet, F.C.; LaRoe, E.T. 1979. *Classification of Wetlands and Deepwater Habitats of the United States.* U.S. Department of the Interior, Fish and Wildlife Service, Washington, DC, 131 pp.
25. Kirwan, M.L.; Guntenspergen, G.R. Influence of Tidal Range on the Stability of Coastal Marshland. *J. Geophys. Res.: Earth Surf.* **2010**, 115 (F2), 1–10.
26. Planet, Planet Surface Reflection Product 1.0. 2018. https://assets.planet.com/marketing/PDF/Planet_Surface_Reflectance_Technical_White_Paper.pdf

Index

A

B

C

D

E

F

H

I

L

M

O

P

R

S

T

W

T - #0621 - 101024 - C0 - 254/178/23 - PB - 9781032474403 - Gloss Lamination